The Synthesis

and

Characterization

of

Inorganic Compounds

The Synthesis
and
Characterization
of
Inorganic Compounds

William L. Jolly

Professor of Chemistry
University of California, Berkeley

WAVELAND
PRESS, INC.
Prospect Heights, Illinois

For information about this book, write or call:

Waveland Press, Inc.
P.O. Box 400
Prospect Heights, Illinois 60070
(708) 634-0081

Preface

Inasmuch as inorganic chemistry encompasses a wide variety of chemical structures and reactions, inorganic synthesis is a difficult and challenging field, in which chemists must be exceptionally versatile. Almost all aspects of chemistry are touched on — the systematization of chemical reactions with thermodynamics and kinetics, the determination and prediction of structure, the correlation and prediction of physical properties, the theory of chemical bonding, chemical analysis, and the design and manipulation of laboratory apparatus. Besides possessing competence in these areas, synthetic chemists must be imaginative and must know how to apply analogies to the solution of problems.

The purpose of this book is to discuss the important principles and techniques applicable to synthetic inorganic chemistry. Because I cannot write authoritatively on all the topics described above, some chapters in this book are necessarily brief and introductory. Nevertheless, I believe that these topics should be discussed in a single book by one author for the following reasons: (1) the level of discussion is consistent throughout the text, (2) relationships between topics can be readily indicated, and (3) the book is made cohesive by the central theme of chemical synthesis. I hope the book can serve as a useful reference and guide for all experimental chemists. I believe it can be used as a text for several kinds of physical and inorganic chemistry courses.

I wish to thank my colleagues who examined parts of the manuscript and offered constructive criticism: Leo Brewer, Robert E. Connick, Rollie J. Myers, Richard E. Powell, and Kenneth N. Raymond.

WILLIAM L. JOLLY

Berkeley, California

Contents

ix

III   COMPOUND CHARACTERIZATION

IV   SYNTHESIS

33 APPENDICES

The Synthesis

and

Characterization

of

Inorganic Compounds

Introduction

1

THE PREPARATION OF NEW COMPOUNDS

Why do chemists prepare new compounds?[1] This question may seem inappropriate in a book the main purpose of which is to describe *how* chemists prepare new compounds. In fact, devotees of pure research may well consider the question irrelevant and trivial. However, it must be remembered that among the readers of this book are students who have not yet engaged in research and who have not enjoyed the thrill of original discovery. Naturally, such students will seek some sort of justification for a scientific endeavor like inorganic synthesis.

Discovery Through Curiosity

Curiosity led Alfred Stock to the discovery of an entirely new class of compounds: the boron hydrides and their derivatives. Stock has stated[2] that in

[1] Some of the material in this chapter first appeared in *J. Chem. Educ.*, **36**, 513 (1959).

[2] A. Stock, *Hydrides of Boron and Silicon*, Cornell University Press, Ithaca, N.Y., 1933.

1912 he chose to study the boron hydrides because he felt that boron, the neighbor of carbon in the periodic system, might be expected to form a much greater variety of interesting compounds than merely boric acid and the borates, which were almost the only ones known at that time. But Stock probably never dreamed that, in the following twenty years, he and his co-workers would prepare such an extensive and fascinating series of compounds as is described in his monograph, "Hydrides of Boron and Silicon."[2] Some of these compounds are shown in Fig. 1.1. In retrospect, Stock stated,

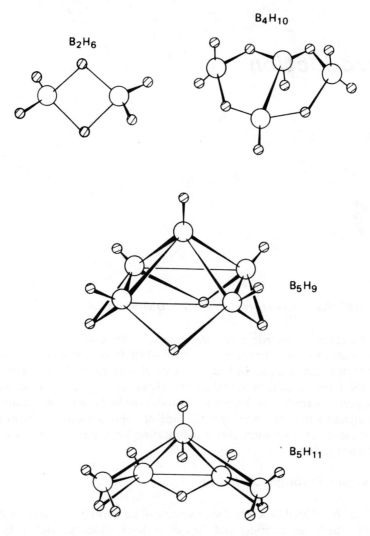

FIG. 1.1. Some compounds discovered by Stock and his co-workers.

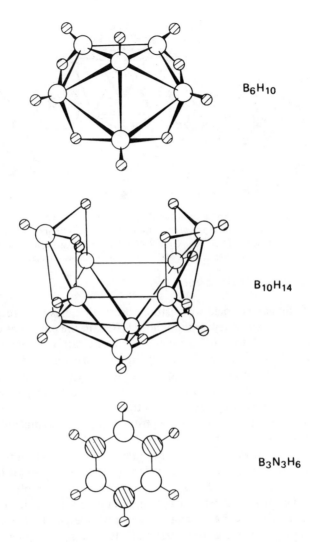

B$_6$H$_{10}$

B$_{10}$H$_{14}$

B$_3$N$_3$H$_6$

"The chemistry of boron has proved unexpectedly rich in results and many-sided in character although, just as in the case of silicon, the mobile portion of its chemistry is confined to the laboratory. In nature boron's dominating affinity for oxygen restricts it to the monotonous role of boric acid and the borates and prevents it from competing with carbon, its neighbor in the periodic system."[2] These words were written in 1932. Surely Stock would have been amazed if he could have known of the rapid development of boron chemistry in the 1960's. Even modern chemists marvel at the synthesis and characterization of polyhedral boranes, such as that shown in Fig. 1.2.[3]

[3] J. N. Francis and M. F. Hawthorne, *J. Am. Chem. Soc.*, **90**, 1663 (1968).

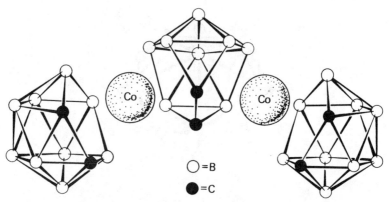

FIG. 1.2. The structure of the red anion[3] $B_{26}C_6H_{32}Co_2{}^{2-}$. All the carbon atoms and boron atoms are bonded to hydrogen atoms (not shown). Structure determined by D. H. Templeton, A. Zalkin, and D. J. St. Clair. Reproduced with permission of the American Chemical Society.

"Accidental" Discovery

Simple chance, coupled with astute observation, has been responsible for the discovery of many new compounds. Often, in the course of what is thought to be a straightforward synthesis or investigation, something completely unexpected happens. Perhaps a precipitate forms, a gas is evolved, a reaction mixture turns an unusual color, or a yield of expected product is very low. Unfortunately, the average chemist usually ignores such phenomena and goes on to work he can understand. But the curious chemist tries to find out what "went wrong" and in the process usually makes a significant—sometimes a spectacular—discovery.

Let us consider the unsuccessful, yet famous, attempt of Kealy and Pauson[4] to prepare dihydro-fulvalene (dicyclopentadienyl) by the coupling of cyclopentadienyl radicals. These investigators attempted the synthesis by the oxidation of cyclopentadienylmagnesium bromide by ferric chloride. (The technique of treating a Grignard reagent with ferric chloride is a well-known method for preparing hydrocarbons—for example, diphenyl from phenyl-magnesium bromide.) However, they obtained no dihydro-fulvalene. Instead, they isolated a remarkably stable organo-iron compound, which they identified as bis(cyclopentadienyl)iron(II). In their reaction, the ferric chloride presumably was first reduced to ferrous chloride by the Grignard reagent and the following reaction then took place:

$$2\,C_5H_5MgBr + FeCl_2 \longrightarrow C_5H_5FeC_5H_5 + MgBr_2 + MgCl_2 \qquad (1.1)$$

Later studies showed the bis(cyclopentadienyl)iron(II) ("ferrocene") to have the illustrated "sandwich" type of structure.

[4] T. J. Kealy and P. L. Pauson, *Nature*, **168**, 1039 (1951).

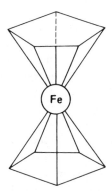

Soon after this discovery, better synthetic methods were developed for preparing ferrocene, and analogous compounds of the other transition metals were prepared.[5]

Testing Theories by Synthesis

There are innumerable examples of compounds that were prepared for the first time to test the validity of a theory. In 1893, Alfred Werner proposed his now famous coordination theory to explain the properties of inorganic "complex compounds."[6] At the time of its proposal it was a revolutionary doctrine, and for many years it met with criticism. Although some of the criticisms were not well founded, others were thoroughly sound and challenged Werner's ingenuity and experimental skill to the utmost. For example, the theory predicted optical isomerism for certain types of hexa-coordinate complexes, and Werner succeeded in resolving the *cis* form of the $Co(en)_2(NH_3)Br^{2+}$ ion into $(+)$ and $(-)$ forms.[7] But his unyielding critics argued that the optical activity centered in the carbon atoms of the ethylenediamine molecules (even though they are optically inactive themselves). Werner[8] provided final proof that the optical activity of such compounds lay in the geometrical configuration about the metal ion by effecting the resolution of a complex containing no carbon, the μ-hexahydroxododecaamminetetracobalt(III) ion:

$$\left[Co \left\{ \begin{matrix} H \\ O \\ \\ O \\ H \end{matrix} Co(NH_3)_4 \right\}_3 \right]^{6+}$$

[5] G. E. Coates, *Organo-Metallic Compounds*, 2nd ed., John Wiley & Sons, Inc., New York, 1960, pp. 233–359.

[6] A. Werner, *Z. anorg. Chem.*, **3**, 267 (1893).

[7] A. Werner, *Ber.*, **44**, 1887 (1911). See Chapter 23 for a discussion of optical activity.

[8] A. Werner, *Ber.*, **47**, 3087 (1914).

Discovery to Fulfill a Need

New compounds are often prepared because they are urgently needed. A problem that arose during the Atomic Bomb Project of World War II was the need to handle the very reactive molten metals uranium and plutonium. In a search for new refractories that could easily be formed and sintered into crucibles, yet would be stable and refractory, it was felt that certain sulfides might be useful. Preliminary thermodynamic calculations and estimations indicated that the most stable sulfides should lie in the periodic table around thorium and the rare earth metals. Accordingly, the sulfides of the most abundant rare earth, cerium, were investigated.[9] The theretofore unknown CeS was found to be a particularly good refractory that is useful whenever one wishes to avoid contamination of very electropositive metals by oxygen.

It is also interesting that most of the impetus for the renaissance of boron hydride chemistry beginning around 1941 was provided by a need for materials of importance to national defense. From 1941 to 1945 a great deal of effort was expended in an attempt to find a volatile uranium compound that could be used in the separation of the natural uranium isotopes, by either gaseous diffusion, centrifugation, thermal diffusion, or distillation. (Uranium hexafluoride was a known compound at the time and was eventually used, but it was feared that it might prove too difficult to prepare and handle on a large scale.) The fact that the hydroborates of aluminum and of beryllium had proved to be the most volatile compounds of these elements suggested the attempt to prepare a hydroborate of uranium. Uranium(IV) hydroborate, $U(BH_4)_4$, as well as the methyl derivatives, $U(BH_4)_3(BH_3CH_3)$ and $U(BH_3CH_3)_4$, were prepared and proved to be the most volatile uranium compounds other than the hexafluoride.[10,11] But of more interest than the preparation of these compounds were the results of efforts to improve the methods of preparing the starting materials and the intermediates required for the preparations. Many new types of reactions were observed, hitherto unknown compounds were discovered, and the chemistry of the boron hydrides was greatly enlarged.[12]

Immediately after World War II, a program for the development of boranes as fuels for air-breathing engines was started.[13] This effort led to the intensive study of many novel alkylated boranes and to the discovery of the remarkably stable class of compounds known as carboranes. Even after

[9] E. D. Eastman, L. Brewer, L. A. Bromley, P. W. Gilles, and N. L. Lofgren, *J. Am. Chem. Soc.*, **72**, 2248 (1950).

[10] H. I. Schlesinger and H. C. Brown, *J. Am. Chem. Soc.*, **75**, 219 (1953).

[11] H. I. Schlesinger, H. C. Brown, L. Horvitz, A. C. Bond, L. D. Tuck, and A. O. Walker, *J. Am. Chem. Soc.*, **75**, 222 (1953).

[12] H. I. Schlesinger and H. C. Brown, *J. Am. Chem. Soc.*, **75**, 186 (1953).

[13] R. T. Holzmann, ed., *Production of the Boranes and Related Research*, Academic Press Inc., New York, 1967.

government support for this project was cut off in the early 1960's, related work was continued in industrial and academic laboratories, with many spectacular discoveries[14] (see Fig. 1.2).

NEW METHODS FOR PREPARING OLD COMPOUNDS

The synthetic chemist is not only concerned with the preparation of new *compounds*; he often seeks new and better *methods* for preparing compounds that have been known for many years. Usually the first method used for the preparation of a compound is inefficient or inconvenient. Thus, the motive for seeking a better synthetic procedure (in contrast to the motive for preparing a new compound) is obvious. There are many inorganic compounds that might change from "laboratory curiosities" to commercially important chemicals if practical syntheses were found for them. The compounds B_2Cl_4 and S_4N_4, for example, challenge chemists' ingenuities. In this chapter we shall briefly examine, from a synthetic point of view, the histories of two compounds (borazine and diamidophosphoric acid) from the times of their discovery to the present.

Borazine

In 1926, Stock and Pohland[15] discovered that when the diammoniate of diborane ($B_2H_6 \cdot 2\,NH_3$) is heated to about 200°, a volatile liquid of composition $B_3N_3H_6$ is formed in yields of about 30 per cent:

$$3\,B_2H_6 \cdot 2\,NH_3 \longrightarrow 2\,B_3N_3H_6 + 12\,H_2 \tag{1.2}$$

They correctly assumed that the $B_3N_3H_6$ molecule is a ring of alternating BH and NH groups (see Fig. 1.1). This compound, now called borazine, was particularly fascinating because of its structural similarity to benzene. But at the time it was truly a laboratory curiosity because its synthesis required the preparation of diborane, B_2H_6, as an intermediate. Diborane itself was a laboratory curiosity because it was made only by the careful thermal decomposition of tetraborane, B_4H_{10}.

In 1930, Stock, Wiberg, and Martini[16] reported a synthetic method that was better because it did not require the preparation of diborane and because

[14] E. L. Muetterties and W. H. Knoth, *Polyhedral Boranes*, Marcel Dekker, Inc., New York, 1968; M. F. Hawthorne, *Acc. Chem. Res.*, **1**, 281 (1968).

[15] A. Stock and E. Pohland, *Ber.*, **59B**, 2215 (1926).

[16] A. Stock, E. Wiberg, and H. Martini, *Ber.*, **63**, 2927 (1930).

yields of 40 per cent were obtained. The method involved heating the tetraammoniate of tetraborane to 180°:

$$3 B_4H_{10} \cdot 4 NH_3 \longrightarrow 4 B_3N_3H_6 + 21 H_2 \tag{1.3}$$

During the next 20 years, the alkali metal hydroborates were prepared and characterized. These offered a new route to borazine. In 1951, Schaeffer, Schaeffer, and Schlesinger[17] reported that borazine can be obtained in 30 to 35 per cent yield by heating a mixture of lithium hydroborate and ammonium chloride from 230 to 300°:

$$3 LiBH_4 + 3 NH_4Cl \longrightarrow B_3N_3H_6 + 9 H_2 + 3 LiCl \tag{1.4}$$

More recently, borazine has been prepared by the reduction of B-trichloroborazine. The methods for the preparation of B-trichloroborazine and for its subsequent reduction have undergone continual improvement.

In 1949, Laubengayer and Brown[18] reported that B-trichloroborazine may be prepared in 35 per cent yield by the reaction of boron trichloride with ammonium chloride at 130 to 175°:

$$3 BCl_3 + 3 NH_4Cl \longrightarrow B_3N_3H_3Cl_3 + 9 HCl \tag{1.5}$$

In 1959, Eméleus and Videla[19] reported that the yield of the latter reaction may be raised to 60 per cent and that temperatures around 100° may be employed if the ammonium chloride is intimately mixed with a cobalt-on-pumice catalyst. In the same year, Leifield and Hohnstedt[20] described the synthesis of B-trichloroborazine by the reaction of an equal-volume mixture of boron trichloride and ammonia at 100°:

$$3 BCl_3 + 3 NH_3 \longrightarrow B_3N_3H_3Cl_3 + 6 HCl \tag{1.6}$$

Although the solid product contains about 35 per cent ammonium chloride by weight, practically all of the boron trichloride that is consumed is converted to B-trichloroborazine. Probably the simplest procedure is that of Rothgery and Hohnstedt,[21] who reported in 1967 that 57-per cent yields of B-trichloroborazine are obtained by refluxing a suspension in chlorobenzene of the acetonitrile adduct of boron trichloride and ammonium chloride:

$$3 CH_3CNBCl_3 + 3 NH_4Cl \longrightarrow B_3N_3H_3Cl_3 + 9 HCl + 3 CH_3CN \tag{1.7}$$

---

[17] G. W. Schaeffer, R. Schaeffer, and H. I. Schlesinger, *J. Am. Chem. Soc.*, **73**, 1612 (1951).

[18] A. W. Laubengayer and C. A. Brown, Paper given at the American Chemical Society meeting, September 19, 1949 [see *J. Am. Chem. Soc.*, **77**, 3699 (1955)].

[19] H. J. Eméleus and G. J. Videla, *J. Chem. Soc.*, 1306 (1959).

[20] R. F. Leifield and L. F. Hohnstedt, Paper given at the American Chemical Society meeting, April 9, 1959.

[21] E. F. Rothgery and L. F. Hohnstedt, *Inorg. Chem.*, **6**, 1065 (1967).

Schlesinger and his co-workers[22] showed that B-trichloroborazine may be hydrogenolyzed by treatment with an ether solution of lithium hydroborate:

$$3\,LiBH_4 + B_3N_3H_3Cl_3 \longrightarrow B_3N_3H_6 + \tfrac{3}{2}B_2H_6 + 3\,LiCl \qquad (1.8)$$

This reduction suffers from the disadvantage that it is very difficult to separate borazine from ether inasmuch as an azeotrope or etherate seems to form. This difficulty was eliminated by Hohnstedt and Haworth,[23] who carried out the reaction using sodium hydroborate as the reducing agent and the dimethyl ether of triethylene glycol as the solvent. Borazine yields in excess of 90 per cent were obtained.

Diamidophosphoric Acid

Stokes[24] first prepared diamidophosphoric acid by the following scheme:

$$POCl_3 \xrightarrow[\Delta]{C_6H_5OH} POCl_2OC_6H_5 \xrightarrow[NH_3]{aq\ or\ alc} PO(NH_2)_2OC_6H_5$$

$$\downarrow{}^{aq}_{OH^-}$$

$$HOPO(NH_2)_2 \xleftarrow[HOAc]{cold\ conc} PO_2(NH_2)_2{}^-$$

The scheme involves the phenolysis of one of the chlorines of phosphorus oxytrichloride followed by the ammonolysis of the remaining chlorines. The phenyl diamidophosphate is then saponified and the free acid is precipitated with cold acetic acid. Audrieth and Toy[25] carried out the ammonolysis in liquid ammonia and thus avoided any hydrolysis of the chlorines. The diamidophosphate is but slightly soluble in liquid ammonia, whereas ammonium chloride is very soluble.

Kirsanov and Abrazhanova[26] reported an interesting, if impractical, method for preparing diamidophosphates. They found that the compound $C_6H_5SO_2N{=}P(NH_2)_3$ undergoes hydrolysis to give benzene sulfonamide and diamidophosphate.

$$NaC_6H_5SO_2NCl \xrightarrow{PCl_3} C_6H_5SO_2N{=}PCl_3 \xrightarrow{NH_3} C_6H_5SO_2N{=}P(NH_2)_3$$

$$C_6H_5SO_2N{=}P(NH_2)_3 \begin{cases} \xrightarrow[\Delta]{H_2O} C_6H_5SO_2NH_2 + NH_4{}^+ + PO_2(NH_2)_2{}^- \\ \xrightarrow{aq\ OH^-} C_6H_5SO_2NH_2 + PO_2(NH_2)_2{}^- \end{cases}$$

[22] R. Schaeffer, M. Steindler, L. Hohnstedt, H. S. Smith, Jr., L. B. Eddy, and H. I. Schlesinger, *J. Am. Chem. Soc.*, **76**, 3303 (1954).

[23] L. F. Hohnstedt and D. T. Haworth, *J. Am. Chem. Soc.*, **82**, 89 (1960).

[24] H. N. Stokes, *Am. Chem. J.*, **15**, 198 (1893).

[25] L. F. Audrieth and A. D. F. Toy, *J. Am. Chem. Soc.*, **63**, 2117 (1941).

[26] A. V. Kirsanov, *Sbornik Statei Obshchei Khim.*, **2**, 1046 (1953); A. V. Kirsanov and E. A. Abrazhanova, *Sbornik Statei Obshchei Khim.*, **2**, 1059 (1953); *cf. Chem. Abstr.*, **49**, 3051, 5406 (1955).

The hydrolysis may involve phosphorus oxytriamide as an intermediate and suggests that the latter might be a useful starting point for the synthesis. Klement and Koch[27] prepared this compound by the reaction of ammonia with phosphorus oxytrichloride in cold chloroform solution and then hydrolyzed it to the diamidophosphate ion:

$$POCl_3 \xrightarrow[\text{in CHCl}_3]{NH_3} PO(NH_2)_3 \xrightarrow{aq\ OH^-} PO_2(NH_2)_2^-$$
$$\downarrow HOAc$$
$$HOPO(NH_2)_2$$

Only a rather impure acid is obtained by acidification of the alkaline solutions of diamidophosphate that are involved in the above procedures. Becke-Goehring and Sambeth[28] avoided the alkaline-solution step by hydrogenation of phenyl diamidophosphate directly to the free acid and cyclohexane.

$$C_6H_5OPO(NH_2)_2 + 4\,H_2 \xrightarrow[\text{catalyst}]{\text{Adams'}} HOPO(NH_2)_2 + C_6H_{12} \tag{1.9}$$

[27] R. Klement and O. Koch, *Ber.*, **87**, 333 (1954).
[28] M. Becke-Goehring and J. Sambeth, *Ber.*, **90**, 2075 (1957).

SYNTHETIC PRINCIPLES

Part

The Application

of

Thermodynamics

to

Synthetic Problems

2

Thermodynamic Relations

A good working knowledge of thermodynamics is invaluable to a synthetic chemist. We shall assume that the reader has already learned the fundamental principles of thermodynamics, and we shall discuss only the application (and not the derivation) of some important thermodynamic relations.

Chemists are particularly interested in quantitative expressions for the driving force of chemical reactions. The driving force can be measured in terms of the free energy change (ΔG), the equilibrium constant (K), the potential of the reaction (E), or a combination of the changes in heat content (ΔH) and entropy (ΔS). The relationships between these thermodynamic functions, and definitions of some of the terms, follow.

$$\Delta G = -nE\mathscr{F} = -RT \ln (K/Q) = \Delta H - T\Delta S \qquad (2.1)$$

13

Here n is the number of faradays of electricity involved in the reaction, and Q is the reaction quotient, or the product of the activities of the resulting substances divided by the product of the activities of the reacting substances, each activity raised to a power equal to the coefficient of the substance in the chemical equation. The activities of pure solid and liquid substances are taken to be unity. As a fair approximation, the activity of a gaseous substance is the partial pressure of that substance, expressed in atmospheres. For a substance in a dilute aqueous solution, the activity is roughly equal to the concentration, expressed either as molality (moles per kilogram of water) or molarity (moles per liter of solution). For concentrated solutions, these approximations cannot be used; activity coefficient data are required. It will be noted that the equilibrium constant K is the value of Q at equilibrium.

Following are the usual units employed and the corresponding values of the constants: ΔG and ΔH, calorie mole^{-1}; ΔS, calorie mole^{-1} degree^{-1}; E, volts. Also, $R = 1.987$ calorie mole^{-1} degree^{-1}; $\mathscr{F} = 23{,}060$ calorie volt^{-1} equivalent^{-1}. At 25°C (298.15°K), $RT\ln(K/Q) = 1364\log(K/Q)$.

When $Q = 1$ (which is the case when all the reactants and products are at unit activity), the thermodynamic functions possess their "standard" values:

$$\Delta G^\circ = -nE^\circ \mathscr{F} = -RT\ln K = \Delta H^\circ - T\,\Delta S^\circ \tag{2.2}$$

Tabulated Data

Equilibrium constants are known, or calculable, for many thousands of reactions. Some of these constants (e.g., solubility products, ionization constants of acids, and complex ion dissociation constants) are tabulated in various compendia. However, most equilibrium constants must be sought in the original literature or must be calculated from other data (such as free energies of formation, to be discussed). One of the important lessons of thermodynamics is that the equilibrium constant of a reaction which is the combination of two or more other reactions can be readily calculated from the equilibrium constants of the other reactions. For example, the solubility product of calcium carbonate may be combined with the first and second ionization constants of carbonic acid to give the equilibrium constant for the dissolution of calcium carbonate in acid:

$$CaCO_3 = Ca^{2+} + CO_3^{2-} \qquad K_1 = 4.8 \times 10^{-9}$$

$$H^+ + HCO_3^- = H_2CO_3 \qquad K_2 = \frac{1}{4.3 \times 10^{-7}}$$

$$H^+ + CO_3^{2-} = HCO_3^- \qquad K_3 = \frac{1}{4.7 \times 10^{-11}}$$

$$CaCO_3 + 2H^+ = H_2CO_3 + Ca^{2+} \qquad K = K_1 K_2 K_3 = 2.4 \times 10^8$$

Equilibrium constants have been determined for the reactions of fluorine with xenon to form XeF_2, XeF_4, and XeF_6.[1] The values at 250° and at 400° are given.

$$Xe(g) + F_2(g) = XeF_2(g) \tag{2.3}$$

$$K_{250} = 8.80 \times 10^4 \qquad K_{400} = 3.60 \times 10^2$$

$$Xe(g) + 2F_2(g) = XeF_4(g) \tag{2.4}$$

$$K_{250} = 1.07 \times 10^8 \qquad K_{400} = 1.98 \times 10^3$$

$$Xe(g) + 3F_2(g) = XeF_6(g) \tag{2.5}$$

$$K_{250} = 1.01 \times 10^8 \qquad K_{400} = 36.0$$

These data are of considerable significance in designing procedures for the preparation of the individual xenon fluorides.

First let us assume that we wish to prepare XeF_2. To avoid contamination of the product with XeF_4, we should use an excess of xenon. But how much of an excess would be required to limit the amount of XeF_4 to less than, say, 1 per cent (i.e., to maintain a ratio $(XeF_2)/(XeF_4) > 99$)? Using the data for reactions (2.3) and (2.4), we calculate the equilibrium constants for the following reaction.

$$Xe + XeF_4 = 2XeF_2 \tag{2.6}$$

$$K_{250} = 72 \qquad K_{400} = 65$$

We see that the required excess pressure of xenon depends on the pressure of XeF_2 present. Let us assume that we start with an amount of fluorine sufficient to produce 1 atm of XeF_2 at the equilibrium temperature. In that case it would be necessary to start with an $Xe:F_2$ ratio greater than about 2.5:1 in order to limit the XeF_4 to 1 per cent.

Second, let us assume that we wish to prepare XeF_6. To minimize the formation of XeF_4, we must employ an excess of fluorine. From the data for reactions (2.4) and (2.5), we calculate the equilibrium constants for the reaction

$$XeF_4 + F_2 = XeF_6 \tag{2.7}$$

$$K_{250} = 0.94 \qquad K_{400} = 0.018$$

Obviously we should use as low a reaction temperature as possible, consistent with a reasonable reaction rate. (The lowest practical reaction temperature is about 250°.) We calculate that at 250° the equilibrium pressure of fluorine must be greater than about 10.5 atm in order that the ratio of XeF_6 to XeF_4 be greater than 10. Obviously extremely high pressures of fluorine would be required to approach a quantitative yield of XeF_6, and any reasonable synthesis would require the separation of some XeF_4 from the product XeF_6.

[1] B. Weinstock, E. E. Weaver, and C. P. Knop, *Inorg. Chem.*, **5**, 2189 (1966).

Finally, let us consider the synthesis of XeF_4. From the data we see that XeF_4 is stable with respect to disproportionation.

$$2 XeF_4 = XeF_2 + XeF_6 \tag{2.8}$$

$$K_{250} = 7.8 \times 10^{-4} \qquad K_{400} = 3.3 \times 10^{-3}$$

There are two reasons for carrying out the synthesis at 400° rather than at a lower temperature, such as 250°. At 400°, the reaction rate is fairly high, and the equilibrium constant for reaction (2.7) is so low that a moderate excess of fluorine can be used without causing much contamination of the product with XeF_6.

A very convenient and concise way for summarizing equilibrium data for oxidation-reduction reactions in solution is by tabulating reduction potentials (or their negative, oxidation potentials) for half-reactions.[2] A short list of reduction potentials for couples useful in synthetic work is presented in Table 2.1. To illustrate the utility of such a table, let us suppose that we wish to prepare periodate $(H_3IO_6^{2-})$ by oxidizing iodate (IO_3^-) with some aqueous oxidizing agent. In Table 2.1 we see that the periodate-iodate couple has a potential of 1.6 v in acid solutions and a potential of 0.7 v in basic solutions. Any oxidizing agent with a reduction potential more positive than 1.6 v in acid solutions or more positive than 0.7 v in basic solutions is thermodynamically capable of oxidizing iodate to periodate. In practice, hypobromite or hypochlorite is usually used, because these oxidizing agents react with reasonable rapidity and are relatively cheap. Either may be conveniently prepared by dissolving the appropriate halogen in an alkaline solution:

$$2 e^- + Cl_2(g) = 2 Cl^- \qquad\qquad E° = +1.36$$

$$Cl^- + 2 OH^- = ClO^- + H_2O + 2 e^- \qquad E° = -0.89$$

$$Cl_2(g) + 2 OH^- = Cl^- + ClO^- + H_2O \qquad E° = +0.47$$

$$K = 8 \times 10^{15}$$

Probably the most generally useful way to summarize equilibrium data for reactions is to tabulate free energies of formation $(\Delta G_f°)$ for individual species. The free energy of formation of a compound is the change in free energy accompanying the formation of the compound from its elements in their standard states. Thus, the free energy of formation of liquid water is the value of $\Delta G°$ for the reaction

$$H_2(g) + \tfrac{1}{2} O_2(g) = H_2O(l) \tag{2.9}$$

To illustrate the application of such free energy data, consider the case of

---

[2] W. M. Latimer, *Oxidation Potentials*, 2nd ed., Prentice-Hall, Inc., Englewood Cliffs, N.J., 1952.

Table 2.1 Some Aqueous Couples Used in Synthetic Chemistry*

Acid Solutions		Basic Solutions	
Couple	$E°$	Couple	$E°$
$Mg^{2+} + 2e^- = Mg$	-2.37	$Mg(OH)_2 + 2e^- = Mg + 2OH^-$	-2.69
$Al^{3+} + 3e^- = Al$	-1.66	$H_2AlO_3^- + H_2O + 3e^- = Al + 4OH^-$	-2.35
$Zn^{2+} + 2e^- = Zn$	-0.76	$B(OH)_4^- + 2H_2 + 4e^- = BH_4^- + 4OH^-$	-1.60
$H_3PO_3 + 2H^+ + 2e^- = H_3PO_2 + H_2O$	-0.50	$HPO_3^{2-} + 2H_2O + 2e^- = H_2PO_2^- + 3OH^-$	-1.57
$H_3BO_3 - 7H^+ + 8e^- = BH_4^- + 3H_2O$	-0.47	$ZnO_2^{2-} + 2H_2O + 2e^- = Zn + 4OH^-$	-1.22
$Sn^{2+} + 2e^- = Sn$	-0.14	$2SO_3^{2-} + 2H_2O + 2e^- = S_2O_4^{2-} + 4OH^-$	-1.12
$2H_2SO_3 + H^+ + 2e^- = HS_2O_4^- + 2H_2O$	-0.08	$SO_4^{2-} + H_2O + 2e^- = SO_3^{2-} + 2OH^-$	-0.93
$2H^+ + 2e^- = H_2(g)$	0.00	$HSnO_2^- + H_2O + 2e^- = Sn + 3OH^-$	-0.91
$HCOOH + 2H^+ + 2e^- = HCHO + H_2O$	0.06	$2H_2O + 2e^- = H_2(g) + 2OH^-$	-0.83
$S + 2H^+ + 2e^- = H_2S$	0.14	$AsO_4^{3-} + 2H_2O + 2e^- = AsO_2^- + 4OH^-$	-0.67
$Sn^{4+} + 2e^- = Sn^{2+}$	0.15	$S + 2e^- = S^{2-}$	-0.48
$SO_4^{2-} + 4H^+ + 2e^- = H_2SO_3 + H_2O$	0.17	$CrO_4^{2-} + 4H_2O + 3e^- = Cr(OH)_3 + 5OH^-$	-0.13
$Fe(CN)_6^{3-} + e^- = Fe(CN)_6^{4-}$	0.36	$O_2(g) + H_2O + 2e^- = HO_2^- + OH^-$	-0.08
$I_3^- + 2e^- = 3I^-$	0.54	$PbO_2 + H_2O + 2e^- = PbO + 2OH^-$	0.25
$H_3AsO_4 + 2H^+ + 2e^- = HAsO_2 + 2H_2O$	0.56	$O_2(g) + 2H_2O + 4e^- = 4OH^-$	0.40
$O_2(g) + 2H^+ + 2e^- = H_2O_2$	0.68	$IO^- + H_2O + 2e^- = I^- + 2OH^-$	0.49
$NO_3^- + 4H^+ + 4e^- = NO(g) + 2H_2O$	0.96	$MnO_4^- + 2H_2O + 3e^- = MnO_2 + 4OH^-$	0.59
$Br_2 + 2e^- = 2Br^-$	1.06	$H_3IO_6^{2-} + 2e^- = IO_3^- + 3OH^-$	0.70
$O_2(g) + 4H^+ + 4e^- = 2H_2O$	1.23	$BrO^- + H_2O + 2e^- = Br^- + 2OH^-$	0.76
$Cr_2O_7^{2-} + 14H^+ + 6e^- = 2Cr^{3+} + 7H_2O$	1.33	$HO_2^- + H_2O + 2e^- = 3OH^-$	0.88
$Cl_2(g) + 2e^- = 2Cl^-$	1.36	$ClO^- + H_2O + 2e^- = Cl^- + 2OH^-$	0.89
$PbO_2 + 4H^+ + 2e^- = Pb^{2+} + 2H_2O$	1.46	$O_3(g) + H_2O + 2e^- = O_2(g) + 2OH^-$	1.24
$MnO_4^- + 8H^+ + 5e^- = Mn^{2+} + 4H_2O$	1.51	$S_2O_8^{2-} + 2e^- = 2SO_4^{2-}$	2.01
$H_5IO_6 + H^+ + 2e^- = IO_3^- + 3H_2O$	1.60		
$MnO_4^- + 4H^+ + 3e^- = MnO_2 + 2H_2O$	1.70		
$H_2O_2 + 2H^+ + 2e^- = 2H_2O$	1.77		
$S_2O_8^{2-} + 2e^- = 2SO_4^{2-}$	2.01		
$O_3(g) + 2H^+ + 2e^- = O_2(g) + H_2O$	2.07		

* Ref. 2.

a chemist who wishes to know whether he can make hydrazine by the oxidation of aqueous ammonia by nitric oxide. From a table of free energies of formation[2], he could obtain the following data:

Species	$\Delta G_f^\circ(25^\circ)$, kcal mole^{-1}
$H_2O(l)$	-56.69
$N_2H_4(aq)$	30.56
$NO(g)$	20.72
$NH_3(aq)$	-6.36

Thus he could calculate $\Delta G^\circ = +48.86$ kcal mole^{-1} and $K = 1.5 \times 10^{-36}$ for the reaction

$$8\,NH_3 + 2\,NO(g) = 5\,N_2H_4 + 2\,H_2O \tag{2.10}$$

He would then know that, under ordinary conditions, the reaction cannot proceed sufficiently to give an appreciable concentration of hydrazine.

Tables of heats of formation for species are also very useful. Often one finds that ΔH° is known for a reaction when ΔG° is not. In some such cases, ΔH° may be taken as a measure of the driving force for the reaction. (This procedure amounts to neglecting the last term in the equation $\Delta G^\circ = \Delta H^\circ - T\,\Delta S^\circ$.) Sometimes ΔH° is known and ΔS° either can be calculated from available entropy data or (as we shall show) can be estimated. Then, of course, ΔG° is, in effect, known.

The Relation between Equilibrium Constants and Temperature

The value of ΔH° for a reaction provides a measure of the temperature coefficient of the equilibrium constant. By differentiating the relation

$$\ln K = \frac{\Delta S^\circ}{R} - \frac{\Delta H^\circ}{RT} \tag{2.11}$$

we obtain the equation

$$\frac{d \ln K}{dT} = \frac{\Delta H^\circ}{RT^2} \tag{2.12}$$

An appreciation of this relation was of importance in the first synthesis of platinum hexafluoride,[3] in which platinum was treated with fluorine. It was presumed that although PtF_6 is probably stable with respect to decomposition to the elements, it would be unstable with respect to dissociation into fluorine and a lower platinum fluoride. It was supposed that ΔH° for the reaction

$$PtF_4 + F_2 = PtF_6 \tag{2.13}$$

would be positive (endothermic) and that the formation of PtF_6 would be

[3] B. Weinstock, H. H. Claassen, and J. G. Malm, *J. Am. Chem. Soc.*, **79**, 5832 (1957); B. Weinstock, J. G. Malm, and E. E. Weaver, *J. Am. Chem. Soc.*, **83**, 4310 (1961).

favored at very high temperatures. The reaction vessel, containing a platinum wire with external electrical terminals, was cooled to $-196°$. A current through the platinum wire started the exothermic reaction, which then continued without the passage of current until the platinum was consumed. The steady-state temperature of the platinum during the reaction was approximately 1000°C. Although PtF_6 is more stable at high temperatures than at low temperatures, it is nevertheless unstable, and it was isolable only because convection currents in the reactor brought the PtF_6 in contact with the cold walls of the reactor (where it froze out) faster than it could decompose.

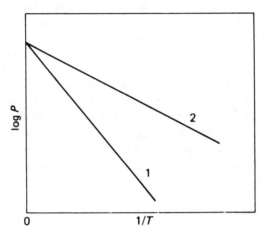

FIG. 2.1. The partial pressures of two species in equilibrium with the same condensed phase. In this example, $\Delta H_1° > \Delta H_2°$.

Consider a gaseous phase containing several types of molecules (say monomers, dimers, etc.) in equilibrium with a liquid or solid phase. How will the molecular composition of the gas change with increasing temperature? For simplicity, let us first consider a gaseous phase containing only two species, each in equilibrium with the same liquid phase. From Eq. (2.12) we find that the logarithms of the partial pressures may be expressed as follows:

$$\frac{d \ln P_1}{dT} = \frac{\Delta H_1°}{RT^2} \tag{2.14}$$

$$\frac{d \ln P_2}{dT} = \frac{\Delta H_2°}{RT^2} \tag{2.15}$$

Using these equations, we may write

$$\frac{d \ln(P_1/P_2)}{dT} = \frac{\Delta H_1° - \Delta H_2°}{RT^2}$$

Hence, we see that the gaseous species corresponding to the higher heat of reaction (more positive $\Delta H°$) will increase in relative importance as the temperature is raised. Now, as a fair approximation, the values of $\Delta S°$ for the reactions forming the two species are equal.[4] Consequently, the partial pressures of the species will approach each other as the temperature becomes very large. In Fig. 2.1 we have plotted the logarithms of the pressures against $1/T$. Obviously we may state, as a general rule for a mixture of vapor species in equilibrium with a common condensed phase, that *whatever species is in relatively small abundance at low temperatures will gain in relative abundance as the temperature is raised.*

In the case of monomeric and dimeric acetic acid vapor in equilibrium with the liquid, $\Delta H°_{monomer} > \Delta H°_{dimer}$. Hence, the vapor is principally dimer at low temperatures, and the fraction of monomer increases with temperature. However, it is more commonly observed that low molecular weight species predominate in saturated vapor at low temperatures, and polymeric species increase in importance with increasing temperature. For example, at low temperatures the main gaseous species in equilibrium with solid molybdenum trioxide is Mo_3O_9, but as the temperature is increased the proportions of Mo_4O_{12} and Mo_5O_{15} steadily increase. Thus, if one seeks a gaseous system containing a large variety of molecular species of complex structure, it is usually advisable to go to the highest possible temperature at which the saturated system can still exist.[6]

Qualitative Data

Thermodynamic reasoning can be applied even to qualitative data. As an example we shall cite the development of practical syntheses of bis(trifluoromethyl)disulfide, CF_3SSCF_3, and trifluoromethanethiol, CF_3SH.[7] It was known that the following two reactions proceed spontaneously.

$$CF_3SH + CF_3SCl \longrightarrow CF_3SSCF_3 + HCl \qquad (2.17)$$

$$CF_3SSCF_3 + 2\,HCl + Hg \longrightarrow 2\,CF_3SH + HgCl_2 \qquad (2.18)$$

If we double Eq. (2.17) and add Eq. (2.18), we obtain

$$2\,CF_3SCl + Hg \longrightarrow CF_3SSCF_3 + HgCl_2 \qquad (2.19)$$

and if we simply add Eqs. (2.17) and (2.18), we obtain

$$CF_3SCl + Hg + HCl \longrightarrow CF_3SH + HgCl_2 \qquad (2.20)$$

---

[4] This condition follows from the facts that the entropy of the condensed phase is small relative to that of the gas and that the major contribution to the entropy of the gas molecules is the translational entropy, which is the same for each molecule. For a discussion of this approximation, see pp. 535–36, ref. 5.

[5] G. N. Lewis, M. Randall, K. S. Pitzer, and L. Brewer, *Thermodynamics*, McGraw-Hill Book Company, New York, 1961.

[6] L. Brewer, *J. Chem. Educ.*, **35**, 153 (1958).

[7] W. R. Cullen and P. S. Dhaliwal, *Can. J. Chem.*, **45**, 379 (1967).

Now the latter two reactions, inasmuch as they are additive combinations of spontaneous reactions, are themselves spontaneous reactions. This fact was recognized by Cullen and Dhaliwal, who showed that the latter reactions were useful synthetic methods for CF_3SSCF_3 and CF_3SH, respectively.

THE ESTIMATION OF THERMODYNAMIC DATA

Estimating Entropies

There are many more compounds whose heats of formation are known than there are compounds whose free energies of formation are known. Inasmuch as free energy data are usually of more interest to chemists than heat data, it is fortunate that fairly reliable methods are available for estimating entropies.

Experimental values exist for the entropies of many gaseous molecules, and the entropies of gaseous molecules for which there are no entropy data can usually be estimated by using the general rule that molecules having similar structure and mass have approximately the same entropy. Periodic table extrapolation and interpolation is generally quite reliable. When this method is not possible, rough estimation can be made using the following empirical equations[8] (where M is the molecular weight):

$$\text{diatomic gases:} \quad S° = 53.8 + 0.043M - 240M^{-1} \tag{2.21}$$

$$\text{polyatomic gases:} \quad S° = 39.0 + 0.34M - 6.2 \times 10^{-4}M^2 \tag{2.22}$$

The entropies of some gaseous molecules of relatively simple structure are given in Table 2.2. Other tabulations should be consulted for more complete listings.

The entropy of an ionic compound at 25° may be estimated by a procedure due to Latimer[9] in which the entropy contributions of the metal cation(s) and the anion(s) are added. The contributions of the metal cations are listed in Table 2.3. The values are a smooth function of atomic weight, and values for metals not listed may be interpolated from the data in the table. The contributions of the anions are listed in Table 2.4. It will be noted that the value to be used for the contribution of a particular anion depends on the charge on the cation in the solid. As an illustration of the method of calculation, let us estimate the entropy of manganese(II) chloride, $MnCl_2$.

	Mn:	10.3	(interpolated value)
	2 Cl:	16.2	
estimated entropy:		26.5	cal deg^{-1} mole^{-1}
experimental value:		28.0	cal deg^{-1} mole^{-1}

[8] O. Kubaschewski and E. L. Evans, *Metallurgical Thermochemistry*, Pergamon, London, 1956.

[9] Latimer, *Oxidation Potentials*, 1952.

Table 2.2   Entropies of Gaseous Molecules at 25°C, cal deg $^{-1}$ mole $^{-1}$

Ne	35.0	CO_2	51.1	CH_4	44.5
Ar	37.0	N_2O	52.6	SiH_4	48.7
Kr	39.2	XeF_2	62.1	CCl_4	74.0
Xe	40.5			$SiCl_4$	79.2
		BF_3	60.7		
H_2	31.2	BCl_3	69.3	SF_6	69.5
D_2	34.6	BBr_3	77.5	XeF_6	87.5
N_2	45.8	SO_3	61.2	UF_6	90.8
O_2	49.0	$COCl_2$	67.5		
F_2	48.5				
Cl_2	53.3	H_2O	45.1		
CO	47.2	H_2S	49.2		
NO	50.3	H_2Se	52.9		
		H_2Te	56.0		
HCl	44.6	F_2O	58.9		
HBr	47.4	SO_2	59.4		
HI	49.3				

Table 2.3   Cation Contributions to Entropies of Solids*

Li	3.5	Fe	10.4	Cd	12.9
Be	4.3	Ni	10.5	Sn	13.1
B	4.9	Cu	10.8	Cs	13.6
Na	7.5	Zn	10.9	Ba	13.7
Mg	7.6	Ge	11.3	La	13.8
Al	8.0	As	11.5	Hf	14.8
Si	8.1	Rb	11.9	Pt	15.2
K	9.2	Sr	12.0	Hg	15.4
Ca	9.3	Y	12.0	Pb	15.5
Sc	9.7	Mo	12.3	Th	15.9
Cr	10.2	Ag	12.8	U	16.0

* Ref. 9.

In hydrated salts, the contribution per mole of water is 9.4 cal deg $^{-1}$ mole $^{-1}$. When the water is strongly coordinated to a cation, the contribution of the anion should be increased by a unit or two. The contribution of NH_4^+ is 13.6 cal deg $^{-1}$ mole $^{-1}$. As another example, we shall show the calculations for ammonium aluminum alum, $NH_4Al(SO_4)_2 \cdot 12 H_2O$. Inasmuch as the average cation charge is $+2$, we shall use the value for sulfate appropriate for a $+2$ cation, but increased by one unit (to give 18.2) because of the strong hydration of the aluminum ion.

$$
\begin{array}{rl}
NH_4: & 13.6 \\
Al: & 8.0 \\
12\,H_2O: & 112.8 \\
2\,SO_4: & \underline{36.4} \\
\text{estimated entropy:} & 170.8 \text{ cal deg}^{-1}\text{ mole}^{-1} \\
\text{experimental value:} & 166.6 \text{ cal deg}^{-1}\text{ mole}^{-1}
\end{array}
$$

Table 2.4   Entropy Contributions of Anions in Solid Compounds* (cal deg^{-1} mole^{-1})

Negative Ion	Charge on Positive Ion			
	+1	+2	+3	+4
F^-	5.5	4.0	3.5	5.0
Cl^-	10.0	8.1	6.9	8.1
Br^-	13.0	10.9	(9.)	(10.)
I^-	14.6	13.6	12.5	13.0
CN^-	7.2	(6.)		
OH^-	5.0	4.5	3.0	
ClO^-	(14.)	(10.)	(8.)	
ClO_2^-	19.2	(17.)	(14.)	
ClO_3^-	24.9	(20.)		
ClO_4^-	26.0	(22.)		
BrO_3^-	26.5	22.9	(19.)	
IO_3^-	25.5	(22.)		
$H_4IO_6^-$	33.9	(30.)		
NO_2^-	17.8	(15.)		
NO_3^-	21.7	17.7	(15.)	(14.)
VO_3^-	20.0	(18.)		
MnO_4^-	31.8	(28.)		
O^{2-}	2.4	0.5	0.5	1.0
S^{2-}	8.2	5.0	1.3	2.5
Se^{2-}	(16.)	11.4	(8.)	
Te^{2-}	(16.5)	12.1	(9.)	
CO_3^{2-}	15.2	11.4	(8.)	
SO_3^{2-}	(19.)	14.9	(11.)	
$C_2O_4^{2-}$	(22.)	17.7	(14.)	
SO_4^{2-}	22	17.2	13.7	(12.)
CrO_4^{2-}	26.2	(21.)		
SiO_4^{2-}	(19.)	13.8	(9.)	7.9
SiO_3^{2-}	16.8	10.5	(7.)	
PO_4^{3-}	(24.)	17.0	(12.)	
HCO_3^-	17.4	(13.)	(10.)	
$H_2PO_4^-$	22.8	(18.)		
$H_2AsO_4^-$	25.1	(21.)		

* Ref. 9.

Sometimes it is necessary to estimate the entropy of an aqueous species in order to calculate, from thermal data, the free energy change for a reaction involving species dissolved in water. The entropy of an aqueous monatomic ion can be estimated with fair accuracy using the empirical relation[10]

$$S^\circ = 47 - \frac{154Z}{r_e^2} \tag{2.23}$$

where Z is the absolute value of the ionic charge, and r_e is the "effective

[10] R. E. Powell, *J. Phys. Chem.*, **58**, 528 (1954).

radius" of the ion (r_e is greater than the Pauling crystal radius[11] by 1.3 Å for cations and by 0.4 Å for anions). The entropy of an aqueous oxy-anion of formula $XO_n{}^{Z-}$ can be estimated using the equation[12]

$$\bar{S}^\circ = 43.5 - 46.5(Z - 0.28n) \tag{2.24}$$

For anions containing hydroxyl groups, the hydroxyl groups are disregarded in counting the number of oxygens, n. The entropy of an aqueous non-electrolyte may be estimated as 14 e.u. plus contributions from the rotational and vibrational entropy. The paper by Powell[10] should be consulted for a discussion of difficulties involved in this latter method. Miller and Hildebrand[13] have shown that the entropy of solution of nonreactive gases to the same mole fraction in water varies linearly with the two-thirds power of their molar volumes at their boiling points. This fact suggests that the entropy change is proportional to the number of hydrogen bonds broken—a quantity that should be proportional to the total surface of the solute.

Empirical Correlations

Let us suppose that we must estimate a thermodynamic function (e.g., $\Delta F_f^\circ, \Delta H_f^\circ, K, E^\circ$, etc.) for a process involving a particular compound. It may be that the compound is a member of a set of compounds, related either by formula or structure, for some of which the thermodynamic function is known. Then by plotting the function either against position in the periodic table or against an appropriate structural parameter, the desired function can be estimated by interpolation or extrapolation.

For example, let us suppose that we wish to estimate the pK value for arsine in aqueous solution. Arsine is one of a family of binary hydrides for most of which the aqueous pK values are known. In Table 2.5, these hydrides

Table 2.5   Aqueous pK Values for Binary Hydrides

CH_4	NH_3	H_2O	HF
45	39	16	3
SiH_4	PH_3	H_2S	HCl
~35	27	7	−7
GeH_4	AsH_3	H_2Se	HBr
~25	?	4	−9
		H_2Te	HI
		3	−10

[11] Selected values are given in Table 2.16.
[12] R. E. Connick and R. E. Powell, *J. Chem. Phys.*, **2**, 2206 (1953).
[13] K. W. Miller and J. H. Hildebrand, *J. Am. Chem. Soc.*, **90**, 3001 (1968).

are arranged according to the periodic table, with the pK values indicated, when known. By considering both the horizontal and vertical trends in this table, we conclude that the pK of arsine probably is about 23. (The reader should verify these extrapolations.) In this case, the reliability of the estimate is very good, and it would be surprising to find (at such time that the pK of arsine is actually determined) that this estimate is in error by more than two pK units.

As a second example, let us suppose that we wish to estimate the heat of formation at 25° of gaseous *n*-tetragermane, Ge_4H_{10}. This compound is the fourth member of an homologous series starting with GeH_4, Ge_2H_6, and Ge_3H_8, for which the heats of formation (gaseous) are 21.6, 38.7, and 53.6 kcal mole^{-1}, respectively.[14] A reasonable estimate for ΔH_f° of *n*-Ge_4H_{10}, obtained by extrapolation from these data, is 67 kcal mole^{-1}.

Considerable caution must be used in making extrapolations of the type just described. For example, properties are not always a smooth function of position in the periodic table. This fact is obvious from a comparison of the properties of a compound of the first main row of the periodic table with the properties of the corresponding compounds in the same vertical family. As an illustration, the dissociation energy of F_2 (36.7 kcal mole^{-1}) is not predictable from the corresponding values for Cl_2, Br_2, and I_2 (57.1, 45.5, and 35.5 kcal mole^{-1}, respectively).

Wilcox and Bromley[15] have shown that heats and free energies of formation for simple ionic solids can be estimated by the following type equation:

$$-\Delta H_f^\circ \quad \text{or} \quad -\Delta G_f^\circ = n_{AB}(X_B - X_A)^2 + n_A Y_A + n_B Y_B + n_{AB}\left(\frac{W_B}{W_A}\right) \quad (2.25)$$

The symbol n_{AB} refers to the "apparent number of single bonds," and n_A and n_B are the number of ions A and B in the "molecule." The parameters X, Y, and W were evaluated for many ions by an iterative least squares treatment of the available literature data. (Of course, different parameters were obtained for $-\Delta H_f^\circ$ and $-\Delta G_f^\circ$.) The average deviation of ΔH_f° obtained for 427 compounds containing monovalent and divalent cations was 1.98 kcal mole^{-1}, and the maximum deviation was 8.73 kcal mole^{-1}. Wilcox and Bromley[15] have tabulated the estimated values of $-\Delta H_f^\circ$ for a large number of compounds for which thermochemical data are unknown.

Estimating Heats of Reaction

Sometimes it is necessary to estimate the value of $\Delta H°$ for a reaction when there are no data available for analogous reactions. In such cases it is usually

[14] S. R. Gunn and L. G. Green, *J. Phys. Chem.*, **65**, 779 (1961); *J. Phys. Chem.*, **68**, 946 (1964).

[15] D. E. Wilcox and L. A. Bromley, *Ind. Eng. Chem.*, **55**, 32 (1963); also see D. E. Wilcox, M.S. Thesis, University of California, Berkeley, UCRL Report 10397, Aug. 1, 1962.

helpful to break up the reaction into a number of simpler processes for each of which the value of $\Delta H°$ is known or able to be estimated. Some important types of processes into which more complicated reactions can often be broken are listed, with the names and symbols of the energies involved, in Table 2.6.

Table 2.6   Some Elementary Processes

Process	Energy	Symbol
$X(g) \rightarrow X^+(g) + e^-(g)$	ionization potential	I
$X(g) + e^-(g) \rightarrow X^-(g)$	electron affinity*	E.A.
$X(s) \rightarrow X(g)$	sublimation energy	S
$X(l) \rightarrow X(g)$	vaporization energy	ΔH_v
$X{-}Y \rightarrow X\cdot + Y\cdot$	dissociation energy	D
$M^+(g) + X^-(g) \rightarrow M^+X^-(c)$	lattice energy*	U
$X^+(g) \rightarrow X^+(soln)$	solvation energy	ΔE_{solv}
$X\;Y \rightarrow X + {:}Y$	coordinate bond energy or dissociation energy of acid-base adduct	

    * Usually taken as a positive quantity when exothermic.

These energies are usually tabulated as changes in internal energy $\Delta E°$ at $0°K$. At this temperature, $\Delta E° = \Delta H°$, and to convert the data to $\Delta H°$ values at $298°K$ we must use experimental or estimated values of $\Delta C_p°$. The correction is usually small (a few kilocalories) compared with other uncertainties involved in the approximate calculations that we shall describe, and we therefore shall assume that $\Delta E°$ and $\Delta H°$ are equivalent and temperature independent.

Ionization Potentials. The ionization potential of an atom or molecule is the energy required to remove an electron from the gaseous species to form the corresponding gaseous ion and gaseous electron. The ionization potentials are known for most of the monatomic elements of the periodic table[16] and are tabulated in Table 2.7. The "first ionization potential" of an element (listed under I in the table) corresponds to the formation of the singly charged ion. The second, third, etc. ionization potentials correspond to the processes

$$M^+(g) \longrightarrow M^{2+}(g) + e^-(g) \tag{2.26}$$

$$M^{2+}(g) \longrightarrow M^{3+}(g) + e^-(g) \quad (etc.) \tag{2.27}$$

The ionization potentials of some simple molecules are listed in Table 2.8.[17]

[16] B. E. Douglas and D. H. McDaniel, *Concepts and Models of Inorganic Chemistry*, Blaisdell Publishing Co., New York, 1965, pp. 34–36.

[17] More extensive tabulations of ionization potentials are to be found in refs. 18 and 19.

[18] R. W. Kiser, *Introduction to Mass Spectrometry and its Applications*, Prentice-Hall, Inc., Englewood Cliffs, N.J., 1965, pp. 308–18.

[19] V. I. Vedeneyev et al., *Bond Energies, Ionization Potentials and Electron Affinities*, E. Arnold Ltd., London, 1966, pp. 151–90.

Table 2.7.   Ionization Potentials of the Elements* (in eV)

Z	Element	I	II	III	IV	V	VI	VII	VIII
1	H	13.595							
2	He	24.581	54.403						
3	Li	5.390	75.619	122.419					
4	Be	9.320	18.206	153.850	217.657				
5	B	8.296	25.149	37.920	259.298	340.127			
6	C	11.256	24.376	47.871	64.476	391.986	489.84		
7	N	14.53	29.593	47.426	77.450	97.863	551.925	666.83	
8	O	13.614	35.108	54.886	77.394	113.873	138.080	739.114	871.12
9	F	17.418	34.98	62.646	87.14	114.214	157.117	185.139	953.60
10	Ne	21.559	41.07	63.5	97.02	126.3	157.91		
11	Na	5.138	47.29	71.65	98.88	138.37	172.09	208.444	264.155
12	Mg	7.644	15.031	80.12	109.29	141.23	186.49	224.90	265.957
13	Al	5.984	18.823	28.44	119.96	153.77	190.42	241.38	284.53
14	Si	8.149	16.34	33.46	45.13	166.73	205.11	246.41	303.07
15	P	10.484	19.72	30.156	51.354	65.007	220.414	263.31	309.26
16	S	10.357	23.4	35.0	47.29	72.5	88.029	280.99	328.80
17	Cl	13.01	23.80	39.90	53.5	67.80	96.7	114.27	348.3
18	Ar	15.755	27.62	40.90	59.79	75.0	91.3	124.0	143.46
19	K	4.339	31.81	46	60.90	82.6	99.7	118	155
20	Ca	6.111	11.868	51.21	67	84.39	109	128	143.3
21	Sc	6.54	12.80	24.75	73.9	92	111	139	159
22	Ti	6.82	13.57	27.47	43.24	99.8	120	141	172
23	V	6.74	14.65	29.31	48	65	129	151	174
24	Cr	6.764	16.49	30.95	50	73	91	161	185
25	Mn	7.432	15.636	33.69	52	76	100	119	196
26	Fe	7.87	16.18	30.643	56.8		103	130	151
27	Co	7.86	17.05	33.49			83.1	133	163
28	Ni	7.633	18.15	35.16					168
29	Cu	7.724	20.29	36.83					
30	Zn	9.391	17.96	39.70					
31	Ga	6.00	20.51	30.70	64.2				
32	Ge	7.88	15.93	34.21	45.7	93.4			
33	As	9.81	18.63	28.34	50.1	62.6	127.5		
34	Se	9.75	21.5	32	43	68	82	155	
35	Br	11.84	21.6	35.9	47.3	59.7	88.6	103	193
36	Kr	13.996	24.56	36.9	52.5	64.7	78.5	111.0	126
37	Rb	4.176	27.5	40	52.6	71.0	84.4	99.2	136
38	Sr	5.692	11.027	43.6	57	71.6	90.8	106	122.3
39	Y	6.38	12.23	20.5	61.8	77	93.0	116	129
40	Zr	6.84	13.13	22.98	34.33	82.3	99	116	139
41	Nb	6.88	14.32	25.04	38.3	50	103	125	141
42	Mo	7.10	16.15	27.13	46.4	61.2	68	126	153
43	Tc	7.28	15.26	29.54					
44	Ru	7.364	16.76	28.46					
45	Rh	7.46	18.07	31.05					
46	Pd	8.33	19.42	32.92					
47	Ag	7.574	21.48	34.82					

Table 2.7 (cont.)

Z	Element	I	II	III	IV	V	VI	VII	VIII
48	Cd	8.991	16.904	37.47					
49	In	5.785	18.86	28.03	54.4				
50	Sn	7.342	14.628	30.49	40.72	72.3			
51	Sb	8.639	16.5	25.3	44.1	56	108	119	
52	Te	9.01	18.6	31	38	60	72	137	
53	I	10.454	19.09	33		71	83	104	170
54	Xe	12.127	21.2	32.1	44	60	83	102	126
55	Cs	3.893	25.1	35				108	122
56	Ba	5.210	10.001	35.5					127
57	La	5.61	11.43	19.17					
72	Hf	7	14.9						
73	Ta	7.88	16.2						
74	W	7.98	17.7						
75	Re	7.87	16.6						
76	Os	8.7	17						
77	Ir	9	17						
78	Pt	9.0	18.56						
79	Au	9.22	20.5						
80	Hg	10.43	18.751	34.2	72	82			
81	Tl	6.106	20.42	29.8	50.7				
82	Pb	7.415	15.028	31.93	42.31	68.8			
83	Bi	7.287	16.68	25.56	45.3	56.0	88.3		
84	Po	8.43	19.4	27.3					
85	At	9.5	20.1	29.3					
86	Rn	10.746	21.4	29.4					
87	Fr	3.83	22.5	33.5					
88	Ra	5.277	10.144						
89	Ac	6.9	12.1	20					
90	Th		11.5	20.0	29.38				

* Ref. 16.

Table 2.8.   Ionization Potentials of Some Molecules (in eV)

H_2	15.427	B_2H_6	11.9	CCl_4	11.47	S_8	8.9
Li_2	4.96	CH_4	12.98	$SiCl_4$	12.06	SF_6	19.3
B_2	12.06	C_2H_6	11.65	$GeCl_4$	11.90	SiF_4	15.4
C_2	12.0	C_3H_8	11.07	$TiCl_4$	11.7	BF_3	15.5
N_2	15.576	NH_3	10.154	NO	9.25	BCl_3	12.0
O_2	12.075	H_2O	12.59	NO_2	9.78	BBr_3	9.7
F_2	15.7	HF	15.77	N_2O	12.89	BI_3	9.0
Cl_2	11.48	HCl	12.74	O_3	12.80	$AlCl_3$	12.8
Br_2	10.55	HBr	11.62	CO_2	13.79	C_6H_6	9.245
I_2	9.28	HI	10.38	CS_2	10.08	$(C_6H_5)_2NH$	7.4
S_2	9.9	H_2S	10.46	OCS	11.17	N_2H_4	9.00

Electron Affinities. The electron affinity of an atom or molecule is the energy released upon adding a gaseous electron to the gaseous species to form the corresponding gaseous anion. An electron affinity is essentially an ionization potential of an anion. Electron affinities are usually more difficult to measure than ionization potentials; consequently, not very many electron affinities are known, and these are generally not known with high precision. Values for some atoms and molecules are given in Table 2.9.[20,21]

Table 2.9   Electron Affinities (kcal mole $^{-1}$)*

H	17.4	O_2	10.6	OH	42
Li	14	O_3	66	SH	54
Na	12	SO_2	64	SeH	47
K	11	SeO_2	53	NH_2	33
Hg	35	NO_2	92	CN	88
F	78.6	NO_3	90	ClO	67
Cl	83.5	ClO_2	79	N_3	72
Br	77.8	ClO_3	91		
I	70.9	ClO_4	134		
$O \rightarrow O^{2-}$	-156	$O_2 \rightarrow O_2^{2-}$	-110		
$O \rightarrow O^-$	34.1	$O_2^- \rightarrow O_2^{2-}$	-130		
$S \rightarrow S^{2-}$	-80				
$S \rightarrow S^-$	48				
$Se \rightarrow Se^{2-}$	-97				
$N \rightarrow N^{3-}$	-547				
$C \rightarrow C^{4-}$	-708				

* See refs. 20, 21.

Heats of Sublimation and Vaporization. Heats of sublimation are known for many elements and compounds. For most of the metals, the heat of sublimation is equivalent to the heat of formation of the gaseous atoms. The latter quantity[22,23] is tabulated in Table 2.10 for most of the elements, including the nonmetals. When direct experimental heats of sublimation are desired but not available, the value often may be calculated by adding the heat of melting and the heat of vaporization of the liquid. In fact, even when these latter quantities are unknown, they may be estimated if the melting point and boiling point are known. From the data in Table 2.11, it will be noted that compounds having similar solid state structures usually have similar entropies of melting, ΔS_M°. In fact, it is much easier to estimate ΔS_M° than ΔH_M°.[24]

[20] H. O. Pritchard, *Chem. Rev.*, **52**, 529 (1953).

[21] R. S. Berry, *Chem. Revs.*, **69**, 533 (1969).

[22] L. Brewer, *Science*, **161**, 115 (1968); L. Brewer, private communication.

[23] L. Pauling, *The Nature of the Chemical Bond*, 3rd ed., Cornell University Press, Ithaca, N.Y., 1960.

[24] Empirical rules for estimating ΔS_M° are given in Kubaschewski and Evans, *Metallurgical Thermochemistry*, 1956, and in Lewis, Randall, Pitzer, and Brewer, *Thermodynamics*, 1961, pp. 518–19.

Table 2.10.   Heats of Formation of the Atoms at 25° (kcal g-atom^{-1})*

Aluminum	78.7	Hydrogen	52.1	Radon	0
Americium	66	Indium	58	Rhenium	187
Antimony	63	Iodine	25.5	Rhodium	133
Argon	0	Iridium	160	Rubidium	19.5
Arsenic	72 ± 3	Iron	99.3	Ruthenium	155.5
Barium	42.5	Krypton	0	Samarium	49.4
Beryllium	77.5	Lanthanum	103	Scandium	90
Bismuth	50.1	Lead	46.6	Selenium	54
Boron	135	Lithium	38.6	Silicon	109
Bromine	26.7	Lutecium	102	Silver	68
Cadmium	26.7	Magnesium	35	Sodium	25.9
Calcium	42.5	Manganese	68	Strontium	39
Carbon	170.9	Mercury	14.5	Sulfur	66.6
Cerium	101 ± 3	Molybdenum	157	Tantalum	187
Cesium	18.7	Neodymium	77	Technetium	158
Chlorine	29.0	Neon	0	Tellurium	48
Chromium	95	Neptunium	105	Terbium	93
Cobalt	102.4	Nickel	102.8	Thallium	44
Copper	80.7	Niobium	172	Thorium	137.5
Dysprosium	71	Nitrogen	113.0	Thulium	55.5
Erbium	75.8 ± 1	Osmium	188	Tin	72
Europium	42.4	Oxygen	59.2	Titanium	112
Fluorine	18.9	Palladium	91	Tungsten	203
Gadolinium	95.0	Phosphorus	80 ± 10	Uranium	125
Gallium	66	Platinum	135	Vanadium	123
Germanium	89.5	Plutonium	84	Xenon	0
Gold	88	Polonium	34.5	Ytterbium	36.4
Hafnium	148	Potassium	21.5	Yttrium	101.5
Helium	0	Praeseodymium	85.0	Zinc	31.2
Holmium	71.9	Protactinium	126	Zirconium	145.5

* See refs. 22 and 23.

Therefore, using the relation $\Delta S_M^\circ = \Delta H_M^\circ / T_{\text{m.p.}}$, the heat of melting can be estimated from the melting point and an estimated value of ΔS_M°. The ΔS_V° data in Table 2.11 nicely bear out Trouton's Rule: The entropy of vaporization of a normal, unassociated liquid is about 21 e.u. Consequently, a fairly good estimate of ΔH_V° can be made from the boiling point by utilizing the relation $\Delta H_V^\circ = 21 T_{\text{b.p.}}$

Dissociation Energies and Bond Energies. The dissociation energy (D) of a bond in a molecule is the energy required to break the bond in the gaseous molecule with formation of two fragments (atoms or radicals). For example, the dissociation energy of the N—N bond in hydrazine is 60 kcal mole^{-1}.[26]

[25] Kubaschewski and Evans, *Metallurgical Thermochemistry*, 1956; National Bureau of Standards Circular 500, *Selected Values of Chemical Thermodynamic Constants*, 1952; N.B.S. Technical Notes 270–1 and 270–2, 1965 and 1966; L. Brewer, private communication.

[26] W. L. Jolly, *The Inorganic Chemistry of Nitrogen*, W. A. Benjamin, Inc., New York, 1964.

Table 2.11 Heats and Entropies of Melting and Vaporization (at b.p. and in kcal mole^{-1} and e.u.)*

	ΔH°_M	ΔS°_M	ΔH°_V	ΔS°_V		ΔH°_M	ΔS°_M	ΔH°_V	ΔS°_V
CF_4	0.167	1.87	3.01	20.7	PF_3			3.43	19.9
CCl_4	0.60	2.4	7.17	20.5	PCl_3			7.28	20.9
CBr_4	0.98	2.7	10.4	22.6	AsF_3	2.49	9.30	8.57	29.3
$SiCl_4$	1.84	9.0	7.0	21.2	$AsCl_3$	2.42	9.42	7.5	18.6
$GeCl_4$			7.9	22.2	$SbCl_3$	3.03	8.74	10.80	21.9
$SnCl_4$	2.19	9.13	8.3	21.5	$BiCl_3$	2.6	5.1	17.35	24.3
F_2	0.372	6.74	1.51	17.7	S_8	2.34	6.0	20.0	28.0
Cl_2	1.531	8.89	4.88	20.4	NaF		6.2		
Br_2	2.52	9.48	7.34	24.6	NaCl		6.4		
I_2	3.74	9.67			KF		6.0		
					KCl		5.8		
					FeO		4.5		
					AgCl		4.25		
					$MgCl_2$		10.4		
					$MnCl_2$		9.75		
					$FeCl_2$		10.8		

* See ref. 25.

For this process we write

$$N_2H_4(g) \longrightarrow 2\,NH_2(g) \qquad D(H_2N{-}NH_2) = 60\,\text{kcal mole}^{-1} \qquad (2.28)$$

When there are two or more equivalent bonds in a molecule, as in ammonia, the successive dissociation energies are usually significantly different from one another.[26]

$$NH_3(g) \longrightarrow NH_2(g) + H(g) \qquad \Delta H^\circ = 104\,\text{kcal mole}^{-1} \qquad (2.29)$$

$$NH_2(g) \longrightarrow NH(g) + H(g) \qquad \Delta H^\circ = 95\,\text{kcal mole}^{-1} \qquad (2.30)$$

$$NH(g) \longrightarrow N(g) + H(g) \qquad \Delta H^\circ = 81\,\text{kcal mole}^{-1} \qquad (2.31)$$

The term "bond energy" refers to the average dissociation energy. Of course, it is not necessary to know each of the successive dissociation energies to calculate the bond energy. Thus, in the case of ammonia, the bond energy is most accurately calculated by dividing the overall atomization energy of ammonia by three.

$$NH_3 \longrightarrow N + 3\,H \qquad (2.32)$$

$$\Delta H^\circ = 280.3\,\text{kcal mole}^{-1} \qquad E(N{-}H) = \frac{280.3}{3} = 93.4\,\text{kcal mole}^{-1}$$

To calculate the N—N *bond energy* in hydrazine, we assume that the average bond energy of the N—H bonds is the same as that in ammonia. Thus,

$$N_2H_4(g) \longrightarrow 2\,N(g) + 4\,H(g) \tag{2.33}$$

$$\Delta H° = 411.6 \text{ kcal mole}^{-1} = E(\text{N—N}) + 4\,E(\text{N—H})$$

$$E(\text{N—N}) = 411.6 - 373.6 = 38.0 \text{ kcal mole}^{-1}$$

We note that the calculated value of $E(\text{N—N})$ is quite different from $D(\text{H}_2\text{N—NH}_2)$. Bond energies may be used to estimate the heats of atomization (and consequently the heats of formation) of molecules for which no thermodynamic data are available. For example, using the values of $E(\text{N—N})$ and $E(\text{N—H})$ we may calculate the heat of formation of triazane, NH_2NHNH_2, from its constituent atoms.

$$3\,N(g) + 5\,H(g) \longrightarrow N_3H_5(g) \tag{2.34}$$

$$\Delta H° = -2\,E(\text{N—N}) - 5\,E(\text{N—H})$$

$$= -76.0 - 467.0 = -543.0 \text{ kcal mole}^{-1}$$

From Table 2.10 we obtain the data required to calculate the heats of the following reactions:

$$\tfrac{3}{2}\,N_2(g) \longrightarrow 3\,N(g) \qquad \Delta H° = 339.0 \text{ kcal mole}^{-1} \tag{2.35}$$

$$\tfrac{5}{2}\,H_2(g) \longrightarrow 5\,H(g) \qquad \Delta H° = 260.5 \text{ kcal mole}^{-1} \tag{2.36}$$

When all three heats are added, we obtain the estimated heat of formation of $N_3H_5(g)$, $\Delta H° = 56.5 \text{ kcal mole}^{-1}$. Obviously this molecule is very unstable.

Table 2.12 contains a list of bond energies. Many more bond energies can be calculated from appropriate thermal data. When the thermal data necessary for calculating a bond energy are lacking, it is possible to make a crude estimate of the bond energy by using the empirical relation[23]

$$E(\text{A—B}) \approx \frac{E(\text{A—A}) + E(\text{B—B})}{2} + 23(X_A - X_B)^2 \tag{2.37}$$

Here X_A and X_B are the electronegativities of atoms A and B, respectively. The electronegativities of the elements are listed in Table 2.13. Equation (2.37) should be used only when the electronegativity difference is small ($|X_A - X_B| < 1.5$) and when $|E(\text{A—A}) - E(\text{B—B})| < 40 \text{ kcal mole}^{-1}$.[27]

As an example of the application of bond energies, let us suppose we wish to estimate $\Delta H°$ for the reaction of 1-butanethiol with arsenic trichloride in carbon tetrachloride solution:

$$3\,\text{BuSH} + \text{AsCl}_3 \longrightarrow \text{As(SBu)}_3 + 3\,\text{HCl} \tag{2.38}$$

[27] R. G. Pearson, *Chem. Commun.,* 65 (1968); J. E. Huheey and R. S. Evans, *ibid.,* 968 (1969).

[28] Taken from ref. 23 and S. R. Gunn and L. G. Green, *J. Phys. Chem.,* **65**, 779 (1961), *J. Phys. Chem.,* **65**, 2173 (1961); E. L. Muetterties and C. W. Tullock, *Preparative Inorganic Reactions,* **2**, 237 (1965). The data in this table may not be entirely consistent with other thermochemical data in this chapter.

Table 2.12 Selected Bond Energies (kcal mole^{-1})*

H—H	104.2	O—H	110.6	O—Cl	48.5
F—F	36.6	S—H	81.1	S—Cl	59.7
Cl—Cl	58.0	Se—H	66.1	N—Cl	47.7
Br—Br	46.1	Te—H	57.5	P—Cl	79.1
I—I	36.1	N—H	93.4	As—Cl	68.9
O—O	33.2	P—H	76.8	C—Cl	78.5
O=O	118.3	As—H	66.8	Si—Cl	85.7
S—S	50.9	Sb—H	60.9	Ge—Cl	97.5
Se—Se	44.0	C—H	98.8		
Te—Te	33	Si—H	76.5	N—O	46
N—N	38.4	Ge—H	69.0	N=O	146
N=N	99	Sn—H	60.4	C—O	84.0
N≡N	226.0	B—H	91	C=O	174
P—P	46.8	BBB	91	Si—O	88.2
As—As	32.1	BHB	105	C=N	147
Sb—Sb	30.2			C≡N	213
Bi—Bi	25	O—F	44.2	C=S	114
C—C	83.1	N—F	64.5		
C=C	147	P—F	116		
C≡C	194	As—F	120		
Si—Si	46.4	Sb—F	107		
Ge—Ge	38.2	C—F	117		
Sn—Sn	34.2	Si—F	143		
B—B	79.3	B—F	153		
Li—Li	26.5				
Na—Na	18.0				
K—K	13.2				
Rb—Rb	12.4				
Cs—Cs	10.7				

* See ref. 28.

Interactions of these species with carbon tetrachloride are small, and we may assume that the heat of reaction is the same in carbon tetrachloride as in the gaseous state. Inasmuch as the BuS— group remains intact during the reaction, we may, to the approximation that the concept of bond energies is valid, rewrite the reaction thus:

$$3 \text{ S—H} + 3 \text{ As—Cl} \longrightarrow 3 \text{ As—S} + 3 \text{ H—Cl} \qquad (2.39)$$

$$\Delta H° = 3[E(\text{S—H}) + E(\text{As—Cl}) - E(\text{As—S}) - E(\text{H—Cl})] \qquad (2.40)$$

All the required bond energies are known except that for the As—S bond. We may estimate $E(\text{As—S})$ using Eq. (2.37):

$$E(\text{As—S}) = \frac{E(\text{As—As}) + E(\text{S—S})}{2} + 23(X_{\text{As}} - X_{\text{S}})^2$$

$$= \frac{32.1 + 50.9}{2} + 23(2.0 - 2.5)^2 = 47$$

Table 2.13   The Electronegativity Scale*

H 2.1																
Li 1.0	Be 1.5											B 2.0	C 2.5	N 3.0	O 3.5	F 4.0
Na 0.9	Mg 1.2											Al 1.5	Si 1.8	P 2.1	S 2.5	Cl 3.0
K 0.8	Ca 1.0	Sc 1.3	Ti 1.5	V 1.6	Cr 1.6	Mn 1.5	Fe 1.8	Co 1.8	Ni 1.8	Cu 1.9	Zn 1.6	Ga 1.6	Ge 1.8	As 2.0	Se 2.4	Br 2.8
Rb 0.8	Sr 1.0	Y 1.2	Zr 1.4	Nb 1.6	Mo 1.8	Tc 1.9	Ru 2.2	Rh 2.2	Pd 2.2	Ag 1.9	Cd 1.7	In 1.7	Sn 1.8	Sb 1.9	Te 2.1	I 2.5
Cs 0.7	Ba 0.9	La–Lu 1.1–1.2	Hf 1.3	Ta 1.5	W 1.7	Re 1.9	Os 2.2	Ir 2.2	Pt 2.2	Au 2.4	Hg 1.9	Tl 1.8	Pb 1.8	Bi 1.9	Po 2.0	At 2.2
Fr 0.7	Ra 0.9	Ac 1.1	Th 1.3	Pa 1.5	U 1.7	Np–No 1.3										

* See ref. 23.

Then, by using Eq. (2.40) and the appropriate values from Table 2.12, we calculate for the reaction in Eq. (2.38)

$$\Delta H° = 3(81 + 69 - 47 - 103) = 0$$

Equation (2.37) leads to a useful generalization: *The most stable combination of covalent bonding for a group of atoms is that combination in which the atom with the highest electronegativity is bonded to the atom with the lowest electronegativity.*[29] There are many examples of spontaneous reactions, such as the following, which bear out this rule.

$$BBr_3 + PCl_3 \longrightarrow BCl_3 + PBr_3 \tag{2.41}$$

$$2\,HI + Cl_2 \longrightarrow 2\,HCl + I_2 \tag{2.42}$$

$$TiCl_4 + 4\,EtOH \longrightarrow Ti(OEt)_4 + 4\,HCl \tag{2.43}$$

The rule should be applied only to reactions in which the number of bonds of each type (covalent, ionic, hydrogen-, three-center, etc.) remains constant. Most apparent exceptions to the electronegativity generalization are reactions in which the type of bonding changes. Thus, the rule correctly predicts a positive free energy change for the following gaseous reaction:

$$ONCl(g) + H_2O(g) = HONO(g) + HCl(g) \qquad \Delta G° = 5.1\ kcal\ mole^{-1} \tag{2.44}$$

However, when nitrosyl chloride is passed into liquid water, hydrolysis occurs because of the strong hydration of the hydrochloric acid that forms:

$$ONCl(g) + H_2O(l) = HONO(aq) + H^+(aq) + Cl^-(aq)$$

$$\Delta G° = -3.4\ kcal\ mole^{-1} \tag{2.45}$$

Consider the following reaction.

$$Et_4Pb + Me_4Pb \longrightarrow 2\,Et_2Me_2Pb \tag{2.46}$$

According to the principle of the additivity of bond energies, the heat of this reaction should be zero. In fact, the heat is very small, and this reaction is just one example of an important class of reactions for which ΔH is usually small and for which the driving force is principally entropy controlled. These reactions are the so-called ligand reorganization reactions in which two or more types of ligands distribute themselves among the various possible combinations with a central electropositive element. When a mixture of tetraethyllead and tetramethyllead is heated in the presence of a catalyst at 80°, reaction (2.46) is not the only reaction that takes place; in fact, an equilibrium mixture

---

[29] The reader may readily prove this for the system of atoms A, B, C, and D for which $X_A > X_B > X_C > X_D$. Assume that there are three bonding combinations: A—B + C—D, A—C + B—D, and A—D + B—C. To prove that the last combination is most stable, it must be shown that $(X_A - X_D)^2 + (X_B - X_C)^2$ is greater than either $(X_A - X_B)^2 + (X_C - X_D)^2$ or $(X_A - X_C)^2 + (X_B - X_D)^2$.

of Et_4Pb, Et_3MePb, Et_2Me_2Pb, $EtMe_3Pb$, and Me_4Pb is formed. The relative amounts of these compounds at equilibrium are very close to those expected for a random distribution of the ethyl and methyl groups.[30] In this particular case, the compounds can be separated from one another without shifting their relative amounts by removing the catalyst and fractionally distilling the mixture at a pressure of 50 mm. Plots of the boiling point vs. the fraction distilled are shown in Fig. 2.2.

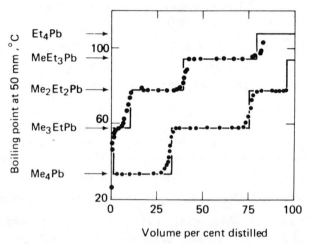

FIG. 2.2. Distillation of the reaction product from Me_4Pb + Et_4Pb. Solid lines calculated for random equilibrium mixtures having Me/Et = 0.776 and 0.345. (From Calingaert and Beatty.[30] Reproduced with permission of the American Chemical Society.)

In general, the technique used for separating the components of an equilibrated mixture must be gentle enough such that redistribution reactions do not occur at an appreciable rate. Otherwise the components of limiting composition will be the principal species isolated. For example, if one attempts to separate the components of an equilibrated mixture of $SnCl_4$ and $SnBr_4$ by fractional distillation, no clean-cut fractions corresponding to $SnCl_3Br$, $SnCl_2Br_2$, and $SnClBr_3$ are collected because of the ease with which reactions of the type

$$4\,SnCl_3Br \longrightarrow 3\,SnCl_4 + SnBr_4 \qquad (2.47)$$

take place during the distillation.[31] A plot of boiling point vs. fraction distilled for a $SnCl_4$—$SnBr_4$ mixture is shown in Fig. 2.3.

Gas chromatography is often used to separate the components of a mixture at a temperature considerably below the boiling points of the

[30] G. Calingaert and H. A. Beatty, *J. Am. Chem. Soc.*, **61**, 2748 (1939).
[31] G. S. Forbes and H. H. Anderson, *J. Am. Chem. Soc.*, **66**, 931 (1944).

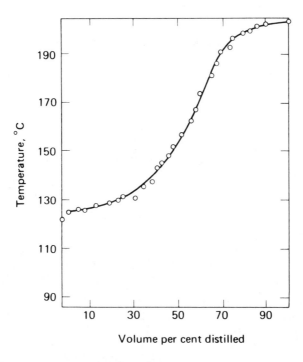

FIG. 2.3. Distillation curve for tin chlorobromides. (From Forbes and Anderson.[31])

components; consequently, this technique can be used for separating components that might undergo reorganization reactions upon distillation. Clark and Brimm[32] used chromatography to analyze the compositions of various mixtures of $Ni(CO)_4$ and $Ni(PF_3)_4$, equilibrated at $75°$. The data, presented in Fig. 2.4, show that the CO and PF_3 groups are essentially randomly distributed.

The reaction of compounds of type MA_n and MB_n to form an intermediate compound MA_iB_j (where $i + j = n$) has a very low value for ΔH whenever the ligands A and B are of the same type (e.g., both halogen or alkoxy or amino or alkyl). When A and B are of different types (e.g., A is a halogen and B is an alkoxy group), sizable exothermic heats are found and the equilibria depart widely from random. These trends may be observed in the data of Table 2.14.

The fact that ΔH differs from zero for reorganization reactions indicates that the principle of the additivity of bond energies is not strictly valid. The deviations can sometimes be rationalized in terms of steric repulsion between large groups, but more often it is helpful to consider the relative preference of groups to bond with orbitals of p (or s) character. For example, consider the

[32] R. J. Clark and E. O. Brimm, *Inorg. Chem.*, **4**, 651 (1965).

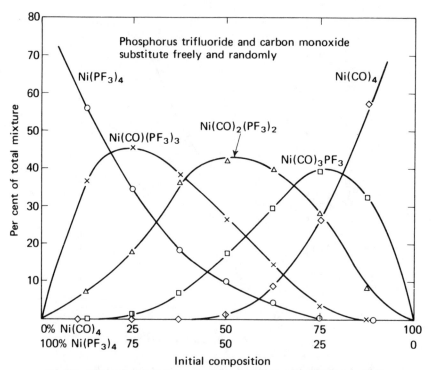

FIG. 2.4. When tetrakis (trifluorophosphine) nickel tetracarbonyl are mixed, the distribution of intermediate substitution compounds after 24 hours at 75°C is almost statistically random. Distribution depends only on the molar ratio of the two ligands. [Figure from *Chem. and Eng. News*, **42**, 52 (Oct. 26, 1964); data from Clark and Brimm.[32] Reproduced with permission of the American Chemical Society.]

Table 2.14   Values of ΔH for Reactions of the Type

$$\frac{i}{n} MA_n + \frac{j}{n} MB_n \rightarrow MA_iB_j*$$

Compound	ΔH (kcal mole^{-1})	Compound	ΔH (kcal mole^{-1})
$BClF_2$	0.49	$Si(OMe)_3OEt$	0.02
BCl_2F	0.7	$Si(OMe)_2(OEt)_2$	0.01
$B(OEt)_2Cl$	−4.0	$Si(OMe)(OEt)_3$	0.05
$B(OEt)Cl_2$	−4.8	$Si(OEt)_3Cl$	−4.3
$B(NMe_2)_2Cl$	−8.9	$Si(NMe_2)_3Cl$	−7.8
$B(NMe_2)Cl_2$	−12.1	$Si(NMe_2)_2Cl_2$	−9.1
		$Si(NMe_2)Cl_3$	−6.5
$PClBr_2$	0.2		
PCl_2Br	−0.8	$MeHgCl$	−6.6
$P(OEt)_2Cl$	−3.2	$MeHgBr$	−4.6
$P(OEt)Cl_2$	−3.2	$MeHgI$	−3.2
$P(NEt_2)_2Cl$	−8.6		
$P(NEt_2)Cl_2$	−9.0		

* See ref. 33.

fact that CH_3HgCl is much more stable with respect to decomposition to $Hg(CH_3)_2$ and $HgCl_2$ than would have been predicted for a random distribution of methyl and chloro groups. Divalent mercury(II) compounds have linear structures, and an s orbital and one p orbital are available on the mercury atom for bonding. The most stable bonding will occur when it is possible for s character to concentrate in an orbital directed toward an electropositive substituent and for p character to concentrate in an orbital directed toward an electronegative substituent.[34] Inasmuch as chlorine is much more electronegative than a methyl group, the combination of bonds in $2\,CH_3HgCl$ would be expected to be more stable than that in $HgCl_2 + Hg(CH_3)_2$.

Bond energies of ionic bonds (as between metals and electronegative elements) can be estimated approximately when the bond distance either is known or can be estimated.[35] The procedure is to calculate the energy of dissociation into ions and then to take account of the ionization potential of the metal and the electron affinity of the electronegative element. The energy required to dissociate a pair of ions separated by a distance R may be estimated using the relation

$$D = -\frac{Z_1 Z_2 e^2}{R} - be^{-aR} + \frac{d}{R^6} \tag{2.48}$$

where Z_1 and Z_2 are the ionic charges (taking account of sign), e is the charge on the electron, and a, b, and d are constants. The first term of Eq. (2.48) is the simple coulombic energy term; the second term is the van der Waals repulsion term, and the third term is the London attractive term. For the case of ion pairs in which each ion has a rare gas electronic configuration, the values of a, b, and d may be taken from Table 2.15. To use the values in Table 2.15, we must take the unit of distance as 0.529 Å, the charge on the electron as unity, and the unit of energy as 627.5 kcal mole^{-1}.

We may illustrate the method for the case of magnesium chloride, for which the energy of dissociation into atoms is known:

$$MgCl_2(g) \longrightarrow Mg(g) + 2\,Cl(g) \qquad \Delta H^\circ = 194\,kcal\,mole^{-1} \tag{2.49}$$

The Mg–Cl distance in gaseous $MgCl_2$ is 2.18 Å; hence, for each Mg^{2+}–Cl^- pair we take $R = 2.18/0.529$. An Mg^{2+}–Cl^- ion pair corresponds to an Ne–Ar rare gas pair; hence, we take $a = 2.18$, $b = 242$, and $d = 30.6$. We calculate, for each Mg^{2+}–Cl^- pair, $D = 290\,kcal\,mole^{-1}$. We assume that $MgCl_2(g)$ is a linear molecule and neglect the second and third terms of Eq. (2.48) for the Cl^-–Cl^- pair. For the Cl^-–Cl^- pair, we calculate $D = -63\,kcal$

[33] J. C. Lockhart, *Chem. Revs.*, **65**, 131 (1965).

[34] H. A. Bent, *Chem. Revs.*, **61**, 275 (1961).

[35] H. B. Gray, *Electrons and Chemical Bonding*, W. A. Benjamin, Inc., New York, 1964, pp. 75–78.

Table 2.15   Van der Waals Energy Parameters*

Interaction Pair	a	b	d
He–He	2.10	6.55	2.39
He–Ne	2.27	33	4.65
He–Ar	2.01	47.9	15.5
He–Kr	1.85	26.1	21.85
He–Xe	1.83	42.4	33.95
Ne–Ne	2.44	167.1	9.09
Ne–Ar	2.18	242	30.6
Ne–Kr	2.02	132	42.5
Ne–Xe	2.00	214	66.1
Ar–Ar	1.92	350	103.0
Ar–Kr	1.76	191	143.7
Ar–Xe	1.74	310	222.1
Kr–Kr	1.61	104	200
Kr–Xe	1.58	169	310
Xe–Xe	1.55	274	480

*See refs. 35 and 36.

$mole^{-1}$. Therefore, for the dissociation of $MgCl_2(g)$ into ions, we estimate $D = 2 \times 290 - 63 = 517 \, kcal \, mole^{-1}$. By combining this value with the overall ionization potential of Mg ($522.8 \, kcal \, mole^{-1}$) and the electron affinity of Cl ($88.2 \, kcal \, mole^{-1}$), we obtain $\Delta H° = 171 \, kcal \, mole^{-1}$ for the atomization of $MgCl_2(g)$.

The problem of estimating relative bond energies arises when one wishes to predict the most stable of several possible structures for a molecule or ion. The following procedure may be employed. First write classical formulas (obeying the Octet Rule when possible) for the various possible structures. Then discard the less stable structures according to the following rules, in the order given, until (it is hoped) only one structure remains.

1. Discard structures that require adjacent atoms to have formal charges of the same sign. For example, we discard the $\overset{+}{N}=\overset{+}{S}-\overset{+}{S}=\overset{-}{N}$ structure in favor of the $\begin{array}{c} \overset{-}{N}-\overset{+}{S} \\ | \quad || \\ S-N \end{array}$ structure for S_2N_2, and we discard the $\overset{-}{O}-\overset{-}{C}=\overset{+}{O}-\overset{-}{O}$ structure in favor of the

$$\begin{array}{c} O \\ || \\ C \\ \diagup \; \diagdown \\ -O \quad\quad O- \end{array}$$

structure for $CO_3{}^{2-}$.

2. Discard structures in which nearby atoms have unlike formal charges contradicting electronegativities, or in which a formal charge resides on an unlikely atom (according to electronegativities). Thus, we discard the

[36] E. A. Mason, *J. Chem. Phys.*, **23**, 49 (1955).

$\overset{-}{N}=\overset{2+}{O}=\overset{-}{N}$ structure in favor of the $N\equiv\overset{+}{N}-\overset{-}{O}$ structure for N_2O.

3. Discard structures with relatively low resonance energy. Thus, we predict nitramide $\left(H_2N-\overset{+}{N}\underset{O^-}{\overset{O}{\diagup}} \right)$ to be stable with respect to hyponitrous acid (HON=NOH).

4. Discard structures that do not maximize the sum of the electronegativity differences of adjacent atoms. For example, the structure H—O—Cl is probably more stable than the structure $H-\overset{+}{Cl}-\overset{-}{O}$. This last rule is simply an application of the generalization discussed on p. 35.

The preceding simple rules are not always effective. Thus they cannot help to decide whether Se_4^{2+} should have a linear, square planar, or tetrahedral structure. In this particular case, simple molecular orbital theory has been used to predict correctly a square planar configuration.[37] Notice that Se_4^{2+} is essentially isoelectronic with S_2N_2 (previously discussed).

Lattice Energies. In the case of ionic crystalline compounds, it is often important to know the energy required to separate the constituent ions in a mole of the crystal to a distance sufficiently great that the interaction energy is negligible. This energy is called the lattice energy. As a rough approximation, the lattice energy, U, may be considered the electrostatic interaction energy less the repulsive energy caused by the crowding of the electronic clouds of the ions. This relation is usually expressed by the equation[38]

$$U = \frac{NMZ_+Z_-e^2}{r}\left(1 - \frac{\rho}{r}\right) \qquad (2.50)$$

where N is Avogadro's number, M is the so-called Madelung constant (the value of which depends on the crystal type), Z_+ and Z_- are the absolute ionic charges, e is the electronic charge, r is the nearest distance between oppositely charged ions, and ρ is a constant, approximately 0.31 Å. The Madelung constant is a geometric factor which takes into account the fact that more than the nearest-neighbor interactions must be considered. The following are Madelung constants for a few important crystal types:

Crystal Type	M
NaCl	1.747
CsCl	1.763
Zinc-blende	1.638
Wurtzite	1.641
CaF_2	2.519
TiO_2	~2.408

[37] I. D. Brown, D. B. Crump, R. J. Gillespie, and D. P. Santry, *Chem. Commun.*, 853 (1968).

[38] K. B. Harvey and G. B. Porter, *Introduction to Physical Inorganic Chemistry*, Addison-Wesley Publishing Co., Inc., Reading, Mass., 1963, pp. 25–36.

When the crystal structure is unknown, a very approximate value of the lattice energy may be calculated using Kapustinskii's formula:[39]

$$U = \frac{287.2 v Z_+ Z_-}{r_+ + r_-}\left(1 - \frac{0.345}{r_+ + r_-}\right) \text{kcal mole}^{-1} \tag{2.51}$$

Here v is the number of ions per chemical formula and r_+ and r_- are the ionic radii. The values of some ionic radii, as calculated by Pauling,[40] are given in Table 2.16.

Qualitative rules, based on lattice energy considerations, have been devised for predicting the sign of $\Delta G°$ for reactions involving only ionic crystals. Thus, one rule states that a reaction can be spontaneous ($\Delta G < 0$) when the overall change in molar volume, ΔV, is negative.[41] For example, $\Delta G° < 0$ for the following reactions (the molar volumes, in cubic centimeters, are given beneath each formula).

Table 2.16 Pauling Ionic Radii (Å)*

Ag^+	1.26	Fe^{2+}	0.76	Pb^{2+}	1.20
Al^{3+}	0.50	Fe^{3+}	0.64	Pb^{4+}	0.84
As^{3-}	2.22	Ga^+	1.13	P^{3-}	2.12
As^{5+}	0.47	Ga^{3+}	0.62	P^{5+}	0.34
Au^+	1.37	Ge^{2+}	0.93	Pd^{2+}	0.86
B^{3+}	0.20	Ge^{4+}	0.53	Ra^{2+}	1.40
Ba^{2+}	1.35	H^-	2.08	Rb^+	1.48
Be^{2+}	0.31	Hf^{4+}	0.81	S^{2-}	1.84
Br^-	1.95	Hg^{2+}	1.10	Sb^{3-}	2.45
C^{4-}	2.60	I^-	2.16	Sc^{3+}	0.81
C^{4+}	0.15	In^+	1.32	Se^{2-}	1.98
Ca^{2+}	0.99	In^{3+}	0.81	Sr^{2+}	1.13
Cd^{2+}	0.97	K^+	1.33	Sn^{2+}	1.12
Ce^{3+}	1.11	La^{3+}	1.15	Sn^{4+}	0.71
Ce^{4+}	1.01	Li^+	0.60	Te^{2-}	2.21
Cl^-	1.81	Lu^{3+}	0.93	Ti^{2+}	0.90
Co^{2+}	0.74	Mg^{2+}	0.65	Ti^{3+}	0.76
Co^{3+}	0.63	Mn^{2+}	0.80	Ti^{4+}	0.68
Cr^{2+}	0.84	Mn^{3+}	0.66	Tl^+	1.40
Cr^{3+}	0.69	Mo^{6+}	0.62	Tl^{3+}	0.95
Cr^{6+}	0.52	N^{3-}	1.71	U^{3+}	1.11
Cs^+	1.69	N^{5+}	0.11	U^{4+}	0.97
Cu^+	0.96	Na^+	0.95	V^{2+}	0.88
Eu^{2+}	1.12	$NH_4{}^+$	1.48	V^{3+}	0.74
Eu^{3+}	1.03	Ni^{2+}	0.72	V^{4+}	0.60
F^-	1.36	O^{2-}	1.40	Y^{3+}	0.93
				Zn^{2+}	0.74

* See ref. 40.

[39] A. F. Kapustinskii, *Quart. Rev.*, **10**, 283 (1956).
[40] Pauling, *The Nature of the Chemical Bond*, 1960.
[41] This rule follows from the fact that U is inversely proportional to r, and V is proportional to r^3.

$$\underset{(22.8)}{KF} + \underset{(19.6)}{MgF_2} \longrightarrow \underset{(38.6)}{KMgF_3} \qquad (2.52)$$

$$\underset{(43.3)}{KBr} + \underset{(21.7)}{AgF} \longrightarrow \underset{(23.4)}{KF} + \underset{(29.0)}{AgBr} \qquad (2.53)$$

Another rule states that $\Delta G° < 0$ for a reaction of the type

$$MX + NY \longrightarrow MY + NX \qquad (2.54)$$

when $r_{M+} < r_{N+}$ and $r_{Y-} < r_{X-}$, or when $r_{N+} < r_{M+}$ and $r_{X-} < r_{Y-}$. That is, the reaction can be spontaneous when the smallest cation combines with the smallest anion.[42] This rule is obeyed in the following two cases:

$$KF + LiBr \longrightarrow KBr + LiF \qquad \Delta H° = -21.8 \text{ kcal mole}^{-1} \qquad (2.55)$$

$$MgCl_2 + 2 NaF \longrightarrow MgF_2 + 2 NaCl \qquad \Delta H° = -34.5 \text{ kcal mole}^{-1} \quad (2.56)$$

However, the rule is not generally valid for cations without rare gas electronic configurations, such as Cu^+, Ag^+, Au^+, Zn^{2+}, and so on. Such cations usually prefer to combine with the larger, more polarizable, anions. Thus, reaction (2.53) is spontaneous even though AgF (the combination of the smallest cation with the smallest anion) is a reactant.

At this point we have discussed all the processes and energies required for an understanding of the Born-Haber cycle. This cycle is based on the fact that the heat of formation of a crystalline compound may be obtained from a direct thermochemical measurement or from the sum of the energies of a series of processes which together are equivalent to the formation step. In the case of an alkali metal chloride, the following Born-Haber cycle may be constructed:

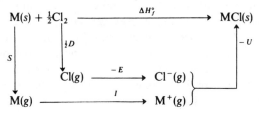

The various energies are related by the equation

$$\Delta H_f° = S + \tfrac{1}{2}D + I - E - U \qquad (2.57)$$

It is instructive to consider which of these terms dominates in establishing trends in heats of formation for the alkali metal halides. From Table 2.17 we see that $-\Delta H_f°$ for the fluorides *increases* on going from CsF to LiF, whereas $-\Delta H_f°$ for the iodides *decreases* on going from CsI to LiI. The only terms in Eq. (2.57) that change in either of these series are S, I, and U. Both S and I

---

[42] The plausibility of the rule can be shown by proving that $1/(r_{M+} + r_{Y-}) + 1/(r_{N+} + r_{X-}) > 1/(r_{M+} + r_{X-}) + 1/(r_{N-1+} + r_{Y-})$ when $r_{M+} < r_{N+}$ and $r_{Y-} < r_{X-}$.

Table 2.17 Heats of Formation of the Alkali Metal Halides (kcal mole^{-1} at 25°)*

	F$^-$	Cl$^-$	Br$^-$	I$^-$
Li$^+$	-147.1	-97.6	-83.9	-64.6
Na$^+$	-137.5	-98.2	-86.3	-68.8
K$^+$	-136.2	-104.4	-94.1	-78.4
Rb$^+$	-133.2	-103.3	-93.6	-78.8
Cs$^+$	-131.7	-106.8	-97.9	-83.7

* See ref. 43.

increase on going from cesium to lithium, whereas U decreases on going from a cesium salt to a lithium salt. Obviously the trend in ΔH_f° for the fluorides is established by the trend in U, which overpowers the opposing trends in S and I. However, the trend in ΔH_f° for the iodides is established by the trends in S and I, which overpower the opposing trend in U. The reason for the relatively small trend in U for the iodides is to be found in the fact that the center-to-center distance between the metal ion and the iodide ion [r in Eq. (2.50)] is principally determined by the larger iodide ion. Consequently the fractional change in r on going from the cesium salt to the lithium salt is relatively small, and the lattice energy does not increase markedly. In the case of the fluorides, the fractional change in r is quite marked on going from one metal to the next; consequently, a large trend in lattice energy is observed.

Of course, in energetic calculations it is completely arbitrary how one chooses, on paper, to break up an ionic solid. The most suitable process is that for which the energy can be either accurately measured or reliably calculated. Fair success has been had in calculating lattice energies (corresponding to the formation of gaseous ions). However, Sanderson[44] has argued that the heat of *atomization*, or total bond energy, is a more logical quantity to use. He has shown that this energy can be calculated as the sum of a covalent contribution and an ionic contribution. The covalent contribution is calculated from homonuclear single bond energies and bond length, and the ionic contribution is calculated from Coulomb's Law. The relative weights of these contributions are evaluated by application of the principle of electronegativity equalization.

Solvation Energies. The solvation energy of a species is the energy absorbed when the species is transferred from the gaseous state to the dissolved state. Thus, the hydration energy for the Na$^+$–Cl$^-$ pair of ions is the energy of the exothermic reaction

$$Na^+(g) + Cl^-(g) \longrightarrow Na^+(aq) + Cl^-(aq) \tag{2.58}$$

$$\Delta E_{solv} = -185 \text{ kcal mole}^{-1}$$

[43] N.B.S. Circular 500, *Selected Values of Chemical Thermodynamic Constants*, 1952; N.B.S. Technical Notes 270–1 and 270–2, 1965 and 1966; L. Brewer, private communication.

[44] R. T. Sanderson, *J. Inorg. Nucl. Chem.*, **28**, 1553 (1966).

Now there is no experimental method known for measuring the solvation energy of a single ion—for example, the Na^+ ion. Only rough estimates of individual ionic solvation energies, obtained by splitting up the solvation energies of oppositely charged pairs of ions according to various theoretical schemes, are available. The simplest and most well-known equation for estimating the solvation energy of an individual ion from first principles is the Born equation:

$$\Delta E_{solv} = -\left(\frac{e^2}{2r}\right)\left(1 - \frac{1}{D}\right) \qquad (2.59)$$

This relation is based on the assumption that the ion is a sphere with a definite radius r (such as the ionic crystal radius) and that it is immersed in a solvent whose effective dielectric constant in the neighborhood of the ion is the ordinary macroscopic dielectric constant D. Neither of these assumptions is valid, and about the only practical application of the equation is its use as a qualitative guide for comparing solvation energies of ions of different radius.

The correlation of ionic radius with solvation energy may be seen in Table 2.18, where the hydration energies for the various alkali metal-chloride

Table 2.18 Hydration Energies of Alkali Metal-Chloride Pairs*

Ions	Hydration Energy (kcal mole^{-1})	Cation Ionic Radius (Å)
$Li^+ + Cl^-$	-209	0.60
$Na^+ + Cl^-$	-185	0.95
$K^+ + Cl^-$	-165	1.33
$Rb^+ + Cl^-$	-159	1.48
$Cs^+ + Cl^-$	-151	1.69

* See ref. 43.

pairs are listed with the corresponding metal ionic radii. We can similarly attempt to observe a correlation of solvent dielectric constant with solvation energy by listing the heats of solution of solid[45] sodium iodide in various solvents with the corresponding solvent dielectric constants. Such data[46-52]

[45] Differences between heats of solution are equal to differences between solvation energies. Thus, both quantities will show exactly the same trends, if any, with various parameters.

[46] Y.-C. Wu and H. L. Friedman, *J. Phys. Chem.*, **70**, 501 (1966).

[47] F. A. Askew, E. Bullock, H. T. Smith, R. K. Tinkler, O. Gatty, and J. H. Wolfenden, *J. Chem. Soc.*, 1368 (1934).

[48] K. P. Mishchenko and A. M. Sukhotin, *Dokl. Akad. Nauk. SSSR*, **98**, 103 (1954).

[49] G. Somsen and J. Coops, *Rec. Trav. Chim.*, **84**, 985 (1965).

[50] R. P. Held and C. M. Criss, *J. Phys. Chem.*, **69**, 2611 (1965).

[51] E. M. Arnett and D. R. McKelvey, *J. Am. Chem. Soc.*, **88**, 2598 (1966).

[52] W. M. Latimer and W. L. Jolly, *J. Am. Chem. Soc.*, **75**, 4147 (1953).

Table 2.19   Heat of Solution of Sodium Iodide in Various Solvents

Solvent	$-\Delta H^{\circ}_{soln}$ (kcal mole^{-1})	Solvent Dielectric Constant	Drago and Wayland E_B Parameter for Solvent*
Water	1.9	78.5	
Propylene carbonate[46]	5.0	65.1	
Ethanol[47]	5.8	24.3	
Acetonitrile[48]	6.8	36	0.53
Formamide[49]	7.4	109.5	
Methanol[47]	7.5	32.6	0.78
N-methylformamide[50]	8.2	182.4	
Methyl ethyl ketone[48]	9.2	18.5	
Acetone[48]	9.6	20.7	0.71
Furfural[48]	10.4	41	
Piperidine[48]	10.4	5.8	
Dimethyl sulfoxide[51]	11.5	48	0.97
Pyridine[48]	12.7	12.3	0.88
Ammonia[52]	14.6	16.9	1.34

* See ref. 53.

are presented in Table 2.19, where it is obvious that no correlation exists at all. One might expect a better correlation to be found between the solvation energy and some parameter, such as Drago and Wayland's E_B parameter,[53] which measures the donor[54] strength of the solvent molecules. The E_B values in the last column of Table 2.19, although only a few are available, do suggest that such a correlation exists.

Acid-Base Reactions

Protonic Acids. A protonic acid is a hydrogen-containing compound that can dissociate to give a proton and an anion:

$$HA = H^+ + A^- \qquad (2.60)$$

The simplest protonic acids are the binary hydrides of the more electronegative elements. These are listed, with their pK values for aqueous solutions, in Table 2.5. It will be noticed that acidity increases (pK decreases) as one moves in the periodic table from left to right in any row or down within any vertical family. The horizontal trend reflects the increase in the electron affinity of the radicals formed by removing hydrogen atoms from the hydrides

[53] R. S. Drago and B. B. Wayland, *J. Am. Chem. Soc.*, **87**, 3571 (1965); R. S. Drago, *Chem. Britain*, **3**, 516 (1967). The source and application of these parameters are discussed on p. 51.

[54] It is generally agreed that solvation energies of cations are usually larger than solvation energies of anions.

on moving from left to right in the table; the vertical trend reflects the decrease in the energy required to remove hydrogen atoms from the hydrides on moving down the table.

Except for the hydrogen halides, each of these binary hydrides may be looked upon as the parent of a whole family of related protonic acids. For example, by replacing one of the hydrogen atoms of water with a series of other groups, we generate the enormous family of hydroxy acids. Similarly, nitrogen acids can be derived from ammonia, carbon acids from methane, and so on. The following are a few nitrogen acids and their corresponding aqueous pK values:

Nitramide, O_2NNH_2	7
Cyanamide, $NCNH_2$	10.5
p-nitroaniline, $O_2NC_6H_4NH_2$	12
Acetamide, CH_3CONH_2	15
Aniline, $C_6H_5NH_2$	27
Ammonia, NH_3	39

Obviously the acidity of the parent hydride is drastically affected by the nature of the substituent. Groups that are more electron withdrawing than hydrogen cause an increase in acidity; groups that are less electron withdrawing cause a decrease in acidity.

Branch and Calvin[55] devised the following equation for estimating the acidity of a hydroxy acid, HOX.

$$pK = 16 - \Sigma I_\alpha \left(\frac{1}{2.8}\right)^i - \Sigma I_c \left(\frac{1}{2.8}\right)^i - \log\frac{n}{m} \qquad (2.61)$$

The summations are evaluated over all of the atoms in the radical X. The value of I_α is a measure of the electron-attracting ability of the atom: values of I_α for some of the important elements are given in Table 2.20. The formal charge[56] on the atom times 12.3 is I_c. The number of atoms separating the

Table 2.20   Values of I_α for use in Branch and Calvin's
Equation*

H	0	Se	2.7
Cl	8.5	N	1.3
Br	7.5	P	1.1
I	6.0	As	1.0
O	4.0	C	−0.4
S	3.4	C_6H_5	2.0
	* See ref. 55.		

[55] G. E. K. Branch and M. Calvin, *The Theory of Organic Chemistry*, Prentice-Hall, Inc., Englewood Cliffs, N.J., 1941.

[56] In a simple equation of J. Ricci [*J. Am. Chem. Soc.*, **70**, 109 (1948)], the pK is essentially a function of only the formal charges of the atom immediately attached to the OH group and of the oxygen atoms attached to it: $pK = 8 - 9m + 4n$, where m is the formal charge on the central atom and n is the number of non-hydroxyl oxygen atoms.

OH group from the atom in question is i. The last term in the equation is a statistical factor; n is the number of acidic hydrogens in the acid, and m is the number of equivalent oxygen atoms in the anion to which a proton can bond. To illustrate the application of this equation, we shall calculate the pK of the dihydrogen phosphate ion,

$$
\begin{array}{c}
\text{O}^{-} \\
| \\
\text{HO}-\overset{+}{\text{P}}-\text{O}^{-} \\
| \\
\text{O} \\
\text{H}
\end{array}
$$

$$\sum I_a \left(\frac{1}{2.8}\right)^i = 1.1\left(\frac{1}{2.8}\right)^0 + 3(4)\left(\frac{1}{2.8}\right)^1 = 5.3_8$$

$$\sum I_c \left(\frac{1}{2.8}\right)^i = 12.3\left(\frac{1}{2.8}\right)^0 - 2(12.3)\left(\frac{1}{2.8}\right)^1 = 3.5_2$$

$$\log \frac{n}{m} = \log \tfrac{2}{3} = -0.1_8$$

$$pK = 16 - 5.3_8 - 3.5_2 + 0.1_8 = 7.3$$

The estimated value of 7.3 is in good agreement with the experimental value of 7.1.

Under normal circumstances, the ionization of a protonic acid does not involve the liberation of free protons. When such a reaction takes place in aqueous solution, the process is best looked upon as the transfer of a proton to a water molecule (or better, a group of water molecules). If we wish to represent the hydrated proton as an oxonium ion, H_3O^+, we write for the ionization

$$\text{HA} + \text{H}_2\text{O} \longrightarrow \text{H}_3\text{O}^+ + \text{A}^- \qquad (2.62)$$

In fact, all reactions of protonic acids may be conveniently thought of as transfers of a proton from one base to another. Brönsted[57] was one of the first to recognize this fact. He pointed out that acids and bases are related by the following "half-reaction":

$$\text{acid} \rightleftharpoons \text{base} + \text{H}^+ \qquad (2.63)$$

That is, an acid is a hydrogen-containing species capable of acting as a proton donor, and a base is a species capable of acting as a proton acceptor. Under this concept, acid-base equilibria are competitions among bases for protons. A typical acid-base reaction may then be written

$$\text{acid}_1 + \text{base}_2 = \text{acid}_2 + \text{base}_1 \qquad (2.64)$$

In this reaction, acid_1 is the conjugate acid of base_1 and base_1 is the conjugate

[57] J. N. Brönsted, *Rec. trav. chim.*, **42**, 718 (1923).

base of $acid_1$. That is, by adding a proton to $base_1$, we obtain $acid_1$. The same formal relationships exist between $acid_2$ and $base_2$.

Lewis Acid-Base Reactions. The Lewis Electronic Theory is the most general of the commonly used acid-base concepts. In a Lewis acid-base reaction, a pair of electrons from one species is used to form a covalent bond to another species. The species that "donates" the electron pair is the base, and the species that "accepts" the electron pair is the acid. The Lewis concept is completely independent of solvent considerations. Indeed, many Lewis acid-base reactions proceed in the gas phase. The following reactions are all Lewis acid-base reactions in which the first reagent is the acid and the second reagent is the base.

$$H^+ + F^- \longrightarrow HF \tag{2.65}$$

$$(CH_3)_3B + NH_3 \longrightarrow (CH_3)_3BNH_3 \tag{2.66}$$

$$Ag^+ + 2I^- \longrightarrow AgI_2^- \tag{2.67}$$

$$SnCl_4 + 2Cl^- \longrightarrow SnCl_6^{2-} \tag{2.68}$$

$$NO_2^+ + Cl^- \longrightarrow NO_2Cl \tag{2.69}$$

$$CO_2 + GeH_3^- \longrightarrow GeH_3CO_2^- \tag{2.70}$$

$$Ni + 4CO \longrightarrow Ni(CO)_4 \tag{2.71}$$

$$\tfrac{1}{2}B_2H_6 + Mn(CO)_5^- \longrightarrow H_3BMn(CO)_5 \tag{2.72}$$

$$I_2 + C_6H_6 \longrightarrow I_2(C_6H_6) \tag{2.73}$$

One of the difficulties in measuring and correlating the relative basicities of Lewis bases is the fact that different trends are observed when different reference acids are used. For example, the coordinating power of the halide ions toward the Al^{3+} ion increases in the order $I^- < Br^- < Cl^- < F^-$. On the other hand, the order is $F^- < Cl^- < Br^- < I^-$ for the Hg^{2+} ion. To resolve this and many other problems, Pearson has proposed the Principle of Hard and Soft Acids and Bases.[58] A soft acid or base is one in which the valence electrons are easily polarized or removed, and a hard acid or base is one which holds its valence electrons tightly and is not easily distorted. The principle is as follows: Hard acids prefer to coordinate with hard bases, and soft acids prefer to coordinate with soft bases. Tables 2.21 and 2.22 list some acids and bases in the categories hard, soft, and borderline. The opposing trends in the reactivity of the halide ions toward Al^{3+} and Hg^{2+} are easily rationalized using this principle. The aluminum ion is a hard acid that prefers to bond to hard bases—such as fluoride—and the mercuric ion is a soft acid that prefers to bond to soft bases—such as iodide.

[58] R. G. Pearson, *Chem. Britain*, **3**, 103 (1967); *Science*, **151**, 172 (1966); *J. Am. Chem. Soc.*, **85**, 3533 (1963).

Table 2.21   The Classification of Some Lewis Acids*

Hard	*Borderline*	*Soft*
H^+, Li^+, Na^+, K^+	Fe^{2+}, Co^{2+}	Cu^+, Ag^+, Au^+, Tl^+, Hg^+
Be^{2+}, Mg^{2+}, Ca^{2+}, Mn^{2+}	Pb^{2+}, Sb^{3+}, Rh^{3+}	Pd^{2+}, Cd^{2+}, Pt^{2+}, Hg^{2+}
Al^{3+}, Sc^{3+}, La^{3+}, N^{3+}	BMe_3, SO_2, NO^+	Pt^{4+}, Tl^{3+}, BH_3
Cr^{3+}, Co^{3+}, As^{3+}	R_3C^+, $C_6H_5{}^+$	I^+, Br^+, HO^+, RO^+
Si^{4+}, Ti^{4+}, Pu^{4+}, Hf^{4+}		I_2, Br_2, ICN, etc.
$UO_2{}^{2+}$, VO^{2+}, MoO^{3+}		O, Cl, Br, I, N
$BeMe_2$, BF_3, $B(OR)_3$		M^0 (metal atoms)
$AlMe_3$, $AlCl_3$, AlH_3		bulk metals
SO_3, I^{7+}, I^{5+}, CO_2		
HX (hydrogen bonding molecules)		

* See ref. 58.

The Principle of Hard and Soft Acids and Bases has many applications in synthetic chemistry. Inasmuch as the hardness of an element increases with increasing oxidation state, in order to stabilize an element in a very high oxidation state, it should be coordinated to hard bases, such as O^{2-}, OH^-, or F^-. To stabilize an element in a low oxidation state, it should be coordinated to very soft bases, such as CO, PR_3, and CN^-. Carbanions, olefins, and aromatic hydrocarbons are soft bases. Consequently, metal coordination compounds containing the latter groups as ligands are made using transition metals in low oxidation states. The low oxidation states are stabilized by soft ligands; thus we find compounds such as $CH_3Mn(CO)_5$ but not $CH_3Mn(H_2O)_5$. By similar reasoning, one concludes that metal-metal bonds should be most stable when the metal atoms are stabilized in low oxidation states. This is the case with most of the known compounds containing metal-metal bonds, such as $Mn_2(CO)_{10}$, $[MoC_5H_5(CO)_3]_2$, and $(C_6H_5)_3PAuCo(CO)_4$.

Eaborn[59] has shown that it is possible to set up a "conversion series" for changing the group X in compounds of the type R_3SiX. In the series $R_3SiI \rightarrow (R_3Si)_2S \rightarrow R_3SiBr \rightarrow R_3SiNC \rightarrow R_3SiCl \rightarrow R_3SiNCS \rightarrow R_3SiNCO$, a compound can be converted into any other on its right by boiling it with the

Table 2.22   The Classification of Some Lewis Bases*

Hard	*Borderline*	*Soft*
H_2O, OH^-, F^-	$C_6H_5NH_2$	R_2S, RSH, RS^-
$CH_3CO_2{}^-$, $PO_4{}^{3-}$, $SO_4{}^{2-}$	C_5H_5N	I^-, SCN^-, $S_2O_3{}^{2-}$
Cl^-, $CO_3{}^{2-}$, $ClO_4{}^-$, $NO_3{}^-$	$N_3{}^-$, Br^-	R_3P, R_3As, $(RO)_3P$
ROH, RO^-, R_2O	$NO_2{}^-$, $SO_3{}^{2-}$	CN^-, RNC, CO
NH_3, RNH_2, N_2H_4		C_2H_4, C_6H_6
		H^-, R^-

* See ref. 58.

59 C. Eaborn, *J. Chem. Soc.*, 3077 (1950).

appropriate silver salt. However, a compound cannot be appreciably converted in this way into a compound on its left. For example, the following reaction proceeds readily only in the forward direction:

$$Et_3SiI + AgBr \longrightarrow Et_3SiBr + AgI \qquad (2.74)$$

MacDiarmid[60] has extended this idea to silyl compounds: $SiH_3I \rightarrow (SiH_3)_2Se$ $\rightarrow (SiH_3)_2S \rightarrow SiH_3Br \rightarrow SiH_3NCSe \rightarrow (SiH_3N)_2C \rightarrow SiH_3Cl \rightarrow SiH_3CN \rightarrow$ $SiH_3NCS \rightarrow SiH_3NCO \rightarrow (SiH_3)_2O \rightarrow SiH_3F$. Note that in these series the substituents are listed in the order of increasing hardness. The soft silver ion prefers to bond to a soft group, and the hard silicon prefers to bond to a hard group.

Drago and Wayland[61] have shown that, for a Lewis acid-base reaction of the type

$$A + B \longrightarrow AB \qquad (2.75)$$

which takes place either in the gas state or in a poorly solvating solvent, the heat of reaction may be calculated from the equation

$$- \Delta H = E_A E_B + C_A C_B \,(\text{kcal mole}^{-1}) \qquad (2.76)$$

where E_A, E_B, C_A, and C_B are empirical constants characteristic of the acid A and the base B. Some values of these constants are given in Tables 2.23 and 2.24. To illustrate this method, we may calculate the heat of the reaction

$$(CH_3)_3B + C_5H_5N \longrightarrow (CH_3)_3BNC_5H_5 \qquad (2.77)$$

$$- \Delta H = 5.77 \times 0.88 + 1.76 \times 6.92 = 17.3 \,\text{kcal mole}^{-1}$$

Table 2.23 Some Acid Parameters

Acid	C_A	E_A	Acid	C_A	E_A
I_2	1.000	1.00	BMe_3	1.76	5.77
C_6H_5OH	0.574	4.70	SO_2	0.726	1.12
ICl	1.61	4.15	$C_2(CN)_4$	1.51	1.68
Me_3SnCl	0.60	6.25	C_6H_5SH	0.174	1.36
Me_3COH	0.095	3.77			

The experimental value is 17.0 kcal mole^{-1}. The constants E_A and E_B have been interpreted as the tendency of the acid and base to form an electrostatic bond; C_A and C_B have been interpreted as the tendency of the acid and base to form a covalent bond. It will be noted that the ratios C_A/E_A and C_B/E_B are proportional to the softness of the species.

[60] A. G. MacDiarmid, *Preparative Inorganic Reactions*, **1**, 165 (1964).

[61] Drago and Wayland, *J. Am. Chem. Soc.*, **87**, 3571 (1965); Drago, *Chem. Britain*, **3**, 516 (1967).

Table 2.24   Some Base Parameters

Base	C_B	E_B	Base	C_B	E_B
C_5H_5N	6.92	0.88	MeC(O)OEt	2.42	0.639
NH_3	3.42	1.34	Me_2CO	0.66	0.706
$MeNH_2$	6.14	1.19	Me_2SO	3.42	0.969
Me_3N	11.61	0.59	$(CH_2)_4SO$	3.30	1.09
Et_3N	11.35	0.65	Et_2O	3.55	0.654
MeCN	1.77	0.533	$(CH_2)_4O$	4.69	0.61
$MeCONMe_2$	3.00	1.00	MeOH	1.12	0.78
$HCONMe_2$	2.73	0.97	Et_2S	7.78	0.041
$(Me_3CO)_3PO$	1.81	1.09	C_6H_6	1.36	0.143

Edwards[62] has shown that equilibrium constants for various Lewis acid-base reactions may be quantitatively correlated by means of the equation

$$\log\left(\frac{K}{K_0}\right) = \alpha E_n + \beta H \tag{2.78}$$

Here K is the equilibrium constant for a base reacting with a particular acid, and K_0 is the constant for a reference base (say, water) reacting with the same acid. The terms α and β are empirical constants characteristic of the acid, and E_n and H are independent properties of the base. E_n is a sort of polarizability factor of the ligand Y^- defined by $E_n = E^\circ + 2.60$, where E° is the standard oxidation potential for the process

$$Y^- = \tfrac{1}{2} Y_2 + e^- \tag{2.79}$$

H is a proton basicity factor defined by $H = 1.74 + pK$, where pK refers to the conjugate acid of Y^-. By using stability constant data for complex ions in aqueous solution, the values of α and β shown in Table 2.25 were evaluated for metal ions.[63] In Table 2.25, the ions are listed in the order of decreasing α/β, and Pearson's classification of hard, soft, and borderline is indicated in the last column. There is a strong correlation that shows the ratio α/β to be a measure of the hardness of the metal ion.

Complications. It would be unfortunate if the reader were to gain the impression that the theory of acid-base reactions is in a highly developed, quantitative state. Actually the oversimplified generalizations of the preceding paragraphs are of rather limited applicability and subject to many exceptions. To show that even relatively simple acid-base reactions can have difficult-to-explain features, let us consider the reactions of ammonia, methylamine, dimethylamine, and trimethylamine with trimethylboron in the gas

[62] J. O. Edwards, *J. Am. Chem. Soc.*, **76**, 1540 (1954); *J. Am. Chem. Soc.*, **78**, 1819 (1956).
[63] A. Yingst and D. H. McDaniel, *Inorg. Chem.*, **6**, 1067 (1967).

Table 2.25 Values of α and β for Metal Ions*

Metal Ion	α	β	α/β	Hardness†
Hg^{2+}	5.786	-0.031	187	S
Cu^+	4.060	0.143	28.4	S
Ag^+	2.812	0.171	16.5	S
Pb^{2+}	1.771	0.110	16.1	B
Sr^{2+}	1.382	0.094	13.0	H
Cd^{2+}	2.132	0.171	12.5	S
Cu^{2+}	2.259	0.233	9.7	B
Mn^{2+}	1.438	0.166	8.7	H
In^{3+}	2.442	0.353	6.9	H
Mg^{2+}	1.402	0.243	5.8	H
Zn^{2+}	1.367	0.252	5.4	B
Ga^{3+}	3.795	0.767	5.0	H
Ba^{2+}	1.786	0.411	4.4	H
Fe^{3+}	1.939	0.523	3.7	H
Ca^{2+}	1.073	0.327	3.5	H
Al^{3+}	-0.749	1.339	0.6	H
H^+	0.000	1.000	0.0	H

* See ref. 63.

† The Pearson classification (see ref. 58): S = soft, H = hard, B = borderline.

phase and with water in aqueous solution:

$$amine(g) + B(CH_3)_3(g) \longrightarrow amine:B(CH_3)_3(g) \tag{2.80}$$

$$amine(aq) + H_2O(l) \longrightarrow amine:H^+(aq) + OH^-(aq) \tag{2.81}$$

The equilibrium constants[64] for these reactions are plotted in Fig. 2.5. For both trimethylboron and the aqueous proton, the amine basicity increases with the introduction of the first methyl group, increases further with the introduction of the second, but drops sharply with the introduction of the third. Explanations of these interesting phenomena have been offered by several authors, and the interested reader is invited to read the summarizing discussion of Hammond.[65]

Lewis bases which can simultaneously donate electrons to a Lewis acid by a σ bond and accept electrons from the acid by a π bond are very difficult to systematize. For example, consider the replacement of "strong" π-acceptor ligands such as CO and PF_3 by the "weak" ligand PH_3 in transition metal complexes. Some examples are the conversions of metal hexacarbonyls into $(CO)_4M(PH_3)_2$,[66] of $HCo(PF_3)_4$ into $HCo(PF_3)_3PH_3$,[67] of $Ni(CO)_4$

[64] H. C. Brown, H. Bartholomay, Jr., and M. D. Taylor, *J. Am. Chem. Soc.*, **66**, 435 (1944).

[65] G. S. Hammond, *Steric Effects in Organic Chemistry*, M. S. Newman, ed., John Wiley & Sons, Inc., 1956, Chap. 9, pp. 454–60.

[66] E. O. Fischer, E. Louis, W. Bathelt, E. Moser, and J. Müller, *J. Organometallic Chem.*, **14**, P9 (1968).

[67] J. M. Campbell and F. G. A. Stone, *Angew. Chem. Intl. Ed.*, **8**, 140 (1969).

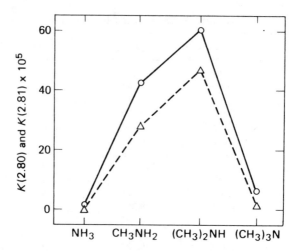

Fig. 2.5. Equilibrium constants for reactions 2.80 ($K_{100°}$, Δ) and 2.81 ($K_{25°} \times 10^5$, O).

into $(CO)_3NiPH_3$,[68] and of $ONCo(CO)_3$ into $ONCo(CO)_2PH_3$.[69] This "strong" behavior of a reputedly weak ligand is probably due to the presence of more than enough π-acceptor ligands for effective employment of all available metal d_π electrons, so that little π-bonding energy is lost by replacing one strong π-acceptor by a weak π-acceptor. Clearly there can be no dependable order of bonding strength for donor-acceptor ligands, even when all are compared by bonding to the same metal.[69]

Nonexistent Compounds

One of the most useful applications of thermodynamic estimation in synthetic chemistry is the estimation of the free energy of formation of compounds that have never been prepared. Such estimated free energy data permit one to decide whether particular synthetic techniques can be used for the preparation of these unknown compounds and whether the compounds will be susceptible to decomposition under ordinary conditions. The book *Nonexistent Compounds* by Dasent should be consulted for interesting discussions of reasons for the nonexistence of, and the likelihood of the synthesis of, a wide variety of compounds that are unknown and of low stability but which "do not offend the simpler rules of valence."[70] Bartlett[71] has shown how thermo-

[68] I. H. Sabherwal and A. B. Burg, *Inorg. Nucl. Chem. Letters*, **5**, 259 (1969).

[69] I. H. Sabherwal and A. B. Burg, *Chem. Commun.*, 853 (1969).

[70] W. E. Dasent, *Nonexistent Compounds*, Marcel Dekker, Inc., New York, 1965; also see D. A. Johnson, *Some Thermodynamic Aspects of Inorganic Chemistry*, University Press, New York, 1968.

[71] N. Bartlett, *Angew. Chem. Intern. Ed. Engl.*, **7**, 433 (1968).

dynamic reasoning was of great importance in planning the preparation of $XePtF_6$ and other compounds formed from reactions of the third transition series hexafluorides.

Problems

1. Although compounds such as carbamic acid (H_2NCOH, with an O double-bonded to C) and aminoguanidine ($H_2NCNHNH_2$, with an NH double-bonded to C) are known, the following compounds are unknown: $HOCOH$ and $HOCNHNH_2$ (with NH double-bonded to C). Explain.

2. Which of the following reactions have equilibrium constants greater than 1 at ordinary temperatures (all species may be assumed to be gaseous, unless otherwise specified)?
   (a) $NH_3 + 3 HOCl = NCl_3 + 3 H_2O$
   (b) $NOCl + H_2O = HNO_2 + HCl$
   (c) $4 ROH + SiCl_4 = 4 HCl + Si(OR)_4$
   (d) $2 COCl_2 + SnBr_4 = 2 COBr_2 + SnCl_4$
   (e) $2 HD = H_2 + D_2$
   (f) $R_3P:BBr_3 + R_3N:BF_3 = R_3P:BF_3 + R_3N:BBr_3$
   (g) $Et_4Pb + Me_4Pb = 2 Et_2Me_2Pb$
   (h) $BF_3 + 3HCl = BCl_3 + 3 HF$
   (i) $(C_6H_5)_3P:SO_2 + (CH_3)_3COH = (C_6H_5)_3P:HOC(CH_3)_3 + SO_2$
   (j) $Ag^+ + 2\,py(pyridine) = Agpy_2^+$ (aqueous)
   (k) $H_2PO_4^- + SO_4^{2-} = HSO_4^- + HPO_4^{2-}$ (aqueous)
   (l) $HOCl + OBr^- = HOBr + OCl^-$ (aqueous)

3. Explain why $NaB(C_6H_5)_4$ is a good reagent for precipitating large cations.

4. Estimate the heat of formation of $AuF(s)$. Do you think it would be possible to prepare $AuF(s)$ from Au and F_2?

5. The reduction potentials at 25° for nickel are
$$Ni(OH)_2(s) + 2\,e^- = Ni(s) + 2\,OH^- \qquad E° = -0.72\ v$$
$$Ni^{2+}(aq) + 2\,e^- = Ni(s) \qquad E° = -0.25\ v$$
For $Ni(NH_3)_4^{2+} = Ni^{2+} + 4\,NH_3(aq)$, $K = 10^{-8}$. It is desired to prepare a solution of $Ni(NH_3)_4^{2+}$ that is just barely unsaturated with respect to $Ni(OH)_2(s)$ by adding 0.1 mole of $Ni(OH)_2$ to a liter of aqueous solution containing $(NH_4)_2SO_4$ and NH_3. If the final solution should be $0.1M$ in NH_3, what should the concentration of ammonium sulfate and ammonia be in the original solution?

6. It is desired to remove the oxygen from an aqueous solution saturated with air by passing a stream of nitrogen through it. The concentration of oxygen in the initial solution is $2.5 \times 10^{-4} M$. The nitrogen contains 0.10 mole per cent oxygen impurity. What concentration of oxygen in the solution can be achieved by this process?

7. A chemist wishes to try to prepare a new form of CuCl in which the crystal lattice contains the molecular units Cu_3Cl_3. (Such a species has been observed in the gas phase in equilibrium with monomeric CuCl molecules.) He plans to quench (cool rapidly) cuprous chloride vapor by passing a stream of vapor equilibrated with liquid CuCl (at about 1300°) over a surface cooled to $-200°$. Will the Cu_3Cl_3/CuCl ratio in the gas phase increase or decrease as he increases the temperature at which the argon is equilibrated with CuCl? Explain. In a separate experiment he plans to prepare the saturated vapor at 1200° and then to allow the gas to pass through a much hotter, empty tube before quenching. How will the composition of the gas change with increasing temperature of the secondary hot tube? Explain.

8. What is wrong with the following syntheses?:
   (a) the preparation of sodium hydride by the reaction of hydrogen with a solution of sodium in liquid ammonia.
   (b) the preparation of ammonium silyl (NH_4SiH_3) by the reaction of NH_3 with SiH_4.
   (c) the preparation of peroxydisulfate $(S_2O_8{}^{2-})$ by the oxidation of $SO_4{}^{2-}$ with $MnO_4{}^-$ in acid solution.
   (d) the preparation of BCl_3 by the treatment of $Cl_3B:N(CH_3)_3$ with BF_3.
   (e) the preparation of sodium phenyl (NaC_6H_5) by the reaction of sodium with chlorobenzene in liquid ammonia.
   (f) the preparation of $HPO_2{}^-$ by the reduction of $HPO_3{}^{2-}$ with tin metal in alkaline solution.

Kinetic Aspects

of

Synthesis

3

THE RATE LAW AND SYNTHESES

When we work with thermodynamic data, it is always important to keep in mind that thermodynamics can tell us whether a reaction is capable of taking place, but it cannot tell us the reaction rate. Thus, a reaction may be thermodynamically favored ($\Delta G < 0$) and yet proceed extremely slowly. Therefore, both kinetics and thermodynamics must be considered.

At a particular temperature, the rate of a reaction may be expressed as some function of the concentration of the reactants and products. For example, the rate of decomposition of aqueous nitrous acid,

$$3\,HNO_2 = H^+ + NO_3{}^- + 2\,NO + H_2O \tag{3.1}$$

may be expressed by the following rate law:[1]

$$\frac{-d(HNO_2)}{dt} = \frac{k(HNO_2)^4}{(NO)^2} \tag{3.2}$$

[1] D. M. Yost and H. Russell, Jr., *Systematic Inorganic Chemistry*, Prentice-Hall, Inc., Englewood Cliffs, N.J., 1946.

Some reactions have rate laws that involve species not appearing in the net reaction. Such species are called catalysts; for example, the reduction of aqueous dichromate by molecular hydrogen is catalyzed by silver ion.[2] The data fit the following rate law:

$$\frac{-d(H_2)}{dt} = k_1(Ag^+)^2(H_2) + \frac{k_2(Ag^+)^2(H_2)}{k_3(H^+) + (Ag^+)} \tag{3.3}$$

The decomposition of arsine to arsenic and hydrogen is catalyzed by arsenic surfaces.[3] The rate law is

$$\frac{d(H_2)}{dt} = kSP_{AsH_3} \tag{3.4}$$

where S is the surface area of arsenic and P_{AsH_3} is the pressure of arsine.

Uncatalyzed heterogeneous reactions are characterized by the presence of reactants or products in more than one phase. Oftentimes the rates of such reactions are affected not only by the concentrations of the reactants, but also by conditions influencing the overall rate of transfer of the reactants to the interface between the phases. Such reactions are "diffusion influenced" and may be accelerated by stirring and by increasing the interfacial area. The reaction of sodium with a solution of chlorobenzene in toluene and the reaction of gaseous arsine with a solution of sodium in ammonia are examples of diffusion-influenced reactions. The rate laws for such reactions are usually quite complicated and are discussed in various chemical engineering texts.[4]

The k's that appear in Eqs. (3.2) through (3.4) are called *rate constants*. Any rate constant may be expressed in terms of the absolute temperature, an entropy of activation $\Delta S^\ddagger$, and a heat of activation $\Delta H^\ddagger$.

$$k_r = \frac{kT}{h} e^{(\Delta S^\ddagger/R)} e^{-\Delta H^\ddagger/RT} \tag{3.5}$$

Here, k_r is the rate constant, k is the Boltzmann constant, and h is Planck's constant. Every reaction path has its characteristic values of $\Delta S^\ddagger$ and $\Delta H^\ddagger$. The exponential term involving the heat of activation is very important in determining the effect of temperature on the rate constant. For example, with a heat of activation of 11 kcal mole^{-1}, a rate constant will approximately double for each 10° rise in temperature near room temperature.

We see that there are two general methods for regulating reaction rates. The concentrations of the species which occur in the rate law may be changed, and the temperature may be changed. Heterogeneous reactions may be

[2] A. H. Webster and J. Halpern, *J. Phys. Chem.*, **61**, 1239 (1957).

[3] K. Tamaru, *J. Phys. Chem.*, **59**, 777 (1955).

[4] See, for example, O. A. Hougen and K. M. Watson, *Chemical Process Principles. Part 3. Kinetics and Catalysis*, John Wiley & Sons, Inc., New York, 1947; E. E. Petersen, *Chemical Reaction Analysis*, Prentice-Hall, Inc., Englewood Cliffs, N.J., 1965; and M. Boudart, *Kinetics of Chemical Processes*, Prentice-Hall, Inc., Englewood Cliffs, N.J., 1968.

speeded up by increasing the surface area at which reaction takes place. If a rate is diffusion influenced, stirring will increase the rate.

Simple rate-law considerations are important when deciding how long to carry out a reaction in order to achieve essentially complete reaction. One must not be as naïve as the chemist who argued: "The reaction is 50 per cent complete in one hour; therefore it will be 100 per cent complete in two hours." This argument would only be correct in the unlikely event that the reaction was zero order; that is, that the rate was independent of the concentrations of reactants and products. Actually most reactions are first-order reactions (or pseudo-first order, as when all the reactants are in large excess except one, the concentration of which appears in the rate law to the first power). In such cases, the reactions theoretically are not complete until an infinite time has passed. One usually compromises by considering the reaction to be practically complete when it is, say, 95 per cent complete. One calculates that the previously cited reaction would be 95 per cent complete in 4.3 hours, assuming first-order kinetics.[5] Several examples of the use of kinetic reasoning in syntheses follow.

Hydroxylamine

Hydroxylamine may be prepared by the formation and subsequent hydrolysis of the hydroxylamine disulfonate ion:

$$H^+ + NO_2^- + 2\,HSO_3^- \longrightarrow HON(SO_3)_2^{2-} + H_2O \tag{3.6}$$

$$HON(SO_3)_2^{2-} + 2\,H_2O \longrightarrow NH_3OH^+ + 2\,SO_4^{2-} + H^+ \tag{3.7}$$

The first of these reactions is acid catalyzed and is very slow if the pH is above 5.3. However, the pH cannot be decreased much below 4 because under such conditions the nitrite ion is protonated to a significant extent (pK $= 3.35$ for HNO_2), and, as seen from Eqs. (3.1) and (3.2), the nitrous acid decomposes. In view of these pH problems, Rollefson and Oldershaw[6] carried out the first step of the synthesis [Eq. (3.6)] in the optimum pH range (around 5) by buffering the reaction solution with acetate and acetic acid. If excess bisulfite is present during the reaction, the hydroxylamine disulfonate ion can react further to form the trisulfonate ion [$N(SO_3)_3^{3-}$], and if excess nitrite is present, the hydroxylamine disulfonate ion reacts to give nitrous oxide. Rollefson and Oldershaw avoided the difficulties associated with excesses of bisulfite and nitrite by using potassium salts in the synthesis. The relatively insoluble salt $K_2HON(SO_3)_2$ then precipitates out before the yield-reducing side reactions can take place. This intermediate can then be isolated and hydrolyzed in acid

---

[5] The rate law $-dx/dt = kx$ integrates to $\log(x_0/x) = (k/2.3)t$, where x_0 is the initial concentration.

[6] G. K. Rollefson and C. F. Oldershaw, *J. Am. Chem. Soc.*, **54**, 977 (1932).

solution to give hydroxylamine [Eq. (3.7)]. The method gives overall yields around 85 per cent.

N,N-Dimethylcyclohexasulfur-1,5-diimide

When trisulfur dichloride and methylamine are allowed to react in an inert solvent, both ring formation and chain polymerization can occur:[7]

$$2\,S_3Cl_2 + 6\,CH_3NH_2 \longrightarrow CH_3N \underset{S-S-S}{\overset{S-S-S}{\diagup\diagdown}} NCH_3 + 4\,CH_3NH_3Cl \qquad (3.8)$$

$$2\,S_3Cl_2 + 6\,CH_3NH_2 \longrightarrow \frac{1}{x}(\overset{CH_3}{\overset{|}{N}}-S-S-S-\overset{CH_3}{\overset{|}{N}}-S-S-S)_x + 4\,CH_3NH_3Cl \qquad (3.9)$$

Let us assume that the cyclization occurs by a process which is first order in the intermediate species $HNCH_3S_3NCH_3S_3Cl$,

$$\overset{CH_3}{\overset{|}{HN}}-S-S-S-\overset{CH_3}{\overset{|}{N}}-S-S-S-Cl \longrightarrow HCl + CH_3N \underset{S-S-S}{\overset{S-S-S}{\diagup\diagdown}} NCH_3 \quad (3.10)$$

[rate of ring formation $\propto (HNCH_3S_3NCH_3S_3Cl)$]

and that the chain formation occurs by a process which is first order in that species and first order in either S_3Cl_2 or CH_3NH_2.[8]

$$\overset{CH_3}{\overset{|}{HN}}-S-S-S-\overset{CH_3}{\overset{|}{N}}-S-S-S-Cl + CH_3NH_2 \longrightarrow$$

$$\overset{CH_3}{\overset{|}{HN}}-S-S-S-\overset{CH_3}{\overset{|}{N}}-S-S-S-\overset{CH_3}{\overset{|}{N}}H + HCl \qquad (3.11)$$

$$S_3Cl_2 + \overset{CH_3}{\overset{|}{HN}}-S-S-S-\overset{CH_3}{\overset{|}{N}}-S-S-S-Cl \longrightarrow$$

$$Cl-S-S-S-\overset{CH_3}{\overset{|}{N}}-S-S-S-\overset{CH_3}{\overset{|}{N}}-S-S-S-Cl + HCl \qquad (3.12)$$

$$\left[\begin{array}{l} \text{rate of chain formation} \propto (HNCH_3S_3NCH_3S_3Cl)(CH_3NH_2) \\ \qquad\qquad\qquad\text{or} \propto (HNCH_3S_3NCH_3S_3Cl)(S_3Cl_2) \end{array} \right]$$

Then the relative rates of the ring- and chain-forming reactions (i.e., the

---

[7] R. C. Brasted and J. S. Pond, *Inorg. Chem.*, **4**, 1163 (1965).

[8] Many other plausible mechanisms can be proposed, each of which will lead to the same conclusion regarding the effect of dilution on the relative yields of rings and chain polymers.

relative yields of rings and chains) are given by the expression

$$\frac{\text{rings}}{\text{chains}} = \frac{(HNCH_3S_3NCH_3S_3Cl)}{(HNCH_3S_3NCH_3S_3Cl)(CH_3NH_2)} = \frac{1}{(CH_3NH_2)}$$

$$\text{or} = \frac{(HNCH_3S_3NCH_3S_3Cl)}{(HNCH_3S_3NCH_3S_3Cl)(S_3Cl_2)} = \frac{1}{(S_3Cl_2)}$$

(3.13)

Thus we understand why Brasted and Pond carried out the synthesis under conditions of very high dilution in order to favor the formation of rings. This concept of using very dilute reagents to favor ring formation can be applied to a wide variety of syntheses.

Diboron Tetrachloride

When a stream of boron trichloride vapor is passed through an electrical discharge, some of the molecules are broken into fragments such as BCl_2, BCl, and Cl. After leaving the discharge, most of these species recombine to form BCl_3. However, some of the species combine to form Cl_2 and B_2Cl_4, perhaps by paths such as the following:

$$2\,Cl + M \longrightarrow Cl_2 + M \tag{3.14}$$

$$BCl + BCl_3 \longrightarrow Cl_2B{-}BCl_2 \tag{3.15}$$

$$2\,BCl_2 \longrightarrow Cl_2B{-}BCl_2 \tag{3.16}$$

Now, molecular chlorine will react with B_2Cl_4 to form boron trichloride, but with appropriate experimental techniques, this reaction may be avoided and the B_2Cl_4 may be isolated. In one method,[9] BCl_3 vapor is passed through a discharge between copper electrodes. The copper serves as a scavenger for chlorine (CuCl is formed) and an appreciable amount of B_2Cl_4 leaves the discharge zone unscathed. In another method, an electrodeless discharge tube is used, and the reaction between chlorine and B_2Cl_4 is retarded by passing the BCl_3 through the discharge very rapidly and at low pressures.[10] The B_2Cl_4 and unreacted BCl_3 are separated from the chlorine by condensation in a $-112°$ trap.

Sodium Amide

Solutions of alkali metals in liquid ammonia are remarkably stable kinetically, even though they are thermodynamically metastable. A solution of, say, sodium at the boiling point of ammonia ($-33.4°$) can be kept in a clean glass

---

[9] T. Wartik, R. Rosenberg, and W. B. Fox, *Inorg. Syn.*, **10**, 118 (1967). A method involving mercury electrodes is described by G. Urry, T. Wartik, R. Moore, and H. Schlesinger, *J. Am. Chem. Soc.*, **76**, 5293 (1954).

[10] J. W. Frazer and R. T. Holzmann, *J. Am. Chem. Soc.*, **80**, 2907 (1958).

vessel for days without undergoing more than a few per cent decomposition. The decomposition reaction yields sodium amide, which is only slightly soluble in ammonia and which precipitates:

$$Na_{(am)} + NH_3 \longrightarrow NaNH_2 + \tfrac{1}{2}H_2 \qquad (3.17)$$

This reaction is catalyzed by a wide variety of solids—in particular, compounds of the transition metals. The catalyzed reaction is convenient for the synthesis of sodium amide. However, the product is naturally contaminated with the catalyst; to minimize this contamination, we must use a small amount of a very efficient catalyst. The usual catalyst for this purpose is the dark material that forms when a solution of iron(III) nitrate reacts with an alkali metal solution (see the synthesis of sodium amide on p. 444).

KINETICS VERSUS THERMODYNAMICS

In the case of an exothermic reaction ($\Delta H° < 0$), care must be taken that the temperature is not raised too far in order to increase the rate. As a rough approximation, we may consider $\Delta H°$ and $\Delta S°$ to be independent of temperature. From Eq. (2.11) we see that if the temperature is increased sufficiently, the equilibrium constant is principally determined by the entropy of the reaction. If $\Delta S°$ is negative, the equilibrium constant will be less than one and any advantage gained by increasing the rate may be offset by reversing the equilibrium.

Ammonia Synthesis

The Haber process for the industrial synthesis of ammonia is a well-known application of both equilibrium and kinetic considerations in synthesis. Hydrogen and nitrogen react at high temperatures and pressures and in the presence of an iron catalyst according to the reversible reaction

$$N_2(g) + 3H_2(g) = 2NH_3(g) \qquad (3.18)$$

The effect of temperature and pressure on the equilibrium is given in Table 3.1. The formation of ammonia is obviously favored by both low temperatures

Table 3.1   Volume Percentage of Ammonia in 3:1 H_2-N_2 Equilibrium Mixtures*

Temperature (°C)	Pressure (atm)				
	10	*50*	*100*	*300*	*1000*
300	14.73	39.41	52.04	70.96	92.55
400	3.85	15.27	25.12	47.00	79.82
500	1.21	5.56	10.61	26.44	57.47
600	0.49	2.26	4.52	13.77	31.43

* See ref. 1.

and high pressures, but if the temperature falls much below 400°, the rate of the catalyzed reaction is too slow for economical production. On the other hand, if the temperature is too high, the equilibrium pressure of ammonia is too low for satisfactory yields. In practice, pressures around 1000 atm and temperatures around 500° are employed.

Diamond Synthesis

Diamond is thermodynamically unstable with respect to graphite at 1 atm pressure at all temperatures. For the process $C_{graphite} \rightarrow C_{diamond}$, $\Delta G° = 692$ cal mole^{-1} at 25°C and $\Delta G° \approx 2400$ cal mole^{-1} at 1200°.[11] Inasmuch as the molar volumes of diamond and graphite are 3.42 and 5.34 cc, respectively, the driving force for the conversion is increased by increasing the pressure. A plot of the pressure required to make $\Delta G = 0$ against temperature is given in Fig. 3.1. Now, for the reaction to proceed at an appreciable

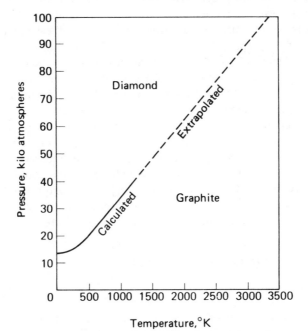

FIG. 3.1. The graphite-diamond equilibrium curve.[11] (Reproduced with permission of the Journal of Chemical Education.)

rate, it is necessary to go to temperatures above approximately 1800°K. Therefore, as shown in Fig. 3.1, pressures greater than about 60,000 atm are required in the synthesis. Another important feature is the use of a suitable

[11] H. T. Hall, *J. Chem. Educ.*, **38**, 484 (1961).

solvent to permit taking apart the graphite lattice atom-by-atom and building the atoms into the diamond lattice. In the first synthesis of a diamond, FeS was the solvent; the temperature was about $1920°K$; the pressure was about 90,000 atm, and the reaction time was 3 min.

In recent years, a number of investigators have succeeded in growing diamonds at low pressures by pyrolyzing the vapors of organic compounds.[12] One process involves passing methane, at a pressure around 0.2 mm, over a diamond seed crystal at about 1050°. After deposition of carbon on the seed crystal, any graphite that has formed is eliminated by reaction with hydrogen at 1033° and 50 atm.

Trimolybdenum Monoxide

There are many compounds which are thermodynamically stable at low temperatures but which disproportionate at higher temperatures. In certain cases, these compounds are stable only at temperatures where the rate of formation is slow. Consequently, they are often difficult to produce and many probably remain undiscovered. However, consider the technique used by Schönberg[13] to prepare Mo_3O, a compound that cannot be produced by conventional methods. If one heats a mixture of molybdenum and MoO_2 at high temperatures, the Mo_3O phase cannot form because it is thermodynamically unstable. If one heats a mixture of molybdenum and MoO_2 at low temperatures, where Mo_3O is stable, the rate of formation is prohibitively slow. To prepare such a phase, one must use a temperature just below the compound's maximum temperature of existence. At this temperature, the rate of formation may be fast enough so that it can be prepared in a reasonable time. Schönberg heated a mixture of molybdenum and MoO_2 in a furnace to 1000° (where Mo_3O is unstable) and very slowly (over a period of days) cooled the sample to room temperature. Upon examining the mixture he found the Mo_3O phase, which had been missed by all previous workers.

NUCLEOPHILIC SUBSTITUTIONS

Nucleophilicity

Lewis bases are sometimes called *nucleophilic reagents* or *nucleophiles*. Hence, the following type of reaction is often called a *nucleophilic substitution* or S_N reaction.[14]

$$Y: + A—X \longrightarrow Y—A + X: \tag{3.19}$$

[12] Recent work has been summarized in *Chem. Eng. News*, July 15, 1968, pp. 44–45.

[13] N. Schönberg, *Acta Chem. Scand.*, **8**, 617 (1954); also see L. Brewer, *J. Chem. Educ.*, **35**, 153 (1958).

[14] C. K. Ingold, *Structure and Mechanism in Organic Chemistry*, Cornell University Press, Ithaca, N.Y., 1953.

The mechanism of such a substitution reaction will generally lie between two limiting extremes: a *dissociative*, S_N1 (lim), mechanism and an *associative*, S_N2 (lim), mechanism.[15,16] An S_N1 (lim) reaction proceeds in two steps. First, a unimolecular dissociation takes place,

$$A—X \longrightarrow A + X: \qquad (3.20)$$

Second, a coordination occurs.

$$Y: + A \longrightarrow A—Y \qquad (3.21)$$

An S_N2 (lim) reaction also has two steps. The first step is the formation of an intermediate in which both X and Y are bound to A. The second step is the loss of X from AXY,

$$Y: + A—X \longrightarrow Y—A—X \longrightarrow A—Y + X: \qquad (3.22)$$

Intermediate mechanisms, in which appreciable amounts of both bond making and bond breaking occur in the rate-determining step, are designated as S_N1 and S_N2. In an S_N1 reaction, the A—X bond is largely broken at the same time that a weak A—Y bond is formed. In an S_N2 reaction, the A—X bond is only slightly broken while a fairly strong A—Y bond is formed. Obviously the borderlines between these mechanism classifications are very fuzzy.

In an S_N1 (lim) reaction in which there is no return of the leaving group, the rate of substitution should be independent of the nucleophilic reagent, because the latter is not involved in the initial dissociation. In an S_N2 (lim) reaction, the rate should be strongly dependent on the nucleophilic character of the incoming ligand. In S_N1 and S_N2 reactions, relatively slight dependence of the rate on the incoming ligand should be observed.

One of the most important problems facing synthetic chemists is that of understanding what properties of a Lewis acid make it susceptible to attack by nucleophiles. When we attempt to correlate the kinetic reactivity of a series of acids or bases, we find the same sort of problem one encounters when attempting to correlate thermodynamic reactivity. Thus, the following ligands are arranged in the order of increasing rate of reaction with *trans*-$Pt(py)_2Cl_2$[15]: $CH_3O^- < C_6H_5NH_2 < NO_2^- < Br^- < C_6H_{11}NC < (C_6H_5)_3As$. The order of kinetic reactivity toward CH_3I for these same ligands is quite different: $C_6H_{11}NC < (C_6H_5)_3As < NO_2^- < C_6H_5NH_2 < Br^- < CH_3O^-$. It appears that soft nucleophiles are most effective toward soft substrates, and hard nucleophiles are most effective toward hard substrates (see pp. 49–52). This rule has been expressed in quantitative terms

[15] F. Basolo and R. G. Pearson, *Mechanisms of Inorganic Reactions*, 2nd ed., John Wiley & Sons, Inc., New York, 1967.

[16] A different nomenclature has been proposed by C. H. Langford and H. B. Gray, *Ligand Substitution Processes*, W. A. Benjamin, Inc., New York, 1965.

by Edwards[17]:

$$\log\left(\frac{k}{k_0}\right) = \alpha E + \beta H \qquad (3.23)$$

Here, k is the rate constant for a nucleophile attacking a particular substrate, and k_0 is the rate constant for water attacking the same substrate. The terms α and β are empirical constants characteristic of the substrate, and E and H are independent properties of the nucleophile. E is a sort of polarizability factor of the ligand Y^- defined by $E = E^\circ + 2.60$, where E° is the standard oxidation potential for the process

$$Y^- = \tfrac{1}{2}Y_2 + e^-$$

H is a proton basicity factor defined by $H = 1.74 + pK$, where pK refers to the conjugate acid of Y^-. Lewis acid substrates that are highly susceptible to attack by hard bases are characterized by large values of β. Substrates that are highly susceptible to attack by soft bases are characterized by large values of α. The ratio α/β is a quantitative measure of the softness of an acid.[18]

Substitution Reactions of Octahedral Complexes

Lability. One of the most important groups of compounds studied by inorganic chemists are the octahedral coordination compounds in which six donor groups are coordinated to a metal ion. Some of these complexes, such as $Co(NH_3)_6^{3+}$ and $Cr(en)_2Cl_2^+$, are relatively inert species the ligands of which can be replaced by other groups only very slowly. Thus, although the following substitution reaction has an enormous driving force,

$$Co(NH_3)_6^{3+} + H^+ + Cl^- \longrightarrow Co(NH_3)_5Cl^{2+} + NH_4^+ \qquad (3.24)$$

the complex $[Co(NH_3)_6]Cl_3$ can be refluxed with 6 M hydrochloric acid for hours without any of the coordinated ammonia being displaced. One must heat the hexaammine complex with concentrated hydrochloric acid in a sealed tube in order to effect reaction. On the other hand, some octahedral complexes, such as $Al(H_2O)_6^{2+}$ and $Cu(NH_3)_4(H_2O)_2^{2+}$, are labile species with ligands that can be rapidly replaced by other groups.

Clearly a synthetic chemist who wishes to make an octahedral complex should be aware of the lability or inertness of the complex in order to choose appropriate experimental conditions for the synthesis. And when a particular geometric or optical isomer of a complex is to be prepared, it is important to know whether the complex is likely to be inert or labile, because a labile

---

[17] J. O. Edwards, *J. Am. Chem. Soc.*, **76**, 1540 (1954); *J. Am. Chem. Soc.*, **78**, 1819 (1956). For a theoretical discussion of this equation, see R. E. Davis, *Organic Sulfur Compounds*, vol. 2, N. Kharasch, ed., Pergamon Press, Inc., New York, 1964, Chap. 1.

[18] A. Yingst and D. H. McDaniel, *Inorg. Chem.*, **6**, 1067 (1967).

complex is liable to isomerize so rapidly as to preclude the isolation of a pure isomer. Thus any generalizations or rules that can help in predicting lability and inertness are of considerable importance.

One important factor in determining the ease with which ligands can be replaced is the charge on the central atom. The higher the positive charge, the more tightly the ligands are held and the more inert the complex is. Thus, a marked decrease in reactivity occurs in the series $AlF_6^{3-} > SiF_6^{2-} > PF_6^- > SF_6$. When base is added to a solution containing AlF_6^{3-}, aluminum hydroxide is precipitated instantly, whereas SF_6 undergoes no detectable reaction with hot concentrated base solutions over long periods of time. Of course, it is probable that part of the trend in reactivity in this series is due to the decrease in size accompanying the increase in charge. That size is an important factor is shown by the rates at which water molecules are replaced in the first coordination sphere of the aqueous ions $Mg(H_2O)_6^{2+}$, $Ca(H_2O)_6^{2+}$, and $Sr(H_2O)_6^{2+}$. The rate constants at 25° are $\sim 10^5$, $\sim 2 \times 10^8$, and $\sim 4 \times 10^8 \sec^{-1}$, respectively.

The lability (or inertness) of transition metal complexes can be correlated with the electronic configurations of the metal ions.[15,19,20] According to a generalization attributable to Taube, any complex with less than three d electrons or with one or more e_g electrons should be labile. All other complexes (i.e., those having 3, 4, 5, or 6 t_{2g} electrons and no e_g electrons) should be inert. This generalization is remarkably accurate. Examples of labile and inert complexes, arranged according to their electronic configurations, are tabulated in Table 3.2. The generalization may be rationalized if we assume that the complexes with d^0, d^1, or d^2 configurations react by $S_N 1$ processes in which the entering group forms a weak bond to the metal as the leaving group breaks its bond to the metal. Such complexes have at least one empty nonbonding d orbital the lobes of which are directed between the attached ligands. An incoming ligand can therefore approach the metal ion from these directions with a minimum of repulsion. The ligands in complexes having e_g electrons are held relatively weakly because of the antibonding character of the e_g electrons; thus, these ligands may be readily displaced.[21]

If one assumes a symmetrical structure for the activated complex of a substitution reaction of an octahedral complex, it is possible to calculate the contribution to the activation energy from the difference in crystal field stabilization energies of the activated complex and the reactant. This calculation has been made for all the d^n configurations, assuming a square pyramid

---

[19] H. Taube, *Chem. Revs.*, **50**, 69 (1952).

[20] L. E. Orgel, *An Introduction to Transition-Metal Chemistry*, 2nd ed., John Wiley & Sons, Inc., New York, 1966.

[21] These rationalizations should not be taken too seriously in view of the similar results obtained from crystal field considerations, assuming different mechanisms. See the following paragraph.

Table 3.2   **Kinetic Classification of**
Complexes

Labile

d^0	$CaEDTA^{2-}$
d^1	$Ti(H_2O)_6{}^{3+}$
d^2	$V(phen)_3{}^{3+}$
d^4 (high spin)	$Cr(H_2O)_6{}^{2+}$
d^5 (high spin)	$Mn(H_2O)_6{}^{2+}$
d^6 (high spin)	$Fe(H_2O)_6{}^{2+}$
d^7	$Co(NH_3)_6{}^{2+}$
d^8	$Ni(H_2O)_6{}^{2+}$
d^9	$Cu(H_2O)_6{}^{2+}$
d^{10}	$Ga(C_2O_4)_3{}^{3-}$

Inert

d^3	$Cr(H_2O)_6{}^{3+}$
d^4 (low spin)	$Cr(CN)_6{}^{4-}$
d^5 (low spin)	$Mn(CN)_6{}^{4-}$
d^6 (low spin)	$Fe(CN)_6{}^{4-}$

activated complex, a pentagonal bipyramid activated complex, and a 7-co-ordinate octahedral wedge activated complex.[15] Positive contributions to the activation energy were assumed to correspond to relative inertness, and negative contributions were assumed to correspond to lability. The results for all the assumed activated complexes were found to be in good agreement with Taube's Rule, except that inertness was predicted for d^8 complexes such as $Ni(H_2O)_6{}^{2+}$. Now although $Ni(H_2O)_6{}^{2+}$ is relatively labile when compared to inert complexes such as $Fe(CN)_6{}^{4-}$, among the aqueous $+2$ ions of the latter half of the first transition series, $Ni(H_2O)_6{}^{2+}$ exchanges its water molecules with the solvent most slowly.

Mechanisms. The available experimental data indicate that almost all of the substitution reactions of octahedral complexes occur by predominantly bond-breaking mechanisms.[15] For some reactions, the S_N1 (lim) mechanism is indicated, and a true 5-coordinate intermediate is probably formed. For most other reactions, the entering ligand or solvent molecule participates to some slight extent in the rate-determining step. The activated complex is 7-coordinate, but the entering and leaving groups are both weakly bound.

The reactions of the $Co(CN)_5H_2O^{2-}$ ion with various ligands are examples of well-substantiated S_N1 (lim) reactions[22]:

$$Co(CN)_5H_2O^{2-} \; \underset{k_2}{\overset{k_1}{\rightleftharpoons}} \; Co(CN)_5{}^{2-} + H_2O \qquad (3.25)$$

$$Co(CN)_5{}^{2-} + X^- \; \underset{k_4}{\overset{k_3}{\rightleftharpoons}} \; Co(CN)_5X^{3-} \qquad (3.26)$$

[22] A. Haim and W. K. Wilmarth, *Inorg. Chem.*, **1**, 573 (1962); A. Haim, R. J. Grassie, and W. K. Wilmarth, *Advan. Chem. Ser.*, **49**, 31 (1965).

Kinetic data for a variety of ligands (Br^-, I^-, SCN^-, and N_3^-) gave the same value for k_1 (1.6×10^{-3} sec^{-1}). In fact, essentially the same value was obtained for the rate constant for the water exchange reaction, using water labeled with ^{18}O. The stability of the 5-coordinate $Co(CN)_5^{2-}$ species is probably attributable to the strong ligand-field stabilization of the cyanide ions.

The reactions of $Ni(H_2O)_6^{2+}$ and $Co(H_2O)_6^{2+}$ with various ligands are examples of S_N1 reactions in which the incoming ligand (probably in an outer-sphere coordination shell) exchanges with one of the water molecules in the inner-sphere coordination shell.[23]

$$M(H_2O)_6^{2+}, L \longrightarrow M(H_2O)_{6-n}L^{2+} + nH_2O \qquad (3.27)$$

The rate constants are given in Table 3.3. The narrow range of values observed shows that only a slight degree of bond formation occurs in the rate-determining step and suggests that the breaking of an $M—OH_2$ bond is the principal process in that step for all the reactions.

The base hydrolysis of the cobalt(III) pentaammine complexes is a class of reactions that has been extensively studied.

$$Co(NH_3)_5X^{2+} + OH^- \longrightarrow Co(NH_3)_5OH^{2+} + X^- \qquad (3.28)$$

The reactions are overall second order (first order in complex and first order in hydroxide ion), and one might suspect a simple S_N1 or S_N2 mechanism; however, an extensive body of information indicates that the following

Table 3.3 Rate Constants for the Reaction of
$Ni(H_2O)_6^{2+}$ and $Co(H_2O)_6^{2+}$ with Various
Ligands at 25°*

Ligand	*Rate Constants (sec^{-1})*	
	$Ni(H_2O)_6^{2+}$	$Co(H_2O)_6^{2+}$
H_2O	2.7×10^4	11.3×10^5
Imidazole	1.6	4.4
SO_4^{2-}	1.5	2
Diglycine	1.2	2.6
$HP_2O_7^{3-}$	1.2	5.3
$HP_3O_{10}^{4-}$	1.2	7.2
Glycine	0.9	2.6
SCN^-	0.6	
$C_2O_4^{2-}$	0.6	
Triglycine	0.5	
$HC_2O_4^-$	0.3	

* See ref. 23.

[23] G. G. Hammes and M. L. Morrell, *J. Am. Chem. Soc.*, **86**, 1497 (1964); G. H. Nancollas and N. Sutin, *Inorg. Chem.*, **3**, 360 (1964).

mechanism is involved[15]:

$$Co(NH_3)_5X^{2+} + OH^- \xrightarrow{\text{fast}} Co(NH_3)_4NH_2X^+ + H_2O \qquad (3.29)$$

$$Co(NH_3)_4NH_2X^+ \xrightarrow{\text{slow}} Co(NH_3)_4NH_2^{2+} + X^- \qquad (3.30)$$

$$Co(NH_3)_4NH_2^{2+} + H_2O \xrightarrow{\text{fast}} Co(NH_3)_5OH^{2+} \qquad (3.31)$$

This is an example of an S_N1CB mechanism (S_N1 conjugate base). Among the evidence for this mechanism are the following facts.[24] The base hydrolysis of various $Co(NH_3)_5X^{2+}$ ions in the presence of various added Y^- ions leads to the formation of some $Co(NH_3)_5Y^{2+}$ ions as well as $Co(NH_3)_5OH^{2+}$ ions. The competition ratio, $Co(NH_3)_5Y^{2+}/Co(NH_3)_5OH^{2+}$, shows little dependence on the leaving group X^-, is independent of the OH^- concentration, but depends on the concentration of Y^- and varies from one Y^- ion to the next. These data strongly suggest the formation of a common intermediate formed by the S_N1CB mechanism.

Synthetic Applications.[25] The S_N1CB mechanism suggests that the synthesis of certain metal complexes might be catalyzed by bases. For example, it explains why the reaction of $CrCl_3$ with liquid ammonia {which ordinarily yields principally $[Cr(NH_3)_5Cl]Cl_2$}, gives good yields of $[Cr(NH_3)_6]Cl_3$ when a trace of KNH_2 is added. The general mechanism suggests that if a good coordinating solvent—such as water—is avoided, the 5-coordinate intermediate should be capable of reacting with suitably-chosen ligands. Thus, in dimethyl sulfoxide the conversion of $Co(en)_2Cl_2^+$ to $Co(en)_2(NO_2)_2$ may be effected by using a catalytic amount of base, such as hydroxide or piperidine.[26] The presumed mechanism for the substitution of the first chloro group follows.

$$Co(en)_2Cl_2^+ + B \xrightarrow{\text{fast}} Co(en)(en-H)Cl_2 + BH^+ \qquad (3.32)$$

$$Co(en)(en-H)Cl_2 \xrightarrow{\text{slow}} Co(en)(en-H)Cl^+ + Cl^- \qquad (3.33)$$

$$Co(en)(en-H)Cl^+ + NO_2^- \xrightarrow{\text{fast}} Co(en)(en-H)NO_2Cl \qquad (3.34)$$

$$Co(en)(en-H)NO_2Cl + BH^+ \xrightarrow{\text{fast}} Co(en)_2NO_2Cl^+ + B \qquad (3.35)$$

Substitution Reactions of Square Planar Complexes

Most metal complexes having coordination numbers less than six react by associative mechanisms, S_N2 and $S_N2(\text{lim})$.[15,27] This result is plausible,

[24] D. A. Buckingham, I. I. Olsen, and A. M. Sargeson, *J. Am. Chem. Soc.*, **88**, 5443 (1966).

[25] J. L. Burmeister and F. Basolo, *Preparative Inorg. Reactions*, **5**, 1 (1968).

[26] R. G. Pearson, H.-H. Schmidtke, and F. Basolo, *J. Am. Chem. Soc.*, **82**, 4434 (1960).

[27] L. Cattalini, R. Ugo, and A. Orio, *J. Am. Chem. Soc.*, **90**, 4800 (1968).

inasmuch as metals with "low" coordination numbers often can readily expand their coordination numbers. Square planar complexes of the low-spin d^8 type, typified by platinum(II) compounds, are relatively "inert," and considerable kinetic and synthetic work has been done with them. The rates of substitution reactions of such compounds are strongly dependent on the entering group. In fact, it is possible to set up a generalized relative rate scale for entering groups in platinum(II) substitutions[28]: $PR_3 > CN^-$, $SC(NH_2)_2$, $SeCN^- > SO_3^{2-}$, SCN^-, $I^- > Br^-$, N_3^-, NO_2^-, py, NH_3, $Cl^- > OR^-$. The relative reactivities of the ligands at opposite ends of this scale are approximately $10^7:1$.

The Trans Effect. [29] In platinum(II) complexes, it has been observed that the reaction rate depends strongly on the nature of the ligand *trans* to the leaving group. We may compare ligands as to their abilities to labilize (make susceptible to substitution) *trans* leaving groups. In the following list, some ligands are listed in the order of decreasing *trans* effect (i.e., decreasing *trans* labilizing ability): CO, CN^-, $C_2H_4 > PR_3$, H^-, $NO > CH_3^-$, $SC(NH_2)_2 > C_6H_5^-$, NO_2^-, I^-, $SCN^- > Br^- > Cl^- > NH_3$, py, RNH_2, $F^- > OH^- > H_2O$.

Consider the reaction of the tetrachloroplatinate(II) ion with two molecules of ammonia. After one molecule of ammonia has reacted, there are two kinds of chloride ions remaining in the complex: those which are *trans* to each other, and that which is *trans* to the ammonia. Inasmuch as chloride is more *trans* directing than ammonia, the chlorides that are *trans* to each other are labilized more than the chloride that is *trans* to the ammonia. Hence, it is one of these labilized chlorides that is next displaced, forming *cis*-$[Pt(NH_3)_2Cl_2]$.

$$\text{Cl-Pt-Cl} \xrightarrow{NH_3} \text{Cl-Pt-NH}_3 \xrightarrow{NH_3} \text{Cl-Pt-NH}_3 \qquad (3.36)$$

Now consider the reaction of the tetraammineplatinum(II) ion with 2 moles of chloride ion. After the displacement of one ammonia molecule, there remain two ammonias *trans* to each other and the ammonia *trans* to the chloride. The latter ammonia molecule is labilized by the chloride and is the next to be displaced, forming *trans*-$[Pt(NH_3)_2Cl_2]$.

$$\text{NH}_3\text{-Pt-NH}_3 \xrightarrow{Cl^-} \text{NH}_3\text{-Pt-Cl} \xrightarrow{Cl^-} \text{NH}_3\text{-Pt-Cl} \qquad (3.37)$$

In a platinum(II) substitution, it is reasonable to assume that the entering group approaches the complex from one side of the plane over the group to be

[28] H. B. Gray and C. H. Langford, *Chem. Eng. News*, **46**, April 1, 1968, p. 68.

[29] For a detailed review of the *trans* effect, the reader is referred to F. Basolo and R. G. Pearson, *Prog. Inorg. Chem.*, **4**, 381 (1962).

replaced. In the activated complex the leaving group will have moved down so that a trigonal bipyramid arrangement is achieved.

$$T\langle\overset{A}{\underset{B}{Pt}}\rangle X + Y \longrightarrow \left[T\underset{B}{\overset{A}{\triangleleft}}\overset{X}{\underset{Y}{}} \right] \longrightarrow T\langle\overset{A}{\underset{B}{Pt}}\rangle Y + X \qquad (3.38)$$

A trigonal bipyramidal activated complex is quite reasonable in view of the fact that a number of stable 5-coordinate platinum(II) complexes having this structure are known (e.g., $Pt(SnCl_3)_5{}^{3-}$). A strong *trans*-directing ligand is one which, relative to other ligands, stabilizes the trigonal bipyramid structure. It is instructive for us to consider the molecular orbitals formed by the overlap of the valence orbitals of the platinum and the ligands. First let us consider the two Pt p orbitals that lie in the plane formed by the entering and leaving groups and the *trans* ligand (see Fig. 3.2). If the *trans* group has a strong σ interaction with the p_x orbital, the bond to the leaving group will be relatively weak in the original complex. Thus, relatively little energy will be expended in moving the leaving group out of the region of strong overlap. In the activated complex, the p_z orbital can be used to help bond both the entering and leaving groups to the platinum atom. The *trans* ligand then can use much more than one-half the p_x orbital in bonding to the platinum. It is known that ligands such as H^-, PR_3, and $CH_3{}^-$ form strong σ bonds and would be expected to overlap well with a $6p$ orbital. Thus, these ligands show what has been referred to as a strong σ-*trans* effect.[16]

Now let us consider the platinum d_{xz} orbital that lies in the plane formed by the entering and leaving groups and the *trans* ligand (see Fig. 3.3). Two lobes of the d_{xz} orbital are directed suitably for π bonding to the *trans* ligand,

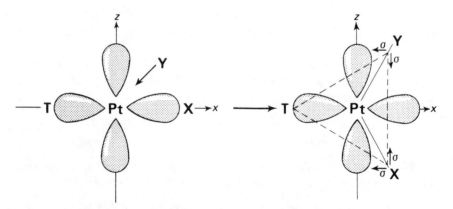

FIG. 3.2. The change in the use of the platinum p_x and p_z orbitals during square-planar substitution.

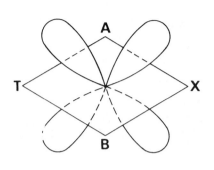

Nonbonding *d* orbital in original complex

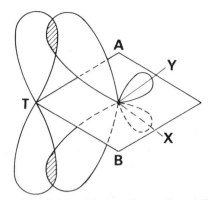

d orbital used in bonding to ligand T in activated complex

FIG. 3.3. d_{xz} orbital of a Pt(II) complex.

and the other two lobes are pointed toward the positions occupied by the entering and leaving groups in the activated complex. The activated complex will be stabilized by a reduction in the density of *d* electrons in these latter positions. This is exactly what happens when the d_{xz} electrons are used in π bonding to the *trans* ligand. The stronger the double bonding, the lower is the electron density in the critical directions and the more *trans* directing is the ligand.[30] Thus it is explained why strongly π-bonding ligands such as CO, CN^-, C_2H_4, etc. are strongly *trans* directing and why ligands that cannot engage in back bonding of this type (NH_3, H_2O, etc.) are weakly *trans* directing.

Synthetic Applications of the Trans Effect. By using the qualitative *trans*-effect series, it is possible to devise syntheses for many isomeric platinum(II) complexes. Often this amounts to simply reversing the order of introduction of groups into $PtCl_4^{2-}$, as in the case of *cis*- and *trans*-$[PtNH_3NO_2Cl_2]^-$:[31]

$$\begin{array}{ccc} \text{Cl} & \text{Cl} \\ & \text{Pt} \\ \text{Cl} & \text{Cl} \end{array} \xrightarrow{NH_3} \begin{array}{ccc} \text{Cl} & \text{NH}_3 \\ & \text{Pt} \\ \text{Cl} & \text{Cl} \end{array} \xrightarrow{NO_2^-} \begin{array}{ccc} \text{Cl} & \text{NH}_3 \\ & \text{Pt} \\ \text{Cl} & \text{NO}_2 \end{array}$$

$$\begin{array}{ccc} \text{Cl} & \text{Cl} \\ & \text{Pt} \\ \text{Cl} & \text{Cl} \end{array} \xrightarrow{NO_2^-} \begin{array}{ccc} \text{Cl} & \text{NO}_2 \\ & \text{Pt} \\ \text{Cl} & \text{Cl} \end{array} \xrightarrow{NH_3} \begin{array}{ccc} \text{Cl} & \text{NO}_2 \\ & \text{Pt} \\ \text{H}_3\text{N} & \text{Cl} \end{array} \qquad (3.40)$$

The *trans* effect can also be used in the synthesis of octahedral complexes.

[30] L. E. Orgel, *J. Inorg. Nucl. Chem.*, **2**, 137 (1956).
[31] I. I. Chernyaev, *Ann. Inst. Platine*, **5**, 102, 118 (1927).

Thus, the reaction of $Mn(CO)_5Br$ with a variety of ligands (pyridine, $\frac{1}{2}$ bipyridine, PR_3, AsR_3, and SbR_3) always leads to the *cis* product.[32–34]

$$
\begin{array}{c}
\text{CO} \\
\text{OC} \diagdown \overset{|}{\underset{|}{\text{Mn}}} \diagup \text{CO} \\
\text{OC} \diagup \ \ \diagdown \text{CO} \\
\text{X}
\end{array}
+ \ 2L \ \longrightarrow \
\begin{array}{c}
\text{CO} \\
\text{OC} \diagdown \overset{|}{\underset{|}{\text{Mn}}} \diagup \text{L} \\
\text{OC} \diagup \ \ \diagdown \text{L} \\
\text{X}
\end{array}
+ \ 2\,\text{CO} \qquad (3.41)
$$

However, the *trans* effect is not as reliable with octahedral substitutions as with square planar substitutions, probably because the 5-coordinate intermediates common in octahedral substitutions can readily lead to rearrangement of ligand positions.[35]

OXIDATION-REDUCTION REACTIONS[36]

Oxidation-reduction reactions generally have either of two mechanisms: *electron transfer*, in which one (or, rarely, two) electrons transfer from one species to another without any simultaneous gross transfer of atoms or groups; or *atom transfer*, in which the change in oxidation state is effected by the transfer of an atom or group from one species to another. The following reaction is an example of a simple 1-equivalent[37] electron transfer reaction:

$$
Fe(CN)_6^{4-} + IrCl_6^{2-} \longrightarrow Fe(CN)_6^{3-} + IrCl_6^{3-} \qquad (3.42)
$$

The reaction of iron(II) with hydrogen peroxide proceeds via two 1-equivalent atom (or group) transfer steps [reactions (3.43) and (3.44)]:

$$
Fe(H_2O)_6^{2+} + H_2O_2 \longrightarrow Fe(H_2O)_5OH^{2+} + H_2O + OH \qquad (3.43)
$$

$$
Fe(H_2O)_6^{2+} + OH \xrightarrow{\text{fast}} Fe(H_2O)_5OH^{2+} + H_2O \qquad (3.44)
$$

$$
2\,H^+ + 2\,Fe(H_2O)_5OH^{2+} \longrightarrow 2\,Fe(H_2O)_6^{3+} \qquad (3.45)
$$

The following reactions are examples of 2-equivalent atom transfer reactions:

$$
CN^- + S_2O_3^{2-} \longrightarrow SCN^- + SO_3^{2-} \qquad (3.46)
$$

$$
NO_2^- + HOCl \longrightarrow H^+ + NO_3^- + Cl^- \qquad (3.47)
$$

Reaction (3.46) can be looked upon as the S_N2 displacement of SO_3^{2-} from $S_2O_3^{2-}$ by the CN^- ion, and Eq. (3.47) is essentially the S_N2 displacement of Cl^- from HOCl by the NO_2^- ion.

[32] R. J. Angelici, F. Basolo, and A. J. Poë, *J. Am. Chem. Soc.*, **85**, 2215 (1963).

[33] E. W. Abel and I. S. Butler, *J. Chem. Soc.*, 434 (1964).

[34] R. J. Angelici, *Inorg. Chem.*, **3**, 1099 (1964).

[35] For example, the base hydrolysis of *trans*-$[Co(en)_2NH_3Cl]^{2+}$ yields *cis*-$[Co(en)_2NH_3OH]^{2+}$.

[36] Some aspects of this topic have been reviewed by N. Sutin, *Acc. Chem. Res.*, **1**, 225 (1968), and A. G. Sykes, *Adv. Inorg. Chem. Radiochem.*, **10**, (1967).

[37] The metal oxidation states change by one unit.

Reactions between Transition Metal Complexes. Reactions of this type generally are of the 1-equivalent type, and may proceed by either an electron transfer or an atom transfer mechanism.

Reaction (3.42) is an example of an *outer*-sphere electron transfer reaction; that is, the coordination shells remain intact throughout the entire reaction. Obviously reactions of this type can be synthetically useful when it is desired to oxidize or reduce a complex without changing the first coordination sphere. Considerable progress has been made in understanding such reactions in recent years. Thus, Marcus[38] has shown that the rate constant for an outer-sphere electron transfer reaction, k_{12}, can be expressed by the relation

$$k_{12} = (k_{11}k_{22}K_{12}f)^{1/2} \tag{3.48}$$

where k_{11} and k_{22} are the rate constants of the *electron exchange* reactions for the species involved, K_{12} is the equilibrium constant, $\log f = (\log K_{12})^2/[4\log(k_{11}k_{22}/Z^2)]$, and Z is the collision frequency for the hypothetically uncharged reactant ions. An *electron exchange* reaction is an electron transfer reaction between two ions that are identical except for charge. The electron exchange between tris(ethylenediamine)cobalt(II) and tris(ethylenediamine)cobalt(III) is an example of such a reaction.

$$\text{Co(en)}_3{}^{2+} + \text{Co(en)}_3{}^{3+} \;\rightleftharpoons\; \text{Co(en)}_3{}^{3+} + \text{Co(en)}_3{}^{2+} \tag{3.49}$$

We can cite an interesting synthetic application of the latter reaction. By the addition of (+)-tartrate to a solution of racemic [Co(en)$_3$]Cl$_3$, it is possible to precipitate essentially half the complex as the salt (+)-[Co(en)$_3$]Cl[(+)-tartrate]. No more than a 50 per cent yield ordinarily can be hoped for, because the racemization of Co(en)$_3{}^{3+}$ is extremely slow. However, both the electron exchange reaction (3.48) and the racemization of Co(en)$_3{}^{2+}$ do proceed at convenient rates. Therefore, Busch[39] was able to resolve Co(en)$_3{}^{3+}$ with yields in excess of 75 per cent by adding a small amount of Co(en)$_3{}^{2+}$ to the solution from which the (+)-[Co(en)$_3$]Cl[(+)-tartrate] was precipitated. The mechanism of the conversion follows:

$$(\pm)\text{-[Co(en)}_3]^{3+} \xrightarrow{\text{(+)-tartrate + Cl}^-} (+)\text{-[Co(en)}_3]\text{Cl[(+)-tartrate]}$$

$$+ \tag{3.50}$$

$$[\text{Co(en)}_3]^{2+} \text{ catalyst} \qquad (-)\text{-[Co(en)}_3]^{3+}$$

The exchange of radioactive chloride ion with *trans*-Pt(en)$_2$Cl$_2{}^{2+}$ in aqueous solution is catalyzed by the presence of Pt(en)$_2{}^{2+}$. The reaction is essentially a 2-equivalent atom transfer reaction.[40]

[38] R. A. Marcus, *Ann. Rev. Phys. Chem.*, **15**, 155 (1964); also see T. W. Newton, *J. Chem. Educ.*, **45**, 571 (1968).

[39] D. H. Busch, *J. Am. Chem. Soc.*, **77**, 2747 (1955).

[40] F. Basolo, M. L. Morris, and R. G. Pearson, *Discussions Faraday Soc.*, **29**, 80 (1960).

$$Pt(en)^{2+} + *Cl^- \xrightleftharpoons{fast} Pt(en)_2 *Cl^+ \qquad (3.51)$$

$$Pt(en)_2 *Cl^+ + Pt(en)_2 Cl_2^{2+} \longrightarrow \begin{bmatrix} & en & & en & \\ *Cl-Pt-Cl-Pt-Cl \\ & en & & en & \end{bmatrix}^{3+} \longrightarrow \qquad (3.52)$$

$$Pt(en)_2 Cl*Cl^{2+} + Pt(en)_2 Cl^+$$

Johnson and Basolo[41] showed that reactions having this type of mechanism have synthetic application by preparing *trans*-$Pt(en)_2 X_2^{2+}$ complexes (X = Br and SCN) from *trans*-$Pt(en)_2 Cl_2^{2+}$ in the presence of catalytic amounts of $Pt(en)_2^{2+}$:

$$trans\text{-}Pt(en)_2 Cl_2^{2+} + 2X^- \xrightarrow{Pt(en)_2^{2+}} trans\text{-}Pt(en)_2 X_2^{2+} + 2\,Cl^- \qquad (3.53)$$

These complexes cannot be prepared by the method used for the preparation of *trans*-$Pt(en)_2 Cl_2^{2+}$ [the oxidation of $Pt(en)_2^{2+}$ with Cl_2] because Br_2 and $(SCN)_2$ are insufficiently strong oxidizing agents.

Taube and his co-workers[42] have shown that, when aqueous chromium(II) is oxidized by a pentaamminecobalt(III) complex, $Co(NH_3)_5 X^{2+}$, the X^- ligand is transferred to the chromium atom. It is believed that the X^- group serves as a path for electron transfer between the metal atoms; in effect, the reactions are 1-equivalent or group transfer reactions:

$$Co(NH_3)_5 X^{2+} + Cr(H_2O)_6^{2+} \longrightarrow (NH_3)_5 Co-X-Cr(H_2O)_5^{4+} \longrightarrow$$

$$Co(H_2O)_6^{2+} + Cr(H_2O)_5 X^{2+} \qquad (3.54)$$

For this type of reaction to be useful in synthesis, it is necessary that the oxidizing complex be inert, that the reducing complex be labile, and that its oxidation product be inert. The following reaction is an interesting application of the method.[43]

$$\begin{bmatrix} (NH_3)_5 CoN \hspace{-0.5em} \diagup \hspace{-0.5em} \bigcirc \hspace{-0.5em} \diagdown \\ \hspace{3em} CONH_2 \end{bmatrix}^{3+} + Cr(H_2O)_6^{2+} + 6H^+ + 5\,H_2O \longrightarrow$$

$$\begin{bmatrix} (H_2O)_5 CrO{=}C{-}\bigcirc \\ H_2N \hspace{2em} N \\ \hspace{4em} H \end{bmatrix}^{4+} + 5\,NH_4^+ + Co(H_2O)_6^{2+} \qquad (3.55)$$

[41] R. C. Johnson and F. Basolo, *J. Inorg. Nucl. Chem.*, **13**, 36 (1960).
[42] See H. Taube, *Adv. Inorg. Chem. Radiochem.*, **1**, 1 (1959).
[43] F. Nordmeyer and H. Taube, *J. Am. Chem. Soc.*, **90**, 1162 (1968).

The product Cr(III)-nicotinamide complex can be isolated, but it undergoes aquation at a measurable rate in solution to give $Cr(H_2O)_6^{3+}$ and nicotinamide. Clearly the complex cannot be prepared by the straightforward reaction of $Cr(H_2O)_6^{3+}$ with nicotinamide. Another application of this ability that $Cr(H_2O)_6^{2+}$ has to capture groups on being oxidized is the synthesis of the $(H_2O)_5CrSH^{2+}$ ion by the action of various sulfur-containing oxidizing agents on aqueous Cr(II). The best yields have been obtained with polysulfide:[44]

$$H^+ + Cr(H_2O)_6^{2+} + HS_2^- \longrightarrow (H_2O)_5CrSH^{2+} + H_2S + H_2O \quad (3.56)$$

Reactions between Nontransition Element Compounds. Reactions of this type generally are of the 2-equivalent variety and involve the transfer of an atom or group.

Among the more commonly used oxidizing agents are oxanions such as BrO_3^-, SO_4^{2-}, OCl^-, and so forth. The reactivities of these reagents are fairly systematic, and their reaction mechanisms have at least one feature in common.[45] The formal charge or oxidation state of the central atom of an oxanion seems to be very important in determining reactivity; the lower the charge or oxidation state, the higher the reactivity. Thus, rates for reactions of the chlorine anions increase in the order $ClO_4^- < ClO_3^- \lesssim ClO_2^- < ClO^-$. The size of the central atom is also important; the larger it is, the higher the reactivity. For example, iodate reactions are fast, chlorate reactions are slow, and bromate reactions are of intermediate rate. One feature shared by most oxanion reactions is acid catalysis. In fact, rate laws with orders of two or more in hydrogen ion concentration are characteristic. The protons effectively labilize the oxygen atoms that break loose from the central atom and, in some cases, are transferred to another species. Thus, in the reaction of chlorate ion with halide ions, the following mechanism is probably involved:

$$2 H^+ + ClO_3^- \; \overset{fast}{\rightleftharpoons} \; H_2ClO_3^+ \quad\quad\quad (3.57)$$

$$X^- + H_2ClO_3^+ \; \overset{slow}{\longrightarrow} \; \text{products (either via HOX + HClO}_2 \text{ or} \\ \text{via XClO}_2 + H_2O) \quad (3.58)$$

The classical synthesis of hydrazine is the reaction of ammonia with hypochlorite in strongly alkaline aqueous solution:

$$2 NH_3 + OCl^- \longrightarrow N_2H_4 + Cl^- + H_2O \quad (3.59)$$

The reaction proceeds in two main steps. First, chloramine is formed,[46]

$$NH_3 + OCl^- \longrightarrow OH^- + NH_2Cl \quad (3.60)$$

---

[44] M. Ardon and H. Taube, *J. Am. Chem. Soc.,* **89**, 3662 (1967).

[45] J. O. Edwards, *Inorganic Reaction Mechanisms*, W. A. Benjamin, Inc., New York, 1964.

[46] This reaction is acid catalyzed, and probably the rate-determining step is the S_N2 displacement of an OH^- ion from HOCl by NH_3. See I. Weil and J. C. Morris, *J. Am. Chem. Soc.,* **71**, 1665 (1949).

which then reacts with ammonia to form hydrazine:

$$NH_3 + NH_2Cl + OH^- \longrightarrow N_2H_4 + Cl^- + H_2O \qquad (3.61)$$

It has been demonstrated that this latter reaction proceeds by two independent paths, a base-independent path and a base-catalyzed path.[47] The base-independent path is believed to be the nucleophilic attack of ammonia on chloramine (S_N2 mechanism):

$$NH_3 + NH_2Cl \xrightarrow{\text{slow}} N_2H_5^+ + Cl^- \qquad (3.62)$$

$$N_2H_5^+ + OH^- \underset{}{\overset{\text{fast}}{\rightleftharpoons}} N_2H_4 + H_2O \qquad (3.63)$$

The base-catalyzed path is believed to involve the preliminary formation of the chloramide ion, followed by attack on this ion by ammonia (S_N2CB mechanism):

$$NH_2Cl + OH^- \underset{}{\overset{\text{fast}}{\rightleftharpoons}} NHCl^- + H_2O \qquad (3.64)$$

$$NH_3 + NHCl^- \xrightarrow{\text{slow}} N_2H_4 + Cl^- \qquad (3.65)$$

Reactions between Transition Metal Complexes and Nontransition Element Compounds. Reactions of this type can be of either the 1-equivalent or 2-equivalent type, although 1-equivalent reactions are more common.

When aqueous solutions of hydrazine are oxidized, the proportions of the products obtained (i.e., the stoichiometry) depend upon whether a 1-equivalent or a 2-equivalent oxidizing agent is used.[46] The *limiting* stoichiometries are represented by the half-reactions

$$N_2H_5^+ = e^- + NH_4^+ + \tfrac{1}{2}N_2 + H^+ \qquad (3.66)$$

$$N_2H_5^+ = 4e^- + N_2 + 5H^+ \qquad (3.67)$$

The same sort of behavior is found in the oxidation of aqueous solutions of sulfur dioxide;[48] the *limiting* stoichiometries are

$$SO_2 + H_2O = e^- + \tfrac{1}{2}S_2O_6^{2-} + H^+ \qquad (3.68)$$

$$SO_2 + 2H_2O = 2e^- + SO_4^{2-} + 4H^+ \qquad (3.69)$$

The *observed* stoichiometries for the oxidation of hydrazine and sulfur dioxide by a variety of oxidizing agents are given in Table 3.4. In this table, the stoichiometry is defined as the number of equivalents of oxidizing agent consumed per mole of hydrazine or sulfur dioxide. For hydrazine, the stoichiometry can range from 1 to 4; for sulfur dioxide, the stoichiometry can range from 1 to 2. It will be noted that the 2-equivalent oxidizing agents almost invariably give the maximum stoichiometry, corresponding to the

---

[47] G. Yagil and M. Anbar, *J. Am. Chem. Soc.*, **84**, 1797 (1962).
[48] W. C. E. Higginson and J. W. Marshall, *J. Chem. Soc.*, 447 (1957).

Table 3.4 Stoichiometries of the Oxidations of Hydrazine and Sulfur Dioxide*

Oxidizing Agent	Hydrazine (pH 0–1)	Sulfur Dioxide (pH 0.5)
1-equiv. oxidants†		
Ce^{4+}	1.05–1.4	1.27–1.44
Co^{3+}	1.03–1.15	1.04–1.37
Fe^{3+}	1.1 –1.35	~1.2
$Mn(H_2P_2O_7)_3^{3-}$	1.05	~1.24
1,2-equiv. oxidants†		
$Cr_2O_7^{2-}$	3–4	1.84–1.95
MnO_4^-	1.45–2.2	1.55–1.80
$PtCl_6^{2-}$	~3.7	2.00
VO_2^+	3.5 –4	1.57
2-equiv. oxidants		
I_2	4.00	2.00
Br_2	4.00	1.98–2.00
Cl_2	4?	1.99–2.00
IO_3^-	4.00	2.00
BrO_3^-	4.00	1.77–1.95
H_2O_2	—	2.00
Tl^{3+}	4.00	2.00

* See ref. 48.

† 1-equivalent oxidants can gain only 1 electron in a given step, whereas 1,2-equivalent oxidants can gain either 1 or 2 electrons in a given step.

proposed mechanisms

$$N_2H_4 \xrightarrow{-2e^-} N_2H_2 \xrightarrow{-2e^-} N_2 \qquad (3.70)$$

and

$$H_2SO_3 \xrightarrow{-2e^-} SO_3 \quad \text{(i.e., } SO_4^{2-}) \qquad (3.71)$$

The 1-equivalent oxidizing agents give intermediate stoichiometry, explained by the mechanisms

$$N_2H_4 \xrightarrow{-e^-} N_2H_3 \longrightarrow \tfrac{1}{2}N_4H_6 \longrightarrow \tfrac{1}{2}N_2 + NH_3$$
$$\phantom{N_2H_4 \xrightarrow{-e^-}} \xrightarrow{-e^-} N_2H_2 \xrightarrow{-e^-} N_2H \xrightarrow{-e^-} N_2 \qquad (3.72)$$

and

$$H_2SO_3 \xrightarrow{-e^-} HSO_3 \longrightarrow \tfrac{1}{2}H_2S_2O_6$$
$$\phantom{H_2SO_3 \xrightarrow{-e^-}} \xrightarrow{-e^-} SO_3 \quad \text{(i.e., } SO_4^{2-}) \qquad (3.73)$$

The 1,2-equivalent oxidizing agents also give intermediate stoichiometry, explained by assuming the simultaneous occurrence of the 1- and 2-equivalent mechanisms. The oxidation of sulfur dioxide is actually used for the preparation of dithionate $(S_2O_6^{2-})$.[49] It is obvious that the choice of oxidizing agent in this synthesis is quite critical.

Many other redox reactions are known, the products of which depend on whether a 1- or 2-equivalent oxidizing or reducing agent is used.[50] When a solution of Co(II) in the presence of excess oxalate is oxidized by the 1-equivalent oxidant Ce(IV), $Co(C_2O_4)_3^{3-}$ is formed.[51] However, the 2-equivalent, oxygen-carrying oxidants, such as H_2O_2 and OCl^-, give the binuclear complex

$$\left[(C_2O_4)_2Co \underset{\underset{H}{O}}{\overset{\overset{H}{O}}{<>}} Co(C_2O_4)_2 \right]^{4-}$$

When the complex $\left[(H_3N)_5CoO_2C-\bigcirc-CHO \right]^{2+}$ is oxidized with a

2-equivalent oxidant, the product is $\left[(H_3N)_5CoO_2C-\bigcirc-CO_2H \right]^{2+}$

(which cannot be made by more straightforward techniques). However, when 1-equivalent oxidants are used, the Co(III) cooperates in the oxidation of the aldehyde group, and the entire complex disintegrates.[52]

$$\left[(H_3N)_5CoO_2C-\bigcirc-CHO \right]^{2+} + Ce^{4+} + 4H^+ + 7H_2O \longrightarrow$$

$$Co(H_2O)_6^{2+} + Ce^{3+} + HO_2C-\bigcirc-CO_2H + 5NH_4^+ \qquad (3.74)$$

Problems

1. When *cis*-$[Pt(NH_3)_2(NH_2OH)_2]^{2+}$ is treated with HCl, $[PtCl_2(NH_3)(NH_2OH)]$ is formed. Would you expect this product to be *cis*- or *trans*? Inasmuch as NH_2OH is a poorer complexing agent than NH_3, why doesn't $[PtCl_2(NH_3)_2]$ form?

[49] Yost and Russell, *Systematic Inorganic Chemistry*, 1946.

[50] Burmeister and Basolo, *Preparative Inorg. Reactions*, **5**, 1 (1968).

[51] A. W. Adamson, H. Ogata, J. Grossman, and R. Newbury, *J. Inorg. Nucl. Chem.*, **6**, 319 (1958).

[52] R. T. M. Fraser and H. Taube, *J. Am. Chem. Soc.*, **82**, 4152 (1960).

2. Using the *trans* effect, predict the product of the reaction of two moles of ethylene-diamine with one mole of $PtCl_6^{2-}$

3. The ammoniation of *cis*-$[Co(en)_2(NH_3)Cl]^{2+}$ to the corresponding diammine complex occurs with almost complete retention of configuration.

   cis-$[Co(en)_2(NH_3)Cl]^{2+} + NH_3 \longrightarrow$ *cis*-$[Co(en)_2(NH_3)_2]^{3+} + Cl^-$

   On the other hand, the ammoniation of *cis*-$[Co(en)_2Cl_2]^+$ to the chloroammine complex gives considerable *trans* product as well as *cis* product.

   cis-$[Co(en)_2Cl_2]^+ + NH_3 \longrightarrow$ *cis*- and *trans*-$[Co(en)_2(NH_3)Cl]^{2+} + Cl^-$

   Explain the different behaviors.

4. When K_2PtCl_4 is treated with ethylene, the salt $K[Pt(C_2H_4)Cl_3]$ is obtained. If this salt is treated with ammonia, compound A—$Pt(C_2H_4)(NH_3)Cl_2$—is obtained. When the salt $K[Pt(NH_3)Cl_3]$ is treated with ethylene, compound B—$Pt(C_2H_4)$-$(NH_3)Cl_2$—is obtained. Draw structures for compounds A and B.

5. Sketch the structure of the complex that forms when:
   (a) two of the chloride ions in $K_2[PtCl_4]$ are displaced by two molecules of $P(C_2H_5)_3$;
   (b) one of the chloride ions in *trans*-$[Pt(NH_3)_3Cl_3]^+$ is displaced by an ammonia molecule.

6. The two isomers of $Pt(NH_3)_2Cl_2$ are treated with thiourea (abbreviated "th"). One isomer (A) forms $Pt(th)_4^{2+}$, whereas the other isomer (B) forms $Pt(NH_3)_2(th)_2^{2+}$. Identify the isomers A and B on the basis of the products formed.

7. Draw structures for the two isomers of $[Pt(NH_3)_2(NO_2)_2]$ and indicate that isomer which you expect to be formed by displacement of NH_3 from $[Pt(NH_3)_3NO_2]^+$ by NO_2^-.

8. There are two isomers of $RuCl_2(H_2O)_4^+$: A and B. Likewise, there are two isomers of $RuCl_3(H_2O)_3$: C and D. When either C or D is hydrolyzed, only A is formed:

$$C \text{ or } D + H_2O \longrightarrow A + Cl^-.$$

   What are the most probable configurations of A and B? Defend your choices, and explain the fact that both C and D hydrolyze to give only A.

9. Given that the diffusion coefficients of substances are of the order of 10^{-5} cm^2 sec^{-1}, show that the mixing of two liquids in a beaker will not occur in a reasonable time by diffusion.

10. For the reaction, *trans*-$Co(en)_2(NCS)Cl^+ + H_2O =$ *trans*-$Co(en)_2(NCS)(H_2O)^{2+}$ $+ Cl^-$, the rate constant at 25° is $k = 3 \times 10^{-6}$ min^{-1}, and the enthalpy of activation is 30.4 kcal. How long would it take for 90 per cent reaction of a 1 M solution at 25° and at 100°?

11. In the reduction step of the preparation of chromous acetate (Synthesis 1, p. 442), there are two simultaneous reactions. Write the net reactions for these. The very fact that chromous chloride can be kept in acidic solution for an hour or so indicates that a particular reaction is slow. What is this slow reaction? There is a fast reaction involving the aqueous chromous ion that is undesirable and is largely avoided during the synthesis. Write the equation for the net reaction.

SYNTHETIC TECHNIQUES

Part

Elementary

Techniques

4

Chemists usually learn most of the elementary operations of synthetic chemistry in a beginning laboratory course in organic chemistry. Some of these operations are briefly described in this chapter; for more details, specialized texts should be consulted.[1-3]

VAPORIZATION PROCESSES

Simple Distillation

The apparatus shown in Fig. 4.1 can be used for the distillation of a liquid compound from a solution of that compound and a nonvolatile material.

[1] K. B. Wiberg, *Laboratory Technique in Organic Chemistry*, McGraw-Hill Book Company, New York, 1960.
[2] J. Cason and H. Rapoport, *Laboratory Text in Organic Chemistry*, 2nd ed., Prentice-Hall, Inc., Englewood Cliffs, N.J., 1962.
[3] Various volumes of A. Weissberger, ed., *The Technique of Organic Chemistry*, Interscience Publishers, New York, 1949–

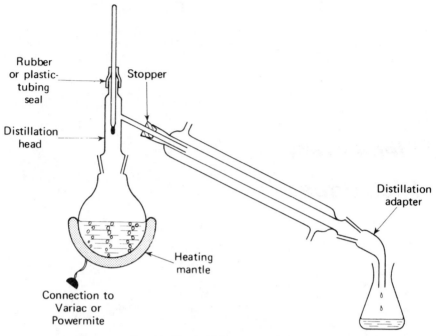

Rubber
or plastic-
tubing
seal

Stopper

Distillation
head

Distillation
adapter

Heating
mantle

Connection to
Variac or
Powermite

FIG. 4.1. Simple distillation apparatus.

Heating mantles are very convenient for supplying heat to the distillation flask, but in the latter stages of a distillation they can cause overheating of the flask walls above the liquid level and thereby decompose any thermally unstable components of the liquid mixture. Therefore, whenever thermally unstable compounds are subjected to distillation, the heat should be supplied to the flask by a liquid bath (water, oil, or molten wax) as shown in Fig. 4.2.

If the compound being distilled is pure, or if it is being distilled from essentially nonvolatile impurities, the operation can be used as a determination of the boiling point. It is important that the bulb of the thermometer lie entirely below the level of the side arm of the distillation head and that, at the time of reading, liquid be condensing on the upper part of the thermometer and flowing back down over the bulb. Of course, if the atmospheric pressure differs much from 760 mm, the boiling point will not be exactly the normal value. If the pressure difference is small, one can make an approximate correction by combining Trouton's Rule ($\Delta H^\circ_{vap} = 21 T_{bp}$) with Eq. (2.11):

$$\frac{T_{bp}(\text{normal})}{T(\text{measured})} = \frac{4.575}{21} \log\left(\frac{760}{P_{mm}}\right) + 1 \tag{4.1}$$

Consider the simple distillation of a mixture of 20 mole per cent benzene (b.p. 80.1°) and 80 mole per cent toluene (b.p. 110.6°).[1] From the vapor-liquid

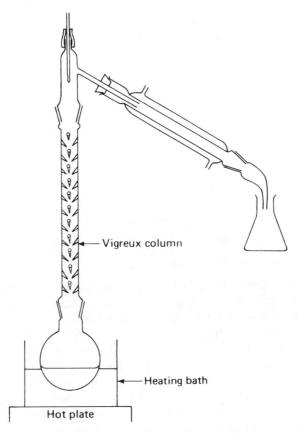

FIG. 4.2. Apparatus for simple fractional distillation.

composition curve shown in Fig. 4.3, we see that the vapor in equilibrium with the boiling mixture contains about 40 mole per cent benzene and 60 mole per cent toluene, which would be the composition of the initial distillate in a simple distillation. As the simple distillation proceeded, both the liquid and vapor would gradually become richer in toluene, and essentially no separation of the substances would be effected.

Fractional Distillation

Now consider the distillation of a 20/80 mole per cent benzene-toluene mixture using the simple fractional distillation apparatus shown in Fig. 4.2. The distillation column is designed so that the rising vapor is equilibrated with condensate which runs back to the distillation flask. In effect, the vapor passes through a series of stages ("theoretical plates") in each of which it partially condenses. From Fig. 4.3 we see that, in the first stage, the vapor

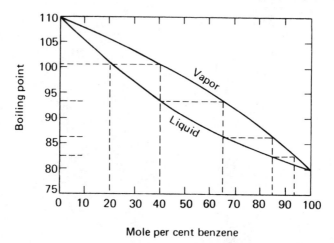

FIG. 4.3. The vapor-liquid composition curve for the benzene-toluene system. The effect of repeated distillation is shown. (Reproduced with permission of the McGraw-Hill Book Company.)

from the flask condenses to form a liquid containing about 40 mole per cent benzene in equilibrium with vapor containing about 65 mole per cent benzene. In successive stages, the vapor would contain 85 mole per cent benzene, 93 mole per cent benzene, and so on. Now, in a fractional distillation with a 30-cm Vigreux column, as shown in Fig. 4.2, about three "theoretical plates" will be realized. Thus one would expect the initial distillate to come over at about 82.5° and to contain about 93 per cent benzene. However, succeeding fractions will be richer in toluene, and thus even this procedure would give a poor separation of the components. More elaborate distillation apparatus, such as described in various texts[1-3] must be used to separate cleanly materials having boiling points as close as 30°.

Vacuum Distillation

Liquids that undergo appreciable decomposition at their normal boiling points must be distilled under reduced pressure. A very simple apparatus for vacuum distillation is shown in Fig. 4.4. The capillary gas inlet[4] allows a small steady stream of bubbles to pass through the boiling liquid and thereby prevents bumping (sudden violent evaporation, which throws liquid up to the condenser). Either an aspirator or an oil pump (protected from volatile chemicals with a cold trap) can be used for evacuation. The pressure is determined with a simple manometer or a rotating McLeod gauge (see p. 144). The pressure can be maintained at some predetermined value by manipulation of a variable air leak; however, special manostats are available.[1]

[4] This capillary should be drawn so fine that a barely perceptible stream of bubbles forms when one blows through the capillary while it is immersed in water.

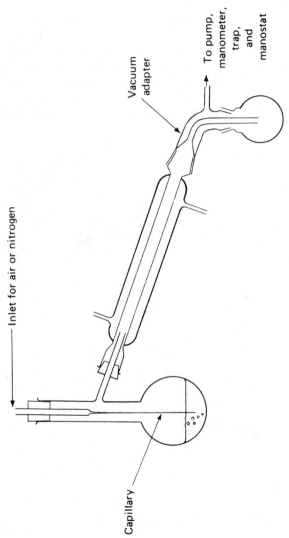

To pump, manometer, trap, and manostat

Vacuum adapter

Inlet for air or nitrogen

Capillary

FIG. 4.4. Apparatus for simple vacuum distillation.

Fractional distillation under reduced pressure is a relatively difficult operation. Figure 32.18 (Synthesis 30) shows a simple setup.

Refluxing

Often it is required to maintain a solution at its boiling point for an extended
period of time in order to carry a reaction to completion. To prevent loss of
the solvent by distillation (and consequent concentration of the solution), the
operation is usually carried out by heating under reflux. Figure 4.5 shows a
typical setup for heating under reflux, using a ground-glass flask and condenser. If it is important to prevent access of moisture to the refluxing liquid,
a drying tube containing a desiccant can be attached to the top of the condenser.

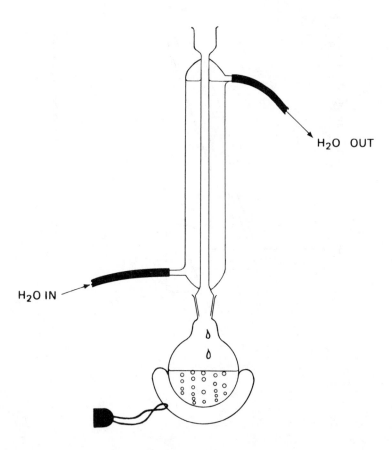

H₂O OUT

H₂O IN

Fig. 4.5. Heating a solution under reflux.

Sublimation

Sublimation is a convenient method for separating a volatile solid from non-volatile solids. In this process, the mixture is usually warmed in an evacuated vessel, and the volatilized solid is condensed on a water-cooled surface. A typical apparatus is shown in Fig. 4.6. Although sublimation strictly refers to the process in which a solid passes directly into the vapor state, the term is also loosely applied when material vaporized from a molten state condenses as a solid. When applicable, sublimation is a valuable purification operation because it involves no solvent that must be removed later.

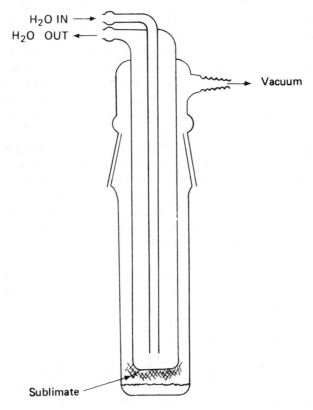

H_2O IN →
H_2O OUT ←

Vacuum

Sublimate

FIG. 4.6. A vacuum sublimer.

RECRYSTALLIZATION

Recrystallization is a procedure used for purifying solids. It usually involves the dissolution of the impure material in a hot solvent (generally at the boiling point) and allowing the material to crystallize as the solution cools. It is important to choose a solvent in which the material being purified has a high temperature coefficient of solubility so as to minimize loss of material. The impurities should be either insoluble (so that they can be separated from the hot solution by filtration) or very soluble (so that they will remain in solution after cooling the solution). The following solvents are commonly used for recrystallizations: acetic acid, acetone, acetonitrile, benzene, carbon tetra-chloride, chloroform, cyclohexane, dioxane, ethanol, ethyl acetate, hexane (or petroleum ether), methanol, toluene, and water. Binary mixtures of these are sometimes used.

The following general procedure is often used for recrystallization.

1. *Preparation of the solution.* The solid is placed in an Erlenmeyer flask and solvent is gradually added, as agitation and heating are maintained, until all the material is dissolved. It is important to wait sufficiently long between additions of solvent to allow saturation to be achieved, otherwise too much solvent will be used. An impurity may be only slightly soluble even in the hot solvent. In this case, just sufficient solvent should be used to dissolve the desired material, and two or three consecutive recrystallizations should be carried out.

2. *Filtration of the hot solution.* If there are no insoluble impurities, this step may be omitted. Otherwise, the hot, almost-saturated solution is filtered *by gravity* through a coarse sintered-glass funnel or a fluted filter paper in a short-stemmed funnel into a wide-mouthed flask or beaker. It is important during the filtration to keep the solution hot enough to prevent crystallization in the funnel. Therefore, the funnel should be preheated in an oven to the approximate temperature of the hot solution.

3. *Cooling.* The solution is cooled slowly, with occasional stirring, to room temperature or 0° in order to cause the formation of reasonably large, readily filtered crystals of the product. Crystals about 2 mm in diameter are usually satisfactory. If the solution is cooled too far, the impurities will begin to precipitate. Some substances crystallize extremely slowly, and several hours or even days at the low temperature may be required to maximize the yield.

4. *Filtration of the cold suspension.* The recrystallized material is usually separated from the supernatant solution by suction filtration through a sintered-glass funnel into a filter flask, as shown in Fig. 4.7. When all the recrystallized material has been collected on the funnel, the crystals are pressed down on the filter with a spatula while suction is applied to remove as much of the adhering solution as possible.

5. *Washing.* The small amount of adhering solution containing soluble impurities is removed by stopping the suction, adding a few milliliters of fresh, cold solvent, stirring up the wet mass of crystals with the spatula, and then reapplying suction.

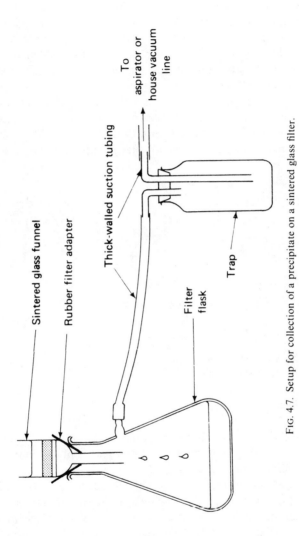

Sintered glass funnel

Rubber filter adapter

Thick-walled suction tubing

To aspirator or house vacuum line

Filter flask

Trap

FIG. 4.7. Setup for collection of a precipitate on a sintered glass filter.

6. *Drying.* After drying the crystals as much as possible on the filter by the application of suction, the crystals are transferred from the filter to a watch glass or large piece of filter paper. Sometimes complete drying can be accomplished by allowing the material to stand in a well-ventilated place for several hours. When a high-boiling solvent has been used, time can be saved by drying the material in a vacuum desiccator connected, through a cold trap, to a vacuum pump. Much more rapid and efficient drying can be accomplished by both heating and evacuating. The Abderhalden drying pistol pictured in Fig. 4.8 may be used for this purpose. A desiccant is placed in the bulb, a boat containing the sample is placed in the drying tube, a vacuum is applied through the stopcock, and the sample is heated by refluxing a liquid through the outer jacket. Some commonly used liquids are acetone (b.p. 56°), methanol (b.p. 65°), benzene (b.p. 80°), water (b.p. 100°), toluene (b.p. 110°), chlorobenzene (b.p. 132°), bromobenzene (b.p. 156°), and nitrobenzene (b.p. 211°).

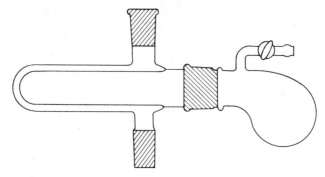

FIG. 4.8. Abderhalden drying pistol.[1] (Reproduced with permission of the McGraw-Hill Book Company.)

STIRRING

The most useful method for stirring reaction mixtures is magnetic stirring, in which a small glass- or plastic-enclosed bar magnet is placed in the bottom of the reaction vessel and is caused to spin by rotation of a larger magnet beneath the reaction vessel with a variable-speed motor. Magnetic stirring has the advantage over most other stirring techniques in that the space directly over the reaction vessel is completely unobstructed. If the magnets are sufficiently strong, it is possible to stir reaction mixtures in round-bottomed flasks fitted with heating mantles and in dewar flasks. Magnetic stirring is ideal for stirring a mixture in a closed or evacuated system. The principal disadvantage of magnetic stirring is that very thick suspensions and viscous solutions are not efficiently stirred. For such reaction mixtures it is necessary to use an overhead stirrer directly coupled to a motor. Two common types of stirrers are the propeller type and paddle type, illustrated in Fig. 4.9. The paddle should be made of Teflon, and it should rotate about its connection to the stirring rod to facilitate insertion through the neck of a round-bottomed flask. These stirrers should be connected by a short length of vacuum tubing to the shaft of the stirring motor, which should have a rating of about $\frac{1}{20}$ hp. The stirring rod itself should pass through a 10-cm length of glass tubing that

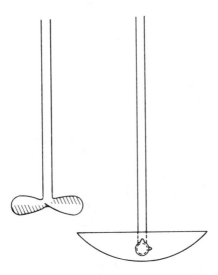

FIG. 4.9. Propeller stirrer and paddle stirrer.

is clamped or passes through a stopper in the reaction vessel and that serves to keep the rod from wobbling. When overhead stirring in a closed system is required, a Truebore[5] rod passing through a ground tube can be used. By lubricating the ground surface, a seal is formed that can prevent gas leakage even under several centimeters pressure differential. Special stirrer seals, in which the shaft passes through a packed gland of the type used in valves, can be purchased.

EXTRACTION OF SOLIDS

If one component of a mixture of solids is highly soluble in a particular solvent, and the other components are essentially insoluble, the soluble material may be separated from the mixture by simply shaking the mixture with the solvent, followed by filtration. If the soluble component is only slightly soluble, or if, because of the form of the mixture, it dissolves slowly, then

[5] Ace Glass Company.

continuous extraction with a Soxhlet extractor (Fig. 4.10) can be used. The mixture, usually contained in a filter-paper thimble, is placed in the central chamber. The solvent is placed in a flask attached to the bottom of the extractor, and a water-cooled condenser is attached to the top. The solvent is continuously heated at a vigorous boil so that a steady trickle of solvent falls

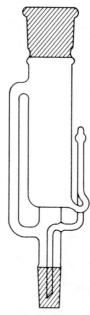

FIG. 4.10. Soxhlet extractor.[1] (Reproduced with permission of the McGraw-Hill Book Company.)

from the condenser into the thimble containing the solid mixture. When the liquid level in the chamber reaches the height of the siphon tube, all of the liquid siphons into the lower flask. As long as heat is supplied to the flask, this extraction cycle is automatically repeated. Obviously it is important that the system contain sufficient solvent so that when the chamber is full of liquid the flask does not go dry.

Solvents

5

There are several reasons why it is often desirable to carry out reactions in a solvent:

1. In a solvent, one may achieve intimate contact between reactants that otherwise would not react.

2. Reactions that are violent in the absence of a solvent may be carried out at a controlled rate in a solvent.

3. It may be possible to separate the product from by-products by utilizing solubility differences.

4. Many reagents are more conveniently handled when in solution than when in the pure state.

5. It is easier to measure out a certain *volume* of solution than to measure out a certain *weight* of material.

Water is the most commonly used solvent because it is cheap, it is readily available in high purity, and it dissolves many compounds. Unfortunately, many chemists behave as if water were the only usable solvent. Reactions that for one reason or another cannot be carried out in water are too often abandoned completely. This devotion to water is not caused so much by an unawareness of the existence of nonaqueous solvents as by a lack of knowledge of the methods of handling them and of the advantages that many of them have in particular experimental problems. These methods and advantages are briefly discussed in this chapter.

Physical Properties. Tables 5.1 and 5.2 list the important physical properties of some protonic and aprotic solvents, respectively. The formulas of the commonly used solvents are printed in boldface type in these tables.

In general, substances which are not liquid at room temperature are impractical solvents. Thus, I_2, IBr, and $HgBr_2$ (with melting points above room temperature) are of only academic interest as solvents. On the other hand, the cheapness and unique solubility properties of NH_3 and SO_2 more than compensate for the fact that these substances boil considerably below room temperature. And, although sodium hydrogen sulfate is a solid at room temperature, its melts[1] are very useful for dissolving various refractory ores.

In general, a low viscosity is a desirable property in a solvent. In solvents with low viscosities, transfer operations—such as pouring and filtering—are rapid, precipitates are easily freed of solvent, and diffusion-limited reactions proceed rapidly. Thus, the high viscosity of sulfuric acid is its major adverse property. Both dissolution and crystallization are very slow in this solvent.

PROTONIC SOLVENTS

For many years it has been recognized that certain solvents resemble water in that they undergo self-ionization to give a solvated proton and an anion, and dissolve salts to give conducting solutions. In these solvents, the protonic concept of acids and bases (see p. 46) is quite applicable. For example, the solvent liquid ammonia undergoes self-ionization as follows:

$$2\,NH_3 \rightleftharpoons NH_4^+ + NH_2^- \qquad K_{-33^\circ} \approx 10^{-30} \tag{5.1}$$

If we look upon the ammonium ion as the acidic species and the amide ion as the basic species, we may define an acid as any species which, when dissolved in ammonia, increases the concentration of ammonium ion and a base as any species which increases the concentration of amide ion. Thus, we may establish a set of ammono acids and bases analogous to corresponding aquo acids and bases (see Table 5.3).

---

[1] Molten salts are important media for chemical reactions but are not discussed in this text. The reader is referred to H. Bloom, *The Chemistry of Molten Salts*, W. A. Benjamin, Inc., New York, 1967, and J. D. Corbett and F. R. Duke, *Technique of Inorganic Chemistry*, vol. 1, H. B. Jonassen and A. Weissberger, eds., Interscience Publishers, New York, 1963, Chap. 3, p. 103.

Table 5.1 Physical Properties of Some Protonic Solvents

Solvent	m.p. (°C)	b.p. (°C)	Dielectric Constant (25°)	Electrical Conductivity, (mho/cm)	$-\log K_{ion}$ (25°)	ΔS_{vap}	1000η
HSO$_3$F	−89.0	162.7	~120	1.1×10^{-4} (25°)	7.4		15.6 (25°)
H$_2$SO$_4$	10.37	330	100	1.04×10^{-2} (20°)	3.4 (10°)		245 (25°)
HNO$_3$	−41.6	86		8.9×10^{-3} (0°)	~1.7 (−40°)		8.9 (20°)
H$_3$PO$_4$	42.35	—	~61	4.6×10^{-2} (25°)	0.9		1780 (25°)
HF	−89.4	19.51	60 (19°)	1.4×10^{-5} (−15°)	≫ 11.7 (0°)	6.1	2.4 (6°)
HCl	−114.6	−84.1	9.28 (−95°)	3.5×10^{-9} (−85°)		20.4	5.1 (−95°)
HOOCH	8.3	100.5	57.9 (20°)	6×10^{-5}	6.2	14.2	18.0 (20°)
HOAc	16.6	118.2	6.19	5×10^{-9} (25°)	14.45	14.9	11.6 (25°)
HOOCCF$_3$	−15.4	71.8					
HCN	−13.2	25.7	106.8	5×10^{-7} (0°)	~18.7 (12°)	20.2	2.0 (20°)
H$_2$S	−85.5	−60.3	10.2 (−60°)	3.7×10^{-11} (−78°)		21.0	4.3 (−60°)
H$_2$O$_2$	−0.9	151.4	93.7	2×10^{-6} (25°)	13	27.3	
H$_2$O	0.0	100.0	78.5	4×10^{-8} (18°)	14.0	26.0	10.1 (20°)
CH$_3$OH	−97.9	64.7	32.6	1.5×10^{-9} (25°)	16.6	25.0	5.45 (25°)
C$_2$H$_5$OH	−114.6	78.5	24.3	1.35×10^{-9} (25°)	18.9	26.2	10.8 (25°)
n-C$_3$H$_7$OH	−127	97.8	19.7				22.5 (20°)
iso-C$_3$H$_7$OH	−85.8	82.5	18.3				17.7 (30°)
C$_6$H$_5$OH	41	182	9.8 (60°)				34.9 (50°)
N$_2$H$_4$	1.5	113.5	51.7	2.3×10^{-6} (25°)	24.7	25.9	9.0 (25°)
NH$_3$	−77.7	−33.38	16.9	1×10^{-11} (−33°)	27	23.3	2.5 (−33°)
iso-C$_3$H$_7$NH$_2$	−101.2	34	5.5 (20°)				
C$_6$H$_5$NH$_2$	−6.2	184.4	6.89 (20°)				37.1 (25°)

Table 5.2 Physical Properties of Some Aprotic Solvents

Solvent	Formula	m.p. (°C)	b.p. (°C)	Dielectric Constant (25°)	1000η
Sulfur dioxide	SO_2	−72.7	−10.2	12.3 (22°)	4.28 (−10°)
Dinitrogen tetraoxide	N_2O_4	−11.2	21.15	2.4 (18°)	
Arsenic trifluoride	AsF_3	−8.5	63	5.7 (< −6°)	
Iodine pentafluoride	IF_5	9.6	98	36.2 (35°)	21.9 (25°)
Bromine trifluoride	BrF_3	8.8	127.6	—	22.2 (25°)
Phosphorus trichloride	PCl_3	−91	76.0	3.43	
Disulfur dichloride	S_2Cl_2	−80	135.6	4.79 (15°)	
Arsenic trichloride	$AsCl_3$	−18	130.2	12.8 (20°)	12.25 (20°)
Antimony trichloride	$SbCl_3$	73	221	33.0 (75°)	33 (95°)
Tin tetrachloride	$SnCl_4$	−30.2	114.1	2.87 (20°)	
Iodine monochloride	ICl	27.2	100	—	41.9 (28°)
Nitrosyl chloride	$NOCl$	−61.5	−5.4	19.7 (−10°)	5.47 (−20°)
Carbonyl chloride	$COCl_2$	−104	8.3	4.3 (22°)	
Thionyl chloride	$SOCl_2$	−104.5	79	9.0 (22°)	
Phosphorus oxychloride	$POCl_3$	1.25	105.3	13.9 (22°)	11.5 (25°)
Selenium oxychloride	$SeOCl_2$	10.9	176.4	46 (20°)	
Antimony tribromide	$SbBr_3$	97	280	20.9 (100°)	68.1 (100°)
Arsenic tribromide	$AsBr_3$	35	220	8.8 (35°)	54.1 (35°)
Iodine monobromide	IBr	41	∼116		
Mercury(II) bromide	$HgBr_2$	238	320	9.8	
Tin tetraiodide	SnI_4	143.5	340		
Iodine	I_2	113.7	184.3	11.1 (118°)	19.8 (116°)
n-Hexane	$n\text{-}C_6H_{14}$	−94.3	69.0	1.90	2.94 (25°)
Isooctane	$iso\text{-}C_8H_{18}$	−111.3	117.2	∼1.94 (20°)	
Kerosene	$\sim C_{12}H_{26}$		210	∼2.0	
Benzene	C_6H_6	5.5	80.1	2.27	6.52 (20°)
Toluene	$C_6H_5CH_3$	−95	110.6	2.38	5.90 (20°)
o-Xylene	$o\text{-}C_6H_4(CH_3)_2$	−25	144.4	2.57 (20°)	8.10 (20°)

Carbon tetrachloride	CCl_4	-22.8	76.8	2.23	9.69 (20°)
Chloroform	$CHCl_3$	-63.5	61.3	4.70	5.42 (25°)
Dichloromethane	CH_2Cl_2	-96.7	40.1	8.9	3.9 (30°)
Ethylene chloride	$ClCH_2CH_2Cl$	-35.3	83.7	10.36	8.0 (20°)
Tetrachloroethane	$Cl_2CHCHCl_2$	-43.8	146.3		18.4 (15°)
Chlorobenzene	C_6H_5Cl	-45.2	132	5.62	7.99 (20°)
Perfluoroheptane	$n\text{-}C_7F_{16}$	-51	82.5		
Diethyl ether	$(C_2H_5)_2O$	-116.3	34.6	4.22	2.22 (25°)
Tetrahydrofuran	$\overline{OCH_2CH_2CH_2CH_2}$		65.4	7.39	
1,2-Dimethoxyethane	$CH_3OCH_2CH_2OCH_3$	-69	85.2	3.5-6.8	11 (20°)
Diglyme	$CH_3(OCH_2CH_2)_2OCH_3$	-64	162.0		20 (20°)
Dioxane	$\overline{OCH_2CH_2OCH_2CH_2}$	11.7	101.5	2.21	
Ethyl acetate	$C_2H_5OCOCH_3$	-83.6	77.1	6.02	4.41 (25°)
Acetone	CH_3COCH_3	-95.4	56.2	20.7	3.16 (25°)
Acetonitrile	CH_3CN	-45.7	81.6	36.2	3.45 (25°)
Pyridine	C_5H_5N	-41.8	115.5	12.3	9.45 (20°)
Nitromethane	CH_3NO_2	-28.5	101.2	38.6 (20°)	6.08 (25°)
Nitrobenzene	$C_6H_5NO_2$	5.7	210.9	34.6	20.3 (20°)
Formamide	$HCONH_2$	2.55	193	109.5	37.6 (20°)
N-Methylformamide	$HCONHCH_3$		111.2	182.4	16.5 (25°)
Dimethylformamide	$HCON(CH_3)_2$	-61	153.0	36.7	7.96 (25°)
N-Methylacetamide	$CH_3CONHCH_3$	29.8	206	178.9 (30°)	38.85 (30°)
Dimethylacetamide	$CH_3CON(CH_3)_2$	-20	166.1	37.8	9.2 (25°)
Carbon disulfide	CS_2	-111.6	46.3	2.64	3.76 (20°)
Dimethyl sulfoxide	$(CH_3)_2SO$	18.55	189.0	47.6 (23°)	19.8 (25°)
Sulfolane	$(CH_2)_4SO_2$	28.37	283	44.0 (30°)	98.7 (30°)
Dimethyl carbonate	$CH_3OCOOCH_3$	5.0	90.4		
Ethylene carbonate	$\overline{OCH_2CH_2OCO}$		244		
Propylene carbonate	$\overline{OCHCH_3CH_2OCO}$	-49.2	241.7	65.1	25.3 (25°)
Tri-*n*-butyl phosphate	$(n\text{-}C_4H_9O)_3PO$	< -80	289 d.		
Hexamethylphosphoryltriamide	$[(CH_3)_2N]_3PO$	7.2	235	30 (20°)	35 (60°)

Table 5.3   Some Ammono and Aquo Compounds

Ammono Compound	Aquo Compound	Compound Type
NH_3	H_2O	solvent
KNH_2	KOH	base
NH_4Cl	HCl	protonic acid
CH_3CONH_2	CH_3COOH	"solvo-acid"
Li_2NH	Li_2O	ansolvous base
NH_2NO_2	HNO_3	"solvo-acid"

Anhydrous sulfuric acid self-ionizes in two ways[2] : by "ionic dehydration"

$$2\,H_2SO_4 = H_3O^+ + HS_2O_7^- \qquad K_{id} = 5.1 \times 10^{-5}\,mole^2\,kg^{-2} \qquad (5.2)$$

and by "autoprotolysis"

or $\qquad H_2SO_4 = H^+ + HSO_4^-$

$$(5.3)$$

$$2\,H_2SO_4 = H_3SO_4^+ + HSO_4^- \qquad K_{ap} = 2.7 \times 10^{-4}\,mole^2\,kg^{-2}$$

If we consider the solvated proton to be the acidic species in sulfuric acid, then the bisulfate ion, HSO_4^-, is the basic species. When water is dissolved in sulfuric acid, the following solvolysis takes place:

$$H_2O + H_2SO_4 = H_3O^+ + HSO_4^- \qquad K = 1\,mole\,kg^{-1} \qquad (5.4)$$

Hence, water is a fairly strong base in sulfuric acid. On the other hand, when sulfur trioxide is dissolved in sulfuric acid, the following solvolysis takes place:

or $\qquad SO_3 + H_2SO_4 = H^+ + HS_2O_7^-$

$$(5.5)$$

$$H_2S_2O_7 = H^+ + HS_2O_7^- \qquad K = 0.014\,mole\,kg^{-1}$$

Hence, sulfur trioxide is an acid in sulfuric acid. The freezing-point diagram for sulfuric acid solutions of water and of SO_3 is given in Fig. 5.1. The marked difference in the slopes of the two arms of the curve reflects the difference in the equilibrium constants for the solvolysis of water and the ionization of disulfuric acid. Freezing-point depression studies have been very useful in elucidating the nature of many sulfuric acid solutions. For example, such studies have shown that nitric acid reacts in sulfuric acid to give 4 moles of species. This result may be explained by reaction (5.6):

$$HONO_2 + 2\,H_2SO_4 = NO_2^+ + H_3O^+ + 2\,HSO_4^- \qquad (5.6)$$

[2] R. J. Gillespie and E. A. Robinson, *Non-Aqueous Solvent Systems*, T. C. Waddington, ed., Academic Press, London, 1965, Chap. 4, p. 117.

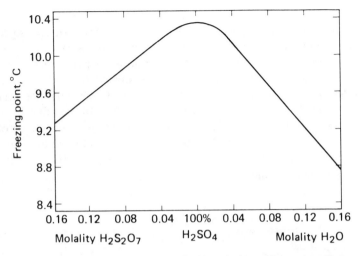

FIG. 5.1. Freezing points of solutions of sulfur trioxide and water in sulfuric acid.

It is believed that ethers, alcohols, and carboxylic acids react as follows:

$$R_2O + H_2SO_4 = R_2OH^+ + HSO_4^- \tag{5.7}$$

$$ROH + 2\,H_2SO_4 = ROSO_3H + H_3O^+ + HSO_4 \tag{5.8}$$

$$RCOOH + 2\,H_2SO_4 = RCO^+ + H_3O^+ + 2\,HSO_4^- \tag{5.9}$$

Anhydrous hydrogen fluoride self-ionizes to the solvated proton and the bifluoride ion

$$2\,HF = H^+ + HF_2^- \tag{5.10}$$

Hence, phosphorus pentafluoride is an acid in hydrogen fluoride:

$$PF_5 + HF = H^+ + PF_6^- \tag{5.11}$$

and water, nitric acid, and potassium nitrate are bases in hydrogen fluoride:

$$H_2O + 2\,HF = H_3O^+ + HF_2^- \tag{5.12}$$

$$HNO_3 + 2\,HF = H_2NO_3^+ + HF_2^- \tag{5.13}$$

$$KNO_3 + 4\,HF = H_2NO_3^+ + K^+ + 2\,HF_2^- \tag{5.14}$$

If it is desired to have a protonic acid and its anion in comparable amounts at equilibrium in water, the pK value of the acid must lie within the range from 0 to 14, or, at the outside, from -2 to 15. If the pK is much less than zero, the acid will be essentially completely ionized even in a strongly acidic solution, and if the pK is much greater than 14, the anion will be completely hydrolyzed even in a strongly basic solution. To have an acid with aqueous pK $\ll$ 0 in a measurable equilibrium with its anion, it is necessary to use a solvent that is less basic than water. To have an acid with aqueous pK $\gg$ 14 in a measurable equilibrium with its anion, it is necessary to use a solvent that is more basic than water. The effective aqueous pH ranges that can be achieved in several protonic solvents are indicated in Fig. 5.2. It will be noted that in an acidic solvent, such as formic acid, one can differentiate the acidities of HCl (pK = -7) and HBr (pK = -9). On the other hand, in a basic solvent, such as ammonia, one can differentiate the acidities of PH_3 (pK $\approx$ 27) and GeH_4 (pK $\approx$ 25). Both dimethyl sulfoxide and ammonia have very low autoprotolysis constants, and consequently these solvents have very wide pH ranges.

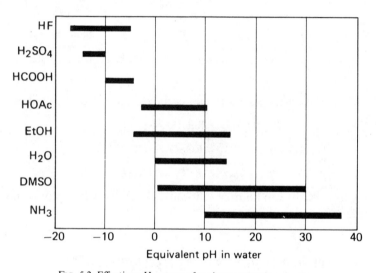

Fig. 5.2. Effective pH ranges of various protonic solvents.

In recent years there has been considerable interest in "superacid" media, which have such high proton activities that extremely weak bases are protonated when dissolved. For example, fluorosulfuric acid, $HOSO_2F$, has been used as such a medium.[3] In fluorosulfuric acid, even acids with pK values as low as -14 are only partially dissociated. The acidity of the solvent can be increased by the addition of SbF_5 and SbF_5-SO_2 mixtures, and the resulting solutions are among the most highly acidic media known. The solutions have been used for the preparation and identification of many new conjugate acid and carbonium ion species. Thus, in an HSO_3F-SbF_5-SO_2 solvent at low temperatures, acetic acid exists as an equilibrium mixture of

[3] R. J. Gillespie, *Acc. Chem. Res.*, **1**, 202 (1968).

the two isomeric protonated species:

$$
\left[CH_3-C \Big\langle \begin{array}{c} O \diagdown H \\[4pt] \diagup \\ O \diagup \\ H \end{array} \right]^{+}
\quad \text{and} \quad
\left[CH_3-C \Big\langle \begin{array}{c} H \diagdown \\ O \diagup \\ \\ O \diagdown \\ H \end{array} \right]^{+}
$$

In aqueous solutions, the activity of the hydrogen ion is very closely proportional to the concentration of strong acid down to about pH 0 and closely proportional to the reciprocal of strong base concentration up to about pH 14. However, beyond these pH limits, the hydrogen ion activity and the effective pH change much more rapidly than one would calculate simply from changes in the concentration of either strong acid or strong base. Nevertheless, Hammett[4] has shown that it is possible to define an acidity function that measures effective pH in solutions where pH loses its ordinary significance. The function H_0 describes the ability of a solution to remove a proton from a cationic acid BH^{+}, and it may be determined for a particular solution by using the relation

$$
H_0 = pK_a + \log \frac{C_B}{C_{BH^+}} \tag{5.15}
$$

where pK_a refers to the ionization constant for the acid, and C_{BH^+} and C_B are the concentrations of the acid and its neutral conjugate base at equilibrium in the solution. When a neutral acid BH and its negative conjugate base B^{-} are used, the function is H_-:

$$
H_- = pK_a + \log \frac{C_{B^-}}{C_{BH}} \tag{5.16}
$$

Some values of H_0 and H_- for various aqueous solutions[5-7] are tabulated in Table 5.4. Also, H_0 or H_- values can be assigned to nonaqueous systems. In Table 5.5, H_- values for ethanol-dimethyl sulfoxide solutions, 10^{-2} M in sodium ethoxide, are given.[8]

When a hydroxy acid is dissolved in a hydroxylic solvent—such as water, an alcohol, acetic acid, and so on—its acidity is enormously enhanced because of strong hydrogen bonding between the anion and the solvent. This effect

---

[4] L. P. Hammett, *Physical Organic Chemistry*, McGraw-Hill Book Company, New York, 1940.

[5] K. Yates and H. Wai, *J. Am. Chem. Soc.*, **86**, 5408 (1964).

[6] M. J. Jorgenson and D. R. Harrter, *J. Am. Chem. Soc.*, **85**, 878 (1963); Gillespie and Robinson, *Non-Aqueous Solvent Systems*, T. C. Waddington, ed., 1965, Chap. 4, p. 117.

[7] G. Schwarzenbach and R. Sulzberger, *Helv. Chim. Acta*, **27**, 348 (1944).

[8] K. Bowden and R. Stewart, *Tetrahedron*, **21**, 261 (1965).

Table 5.4 H_0 and H_- **Values for Aqueous Solutions of Some Strong Acids and Bases**

Concentration of $HClO_4$ or H_2SO_4 (wt. %)	$-H_0$		Molarity of NaOH or KOH	H_-	
	$HClO_4{}^5$	$H_2SO_4{}^6$		$NaOH^7$	KOH^7
10	0.35	0.31	1	13.9	14.0
15		0.66	2	14.4	14.6
20	0.98	1.01	3	14.7	14.9
25		1.37	4	14.9	15.2
30	1.60	1.72	5	15.2	15.5
35		2.06	6	15.4	15.8
40	2.40	2.41	7	15.6	16.2
45		2.85	8	15.8	16.5
50	3.48	3.38	9	16.0	16.8
55	4.23	3.91	10	16.2	17.3
60	5.27	4.46	11	16.5	17.7
65	6.45	5.07	12	16.8	18.2
70	7.75	5.80	13	17.1	
75	9.21	6.56	14	17.5	
80	10.75	7.34	15	17.8	
85		8.14	16	18.1	
90		8.92			
95		9.85			
96		10.03			
97		10.21			
98		10.40			
99		10.72			
100		12.08			
101		12.74			
102		12.92			
104		13.20			
106		13.50			

Table 5.5 H_- **Function for Ethanolic Dimethyl Sulfoxide Containing** 10^{-2} M **Sodium Ethoxide***

Dimethyl Sulfoxide (mole %)	H_-	Dimethyl Sulfoxide (mole %)	H_-
0	13.99	50	17.03
1	14.07	60	17.75
5	14.25	70	18.45
10	14.45	80	18.97
20	14.92	90	19.68
30	15.40	95	20.68
40	16.11		

* See ref. 8.

is particularly noticeable in the acidity of water itself. Thus, the pK of water in water is 16, whereas in the much more basic solvent, liquid ammonia, its pK is 18. If it were not for the abnormal stabilization of the hydroxide ion in water, the aqueous pK of water would probably be about 28. Thus, the hydroxide ion is intrinsically a very basic species whose basicity can only be fully utilized by using nonhydroxylic solvents.[9] A suspension of potassium hydroxide in a solvent such as dimethyl sulfoxide can be used to quantitatively deprotonate extremely weak acids, such as triphenylmethane (aqueous pK $\approx$ 30), phosphine (aqueous pK $\approx$ 27), germane (aqueous pK $\approx$ 25), and indene (aqueous pK $\approx$ 20).[10]

$$2\,KOH_{(s)} + HA = K^+ + A^- + KOH{\cdot}H_2O_{(s)} \qquad K = 10^{31-pK} \qquad (5.17)$$

An application of this technique is described in the synthesis of ferrocene (p. 484).

APROTIC SOLVENTS

The Solvent Concept

There are some solvents that undergo self-ionization, whose salt solutions are electrically conducting, and yet contain no hydrogen. Obviously, Brönsted's protonic concept of acids and bases cannot be used to identify acid-base reactions in such solvents. In such solvents, the usual procedure is to consider the cation formed in the self-ionization as the acidic species and to consider the anion formed in the self-ionization as the basic species. This procedure is the basis of the *"solvent" concept of acids and bases.*

Pure bromine trifluoride is a liquid with a rather high electrical conductivity ($\kappa = 8.1 \times 10^{-3}\,ohm^{-1}\,cm^{-1}$ at 15°).[11] The conductivity is explained by assuming the following self-ionization:

$$2\,BrF_3 = BrF_2{}^+ + BrF_4{}^- \qquad (5.18)$$

The d.c. electrolysis of bromine trifluoride involves the formation of BrF at the cathode and the formation of BrF_5 at the anode.

$$2\,e^- + BrF_3 + BrF_2{}^+ = BrF_4{}^- + BrF \qquad (5.19)$$

$$BrF_3 + BrF_4{}^- = BrF_2{}^+ + BrF_5 + 2\,e^- \qquad (5.20)$$

At the boundary between the anode and cathode compartments, the recombination reaction occurs:

$$BrF + BrF_5 = 2\,BrF_3 \qquad (5.21)$$

[9] W. L. Jolly, *J. Chem. Educ.*, **44**, 304 (1967); *Inorg. Chem.*, **6**, 1435 (1967).

[10] Applications of this general technique are described in *Inorg. Syn.*, **11**, 113–30 (1968).

[11] A. Banks, H. Emeléus, and A. Woolf, *J. Chem. Soc.*, 2861 (1949).

The conductivity is achieved by a countercurrent migration of the BrF_2^+ and BrF_4^- ions.

Bromine trifluoride forms adducts with many metal fluorides. Potassium fluoride reacts with BrF_3 to form potassium tetrafluorobromate, $KBrF_4$. Antimony pentafluoride reacts to form difluorobromine(III) hexafluoroantimonate(V), BrF_2SbF_6. According to the "solvent" concept, $KBrF_4$ is a base and BrF_2SbF_6 is an acid in bromine trifluoride. Thus, when solutions of these reagents are mixed, a typical neutralization reaction takes place:

$$BrF_2^+ + BrF_4^- = 2\,BrF_3 \qquad (5.22)$$

Liquid sulfur dioxide dissolves many substances to form conducting solutions. Certain rapid reactions, such as the reaction between thionyl chloride and cesium sulfite to give cesium chloride, have been interpreted as neutralization reactions.[12] It was supposed that thionyl compounds ionize to give the acidic SO^{2+} ion and that sulfites ionize to give the basic SO_3^{2-} ion. Neutralizations were then presumed to proceed by the reaction

$$SO^{2+} + SO_3^{2-} = 2\,SO_2 \qquad (5.23)$$

The reverse reaction is a plausible self-ionization reaction for sulfur dioxide. However, experiments with radioactive sulfur have shown that although sulfur is exchanged rapidly between the solvent and dissolved sulfites, essentially no exchange occurs between thionyl halides and the solvent.[13] Hence, Eq. (5.23) is an unlikely mechanism for neutralizations. Perhaps some mechanism like the following, involving oxide ion transfer, is operative:

$$SOCl_2 = SOCl^+ + Cl^- \qquad (5.24)$$

$$SOCl^+ + SO_3^{2-} \longrightarrow SO_2 + SO_2Cl^- \qquad (5.25)$$

$$SO_2Cl^- \longrightarrow SO_2 + Cl^- \qquad (5.26)$$

This example shows that caution must be used in applying the solvent concept of acids and bases. Another place where the solvent concept can cause difficulty is in the interpretation of the acidic behavior of metal halides dissolved in nonmetal halides, such as $POCl_3$. According to the concept, $POCl_3$ ionizes as follows:

$$POCl_3 \rightleftharpoons POCl_2^+ + Cl^- \qquad (5.27)$$

Consequently, we would identify $POCl_2^+$ as the acidic species and Cl^- as the basic species. It has been shown[14] that addition of $FeCl_3$ to $POCl_3$ yields $FeCl_4^-$ (and presumably $POCl_2^+$); hence, $FeCl_3$ is an acid. Conductimetric titrations of solutions of $FeCl_3$ in $POCl_3$ with ionic halides[15]

[12] G. Jander, *Die Chemie in Wasserähnlichen Lösungsmitteln*, Springer-Verlag, Berlin, 1949.

[13] R. E. Johnson, T. H. Norris, and J. L. Huston, *J. Am. Chem. Soc.*, **73**, 3052 (1951).

[14] M. Baaz, V. Gutmann, and L. Hubner, *Monatsh.*, **91**, 537 (1960).

[15] V. Gutmann and M. Baaz, *Monatsh.*, **90**, 729 (1959).

give sharp breaks at a $Cl^- : FeCl_3$ ratio of $1:1$. The net reaction was presumed to be

$$POCl_2^+ + Cl^- \longrightarrow POCl_3 \tag{5.28}$$

However, spectral studies show that when $FeCl_3$ is dissolved in triethyl phosphate, $FeCl_4^-$ is formed to essentially the same extent as when it is dissolved in $POCl_3$. Meek and Drago[16] have therefore proposed that the reaction of $FeCl_3$ in both solvents is the same—namely,

$$(1 + x)FeCl_3 + (1 + x)Y_3PO \rightleftharpoons (1 + x)FeCl_3OPY_3$$
$$\rightleftharpoons FeCl_{3-x}(OPY_3)_{1+x}{}^{x+} + xFeCl_4^- \tag{5.29}$$

The conductimetric titration results are then explained by the reaction

$$(1 + x)Cl^- + FeCl_{3-x}(OPY_3)_{1+x}{}^{x+} \longrightarrow FeCl_4^- + (1 + x)OPY_3 \tag{5.30}$$

Solvation

The preceding results in $POCl_3$ point out that the donor property of a solvent (the tendency of a solvent to coordinate to a solute) is very important. The classification of solvents as to their solvating ability is a difficult problem. No simple physical property—such as dielectric constant, dipole moment, or refractive index—seems to correlate well with solvating ability. (Thus, in Chap. 2 we have seen that there is no correlation between solvent dielectric constant and ionic solvation energies.) Probably the most practical method for measuring relative solvating ability is by experimentally determining the effect of various solvents on some solvent-dependent standard process, such as the rate of a chemical reaction or the absorption of light by a solvatochromic dye.[17] Smith, Fainberg, and Winstein[18] used the ionization of *p*-methoxyneophyl *p*-toluenesulfonate at 75° as a standard reaction. They found that the rate constants correlated well with the rates of other solvent-sensitive reactions. Other investigators have used the energies of solvent-sensitive charge transfer absorption bands in certain dyes to correlate solvents.[19,20] The rate constants and three different sets of dye transition energies[21] for various solvents are presented in Table 5.6. Data for both protonic and aprotic solvents are given; the solvents are listed in order of decreasing E_T values. It will be noticed that approximately the same ordering of solvents is found using the $\log k_{ion}$ parameters, the E_T values, and the X_B

[16] D. Meek and R. S. Drago, *J. Am. Chem. Soc.*, **83**, 4322 (1961); also see R. S. Drago and K. F. Purcell, *Non-Aqueous Solvent Systems*, T. C. Waddington, ed., Academic Press, London, 1965, Chap. 5, p. 211, and J. E. Huheey, *J. Inorg. Nucl. Chem.*, **24**, 1011 (1962).

[17] C. Reichardt, *Angew. Chem. Intern. Ed. Engl.*, **4**, 29 (1965).

[18] S. G. Smith, A. H. Fainberg, and S. Winstein, *J. Am. Chem. Soc.*, **83**, 618 (1961).

[19] K. Dimroth, C. Reichardt, T. Siepmann, and F. Bohlmann, data quoted in Ref. 17.

[20] L. G. S. Brooker, A. C. Craig, D. W. Heseltine, P. W. Jenkins, and L. L. Lincoln, *J. Am. Chem. Soc.*, **87**, 2443 (1965).

[21] The E_T values are for a pyridinium N-phenolbetaine[19] and the X_R and X_B values are for merocyanine dyes.[20]

Table 5.6   Empirical Parameters of Solvents for Measuring "Solvating Ability"*

Solvent	$-\log k_{ion}$ (sec^{-1})	E_T (kcal)	X_R (kcal)	X_B (kcal)
H_2O	1.18	63.1		68.9
CH_3OH	2.80	55.5	43.1	63.0
C_2H_5OH	3.20	51.9	43.9	60.4
CH_3COOH	2.77	51.9		
$CH_3CH_2CH_2OH$		50.7	44.1	
$CH_3CH_2CH_2CH_2OH$		50.2	44.5	56.8
$(CH_3)_2CHOH$		48.6	44.5	56.1
$\overline{OCHCH_3CH_2OCO}$		46.6		
CH_3NO_2	3.92	46.3	44.0	
CH_3CN	4.22	46.0	45.7	53.7
CH_3SOCH_3	3.74	45.0	42.0	
$C_6H_5NH_2$		44.3	41.1	
Sulfolane		44.0		
$HCON(CH_3)_2$	4.30	43.8	43.7	51.5
CH_3COCH_3	5.07	42.2	45.7	50.1
$C_6H_5NO_2$		42.0	42.6	
CH_2Cl_2		41.1	44.9	47.5
C_5H_5N	4.67	40.2	43.9	50.0
$CHCl_3$		39.1	44.2	
$CH_3COOC_2H_5$	5.95	38.1	47.2	
C_6H_5Cl		37.5	45.2	
$\underline{CH_2CH_2CH_2CH_2}\text{-O}$	6.07	37.4	46.6	
$\underline{O\text{-}CH_2CH_2\text{-}O\text{-}CH_2CH_2}$		36.0	48.4	
$(C_2H_5)_2O$	7.3	34.6	48.3	
C_6H_6		34.5	46.9	
$C_6H_5CH_3$		33.9	47.2	41.7
CS_2		32.6		
CCl_4		32.5	48.7	
n-Hexane		30.9	50.9	

* See refs. 18–20.

values. These parameters can be used to correlate reactions in rather polar solvents. On the other hand, the X_R values have been found to correlate fairly well with reactions in predominately nonpolar solvents.[20]

SOLUBILITIES OF NONELECTROLYTES

Ideal Solutions

Two liquids will form an ideal solution if they have zero heat of mixing and if the molecules in the mixture arrange themselves completely randomly. The latter clause is equivalent to saying that the increase in entropy for a component of an ideal solution is $-R \ln x$, where x is the mole fraction. Thus we calculate, for the dilution of a pure liquid to form an ideal solution,

$$\Delta G = RT \ln x$$

Obviously two liquids that form an ideal solution are completely miscible, and it can be shown that the partial pressure of each component over a solution is equal to the vapor pressure of the pure liquid times its mole fraction (Raoult's Law). Many pairs of liquids—for example, benzene and toluene—form ideal, or nearly ideal, solutions.

Consider the dissolution of a solid in a liquid. The process may be broken into two steps:

$$\text{solid solute} \longrightarrow \text{liquid solute} \longrightarrow \text{solute in solution}$$

If the hypothetical liquid solute forms an ideal solution with the solvent, we may write, for the dissolution of the solid in a saturated solution,

$$\Delta G = 0 = \Delta G_{fus} + \Delta G_{soln}$$

$$= \Delta H_{fus} - T\,\Delta S_{fus} + RT \ln x \tag{5.31}$$

$$\ln x = \frac{-\Delta H_{fus}}{RT} + \frac{\Delta H_{fus}}{RT_{mp}} \tag{5.32}$$

where x is the mole fraction of the solute and T_{mp} is the absolute melting temperature. Thus, one can calculate the solubility of a solid at any temperature below its melting point without any information about the solvent except that it forms an ideal solution with the hypothetical supercooled molten solute. For example, naphthalene forms essentially ideal solutions with benzene, toluene, and nitrobenzene, and its solubility in these solvents can be closely estimated simply from its melting point and heat of fusion.

Regular Solutions

Hildebrand[22] has shown that there is a class of solutions, far larger than the class of ideal solutions, for which there is a finite heat of mixing, but for which the entropy is the same as in ideal solutions. These are regular solutions. They are formed by most liquid pairs in which dipole-dipole interactions, hydrogen bonding, and chemical complexing are absent. It has been shown[23] that the heat absorbed upon dilution of a pure liquid to form a regular

[22] J. H. Hildebrand, *J. Am. Chem. Soc.*, **51**, 66 (1929); J. H. Hildebrand and R. L. Scott, *The Solubility of Nonelectrolytes*, 3rd ed., Reinhold Publishing Corp., New York, 1950; J. H. Hildebrand and R. L. Scott, *Regular Solutions*, Prentice-Hall, Inc., Englewood Cliffs, N.J., 1962.

[23] G. Scatchard, *Chem. Revs.*, **8**, 321 (1931); J. H. Hildebrand and S. E. Wood, *J. Chem. Phys.*, **1**, 817 (1933).

solution is

$$\Delta H = V_2 \phi_1{}^2 \left[\left(\frac{E_1}{V_1} \right)^{1/2} - \left(\frac{E_2}{V_2} \right)^{1/2} \right]^2 = V_2 \phi_1{}^2 (\delta_1 - \delta_2)^2 \qquad (5.33)$$

where V is the molar volume, ϕ is the volume fraction, E is the molar energy of vaporization, δ is the so-called solubility parameter, and the subscripts 1 and 2 refer to the solvent and solute, respectively. One then calculates, for the solubility (S) of a liquid,

$$\ln S = - \frac{V_2 \phi_1{}^2}{RT} (\delta_1 - \delta_2)^2 \qquad (5.34)$$

and for a solid,

$$\ln S = - \frac{V_2 \phi_1{}^2}{RT} (\delta_1 - \delta_2)^2 - \frac{\Delta H_{fus}}{RT} + \frac{\Delta H_{fus}}{RT_{mp}} \qquad (5.35)$$

A test of this equation appears in Fig. 5.3, where the solubility of iodine at 25° in a wide variety of solvents is plotted[24] vs. $(\delta_1 - \delta_2)^2$. As expected from Eq.

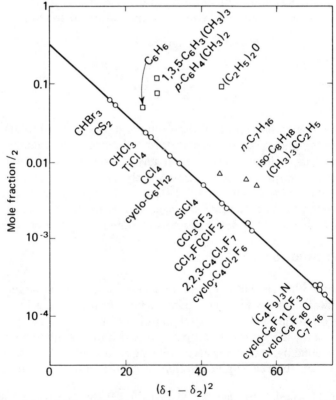

FIG. 5.3. Solubility of iodine at 25°.[24] (Reproduced with permission of The Macmillan Company.)

[24] J. H. Hildebrand and R. E. Powell, *Principles of Chemistry*, 7th ed., The Macmillan Company, New York, 1964, p. 286.

(5.35), most of the points fall on a straight line. Some of the points that lie above the line correspond to liquids which form complexes with iodine, and therefore would not be expected to form regular solutions.

Equations (5.34) and (5.35) permit us to make a generalization having considerable practical value: The more nearly equal are the solubility parameters of two materials, the more ideal are their solutions and the greater are their mutual solubilities. Values of the solubility parameter, δ, for a variety of solvents[25] are given in Table 5.7. When estimating solubilities by means of solubility parameters, it is important to remember that the theory is applicable only to mixtures in which there are no chemical reactions or

Table 5.7   Solubility Parameters*

Solvent	δ	Solvent	δ
Water	23.4	Dimethyl carbonate	9.9
N-methylformamide	16.1	Dioxane	9.9
Ethylene carbonate	14.7	Ethylene dichloride	9.8
N-methylacetamide	14.6	1,1,2,2-Tetrachloroethane	9.7
Methanol	14.5	Dichloromethane	9.7
Ethylene glycol	14.2	Chlorobenzene	9.5
Propylene carbonate	13.3	Chloroform	9.3
Dimethyl sulfoxide	12.8	Benzene	9.2
Ethanol	12.7	Tetrahydrofuran	9.1
Nitromethane	12.7	Ethyl acetate	9.1
Dimethylformamide	12.1	Toluene	8.9
n-Propanol	11.9	Xylene	8.8
Acetonitrile	11.9	Carbon tetrachloride	8.6
Isopropanol	11.5	Benzonitrile	8.4
Nitrobenzene	10.9	Cyclohexane	8.2
Pyridine	10.7	*n*-Octane	7.6
t-Butanol	10.6	*n*-Heptane	7.4
Acetic anhydride	10.3	Diethyl ether	7.4
Nitrobenzene	10.0	*n*-hexane	7.3
Carbon disulfide	10.0	Tetramethylsilane	6.2
Acetone	10.0	Perfluoroheptane	5.8
Iso-amyl alcohol	10.0	Silicones	5.5

* See ref. 25.

solvation effects. Thus, although water and pyridine have quite different parameters, they are infinitely miscible in one another. On the other hand, the solubility trends of saturated hydrocarbons (which are chemically inert to most solvents) are fairly predictable from solubility parameters. For

[25] H. Burrell, *Official Digest, Fed. Paint and Varnish Prod. Clubs*, **27**, 726 (1955); R. F. Weimer and J. M. Prausnitz, *Hydrocarbon Proc.*, **44**, 237 (1965).

cyclohexane, $\delta = 8.2$, and, as one might expect, this liquid is immiscible with water ($\delta = 23.4$) and ethylene glycol ($\delta = 14.2$). However, it is partially miscible with acetonitrile ($\delta = 11.9$) and miscible with iso-amyl alcohol ($\delta = 10.0$) and almost all liquids for which $6 < \delta < 10$. Clearly solubility parameters can be of help when choosing solvents for liquid-liquid extraction.

SOLVENT PURIFICATION

Drying Agents

Frequently it is necessary to purify a solvent before use to avoid a deleterious side reaction caused by an impurity. Usually water is the most significant impurity in a solvent, and therefore most purification schemes are essentially schemes for the removal of water. In the following paragraphs, some important drying agents are listed in the approximate order of decreasing efficiency together with remarks concerning their use.

Sodium. Sodium is very commonly used as a drying agent. It is extremely effective when it can dissolve (as in liquid ammonia and some amines), but is rather slow acting when (as in most cases) it must be used in the form of chunks or wire. A more reactive form of solid sodium is achieved by making a fine dispersion of the metal, as described on p. 476. Ethers may be dried by dissolving about 5 g of benzophenone per liter and warming with sodium. After about an hour, the violet color of the ketyl forms:

$$(C_6H_5)_2CO + Na \longrightarrow (C_6H_5)_2\dot{C}-O^-Na^+$$

This reagent reacts vigorously with any water present. Thus, as long as the solution is violet, dryness is assured, and the ether can be distilled from the solution into a reaction vessel. However, the solution must be stored under an inert atmosphere or in an evacuated vessel.

CAUTION: Never use sodium to dry acidic solvents or solvents having oxidizing power. Chlorinated hydrocarbons must never be treated with sodium.

Lithium aluminum hydride ($LiAlH_4$). It is an inflammable solid that reacts vigorously with protonic species and effectively removes these species from solvents in which it dissolves (such as ethers). As with sodium, oxidizing solvents must be avoided. Lithium aluminum hydride should never be heated with a solvent over 100° because of the danger of explosion. Solvents highly contaminated with protonic species or oxidizing agents (such as peroxides) should be prepurified before treatment with $LiAlH_4$.

Calcium Hydride (CaH_2). It is insoluble in all solvents with which it does not react. Protonic impurities are very efficiently removed by refluxing a solvent with the powdered reagent.

Sodium-lead Alloy (Dri–Na). Available in granular form, it is a relatively safe reagent that should be refluxed with the solvent.

Phosphorus Pentoxide (P_4O_{10}). This reagent reacts vigorously with water to form syrupy phosphoric acid. It is principally used to dry hydrocarbons and their halogenated derivatives.

Barium Oxide (BaO). Obviously, it cannot be used to dry acidic species; it is often used to dry amines and, sometimes, alcohols. The effectiveness of the reagent can be tested by adding a few drops of water to a small lump of the oxide, whereupon a vigorous exothermic reaction will occur if the material is of good quality.

Linde Molecular Sieves. They are very effective for solvent purification when used in the form of a column. A convenient method is to store the solvent in a large bulb that surmounts a column (about 3×60 cm) containing the molecular sieve granules and from the bottom of which the pure solvent is drawn through a Teflon stopcock when needed.

Silica Gel (SiO_2) *and* ***Active Alumina*** (Al_2O_3). Of approximately equal drying efficiency, they are essentially chemically inert and can be regenerated after exhaustion by heating ($\sim 300°$ for SiO_2 and $\sim 500°$ for Al_2O_3).

Potassium Hydroxide (KOH). It is a good drying agent for amines.

Calcium Sulfate ($CaSO_4$; Drierite). A good, convenient drying agent of rather low capacity, it is available as "indicating" granules (blue when fresh, pink when exhausted).

Calcium Oxide (CaO). It has the same characteristics as barium oxide, but it is less efficient.

Calcium Chloride ($CaCl_2$). This fairly good desiccant must be used with discretion, because it forms adducts with alcohols, amines, and many other polar compounds.

Peroxide Removal

Ethers (such as diethyl ether, dioxane, tetrahydrofuran, etc.) are hazardous solvents because, on standing in the presence of air and light, they form peroxides that can explode upon heating. The peroxides are less volatile than the ethers; therefore, they are concentrated during distillations, and the superheating that generally occurs at the end of a distillation is sufficient to cause explosion. It is important to test ethers for the presence of peroxides, and, if more than traces of peroxides are found to be present, they should be destroyed before working with the ethers.

A water-soluble ether may be tested for peroxide by adding a portion to an acidic aqueous iodide solution. A brown color indicates the presence of peroxide. A convenient method for removing peroxide from a water-soluble ether is to reflux the ether with about 0.5 per cent by weight of copper(I) chloride and finally to distill. This peroxide removal should precede any drying operation.

A water-insoluble ether may be tested for peroxide as follows. Dissolve 1 mg of sodium or potassium dichromate in 1 ml of water in a test tube; add a drop of dilute sulfuric acid and about 10 ml of the ether, and shake. A blue color in the ether layer indicates the presence of peroxide which can be removed by shaking with an aqueous 5 per cent iron(II) sulfate solution weakly acidified with sulfuric acid.

Purification and Handling Procedures

Acetic Acid. It is available as the 99.5 per cent reagent. The water may be removed by treatment with triacetyl borate. Triacetyl borate is prepared[26] by heating one part of boric acid (by weight) with five parts of acetic anhydride to 60°; the crystals of product that precipitate on cooling are filtered off. The acetic acid is treated with a calculated three-fold excess of the triacetyl borate; the mixture is refluxed for an hour and is finally distilled.

Acetone. The water content may be reduced to 0.001 per cent by treatment with Drierite. If reducing impurities must be removed, the acetone should first be heated with a little potassium permanganate and then distilled. Very pure material may be prepared by the preparation and subsequent decomposition of the sodium iodide addition compound, $(CH_3)_2CO \cdot NaI$.[27]

Acetonitrile (Methyl Cyanide). Acetonitrile may be dried by refluxing with phosphorus pentoxide, followed by distillation.

Ammonia. Ammonia is available in cylinders as refrigeration-grade material containing less than 0.05 per cent water as the principal impurity. For many purposes such ammonia is sufficiently pure, in which case the ammonia may be passed directly from the cylinder into the reaction vessel, which may be a Dewar, a refrigerated glass tube, and so on. Most large ammonia cylinders have a "dip pipe" running from the base of the valve to the inside cylinder wall near the shoulder of the cylinder. To withdraw gas from the cylinder, one must keep it standing upright; but to withdraw liquid ammonia, one must place the cylinder so that its dip pipe is immersed in the liquid. To do this, lay the cylinder down with the butt end about 2 in. higher than the valve end. The dip pipe will be immersed when the main valve outlet

[26] A. Pictet and A. Geleznoff, *Ber.*, **36**, 2219 (1903).
[27] R. Livingston, *J. Am. Chem. Soc.*, **69**, 1220 (1947).

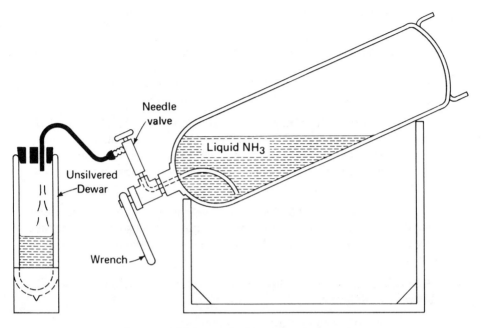

FIG. 5.4. Withdrawing liquid ammonia from a cylinder.

points up. The liquid is generally withdrawn into an unsilvered Dewar so that the rate of withdrawal may be readily observed (see Fig. 5.4 for a sketch of a typical setup).

Ordinarily, a reaction vessel containing liquid ammonia should be protected from the moisture and carbon dioxide of the atmosphere, but even this precaution may be neglected if these impurities are harmless. In fact, ammonia may be handled in ordinary glassware (beakers, flasks, Büchner funnels, etc.) if a good hood is available.

When even a trace of water would be detrimental, ammonia must be dried by dissolving sodium in it, followed by distillation. Such work is usually performed in a vacuum line, although a metal high-pressure system would function equally well.

Ammonia has a vapor pressure near 10 atm at room temperature. Sealed glass tubes of less than 20 mm diameter will seldom burst under such a pressure; nevertheless, any glass tube containing liquid ammonia at room temperature should be handled with the greatest of care (face shield and thick gloves). The reader is referred to Franklin[28] for a discussion of the types of reactions that may be performed by the use of sealed glass tubes containing liquid ammonia. By following the procedure described in Chap. 14 (under

[28] E. C. Franklin, *The Nitrogen System of Compounds*, Reinhold Publishing Corp., New York, 1935, pp. 317–30.

Sealed-Tube Reactions), we may heat sealed tubes containing ammonia to as high as 150°.

Benzene. Benzene may be dried by distilling slowly until the distillate is no longer cloudy. The residue is extremely dry. If thiophene is present and it is necessary to remove it, the benzene should be agitated with concentrated sulfuric acid, washed with water, and then distilled.

Carbon Disulfide. It is an extremely inflammable and toxic liquid. It should be distilled with great care, using a pan of water heated on a steam bath in a hood. The solvent may be purified by shaking, successively with mercury, a mercuric chloride solution, and a potassium permanganate solution. Finally the material is dried over phosphorus pentoxide and distilled.

Carbon Tetrachloride. Carbon tetrachloride is very toxic and should be handled accordingly. It may be dried by refluxing over phosphorus pentoxide, followed by distillation. Never dry with sodium because of the danger of an explosion.

Chlorobenzene. It may be dried by treatment with phosphorus pentoxide, followed by distillation.

Chloroform. Chloroform usually contains about 1 per cent alcohol as a stabilizer, which may be removed by shaking five or six times with one-half volume of water. The alcoholfree material is dried—first with calcium chloride and then with phosphorus pentoxide—and finally distilled. Pure chloroform reacts at an appreciable rate with oxygen in the presence of light to give highly poisonous phosgene ($COCl_2$). Chloroform should never be dried over sodium because of the danger of explosion.

Diglyme. This dimethyl ether of diethylene glycol may be dried by refluxing with calcium hydride or sodium-lead alloy, and then distilled. Lithium aluminum hydride may be used as a drying agent at room temperature, but diglyme should never be heated with this reagent because of the danger of explosion.

Dimethylacetamide. Dimethylacetamide and N-methylacetamide have been purified by vacuum fractional distillation.[29]

Dimethylformamide. Dimethylformamide is dried by treatment with potassium hydroxide, calcium oxide, or barium oxide, followed by vacuum distillation. N-methylformamide is purified by agitation with P_2O_5, followed by filtration and vacuum distillation.[29]

Dimethyl Sulfoxide. It generally contains about 0.5 per cent water and

[29] J. W. Vaughn, *The Chemistry of Non-Aqueous Solvents*, J. J. Lagowski, ed., Academic Press, New York, 1967, p. 191.

small quantities of dimethyl sulfide and dimethyl sulfone. Fractional vacuum distillation from barium oxide is the most commonly used method of purification.

Dioxane. Dioxane may be purified by treatment with potassium hydroxide followed by refluxing for one or two days with sodium-lead alloy, and, finally, by distillation.

Ethanol. Ethanol containing less than 0.5 per cent water is available as the commercial "absolute" or "200 proof" alcohol. When necessary, the remaining water may be removed by the following procedure.[30] About 5 g of magnesium turnings, a few drops of carbon tetrachloride or chloroform, and 60 ml of absolute ethanol are refluxed in a 2-l flask until almost all the magnesium has been converted to the ethoxide. About 900 ml of absolute ethanol is added; the mixture is refluxed for an hour and is then distilled. When halogen compounds must be excluded, ethyl bromide can be used as the catalyst; in that case, the ethyl bromide comes over in the first few milliliters of distillate. Great care must be taken when transferring the solvent to avoid absorption of moisture.

Ether (Diethyl). It is commercially available in an "anhydrous" form containing less than 0.1 per cent water. This water may be removed by refluxing with calcium hydride, lithium aluminum hydride, or sodium-lead alloy, followed by distillation.

Ethyl Acetate. Ethyl acetate may be purified by treatment with anhydrous potassium carbonate, followed by distillation from phosphorus pentoxide.

Ethylene Dichloride. It may be dried by refluxing with phosphorus pentoxide, followed by distillation.

Formic Acid. Formic acid is available as the 88 per cent reagent. It may be dried by storage for several days over boric oxide[31] or three parts of phthalic anhydride,[32] followed by distillation under reduced pressure.

n-Hexane. It may be dried by treatment with Drierite. If the solvent must be freed of aromatic and unsaturated compounds (as when it is used in ultraviolet spectroscopy), it is shaken with a mixture of concentrated sulfuric and nitric acids, washed with water, distilled, and then passed through a column of activated alumina.

Methanol. Methanol may be dried by fractional distillation, followed by treatment with Drierite for a period of several days.

[30] H. Lund and J. Bjerrum, *Ber.*, **64**, 210 (1931); H. Lund, *J. Am. Chem. Soc.*, **74**, 3188 (1952).

[31] H. I. Schlessinger and A. W. Martin, *J. Am. Chem. Soc.*, **36**, 1589 (1914).

[32] Schering-Kahlbaum, A.-G., British Patent No. 308,731; *Chem. Abstr.*, **24**, 380 (1930).

Methylene Chloride (Dichloromethane). It may be dried by refluxing with phosphorus pentoxide, followed by distillation.

Monoglyme. It is also known as glyme, 1,2-dimethoxyethane and dimethyl ether of ethylene glycol. It may be purified by refluxing with sodium, or sodium-potassium alloy, followed by distillation.

Nitrobenzene. Nitrobenzene may be dried over phosphorus pentoxide and then vacuum distilled.

Nitromethane. It may be dried over calcium chloride and then fractionally distilled; the initial distillate is discarded. The main distillate should be stored over Drierite. The literature may be consulted for a method for preparing high-purity nitromethane.[33]

Propylene Carbonate. It has been purified by vacuum distillation.[34]

Pyridine. Pyridine may be dried by allowing it to stand over potassium hydroxide for several days, followed by refluxing with barium oxide and fractional distillation.

Sulfolane. Sulfolane has been purified by repeated vacuum distillation from sodium hydroxide pellets until a 1-ml sample did not become colored within 5 min after the addition of an equal volume of 100 per cent sulfuric acid.[35] A final distillation from calcium hydride removes the last traces of water.

Sulfur Dioxide. Commercial sulfur dioxide contains trace amounts of sulfur trioxide and as much as 0.1 per cent water. Passage of the gas through concentrated sulfuric acid removes both of these impurities. The liquid may be handled in much the same way as ammonia. If small amounts of water are acceptable, ordinary Dewars and flasks may be used to contain the liquid. Otherwise, the vacuum line or closed tubes are recommended.

Sulfuric Acid. The most convenient way to make 100 per cent sulfuric acid is to add just enough fuming sulfuric acid (sulfuric acid containing excess SO_3) to ordinary 96 per cent sulfuric acid to convert all the water to sulfuric acid. The "fair-and-foggy" method[36] is used to determine the endpoint. In this remarkably simple yet delicate test, a gust of moist air is blown across the acid with a small rubber syringe. If the acid is more than 100 per cent pure (i.e., contains excess SO_3), a fog will appear. If the acid is less than 100 per cent pure, no fog will appear. By this procedure, one may adjust the composition to within 0.02 per cent of pure sulfuric acid.

[33] C. J. Thompson, H. J. Coleman, and R. V. Helm, *J. Am. Chem. Soc.*, **76**, 3445 (1954).

[34] P. L. Kronick and R. M. Fuoss, *J. Am. Chem. Soc.*, **77**, 6114 (1955); Y. Wu and H. L. Friedman, *J. Phys. Chem.*, **70**, 501 (1966).

[35] E. M. Arnett and C. F. Douty, *J. Am. Chem. Soc.*, **86**, 409 (1964).

[36] J. E. Kunzler, *Anal. Chem.*, **25**, 93 (1953).

Sulfuric acid is extremely hygroscopic. If it is important that it remain anhydrous, it should be suitably protected from the atmosphere. It also should be remembered that hot sulfuric acid is a powerful oxidizing and dehydrating agent. Whenever more than a few milliliters of the hot acid must be handled, special protective shielding should be provided.

1,1,2,2-Tetrachloroethane. It is purified by agitation with concentrated sulfuric acid for about two hours, followed by washing with water, drying with calcium chloride, and distillation. The purified material may be stored over Drierite.

Tetrahydrofuran. It is often grossly contaminated with peroxides; such material should be discarded. If a peroxide test (p. 116) indicates the presence of only traces of peroxides, these should be removed by treatment with copper(I) chloride[37] before the solvent is dried or used in any way. If the peroxide-free tetrahydrofuran is very wet, it may be partially dried by treatment with potassium hydroxide pellets. The last traces of water are removed by refluxing with calcium hydride, lithium aluminum hydride, or sodium-lead alloy, followed by distillation.

Toluene. Toluene is dried by distillation; the initial cloudy distillate is discarded.

Tributyl Phosphate. It may be purified by washing with aqueous sodium hydroxide and then with water. After drying over Drierite, the solvent is vacuum distilled.[38]

Xylene. Xylene is generally available as a mixture of the three isomers, which are very difficult to separate by distillation. The solvent may be purified by shaking first with sulfuric acid and then water, followed by drying over phosphorus pentoxide and distillation.

Problems

1. From the equilibrium constants for Eqs. (5.2), (5.3), and (5.4), calculate the concentrations of the four major ionic species in 100 per cent sulfuric acid and in a sulfuric acid solution that is nominally 0.02 molal in H_2O.

2. What is the net reaction that takes place when nitryl chloride, NO_2Cl, is formed by bubbling HCl into a solution of nitric acid in sulfuric acid?

---

[37] A 0.5 per cent suspension of CuCl in tetrahydrofuran is refluxed for 30 min and then distilled before further treatment.

[38] L. L. Burger, *The Chemistry of Tributyl Phosphate. A Review*, HW 40910, Hanford Atomic Products Operation Report, 1955.

3. Six moles of dissolved species are formed when either one mole of boric acid (H_3BO_3) or one mole of boric oxide (B_2O_3) is dissolved in sulfuric acid. Write the net reactions for these dissolution processes.

4. Explain how potassium acid phthalate can be a primary standard acid for aqueous solutions, and a primary standard base for titrimetry in anhydrous acetic acid.

5.

Compound	Molecular Weight	Density	Temperature at Which Vapor Pressure is 1 atm (°K)
PCl_5	208	1.6	433
IF_7	222	3.5	371
$SnCl_4$	261	2.2	387

In which solvent, IF_7 or $SnCl_4$, would a solution containing a mole fraction of PCl_5 of 10^{-2} have the highest vapor pressure of PCl_5 (assume regular solutions and Trouton's Rule)?

6. At 17°, WF_6 boils and has a density of 3.42 g cc^{-1}. If CCl_4 and CS_2 are each saturated with WF_6 gas at 0.1 atm, which liquid will have the higher concentration of WF_6?

The Maintenance

and

Measurement

of

Low Temperatures

6

Often it is necessary to carry out a reaction at a low temperature when the temperature need not be controlled any closer than $\pm 3°$ or more. Several methods, each relying upon the occasional attention of the experimenter for the maintenance of the low temperature, are commonly used in such cases.

In the interval between room temperature and about 12°, a bath of flowing tap water is usually adequate. From this point to 0°, a stirred water bath to which crushed ice is occasionally added can be used. From 0 to about $-25°$, a stirred mixture of crushed ice, water, and salt (NaCl) is commonly used. The temperature is maintained by occasionally siphoning off some of the salt solution and adding fresh ice and salt. For the same temperature interval, a mixture of ice and aqueous hydrochloric acid is also used. The temperature is maintained by occasionally siphoning off some of the dilute HCl solution and adding fresh ice and 12 M HCl.

123

Temperatures between about -10 and $-78°$ are conveniently maintained by the occasional addition of dry ice chunks to a stirred bath of alcohol or acetone. A bath of this type has an advantage over the ice-NaCl and ice-HCl baths in that siphoning is not necessary. Baths below $-78°$ require cooling of a liquid with liquid nitrogen. If a piece of apparatus or a reaction vessel is immersed in such a bath, maintenance of the temperature by the addition of further liquid nitrogen to the stirred liquid is very awkward and causes large temperature gradients in the bath. Consequently, constant-temperature slush baths (to be described) are generally used for maintaining temperatures below $-78°$.

It is quite possible, in the case of a very exothermic reaction, for the reaction mixture to be much warmer than the cold bath; in such cases it is obviously important to indicate whether the specified temperature refers to the reaction mixture or the cold bath. Sometimes the temperature of an exothermic reaction mixture is most conveniently controlled by regulating the rate of addition of one of the reagents. Thus, an exothermic reaction mixture could be held at $10°$ by placing the reaction vessel in a $0°$ ice bath and mixing the reagents at the appropriate rate.

CONSTANT-TEMPERATURE COLD BATHS

Simple constant-temperature cold baths are of two types: equilibrated mixtures of pure substances in the liquid and solid states (slush baths), and boiling pure liquids. The most useful baths of these types are listed in Table 6.1.[1] The baths listed in boldface type are useful as secondary standards of temperature.

Except for the ice-water bath, slush baths are prepared in a hood by *slowly* adding liquid nitrogen (never liquid air or liquid oxygen!) to the stirred liquids in Dewars until the consistency of a thick milk shake[2] is achieved. Care must be taken not to add too much liquid nitrogen, or a difficult-to-melt solid mass may form. If the liquid nitrogen is added too rapidly at the beginning, large amounts of the substance being cooled will be thrown out of the Dewar.

The dry ice bath is unique. It is strictly not a slush bath, and it is made by slowly adding crushed dry ice and a liquid, such as 95 per cent ethanol, to a Dewar. If the dry ice is added too rapidly, or if too much liquid is present, the liquid may be thrown out of the Dewar by the violent evolution of CO_2. The final bath should consist of a pile of dry ice chunks, with just enough

[1] A more extensive list of slush baths is given by R. E. Rondeau, *J. Chem. Eng. Data*, **11**, 124 (1966).

[2] Applesauce is a better example for those unfortunates who are unfamiliar with properly made milk shakes (which contain more ice cream than milk).

Table 6.1 Cold Baths

Bath	Temperature (°C)
Ice-water slush	**0**
Carbon tetrachloride slush	**−22.8**
Liquid ammonia*	−33 to −45
Chlorobenzene slush	**−45.2**
Chloroform slush	**−63.5**
Dry ice bath	**−78.5†**
Ethyl acetate slush	**−83.6**
Toluene slush	−95
Carbon disulfide slush	**−111.6**
Methylcyclohexane slush	**−126.3**
n-Pentane slush	−130
iso-Pentane slush	−160.5
Liquid oxygen	−183
Liquid nitrogen	−196

* It must be used in a hood or a very well-ventilated room. When liquid ammonia is freshly drawn from a cylinder, its temperature is usually far below the normal boiling point (−33.4°). It is necessary to keep air from the surface of the liquid and to apply heat in order to reach this temperature.

† The temperature of this bath is a function of the pressure of carbon dioxide vapor in equilibrium with the dry ice, and hence of atmospheric pressure (see Table 6.2).

liquid present to bring the liquid level within 1 or 2 cm of the top of the pile. It is difficult to submerge a reaction tube or other apparatus in a dry ice bath after the bath has been prepared; therefore, it is best to place the apparatus in the Dewar before preparing the bath. As the dry ice sublimes, the pile of chunks will gradually fall down; fresh chunks should occasionally be added to the top to maintain the bath. The liquid serves only as a heat-conducting medium. Other low-freezing liquids—such as acetone, isopropanol, and cellosolve—can be used as well as ethanol. In a properly made bath, the temperature is independent of the heat-conducting liquid used. Oftentimes freshly prepared baths, or baths made from dry ice that has been too finely powdered, are colder than the temperature at which solid CO_2 is in equilibrium with CO_2 gas at atmospheric pressure. The simplest remedy for a supercooled bath is to wait for the temperature to rise to the equilibrium value. An equilibrated dry ice bath is characterized by moderately vigorous bubbling due to the CO_2 evolution.

Liquid oxygen is very hazardous because many materials, such as organic compounds and finely divided metals, can react explosively with it. Mixtures of reducing agents with liquid oxygen can be detonated with sparks, friction, and concussion. Liquid oxygen should never be used to cool a glass vessel containing an oxidizable species, because of the possibility of the glass vessel breaking. Needless to say, liquid oxygen and liquid air should never be used to prepare slush baths.

CRYOSTATS

A cryostat is a device that can automatically hold an apparatus at a specified low temperature for extended periods of time. A very simple liquid bath cryostat, useful for temperatures down to $-70°$, is illustrated in Fig. 6.1. Cooling is provided by a copper rod, one end of which is in contact with liquid nitrogen.[3] The depth of immersion of the copper rod is adjusted so that the temperature is about 5° below the desired temperature. A thermoswitch-controlled heater then maintains the temperature constant to $\pm 0.1°$.

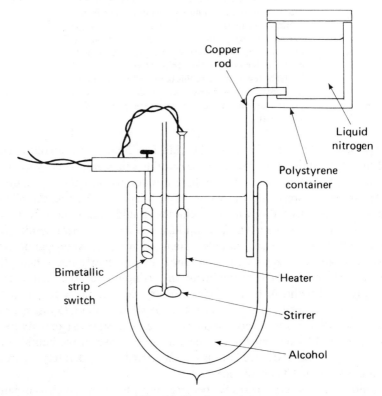

FIG. 6.1. Simple cryostat.

[3] S. R. Gunn, *Rev. Sci. Instr.*, **33**, 880 (1962).

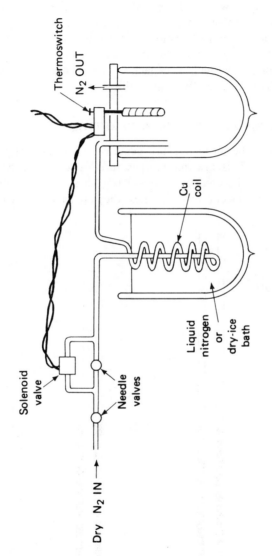

FIG. 6.2. Nitrogen gas bath cryostat.

A simple gas bath cryostat, useful for temperatures to $-150°$, is illustrated in Fig. 6.2. Cooling is provided by a stream of cold nitrogen gas. When the temperature is lower than the desired temperature, the solenoid valve is closed, a relatively slow flow of gas passes through the needle valves, and the cryostat temperature rises. When the temperature is greater than the desired temperature, the bimetallic thermoswitch actuates the solenoid valve, and a relatively fast flow of gas cools the apparatus. This type of cryostat has the disadvantage of a very low heat capacity; a long time is required for objects placed in the bath to reach thermal equilibrium.

LOW-TEMPERATURE THERMOMETERS

Mercury-in-glass thermometers are generally fairly accurate and may be used to about $-30°$. Below that temperature, various hydrocarbon-in-glass thermometers may be used, even to $-200°$. These thermometers, however, are generally very inaccurate, and, unless they are frequently calibrated, they

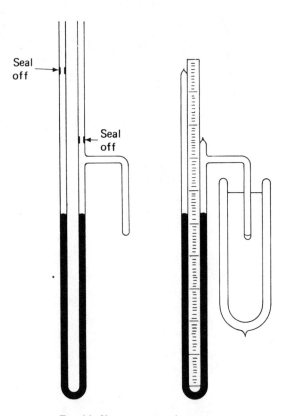

FIG. 6.3. Vapor pressure thermometer.

cannot be trusted to better than $\pm 5°$. In addition, they often give trouble through breaking of the liquid threads in the capillaries.

One of the most convenient and accurate methods for measuring a low temperature is to measure the vapor pressure of an appropriate substance at the temperature in question. A device for accomplishing this (a "vapor-pressure thermometer") is shown in Fig. 6.3. To prepare the thermometer, the mercury in the manometer is first heated in vacuo to eliminate volatile impurities, the pure substance is then condensed in the tip of the thermometer, and finally the thermometer is sealed off where indicated. The vapor pressures of substances suitable for vapor-pressure thermometers in the range from -10 to $-200°$ are given in Table 6.2. The data of Table 6.2 above $-183°$ are those of Stock, Henning, and Kuss,[4] those for $-183°$ and below are from Melville and Gowenlock,[5] and have not been corrected to $0°C = 273.15°K$. The uncorrected data are probably good for determining temperatures to $\pm 0.1°$. For more accurate work, it is suggested that the temperature correction be made and that the more recent literature be consulted.[6]

[4] A. Stock, F. Henning, and E. Kuss, *Ber.*, **54**, 1119 (1921).

[5] H. Melville and B. G. Gowenlock, *Experimental Methods in Gas Reactions*, Macmillan & Co., Ltd., London, 1964, p. 125.

[6] For SO_2, see W. F. Giauque and C. C. Stephenson, *J. Am. Chem. Soc.*, **60**, 1389 (1938); for NH_3, see G. T. Armstrong, *A Critical Review of the Literature Relating to the Vapor Pressure of Ammonia and Trideuteroammonia*, National Bureau of Standards Report No. 2626; for CO_2, see W. F. Giauque and C. J. Egan, *J. Chem. Phys.*, **5**, 45 (1937); for C_2H_4, see A. B. Lamb and E. E. Roper, *J. Am. Chem. Soc.*, **62**, 806 (1940).

Table 6.2 Vapor Pressure Thermometer Data*

°C	SO₂	NH₃	CO₂	C₂H₄	CH₄	O₂
−10	759.8					
−11	726.6					
−12	694.6					
−13	663.7					
−14	633.9					
−15	605.3					
−16	577.7					
−17	551.0					
−18	525.3					
−19	500.6					
−20	476.7					
−21	453.9					
−22	432.1					
−23	411.2					
−24	390.9					
−25	371.3					
−26	352.6					
−27	334.7					
−28	317.7					
−29	301.4					
−30	285.8					
−31	270.8					
−32	256.5					
−33	242.8	773.7				
−34	229.7	736.0				
−35	217.1	699.6				
−36	205.1	664.6				
−37		631.0				
−38		598.9				
−39		568.2				
−40		538.7				
−41		510.5				

°C	SO₂	NH₃	CO₂	C₂H₄	CH₄	O₂
−78			792.9			
−79			730.5			
−80			672.6			
−81			618.7			
−82			568.4			
−83			521.7			
−84			478.6			
−85			438.7			
−86			401.6			
−87			367.4			
−88			335.8			
−89			306.5			
−90			279.4			
−91			254.5			
−92			231.7			
−93			210.7			
−94			191.4			
−95			173.7			
−96			157.4			
−97			142.5			
−98			128.8			
−99			116.3			
−100			104.9			
−101			94.5			
−102			85.0			
−103			76.3	792.0		
−104			68.4	747.5		
−105			61.3	705.1		
−106			54.8	664.5		
−107			49.0	625.6		
−108			43.7	588.4		
−109				552.8		

°C	SO₂	NH₃	CO₂	C₂H₄	CH₄	O₂
−146				23.9		
−147				21.3		
−148				19.0		
−149				16.9		
−150				14.9	1720	
−151					1613	
−152					1510	
−153					1412	
−154					1318	
−155					1229	
−156					1145	
−157					1064	
−158					988	
−159					916	
−160					848	
−161					783.0	
−162					722.4	
−163					663.8	
−164					609.6	
−165					559.3	
−166					512.2	
−167					468.2	
−168					427.2	
−169					388.8	
−170					353.2	
−171					320.4	
−172					290.0	
−173					262.0	
−174					236.1	
−175					212.1	
−176					190.0	
−177					169.6	

Temp	VP		Temp	VP		Temp	VP	VP
−42	483.5		−110	518.8		−178	151.1	
−43	457.7		−111	486.3		−179	134.4	
−44	433.2		−112	455.4		−180	118.9	
−45	409.7		−113	426.0		−181	105.1	
−46	387.2		−114	398.2		−182	92.6	
−47	365.7		−115	371.9		−183	81.1	757.2
−48	345.2		−116	347.0		−184	71.0	681.4
−49	325.7		−117	323.5		−185		611.6
−50	307.1		−118	301.3		−186		547.4
−51	289.4		−119	280.3		−187		488.5
−52	272.5		−120	260.5		−188		434.6
−53	256.5		−121	241.9		−189		385.5
−54	241.3		−122	224.4		−190		340.7
−55	226.8		−123	207.9		−191		300.2
−56	213.0		−124	192.4		−192		263.6
−57	199.9		−125	177.8		−193		230.6
−58	187.5		−126	164.1		−194		200.9
−59	175.8		−127	151.2		−195		174.4
−60	164.7		−128	139.1		−196		150.9
−61	154.2		−129	127.8		−197		129.9
−62	144.2		−130	117.2		−198		111.3
−63	134.8		−131	107.3		−199		95.0
−64	125.9		−132	98.1		−200		80.7
−65	117.5		−133	89.6				
−66	109.5		−134	81.7				
−67	102.0		−135	74.4				
−68	95.0		−136	67.7				
−69	88.4		−137	61.5				
−70	82.2		−138	55.7				
−71	76.3		−139	50.4				
−72	70.8		−140	45.6				
−73	65.6		−141	41.2				
−74	60.7		−142	37.1				
−75	56.2		−143	33.3				
−76	52.0		−144	29.9				
−77	48.0	860.0	−145	26.7				

* Vapor pressures are given in mm Hg.

Manipulation

in

an Inert Atmosphere

7

The chemist has a variety of techniques at his disposal for handling air-sensitive materials. These differ as to their effectiveness in preventing access of air, their ease in setting up, and their convenience in use. In the following paragraphs some of these techniques are described so that the reader can choose the technique most suited to a particular purpose. Chapter 8 is completely devoted to one important "inert atmosphere" technique: the vacuum line.

SIMPLE PURGING WITH AN INERT GAS

A convenient method for handling air-sensitive materials is to use reaction flasks, distillation apparatus, filters, and so forth, from which the air is flushed by a stream of nitrogen or argon. Ground-glass equipment is convenient for such work, but not necessary. A cylinder containing the inert gas is connected by tubing to one arm of a T tube, the main leg of which dips into a pool of mercury in a large test tube or small flask and the other arm of which is connected by flexible tubing to the apparatus to be purged. A joint of the apparatus is loosened,[1] and a stream of inert gas is passed through the apparatus until it is judged that practically all the air has been flushed out. Then the joint is tightened, causing the inert gas to bubble through the mercury "safety valve." The apparatus is then under a slight positive pressure of relatively static inert gas and is ready for operation. After the purging operation, the flow of gas through the bubbler can be adjusted to a low value to minimize wastage. This type of purging and inert atmosphere maintenance can be employed during a variety of operations, such as distillation, refluxing, filtration, and the like.

SCHLENK VESSELS

Herzog and Dehnert[2] have described methods for effecting a wide variety of laboratory operations using Schlenk vessels and similar apparatus equipped with greased ground-glass joints and stopcocks (cf. Figs. 7.1 and 7.2). The apparatus is purged of air by alternately evacuating and introducing inert gas through a two-way stopcock. Even a relatively poor vacuum source (such as an aspirator) suffices if this purging operation is repeated several times in succession. As an illustration of the general technique, we shall describe the operations involved in precipitation, filtration, and washing.

Let us suppose we have an air-sensitive solution in the Schlenk vessel S_1 pictured in Fig. 7.1, to which we wish to add a solution from a dropping funnel to precipitate a solid. The side arm of the Schlenk vessel is connected to an inert gas source by means of tubing that has been purged of air, and the stopcock is opened. The stopper is then removed, and, while a fast stream of inert gas is passing out through the joint, the dropping funnel F_1 is quickly attached. After allowing a little time for flushing traces of air out of the region immediately below the stopcock of the dropping funnel, that stopcock is closed. The solution to be added is placed in the dropping funnel, the dropping funnel is recapped, and the dropping funnel is purged of air. If desired, the solution to be added can be deaerated by partially opening the dropping funnel stopcock and allowing inert gas to bubble through the solution for a while. Finally, both side arms are connected to the same inert gas source, and the upper solution is added to the lower solution.

The Schlenk vessel containing the precipitate and mother liquor is disconnected from the dropping funnel, and, while a fast stream of inert gas is passing out through the joint, the flask is quickly connected to flask S_2 via a

---

[1] Alternatively, a stopper can be loosened or a stopcock can be opened.

[2] S. Herzog and J. Dehnert, *Z. Chem.*, **4**, 1 (1964); S. Herzog, J. Dehnert, and K. Lühder, in *Technique of Inorganic Chemistry*, vol. 7, H. B. Jonassen and A. Weissberger, eds., Interscience Publishers, New York, 1968, p. 119.

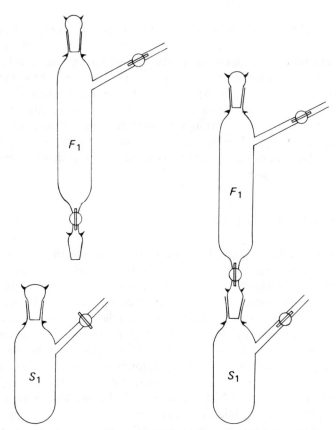

FIG. 7.1. Apparatus for anaerobically adding one solution to another.

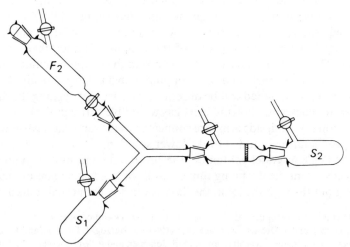

FIG. 7.2. Apparatus for anaerobic filtration and washing.

sintered glass filter and Y tube, as shown in Fig. 7.2. After the entire apparatus is purged, it is tilted so that the precipitate is deposited on the filter. The vessel S_1 and the precipitate are then washed with solvent or other washing solution admitted through dropping funnel F_2.

NEEDLE-PUNCTURE STOPPERS AND HYPODERMIC NEEDLES

A convenient method for transferring a liquid to or from a vessel without introducing air is by means of a hypodermic needle injected through a special rubber stopper, such as that used for serum bottles. A typical needle-puncture stopper is shown in Fig. 7.3. The plug of the stopper is hollow to within a short distance of the top. The sleevelike extension folds down over the neck of the glass tubing and holds the stopper securely in position during handling. The diaphragm can be easily punctured with a syringe needle and seals automatically after the needle is withdrawn.

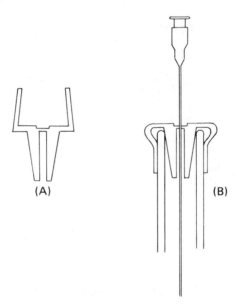

(A)   (B)

FIG. 7.3. Side view of needle-puncture stopper, unattached (A), and attached, with needle inserted (B).

Glass tubing for accepting needle-puncture stoppers can be sealed onto ordinary flasks or Schlenk vessels, and the techniques previously described can be modified to include withdrawal and injection of liquids with syringe needles. Hypodermic needles are available in a variety of lengths and gauges;

a 6-in., 20-gauge needle is of very general applicability. Syringes of the barrel type are provided with a sleeve joint for connection to the needle and are usually calibrated in cubic centimeters; a variety of sizes is available.

A liquid that has been drawn into a syringe through a needle has only very limited access to air through the fine tip of the needle. When even this small access to air must be avoided, the tip of the needle can be kept from the atmosphere by using small capsules closed on opposite sides with needle-puncture stoppers.[3]

THE GLOVE BAG

A very popular method for providing an inert atmosphere for chemical manipulations is to enclose the entire apparatus in a transparent polyethylene bag that is flushed with an inert gas. The bag is provided with inward-pointing gloves that facilitate the handling of objects within the bag[4] (see Fig. 7.4).

Inert gas inlet

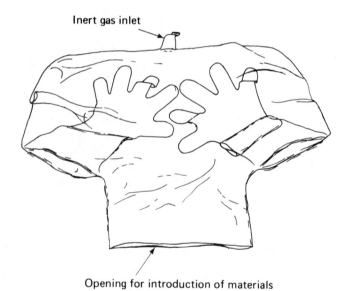

Opening for introduction of materials

FIG. 7.4. Polyethylene glove bag.

[3] For a description of these *Schutzröhrchen*, see F. Fehér, G. Kuhlbörsch, and H. Luhleich, *Z. Naturforsch.*, **14b**, 466 (1959).

[4] L. R. Ocone and J. Simkin, *J. Chem. Educ.*, **39**, 463 (1962); A. J. Franklin and S. E. Voltz, *Anal. Chem.*, **27**, 865 (1955). The type of bag described by these investigators is available from Instruments for Research and Industry, Cheltenham, Pa.

A nitrogen or argon inlet tube is inserted into the narrow opening at the top of the bag and secured with a rubber band. The materials to be handled are introduced through the front flap, and the bag is purged of air by a strong flow of gas for about 10 min. The flap is then loosely folded and clipped, and the gas flow is adjusted so as to maintain the bag in a barely inflated state. The operator then inserts his hands in the gloves and carries out the required manipulations.

THE GLOVE BOX

A glove box is one of the most convenient devices for handling materials in an inert atmosphere, and it is potentially capable of providing an extremely "clean" (i.e., free of oxygen, water, etc.) atmosphere. Large objects that cannot be conveniently handled in a glove bag often can be handled in a glove box[5]; however, a considerable amount of effort must be expended to maintain a glove box in satisfactory working condition.

A wide variety of glove boxes are commercially available.[6] We shall describe the operation of a typically designed box, illustrated in Fig. 7.5. The apparatus consists of two main parts: a large box, fitted with a window and gloves, in which manipulations are carried out; and a lock through which material is moved in or out of the box. The main box is fitted with a manostat that maintains the pressure slightly above atmospheric pressure by allowing pure inert gas to enter when the pressure falls to too low a value and by opening a connection to a vacuum pump when the pressure rises to too high a value. The gas in the box is constantly circulated through a purification train of hot activated copper (to remove oxygen) and molecular sieves (to remove condensible vapors).

Material to be introduced to the box is placed in the lock while the inner door is clamped shut. The outer door is then closed, and the lock is evacuated to a pressure less than 1 mm. About 0.5 atm of inert gas (e.g., nitrogen or argon) is then introduced to the lock from the main box via valve A, and the lock is re-evacuated. Finally, the lock is filled with inert gas at 1 atm pressure, and the inner door is opened so that the material in the lock can be transferred to the box. Material to be removed from the box is placed in the lock; the inner door is closed, and then the outer door is opened.

The principal source of contamination from oxygen and water vapor in a glove box are the gloves. These are best made of butyl rubber and should

[5] For a detailed discussion of glove box techniques, see C. J. Barton, *Technique of Inorganic Chemistry*, vol. 3, H. Jonassen and A. Weissberger, eds., Interscience Publishers, N.Y., 1963, p. 259.

[6] Some of the manufacturers are Vacuum Atmospheres Corp., No. Hollywood, Calif.; Labconco, Kansas City, Mo.; S. Blickman, Inc., Weehawken, N.J.; and Kewaunee Scientific Equipment, Adrian, Mich.

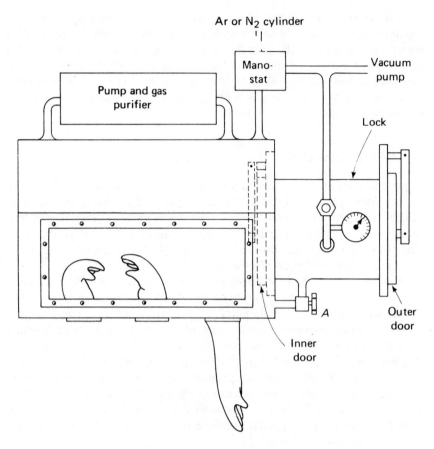

Fɪɢ. 7.5. Inert atmosphere box.

be frequently examined for holes. It is helpful to apply talcum powder to the hands and gloves before working in the box, and to wear an auxiliary pair of tight-fitting rubber gloves to keep the main gloves free from perspiration.

It is best not to work with volatile materials in the dry box; however, if the circulating purification train is efficient enough, contamination from solvents and the like can be eliminated fairly quickly. The interior of the box should be kept clean. It is useful to take a large piece of aluminum foil into the box and to carry out operations that might involve scattering crystals, etc., over the aluminum foil. Afterward, the debris can be wrapped up in the foil and removed through the lock.

The Vacuum Line

8

Manipulation in a chemical vacuum line is the standard technique for handling volatile materials that are poisonous or air sensitive. However, the technique is often used even when the compounds handled do not have these characteristics—for example, whenever only very small amounts of material are available. (As little as 5 μmoles of gas can be easily measured with an accuracy of ± 1 per cent). Inasmuch as vacuum manipulation is carried out in an essentially closed system, loss of material is almost impossible, and quantitative determination of reaction stoichiometry is readily achieved.

Materials of Construction

The preferred material of construction for vacuum lines is a borosilicate glass such as Pyrex[1] or Kimax.[2] Only when it is necessary to handle materials that are reactive toward glass (such as HF and a few exceedingly strong fluorinating agents) does one use, reluctantly, a plastic or metal system. In fact, even then

[1] Copyright Corning Glass Works, Corning, N.Y.
[2] Owens-Illinois Glass Co., Toledo, Ohio.

one often uses metal reaction vessels attached to a glass system, the latter being used for rapidly transferring the corrosive materials.

The properties of glass that are advantageous for vacuum lines are its transparency, strength, relative chemical inertness, and easy workability. The electrical insulating property of glass is a definite advantage when looking for leaks in a vacuum system. For this purpose, one uses a Tesla coil with a tapered metallic output electrode. This device emits an electric discharge in all directions from the end of the electrode. The electrode is moved about the surface of the glass vacuum system while it is being pumped on, and whenever the electrode comes within a centimeter or two of a pinhole in the system, the discharge converges into a single spark that goes directly to the hole. The method cannot be used to detect leaks in the vicinity of metal objects or in metallic systems because the spark is preferentially attracted to metal objects.[3]

Pumps

A mechanical rotary oil pump is the heart of every vacuum system. This pump operates by continuously sweeping gas from a low-pressure inlet to an atmospheric-pressure outlet under a seal of oil (see Fig. 8.1). A wide variety of mechanical oil pumps of different manufacture are available which are satisfactory for use with a vacuum line. For most preparative vacuum lines, a small pump (free air displacement 20 to 50 l min^{-1}) with an ultimate vacuum

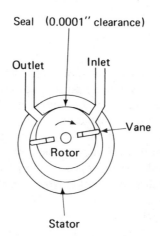

FIG. 8.1. Principal parts of the Welch Duo-Seal vacuum pump. As a vane leaves the seal, moving in a clockwise direction, it sweeps through the crescent-shaped space, drawing air through the inlet and forcing air out the outlet. The outlet is provided with a check valve.

[3] The systematic search for a leak in a vacuum system is discussed by E. B. Wilson, Jr., *An Introduction to Scientific Research*, McGraw-Hill Book Company, New York, 1952, p. 145.

of approximately 10^{-4} mm serves quite well.[4] A pump that is not mistreated (by allowing corrosive gases to pass through it) can operate continuously.

Whenever a corrosive material (such as water, an acid, or a compound that can hydrolyze to give an acid) enters the pump through an oversight, the pump oil should be immediately changed. This change is usually accomplished by turning off the pump, draining the oil from a special outlet near the bottom of the pump chamber, and adding fresh oil at the gas outlet. Normally the pump oil requires changing only every six months or so.

A good pump can evacuate an ordinary vacuum line, initially at atmospheric pressure, to a pressure of 10^{-3} mm in 10 min or less. This operating capability is satisfactory for most preparative work. However, if it is desired to hasten the pumpdown time at pressures below 1 mm and to obtain easily pressures below 10^{-3} mm, a diffusion pump should be used between the

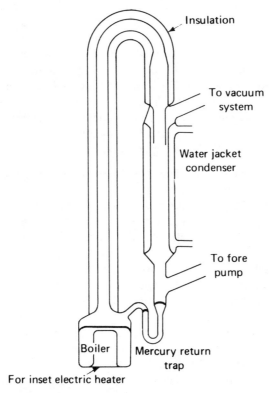

FIG. 8.2. Mercury diffusion pump. Streaming vapors of mercury enter the water-jacketed tube, entraining gas from the vacuum system. The mercury condenses and returns to the boiler via the return trap, and the gas is pumped away by the mechanical fore pump.

[4] The Welch Model 1400B Duo-Seal Pump (Welch Scientific Co., Skokie, Ill. 60076) is quiet operating and commonly used.

mechanical pump and the vacuum line. A diffusion pump operates by entraining gas in a stream of vapor. The vapor is condensed to the liquid, which returns to a boiler where it is evaporated again, permitting continuous cycling of the operating fluid. A typical diffusion pump, using mercury as the operating fluid, is illustrated in Fig. 8.2.

A liquid nitrogen-cooled trap should always be interposed between the vacuum system and the mechanical pump (or the diffusion pump if one is used) to protect the pump from condensible vapors and gases. These materials not only might corrode the pump, but also would dissolve in the pump oil and thereby seriously limit the ultimate vacuum attainable.

Low-Pressure Gauges

A low-pressure gauge can be used to determine when a system has been satisfactorily evacuated, to determine whether or not a large leak is present, and, when a reaction system is being directly evacuated, to determine whether a volatile compound is being evolved. For most synthetic purposes, it is unnecessary to have pressures lower than $1\,\mu$ (10^{-3} mm), and it is seldom necessary to know a low pressure with an accuracy better than a factor of two. Ruggedness and simplicity are also desirable. With these factors in mind, four types of vacuum gauge come under consideration: the thermocouple gauge, the McLeod gauge, the Pirani gauge, and the discharge gauge. These are discussed in the following paragraphs.

The Thermocouple Gauge. The thermocouple vacuum gauge is based on the fact that the thermal conductivity of a rarefied gas increases with increasing pressure. The gauge consists of an electrically heated wire to which a thermocouple is attached. At a given gas pressure, the wire and thermocouple quickly reach a steady-state temperature, at which point the heat loss by gaseous conduction to the walls of the tube approximately equals the electrical heat input of the wire. The greater the pressure of the gas (corresponding to greater heat conductivity), the lower will be the wire temperature and the lower will be the thermocouple current.

Figure 8.3 shows the internal construction of a National Research Corporation[5] Type 501 thermocouple gauge. Figure 8.4 gives the associated circuit diagram, and Table 8.1 gives the thermocouple current as a function of the pressure of dry air. The data in Table 8.1 are only approximate. Old gauges, particularly those which have been subjected to the vapors of thermally unstable compounds, usually give current readings that are too low (corresponding to too high pressure readings). The gauge is of little use at pressures outside the range from 1 to $500\,\mu$. However, inasmuch as the gauge is inexpensive and permits continuous monitoring of the pressure, it is the favorite low-pressure gauge in synthetic vacuum lines.

[5] NRC Equipment Corporation, Newton Highlands, Mass.

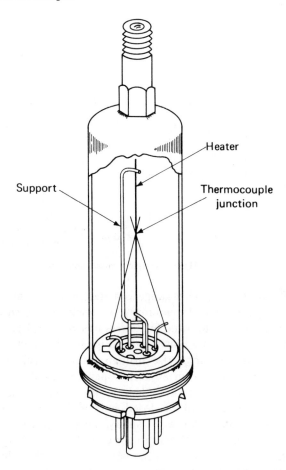

Heater

Support

Thermocouple
junction

FIG. 8.3. The internal construction of an NRC thermocouple gauge.

**Table 8.1   Thermocouple Gauge
Calibration for Dry Air**

Current (μamp.)	Pressure (μ)
20	750
30	300
40	200
60	120
80	70
100	45
120	30
140	20
160	13
180	5
190	1

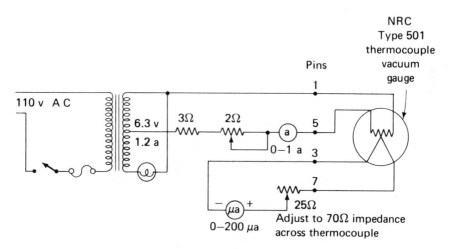

FIG. 8.4. Thermocouple vacuum gauge circuit.

The McLeod Gauge. This gauge is a device for compressing a known volume V of gas to a smaller volume v in which the pressure is measured. When p_0 and p are the initial and final pressures, $p_0 = (v/V)\,p$. Two common types of McLeod gauge are illustrated in Fig. 8.5. The method of operation consists of allowing mercury to rise through a bulb into a closed-end capillary until the mercury level in an adjacent open-end capillary has reached a predetermined mark. From a knowledge of the heights of the mercury in the capillaries, the dimensions of the closed-end capillary, and the volume of the bulb, it is possible to calculate p_0. Usually, the gauge is calibrated so that p_0 may be read from a scale adjacent to the closed-end capillary. Values of p_0 ranging from about 1 to 10^{-6} mm can be measured with a McLeod gauge of the type in Fig. 8.5A. The gauge depends on the validity of Boyle's Law up to the pressures reached during compression. Obviously the technique will not work for vapors condensing under these conditions or for imperfect gases with boiling points near room temperature. However, the McLeod gauge, when used with gases such as nitrogen and oxygen, can be a very accurate absolute manometer and is used to calibrate other types of gauges, such as the thermocouple gauge and the like. Because of the inconvenience of constructing and using the gauge, because of the fact that continuous reading of pressure is not possible, and because of the aforementioned problem with condensible vapors, the gauge would not seem to be appropriate for a synthetic vacuum line. However, for very crude vacuum lines and for bench-top vacuum distillations (where pressure readings lower than 10^{-2} mm are not required), the rotary type McLeod gauge illustrated in Fig. 8.5B is useful.

The Pirani Gauge. This gauge, like the thermocouple gauge, is based on thermal conductivity from a hot wire by the gas. Changes in pressure cause

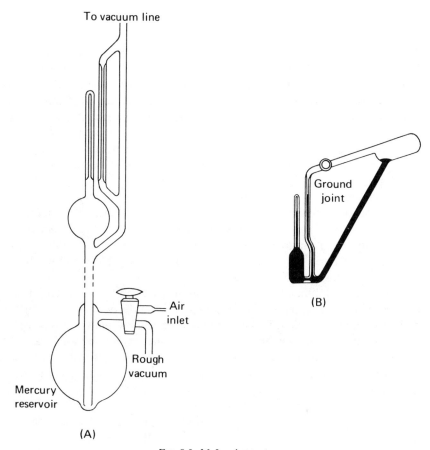

FIG. 8.5. McLeod gauges.

changes in the temperature of, and, consequently, the electrical resistance of, the wire. Most Pirani gauges employ a Wheatstone bridge circuit in which the Pirani wire forms one of the arms of the bridge. The bridge is in balance when the wire is in very high vacuum, and an increase in pressure causes an imbalance which can be measured with a galvanometer. Pirani gauges are useful over the approximate range from 10^{-4} to 50 mm pressure.

The Discharge Gauge. The principle employed in this gauge is the measurement of an ion current produced by a high-voltage discharge. Electrons from a cold cathode are caused to spiral as they move through a magnetic field to an anode. Collisions of electrons with molecules cause positive ions and more electrons to form, with a resultant increase in ion current. This ion current is fed to a microammeter for pressure indication. The gauge covers the range from 10^{-7} mm to 25 μ, and the discharge tube is of simple construction and can be easily cleaned.

Connections

The various parts of a vacuum line and the apparatus connected to it are joined together in different ways depending on the frequency with which the joints must be broken, the reactivity of the chemicals to be exposed to the joints, and the temperatures to which the joints might be heated. Connections that either are semipermanent or must be subjected to elevated temperatures are best made entirely of glass by glass blowing. Some experimenters prefer to make as many connections as possible in this way, because thereby joint leakage is essentially eliminated (assuming, of course, that a good glass-blowing job is done). Sometimes it is necessary to join quartz tubing to Pyrex glass tubing. This procedure cannot be done by making a direct seal because the linear coefficients of thermal expansion of quartz (8×10^{-7}) and Pyrex (32.5×10^{-7}) are sufficiently different that the seal breaks when it is cooled from the fusion temperature to room temperature. It is necessary to join quartz to Pyrex through an intermediate series of glasses of gradually changing composition. Such "graded seals" may be purchased.

The principal difficulty in directly joining glass and metal tubing is the difference in the coefficients of thermal expansion of the two substances. This difficulty may be overcome in the case of copper-to-glass seals (Housekeeper[6] seals) by making the copper thin so that it can readily deform to absorb differences between its expansion and that of the glass. Unfortunately, copper-to-glass seals are not very robust and must be handled with considerable care. The alloy Kovar (which may be soldered and spot welded) has thermal expansion characteristics very much like those of some commercial glasses, and sturdy Kovar-to-glass tube connections may be purchased.[7]

One of the strongest and most convenient ways to join a glass tube to a metal apparatus (tubing, valve, pipe, etc.) is by means of a Swagelok[8] tube fitting. A Swagelok nut, a back ferrule of metal or Zytel,[9] and a front ferrule of Teflon[9] are slipped around the glass tube, as shown in Fig. 8.6A. The glass tubing is inserted against the shoulder of the metal fitting. The nut is turned finger tight and then, carefully, snug tight with a wrench (Fig. 8.6B). Internal pressures greater than 1 atm can be used if the glass tubing is slightly etched with HF before making the connection. Incidentally, Swagelok fittings can be used for making connections to almost all kinds of metal and plastic tubing; therefore, they are very useful in general laboratory work.

When a connection must be made and broken frequently, and when the materials being handled are unreactive toward grease, it is convenient to use a

---

[6] W. G. Housekeeper, *Elec. Eng.*, **42**, 954 (1923).

[7] Stupakoff Ceramic and Manufacturing Co., Latrobe, Pa.

[8] Trademark of the Crawford Fitting Company, Cleveland, Ohio.

[9] Trademark of the E. I. duPont de Nemours and Co.

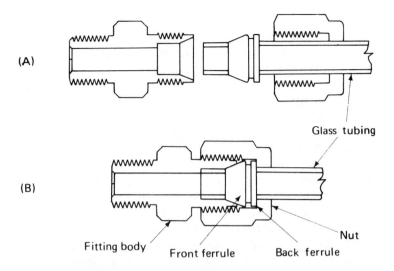

FIG. 8.6. Attaching glass tubing to a metal system with a Swageolok fitting.

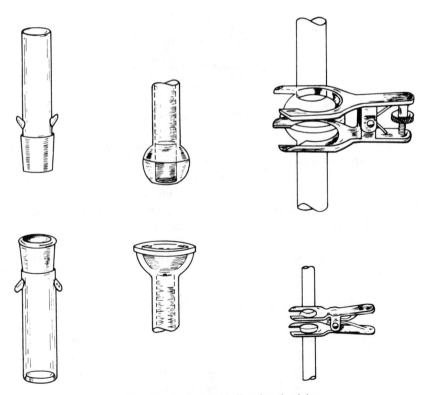

FIG. 8.7. Conical and ball-and-socket joints.

ground joint. There are two main types of ground joints—conical joints and ball-and-socket joints (see Fig. 8.7). Conical joints are used whenever bending at the joint cannot be tolerated; however, the flexibility of the ball-and-socket joint is generally an advantage, and the latter joints are more popular on vacuum lines. Both types must be sealed with grease or wax to render them gas tight. The "ears" on the illustrated conical joint are optional; they are used for holding the joint together with springs or rubber bands. Ball-and-socket joints must be clamped together, best accomplished with spring clamps, as illustrated in Fig. 8.7.

When a connection must be made and broken frequently and when grease cannot be tolerated, O-ring joints are often used. One-half of an O-ring joint is illustrated in Fig. 8.8. A connection is made by clamping together two such parts with an O-ring in between, seated in the circular grooves of each half of the joint. Ordinary Buna-N industrial O-rings can be used at room temperature when unsaturated halogen compounds and oxygenated solvents, such as acetone, are not being used. When greater chemical and thermal resistance is required, Viton-A or Teflon O-rings can be used.

FIG. 8.8. One half of an O-ring joint, with an O ring.

A glass tube may be sealed to another tube (of glass, metal, plastic, etc.) by application of sealing wax. One of the tubes to be joined should fit snugly inside the other, and the wax should be applied on the outside, where the tubes overlap. A very simple (although not completely vacuum tight) method for joining tubing is to make the connection with flexible tubing of rubber, Tygon, or other plastic. Such connections are usually used to join the vacuum system to the mechanical pump and to gas cylinders. For minimal leakage, it is best to make flexible tubing connections as short as possible.

Methods for Opening and Closing Connections

The commonest type of stopcock used in high-vacuum work is illustrated in Fig. 8.9A. This vacuum stopcock requires lubrication with grease and has provision for evacuation of the barrel of the stopcock. When properly lubricated and evacuated, such a stopcock will not pop open even when subjected to pressures as high as 2 atm. When stopcock grease cannot be

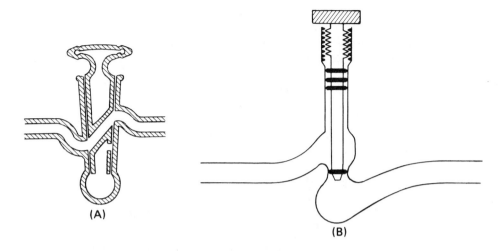

FIG. 8.9. High-vacuum stopcocks.

tolerated, a greaseless stopcock,[10] like the one illustrated in Fig. 8.9B, can be used. This stopcock depends upon O-ring closures to prevent leakage from the atmosphere and to seal one side of the cock from the other. Buna-N and Viton-A rings are commonly used.

One of the most reliable, although inconvenient, substitutes for a greased stopcock is the mercury float valve. A mercury float valve (see Fig. 8.10) consists of a small U tube connected between the parts of the apparatus to be closed from each other. Each arm of the U tube contains a glass float with a ground joint in the arm. When the valve is open, the floats rest on glass supports formed by indenting the tube. The valve is closed by admitting mercury from the reservoir into the U tube, where it forces the floats into place. The ground joints do not permit leakage of mercury under pressure differentials of 1 atm or less.

If the chemicals being studied are incompatible with greases, O-rings, and mercury, or if the apparatus must be subjected to very high temperatures, closures and openings can be made entirely of glass, as shown in Fig. 8.11. Here the reaction vessel can be sealed off with a flame at *A* and later opened by allowing the glass-enclosed iron bar to fall on the fragile tip of the break seal, *B*. All glass-sealed tubes with capillary tips may be opened to the vacuum line by breaking the tips with "tube openers," such as those shown in Fig. 8.12.

---

[10] Greaseless stopcocks similar to that illustrated are manufactured by Fusion, Inc., Rosemont, Ill.; Fischer and Porter Co., Glass Products Division, Warminster, Pa.; Ace Glass, Inc., Vineland, N.J.; Kontes Glass Co., Vineland, N.J.; and Delmar Scientific Laboratories, Inc., Maywood, Ill. A stopcock with a Viton-A or neoprene diaphragm is available from Electronic Space Products, Inc., Los Angeles, Calif., and a stopcock with a metal bellows is available from Delphic Industries, Inc., Livermore, Calif.

Rubber septa of the type used to cap vaccine bottles can be used with the vacuum line.[11,12] However, these seals are not completely vacuum tight and should be used mainly with systems that can be held near atmospheric pressure. The vessel shown in Fig. 8.13 may be used to remove an air-

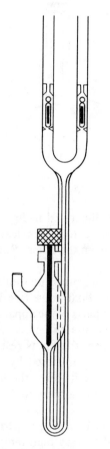

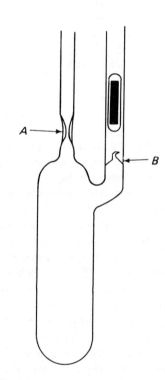

FIG. 8.10. Mercury float valve of the Delmar-Urry type.

FIG. 8.11. All-glass reaction vessel with break-seal.

sensitive liquid from a vacuum line by means of a hypodermic syringe and needle.[12] After the liquid is distilled into the vessel in vacuo, nitrogen is admitted to 1 atm pressure; then the hypodermic needle is inserted through the septum and the sample is withdrawn.

[11] F. Fehér, G. Kuhlbörsch, and H. Luhleich, *Z. Naturforsch.*, **14b**, 466 (1959).
[12] S. D. Gokhale, J. E. Drake, and W. L. Jolly, *J. Inorg. Nucl. Chem.*, **27**, 1911 (1965).

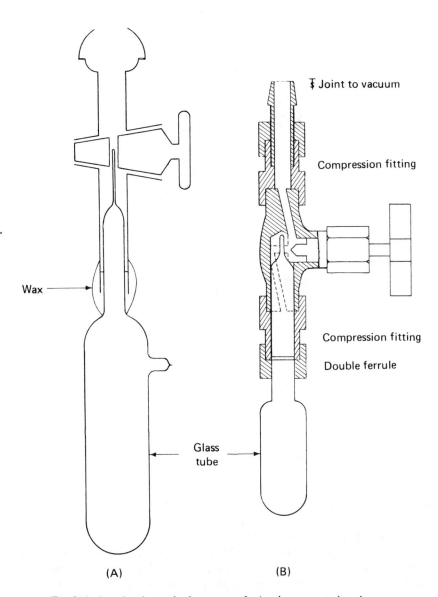

FIG. 8.12. Sample tubes and tube openers. In A, a large stopcock and a wax seal are used. In B, Swagelok fittings and a bored-out needle valve are used. [W. Mahler and H. V. Felmey, *Rev. Sci. Instr.*, **33**, 1127 (1962). Reproduced with permission.]

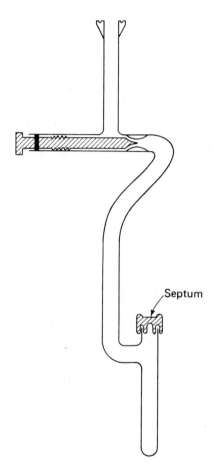

FIG. 8.13. Tube permitting removal of a liquid from a vacuum line with a hypodermic needle and syringe.

Greases and Waxes

Ground glass joints that are to be frequently disconnected or turned and all-glass stopcocks require lubrication with grease. Apiezon greases[13] are most commonly used for these purposes. These are paraffin oil residues that have been freed of all but extremely low vapor-pressure constituents. Apiezon M grease is a good, all-purpose, vacuum-line grease for use at room temperature. Grease N (rather expensive) is especially suited for stopcocks, and grease T is for use in apparatus subject to temperatures as high as 110°. These greases have room-temperature vapor pressures in the neighborhood of 10^{-8} mm

[13] James G. Biddle Co., Plymouth Meeting, Pa.

and are recommended when the chemicals being studied are inert toward, and not highly soluble in, paraffin hydrocarbons. Dow-Corning silicone grease is a lubricant that can be used over a much wider temperature interval (-40 to $+200°$) than paraffin greases can. Silicone grease is also somewhat more resistant toward dissolution of and by organic compounds. Kel–F #90 grease[14] contains polymeric molecules having the general structure $Cl(CF_2CFCl)_xCl$. The grease is inert toward many strongly oxidizing and dehydrating agents—such as fuming nitric and sulfuric acids, halogens, phosphorus oxychloride, and chromyl chloride—but it is very soluble in organic solvents. A working temperature range of -18 to $+175°$ is claimed.

The most useful general purpose sealing wax is Apiezon wax W, which is a hard black wax with a softening point of 80 to 90° and a negligible vapor pressure below 80°. When a wax with high chemical resistance is required, Kel–F #200 wax is recommended.

Layouts of Vacuum Lines

The layout of a simple, very useful vacuum line is shown in Fig. 8.14. Trap A is cooled with liquid nitrogen and serves to protect the pump from obnoxious vapors. Stopcocks 1 and 11 connect the main manifold with the working part of the line. The sockets below cocks 5, 6, and 7 are used to connect various pieces of apparatus, bulbs, gas inlet tubes, and so on, to the line. Traps B, C, and D are used principally for fractional condensation (to be discussed).

A more complicated but highly versatile vacuum line[15] is illustrated in Fig. 8.15. This unit includes a diffusion pump, a McLeod gauge, a Toepler pump, and a copper oxide combustion train.

Transfer of Chemicals in the Vacuum Line

Volatile materials can be quantitatively moved from one place to another in an otherwise evacuated system by simply cooling the receiving vessel to a temperature at which the vapor pressure of the material is negligible (for practical purposes, less than 10^{-2} mm). Liquid nitrogen is an effective coolant for most substances with boiling points above $-110°$[16]; a dry-ice bath can be used for substances boiling above $+100°$. Distillations and sublimations in a closed system are markedly slowed by the presence of a noncondensible gas, even when the latter is present at partial pressures as low as 10^{-2} mm. This effect is most noticeable when distilling or subliming

[14] Chemical Division, 3M Company, St. Paul, Minn. 55119.

[15] Available from Delmar Scientific Laboratories, Inc., Maywood, Ill.

[16] Substances with boiling points in the range from -110 to $-125°$ are trapped at liquid nitrogen temperature, but they exert enough vapor pressure at that temperature so that appreciable losses are suffered when the trap is opened to a pump.

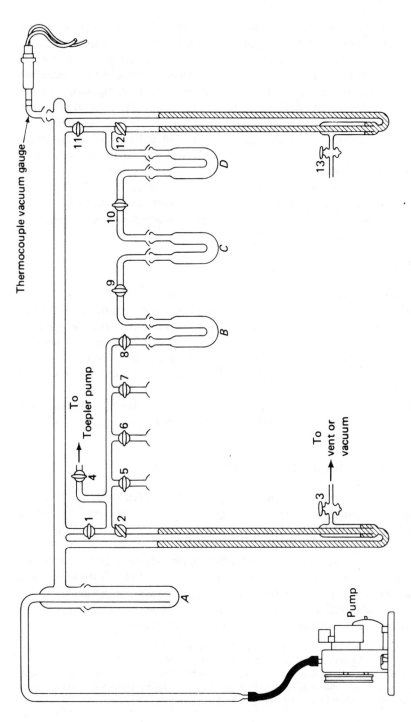

FIG. 8.14. A simple chemical vacuum line.

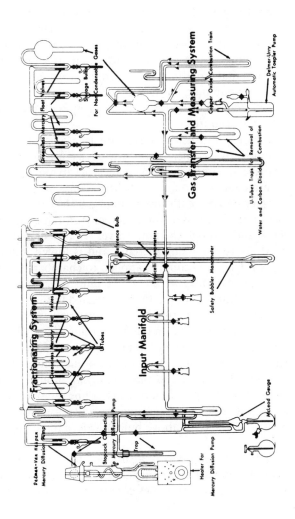

Fig. 8.15. Layout of the Delmar-Urry vacuum line. (Reproduced courtesy of Delmar Scientific Glass Products, Maywood, Ill.)

a compound having a low vapor pressure at room temperature. In such a case, it is helpful to use a U tube for the receiving vessel and to pump continuously on the side of the U tube opposite that from which the vapor is condensing. The U tubes used for trapping out materials in a stream of vapor or gas that is not condensible at the trap temperature must be constructed of large-diameter tubing so that the gas velocity in the condensation zone is as slow as possible. Otherwise, the material to be trapped will condense as a cloud of fine particles that will be blown through the trap and then vaporize again.

In view of the preceding discussion, it is clear that it is important to deaerate volatile materials that are being introduced to the vacuum line. This deaeration can be done by pumping on the material at room temperature if some wastage can be afforded. Distillation into a U trap with pumping is a very efficient deaeration procedure.

Substances with boiling points below $-125°$ are generally transferred by means of a Toepler pump. A Toepler pump is a device for pushing a gas completely from one place to another by causing mercury to move repeatedly up and down through a chamber connected to the source of the gas. Two common types of Toepler pump are illustrated in Fig. 8.16. To put either of these pumps into operation, air is passed into chamber W, forcing the mercury up through chamber X and through the float valve. (In this step, the gas in chamber X is trapped above the rising mercury and is forced irreversibly through the float valve.) Then a vacuum is applied to chamber W, and the mercury is pulled down again (more gas from the vacuum line enters chamber X). This sequence of operations is automatically repeated until 99.9 per cent or more of the gas has been forced through the float valve into bulb Y. Finally, the mercury is raised to a mark on the tubing immediately below the calibrated bulb Y. The pressure of the gas in bulb Y is read directly from a rule next to the inlet tube of the pump. The volume of the collected gas can easily be increased to include bulb Z and the connecting tubing. Bulb Z is detachable, so a sample of the Toepler-pumped gas can be removed from the line for purposes such as vapor density determination and mass spectrometry. Care must be taken to close the cock above bulb Y before removing bulb Z from the system.

Solids may be added, either gradually or all at once, to a reaction mixture within a closed system by tipping and tapping either the entire apparatus (as with the reactor of Fig. 33.16) or the vessel containing the material to be added (as with the reactor of Fig. 32.14).

Liquids can be added to a reaction vessel at a controlled rate by means of a dropping funnel. Two different dropping funnels are illustrated in Fig. 8.17. The funnel pictured in Fig. 8.17A requires that the pressure in the reaction vessel be equal to or less than atmospheric; the funnel pictured in Fig. 8.17B will operate even when the pressure in the reaction vessel slightly exceeds atmospheric pressure.

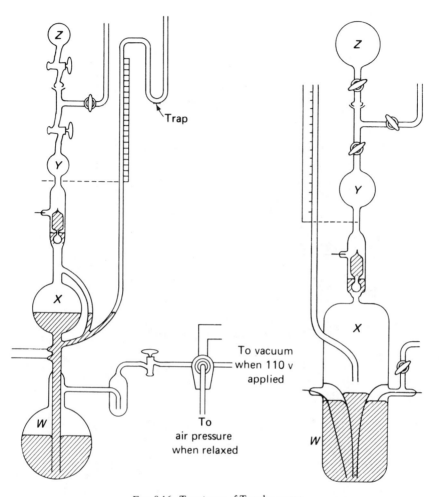

FIG. 8.16. Two types of Toepler pump.

Both liquids and solids can be added to a reaction vessel by crushing a fragile bulb containing the substance to be added. This method is useful when it is necessary to have the added substance thermally equilibrated with the reaction mixture before the addition (as in a kinetic study) or when the substance is essentially nonvolatile and air sensitive (as in the case of metallic sodium). Examples are shown in Fig. 8.18.

Reactions in solvents having boiling points below room temperature and critical points above room temperature can be carried out at room temperature using reactors of the type shown in Fig. 8.19. A solution can be gradually tipped from one leg of the reactor into the other to effect a crude titration. A precipitate may be separated from its mother liquor by decantation, and it

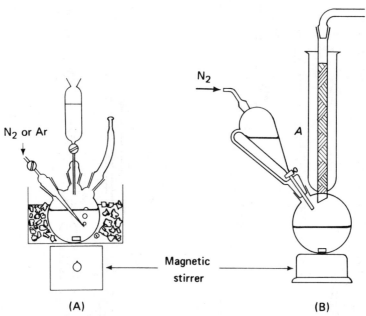

FIG. 8.17. Two types of dropping funnel.

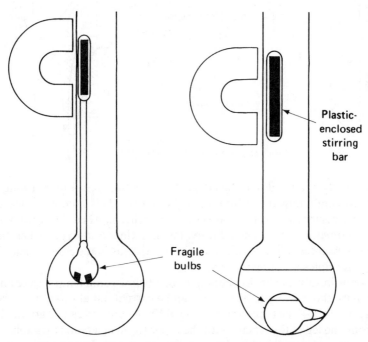

FIG. 8.18. The introduction of reactants by crushing fragile bulbs.

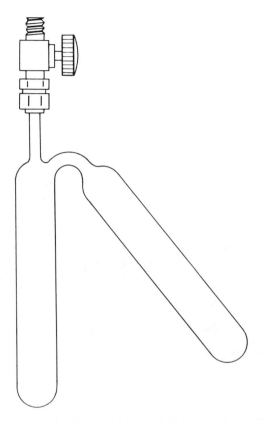

Fig. 8.19. U-tube high-pressure reactor.

may be washed by distilling back some of the solvent onto the precipitate and again decanting.

Reaction vessels can be designed for carrying out a wide variety of reactions involving a succession of processes such as dissolution, stirring, filtration, precipitate washing, and product isolation. For example, the apparatus shown in Fig. 8.20 can be used for the preparation of potassium germyl, $KGeH_3$, by the reaction of potassium hydroxide with germane. Powdered potassium hydroxide and a magnetic stirring bar are introduced into vessel A through the side tube, which is closed off by sealing with a flame at B. The apparatus is evacuated; 1,2-dimethoxyethane and germane are distilled into vessel A, and the mixture is magnetically stirred for an hour or so at room temperature. The apparatus is inverted, and vessel D is cooled slightly to create a pressure differential to aid the filtration. Vessel D is then brought to room temperature and vessel A is cooled with dry ice to distill the solvent back through the fritted disk. The solvent is again sucked into vessel D by slightly cooling the latter and warming vessel A to room temperature. The

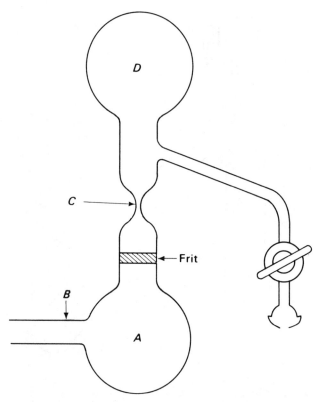

Fig. 8.20. Apparatus for isolating a product dissolved in a filtrate. (See text.)

latter sequence of operations is repeated as many times as necessary to extract all the $KGeH_3$ from the mixture of KOH and $KOH \cdot H_2O$. Finally, with vessel *D* still at the bottom, the region around *C* is thoroughly rinsed by causing solvent to condense there. Then, after the solution of $KGeH_3$ is cooled to liquid nitrogen temperature and the system is pumped out, vessel *A* and the frit are separated by fusing at point *C* with a flame. Vessel *D* now contains a solution of $KGeH_3$ with very little contamination by potassium hydroxide.

Measuring out Volatile Compounds

Compounds with normal boiling points in the vicinity of room temperature or lower can be measured by determining the temperature and pressure of the vapor in a system of known volume. It is usually assumed that the vapor obeys the ideal gas law. In the case of compounds that boil near room temperature, the vapors are very non-ideal at pressures near atmospheric, and to have reasonable accuracy in determining the number of moles of such compounds present, low pressures must be used. Some idea of the magnitude of the errors that can be made by relying on the ideal gas law in volumetric measurements of gases can be seen from Table 8.2, where the values at 25°C of PV/RT for molar quantities of several gases are given for $P = 1$ atm and $P = 0.01$ atm.

Table 8.2 PV/RT **for Several Substances at 25°C***

Substance	b.p. (°C)	PV/RT (1 atm)	PV/RT (0.01 atm)
N_2	−196	1.000	1.000
Xe	−108	0.996	1.000
CO_2	−78.5 (sub.)	0.993	1.000
NH_3	−33.4	0.993	1.000
SO_2	−10.2	0.985	1.000
n-Butane	−0.5	0.965	1.000
$(C_2H_5)_2O$	34.6		0.999

* Calculated from the chart given by R. C. Reid and T. K. Sherwood, *The Properties of Gases and Liquids*, McGraw-Hill Book Company, 1958, p. 43.

A very convenient manometer for use in chemical vacuum lines was shown in Fig. 8.14. This manometer has the following convenient features.

1. It has two arms, so that the pressure can be simply determined by measuring the difference in the heights of the mercury columns.

2. Because of the large mercury reservoir, the height of the mercury column in the arm connected to the vacuum manifold does not change much with changing pressure. Consequently, pressure changes can be quickly estimated by simply watching the mercury in the arm connected to the measuring system.

3. When the pressure in the measuring system exceeds atmospheric pressure by more than a few centimeters, the gas escapes by bubbling through the mercury reservoir.

4. The heights of the mercury columns are readily adjusted by applying either vacuum or pressure to the mercury reservoir. Thus, when one is measuring a gas, the mercury level in the arm connected to the measuring system is brought to a predetermined mark (or "zero") so that the volume of the measuring system is the same as that determined in a previous calibration.

5. The stopcock near the top of the arm connected to the measuring system can be closed when working with vapors that react with mercury or when it is desired to keep mercury vapor out of the system.

It is a good idea to determine the volumes of the isolable sections of a vacuum line (e.g., in the line illustrated in Fig. 8.14, the sections between cocks 1 and 8, 8 and 9, 9 and 10, and 10 and 11). Then it is possible to measure a gas with various manometers and in any of several possible volumes. When large volumes of gas are to be measured, a large bulb of known volume may be connected to the system and included with one or more of the calibrated internal volumes. The volume of a bulb may be determined from the difference in weights of the evacuated bulb and the bulb filled with water (the volume between the stopcock and the joint must also be measured). The volume of an internal section of a vacuum line may be determined by attaching a bulb of known volume containing air at a known pressure to the section. The air is allowed to expand into the evacuated section and the new pressure is determined. By using the relation $pV = p'V'$, the volume of the section can be readily calculated (again, a correction must be made for the volumes between joints and stopcocks).

When it is necessary to measure the pressure of a gas that reacts with mercury, the expedient of covering the exposed mercury surface of the manometer with a centimeter or two of an inert oil (such as silicone or Kel–F oil) can sometimes be used. However, a better solution is to interpose a glass spiral manometer (as in Fig. 8.21) between the system and the mercury manometer. The spiral itself is connected to the system under study, and the jacket around the spiral is connected to a mercury manometer and connections to vacuum and the atmosphere. By measuring the deflection of a light beam by the small mirror attached to the pointer of the spiral, differential pressures can be determined. Alternatively, the device can be used as a null detector, and the pressure can be read directly with the mercury manometer.

When a measured amount of a liquid having a low vapor pressure at room temperature must be introduced to the vacuum line, the volumetric method is impractical. In such cases, the amount of material introduced is determined from the change in weight of a vessel from which the material is distilled.

Separation of Volatile Compounds

Fractional Condensation. The classical method for separating the components of a volatile mixture in the vacuum line is to allow the vapors of the mixture to diffuse in vacuo into a series of traps cooled to successively lower temperatures. The idea is that each compound will diffuse through the traps until it reaches one where its vapor pressure is "negligible," and where it condenses out. Thus, if the compounds in the mixture differ sufficiently in their volatilities, and if the trap temperatures are properly chosen, the compounds will collect in separate traps. A liquid nitrogen bath ($-196°$) is the coldest bath generally used; therefore, compounds that boil below $-110°$ must be separated from higher-boiling compounds either by Toepler pumping or by adsorbing them in a trap containing a high surface-area material, such as charcoal or molecular sieves, at liquid nitrogen temperature (see p. 172).

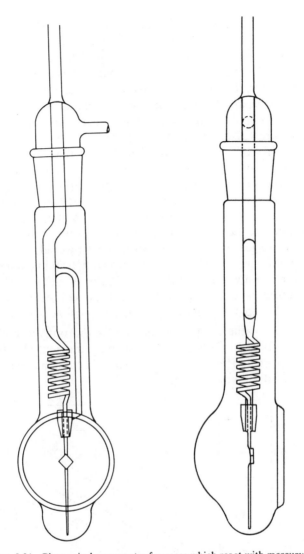

FIG. 8.21. Glass spiral manometer for gases which react with mercury.

For the purposes of fractional condensation, a compound's vapor pressure is "negligible" when its vapor pressure is about 10^{-2} mm or less. That is, a compound can be essentially quantitatively trapped out in a trap cooled to a temperature where the vapor pressure is that low. On the other hand, a compound will pass through a trap cooled to a temperature at which the compound's vapor pressure is greater than about 1 mm. Traps corresponding to intermediate vapor pressures are difficult cases where either very inefficient condensation occurs (i.e., much of the condensing material passes through the trap) or no condensation occurs (as when the distillation is carried out over a long period of time with a small amount of material).

Frequently, temperature-vapor pressure data for pressures in the neighborhood of 10^{-2} mm are unavailable for compounds of interest, and are not quickly calculable. However, it is possible to decide upon a suitable trap temperature for separating two compounds from a knowledge of their boiling points or 1-atm sublimation points (whether actually measured or extrapolated). Figure 8.22 can be used in such cases. For example, let us suppose we wish to separate a mixture of diethyl ether (b.p. = 34.6°) and stibine (b.p. = −18.4°). We seek a trap that will quantitatively condense the ether but will let the stibine pass through. Reading up on the graph from the boiling point of ether (34.6°) to a point slightly above the "fair trapping" line (corresponding to Δb.p. ≈ 53°), we see that a trap at approximately −100° should be used to condense out the ether. By consulting the list of cold baths in Table 6.1, we see that a toluene slush (−95°) would work very well. A carbon disulfide slush (−111.6°) would probably trap out some stibine and would be usable only if the distillation were carried out very slowly.

If the mixed vapors are passed through the cold trap too rapidly, the efficiency of the separation may suffer for two possible reasons.

1. The less volatile component may not be thoroughly condensed in the cold trap; it may be partially carried through the trap (entrained) by the more volatile component. Hence, the more volatile fraction may be contaminated with the less volatile component.

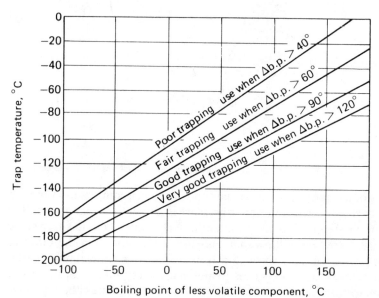

FIG. 8.22. Suggested cold-trap temperatures for separating volatile mixtures.

2. Inasmuch as the pressure in the system is relatively high, the more volatile component may partially condense in the cold trap. Hence, the less volatile fraction may be contaminated by the more volatile component.

If the mixture is distilled too slowly, some of the less volatile component may leave the cold trap by simple evaporation. (Even the less volatile component has a finite vapor pressure at the temperature of the trap). Hence, the more volatile fraction may be contaminated by the less volatile component.

As a rough rule, we may say that an efficient separation will be effected if the materials are distilled at a rate of approximately 1 mmole min.$^{-1}$ Some high-boiling liquids will distill from an ordinary 2-cm-diameter reaction tube much more slowly. In fact, water will usually freeze and then sublime very slowly. In such cases, it is wise to surround the tube with a bath of warm water. When liquids are distilled having boiling points below 100°, the liquid is first frozen to a solid button in the bottom of the distillation tube. During the distillation, the tube is allowed to warm up gradually in the air. If the material boils below −50°, the same procedure is followed, except that the distillation tube is placed in an empty Dewar to reduce the rate of sublimation.

An experiment performed by Schaeffer, Schaeffer, and Schlesinger[17] admirably illustrates the effectiveness of a simple fractionating train. A mixture of 82.3 cc (S.T.P.) of borazine, $B_3N_3H_6$, and 23.8 cc (S.T.P.) of boron trichloride was allowed to react at room temperature for 116 hours. The material remaining consisted of B-dichloroborazine, B-monochloroborazine, unreacted borazine, diborane, and hydrogen. These materials were separated by pumping through four traps. A schematic flow diagram is shown in Fig. 8.23.

A mixture of compounds with boiling points closer than about 40° cannot be quantitatively separated by fractional condensation; however, the isolation of fairly pure samples of the components of a close-boiling mixture is sometimes possible by repeated fractional condensation if low recovery can be tolerated. The fractionation scheme schematically indicated in Fig. 8.24 involves ten different fractional condensation steps. This scheme can be used when Δb.p. is between 20 and 50°. The mixtures that emerge from the intermediate stages of the scheme can be either discarded or combined and put through the fractionation scheme again. Obviously, the scheme of Fig. 8.24 can be greatly simplified if only one of the components of the mixture is sought.

Fractional Vacuum Vaporization. Probably the simplest (but a very inefficient) method for separating two volatile substances is to hold the vessel containing the mixture at a low temperature such that the vapor pressure of the more volatile substance is between 0.5 and 5 mm and to pump on the vessel. Substances boiling above −110° can be "pumped" by distillation

[17] G. W. Schaeffer, R. Schaeffer, and H. I. Schlesinger, *J. Am. Chem. Soc.*, **73**, 1612 (1951).

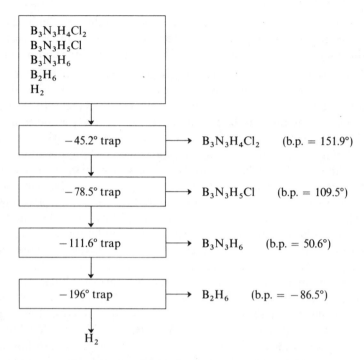

$$B_3N_3H_4Cl_2$$
$$B_3N_3H_5Cl$$
$$B_3N_3H_6$$
$$B_2H_6$$
$$H_2$$

$-45.2°$ trap $\longrightarrow$ $B_3N_3H_4Cl_2$ (b.p. = 151.9°)

$-78.5°$ trap $\longrightarrow$ $B_3N_3H_5Cl$ (b.p. = 109.5°)

$-111.6°$ trap $\longrightarrow$ $B_3N_3H_6$ (b.p. = 50.6°)

$-196°$ trap $\longrightarrow$ B_2H_6 (b.p. = $-86.5°$)

H_2

FIG. 8.23. A separation by fractional condensation.[17]

into a liquid nitrogen-cooled trap; substances boiling below $-110°$ must be Toepler pumped. The method is based on the assumption that the less volatile component of the mixture is essentially nonvolatile and is left behind after all the more volatile substance has been pumped away. In actual practice, the method is effective only if the boiling points of the separated materials differ by more than about 80°. Ordinarily, the low temperature is provided by a slush bath or dry ice bath. However, in a variant of the technique, the mixture is condensed with a liquid nitrogen bath in the bottom of a 20-mm O.D. tube about 30 cm long. The tube is then placed in a Dewar from which liquid nitrogen was just poured, and the vapors of the mixture are successively trapped in separate liquid nitrogen traps as the tube gradually warms up. A thermocouple gauge in the system can be used to monitor the vapors as they emerge from the tube.

The Le Roy still,[18] pictured in Fig. 8.25, represents a refinement of the separation techniques described in the preceding paragraphs. The still is essentially a trap the temperature of which can be varied at will and from which the components of a mixture are successively distilled at a take-off

[18] D. J. Le Roy, *Can. J. Res.*, **28**, 492 (1950); also see A. S. Buchanan and F. Creutzberg, *Australian J. Chem.*, **14**, 527 (1961).

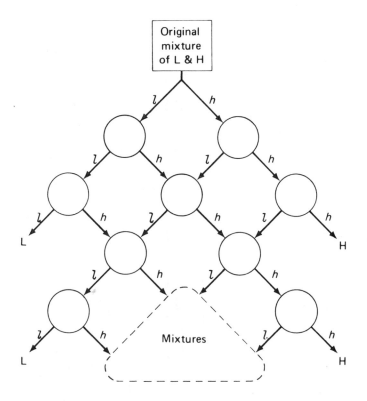

FIG. 8.24. Scheme for the separation by repeated fractional condensation of a mixture of two compounds: a lower boiling compound (L) and a higher boiling compound (H). Each step involves the separation of a mixture into a low-temperature trap (l) and a higher-temperature trap (h).

pressure around 0.5 mm. The outer surface of tube *B* is wound with metal foil and with insulated heating wire. The space between tubes *A* and *B* contains air at a variable reduced pressure, and the entire apparatus is immersed in a liquid nitrogen bath. The mixture of materials to be separated is distilled into the still through tube *C* while the air pressure between *A* and *B* is fairly high (about $\frac{1}{2}$ atm) and tube *B* is essentially at liquid nitrogen temperature. Then the air pressure is reduced, and the column is electrically heated until the pressure in the column is about 0.5 mm. The vapors are drawn off into a liquid nitrogen-cooled trap until the pressure has been reduced to 10^{-2} to 10^{-3} mm. The still outlet is then connected to another liquid nitrogen-cooled trap, and the heater current is again increased until the pressure is once more about 0.5 mm. This fractional vaporization process is repeated until the entire mixture has been distilled. The technique is capable of separating compounds that differ in boiling point by about 40°.

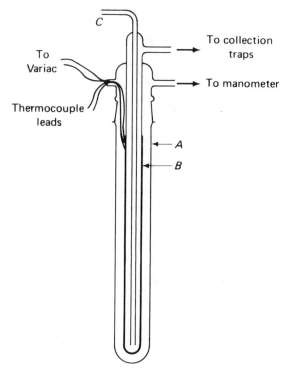

FIG. 8.25. The Le Roy still.[18]

The apparatus[19] pictured in Fig. 8.26 can be used to fractionate volatile mixtures with about the same efficiency as the Le Roy still. It is simpler to fabricate than the Le Roy still and not much more difficult to operate. The mixture to be separated is distilled into the bulb by immersing it in liquid nitrogen. A steady stream of cold nitrogen gas[20] is then passed through the vacuum-jacketed section at a rate sufficient to cool that section as low as possible (about −130°). The mixture is allowed to warm up and reflux. By carefully increasing the flow of cold nitrogen, the temperature of the jacketed section is increased until the first volatile fraction can be pumped off. A series of fractions is collected corresponding to a set of increasing jacket temperatures until the entire sample is vaporized.

Much more efficient separation of mixtures is possible if the vaporized material passes through a cold reflux column through which a thermal

[19] Among the first to use this type of apparatus were C. H. Van Dyke and A. G. MacDiarmid. The apparatus is described in the Ph.D. thesis of C. H. Van Dyke (University of Pennsylvania, *ca.* 1964) and in A. G. MacDiarmid, *Preparative Inorganic Reactions*, **1**, 165 (1964).

[20] The nitrogen (from a cylinder or house line) can be cooled by passing it through a copper coil immersed in liquid nitrogen, or the cold nitrogen can be generated by passing current through heating wire immersed in liquid nitrogen.

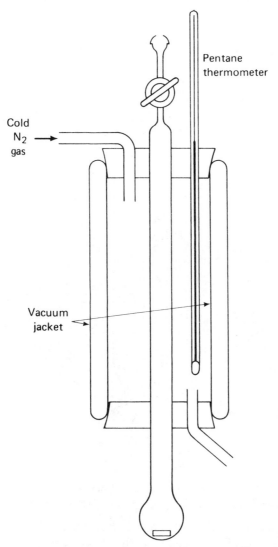

Pentane
thermometer

Cold
N$_2$
gas

Vacuum
jacket

FIG. 8.26. Low-pressure fractionation column.[19]

gradient is maintained. A very simple device for obtaining a reflux column with a low-temperature thermal gradient is shown in Fig. 8.27. This column has a spiral wire packing. In the separation of B$_5$H$_9$ from B$_5$H$_{11}$ with this column, Spielman and Burg[21] maintained a nearly total reflux under high vacuum. A dry ice bath was kept in the large outer chamber, from which the inner column was air-gap insulated to aid development of a thermal gradient favoring separation. Glass wool insulation surrounded the outer chamber.

[21] J. R. Spielman and A. B. Burg, *Inorg. Chem.*, **2**, 1139 (1963).

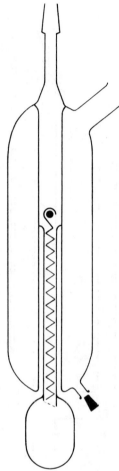

FIG. 8.27. Low-pressure fractionating column with thermal gradient.[21]
(Reproduced with permission of the American Chemical Society.)

The all-glass distillation apparatus[22] pictured in Fig. 8.28 is capable of separating compounds with boiling points differing by as little as 15°. Cold nitrogen gas is passed down the central tube, thereby maintaining the reflux column at a low temperature. A thermal gradient forms as a consequence of thermal and radiative heat conduction to the central tube. The volatile material refluxes under high vacuum in the annular space (1-mm gap) between the central tube and the outer vacuum jacket. As the average column temperature is increased stepwise, fractions are pumped off from the top of the column.

[22] Developed by J. Dobson, A. D. Norman, and others in the laboratory of R. O. Schaeffer, Indiana University, Bloomington, Ind.

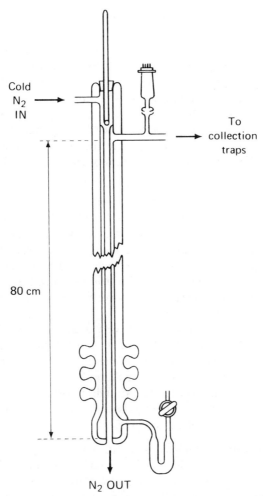

Fig. 8.28. Low-pressure fractionating column with thermal gradient.[22]

Fractional Vaporization in a Carrier Gas. Cady and Siegwarth[23] devised a process for separating volatile substances by a carrier-gas distillation in which the separation is dependent upon liquid-vapor (or solid-vapor) equilibria, a temperature gradient in a packed column, and a rising temperature of the column as a whole. The column is a U tube made from a 60-cm length of $\frac{1}{4}$-in. copper tubing packed with magnesium powder. A steady flow of helium is passed through the tube. The exit gas passes through a katharometer connected to a strip chart recorder and a series of traps for fraction collection.[24] The column is surrounded by liquid nitrogen in a Dewar, and

[23] G. H. Cady and D. P. Siegwarth, *Anal. Chem.*, **31**, 618 (1959).
[24] The apparatus is similar to that illustrated in Fig. 9.5.

the sample is introduced. Then the Dewar is removed, emptied, and immediately replaced around the column. As the column warms, the components
move through the column and emerge in the order of decreasing volatility.
Compounds with boiling points 15° apart can be separated on this column.

Gas chromatography (discussed in Chap. 9) is the most efficient method
for separating volatile compounds. Chromatographic apparatus is more
complicated than that used in the preceding separation techniques, but the
extra trouble involved can be worthwhile when close-boiling (as close as 5°)
compounds must be separated. The sample introduction system, the column,
the detector, and the fraction collecting traps can be made integral parts of
the vacuum line.

Selective Adsorption. A material of very high surface area—such as
activated charcoal or a Linde molecular sieve[25] when cooled to liquid
nitrogen temperature, can trap by adsorption materials that boil as low as
−220°. Thus, substances such as nitrogen, argon, and carbon monoxide can
be separated from substances such as hydrogen, helium, and neon by Toepler
pumping the mixture through a liquid nitrogen-cooled trap containing the
adsorbent. The heavier molecules are adsorbed, and the lighter molecules
collect in the Toepler pump buret.

Linde molecular sieves have another very useful function, based on their
selectivity. This selectivity is attributable to the presence of many pores of
uniform size in the crystalline structure of the sieves. Type 4A sieve has pores
that will permit only molecules smaller than about 4 Å to enter and be adsorbed. Type 5A sieve has pores that will admit molecules up to about 5 Å
in diameter.[26] Table 8.3 shows the kinds of molecules that can be adsorbed
at room temperature by molecular sieves.

Chemical Separation Methods. Sometimes two compounds that cannot
easily be separated on the basis of their different volatilities can be separated
by treating the mixture with an excess of a third compound that forms a
relatively nonvolatile adduct with only one of the two compounds. After
removal of the component of the original mixture, which is uncomplexed,
the adduct is then destroyed by addition of an excess of a fourth compound,
which forms a stronger nonvolatile adduct with the third compound and
completely displaces the other component of the original mixture. Obviously
the volatilities of the added compounds must be such as to permit separation
from the uncomplexed compounds. Examples of such chemical separation
schemes are given in Figs. 8.29 and 8.30.

Sometimes the adduct formed is so weakly associated that its dissociation
products can be partly separated by fast fractional condensation. Although

[25] Linde Air Products Co., A Division of Union Carbide and Carbon Corporation.

[26] This explanation of the selectivity of molecular sieves has been criticized by S. W. Benson
and J. W. King, Jr., *Science*, **150**, 1710 (1965).

Table 8.3   Molecules Adsorbed by Linde Molecular Sieves

Adsorbed by 4A and 5A	Adsorbed by 5A but not 4A	Not Adsorbed by 4A or 5A
H_2O	propane and higher	isobutane and all iso-paraffins
CO_2	n-paraffins	isopropanol and all iso-, secondary, and
$CO*$	butene and higher n-olefins	tertiary alcohols
H_2S	n-butanol and higher n-alcohols	benzene and all aromatics
SO_2	cyclopropane	carbon tetrachloride
NH_3	freon-12	freon-114
N_2*		freon-11
O_2*		SF_6
CH_4*		BF_3
CH_3OH		molecules larger than 5.0 Å
C_2H_6		
C_2H_5OH		
C_2H_4		
C_2H_2		
C_3H_6		
n-C_3H_7OH		
C_2H_4O		

* Adsorbed appreciably only at low temperatures.

this procedure does not permit a quantitative separation of a mixture, it is often useful for eliminating impurities. For example, ethane (b.p. = $-89°$) is sometimes an impurity in diborane (b.p. = $-93°$). If the mixture is added to an excess of dimethyl ether (b.p. = $-24°$), the adduct $(CH_3)_2O\cdot BH_3$ (low volatility at $-130°$) forms, permitting the complete removal of the ethane

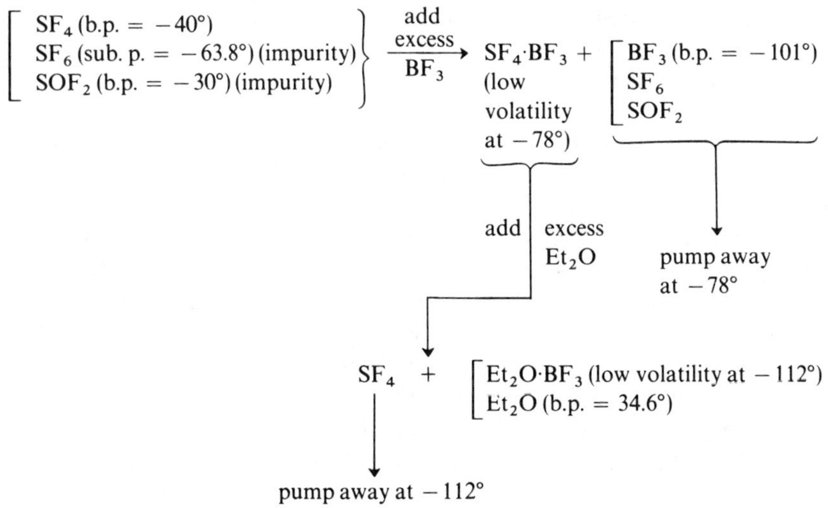

$$
\begin{bmatrix} SF_4 \text{ (b.p.} = -40°) \\ SF_6 \text{ (sub. p.} = -63.8°) \text{ (impurity)} \\ SOF_2 \text{ (b.p.} = -30°) \text{ (impurity)} \end{bmatrix}
\xrightarrow[\text{BF}_3]{\substack{\text{add} \\ \text{excess}}}
SF_4\cdot BF_3 +
\begin{bmatrix} BF_3 \text{ (b.p.} = -101°) \\ SF_6 \\ SOF_2 \end{bmatrix}
$$

(low volatility at $-78°$)

add excess Et_2O pump away at $-78°$

$$
SF_4 + \begin{bmatrix} Et_2O\cdot BF_3 \text{ (low volatility at } -112°) \\ Et_2O \text{ (b.p.} = 34.6°) \end{bmatrix}
$$

pump away at $-112°$

FIG. 8.29. Purification scheme for SF_4. [From N. Bartlett and P. L. Robinson, *J. Chem. Soc.*, 3417 (1961).]

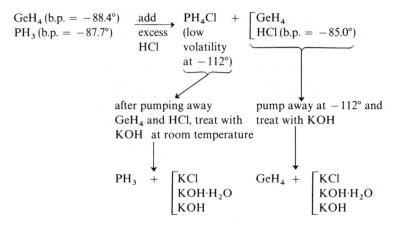

FIG. 8.30. Scheme for separating GeH_4 and PH_3.

impurity. Most of the diborane can be separated from the dimethyl ether by rapid passage through a $-130°$ trap [which removes the $(CH_3)_2O$ and a small amount of $(CH_3)_2O \cdot BH_3$]. By similar procedures, boron trichloride can be purified by treatment with nitrobenzene[27] and boron trimethyl can be purified by treatment with triethylamine.[28]

Sometimes it is necessary to remove a compound from a volatile mixture without the necessity of recovering the removed compound. There are many examples of such separations—for example, the removal of CO_2 from GeH_4 with Ascarite; the removal of B_2H_6 from GeH_4 by treatment with water; the removal of hydrogen from helium by oxidation with CuO; the removal of HCl from CO_2 by treatment with yellow HgO; and so on.

Physical Measurements

Vapor pressures of substances below room temperature may be determined using a manometer of the type shown in Fig. 8.14. The substance is contained in a tube or trap immersed in a suitable constant-temperature cold bath (such as one of the "secondary temperature standards" listed in Table 6.1). Vapor pressure determinations of this type are very useful for identifying and judging the purity of an already-characterized compound. One simply compares the measured vapor pressure with a previously determined value. When the vapor pressure of a new compound is being determined, there is always a question as to whether the compound is pure. Dissolved impurities having vapor pressures different from that of the pure substance can generally be detected by measuring the vapor pressure of the sample for different extents of vaporization (i.e., for different ratios of vaporized sample : liquid sample).

[27] H. C. Brown and R. R. Holmes, *J. Am. Chem. Soc.*, **78**, 2173 (1956).
[28] H. C. Brown, *J. Am. Chem. Soc.*, **67**, 374 (1945).

A pure substance will show no change in vapor pressure upon changing the extent of vaporization. When it is desired to obtain vapor pressures over a range of temperatures, a vapor pressure thermometer and the vessel containing the substance are placed in a Dewar containing a stirred cold liquid (e.g., alcohol). Pressure readings are taken as the bath slowly warms up.

Vapor pressures above room temperature must be determined with an apparatus in which the sample and the entire system containing the vapor are held in a bath. The apparatus shown in Fig. 8.31 is typical. The little manometer in the bath is used as a null detector to indicate when the vapor pressure of the sample on one side and the adjustable pressure of air on the other side are equal.

The melting point of a volatile substance melting below 0° can be determined using the apparatus pictured in Fig. 8.32. The inner tube (having a glass cross at the bottom and a sealed-in iron core) is held up magnetically while a solid ring of the substance is condensed in the outer tube as indicated. The apparatus is placed next to a vapor pressure thermometer in a bath cooled below the melting point of the sample, and the inner tube is carefully lowered

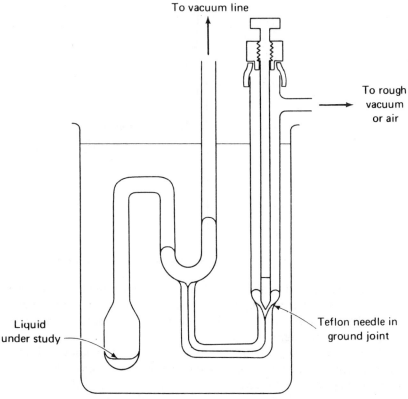

FIG. 8.31. Apparatus for measuring vapor pressures above room temperature.

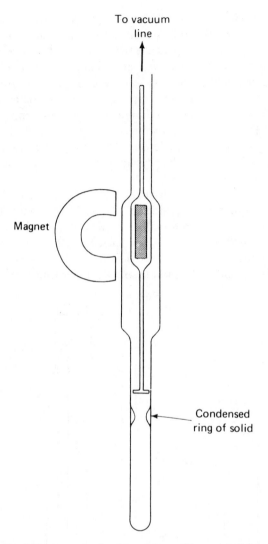

To vacuum
line

Magnet

Condensed
ring of solid

FIG. 8.32. Apparatus for determining melting points below 0°.

so that it is supported by the ring of substance. Then the bath is allowed to warm up slowly. The melting point is indicated by the fall of the inner tube. The method has the advantage that it is not necessary to see the sample during the melting point determination.

When the melting point of a mixture of volatile compounds is desired (as when determining a phase diagram), special apparatus such as that described by Booth and Martin[29] is required.

[29] H. S. Booth and D. R. Martin, *J. Am. Chem. Soc.*, **64**, 2198 (1942).

Several methods are available for determining molecular weights.[30] Here we shall discuss three methods that are most conveniently applied by using a vacuum line. If the sample is sufficiently volatile, the average molecular weight of the vapor molecules can be determined from a measurement of the vapor density at room temperature, assuming the validity of the ideal gas law. This determination can be accomplished by weighing a 100-ml bulb with a stopcock when evacuated and when filled with the vapor at a measured temperature and pressure. If the difference in weight can be determined to ± 0.3 mg, and if the pressure is about 500 mm, the molecular weight can be determined with a precision of ± 0.1 mass unit. The molecular weight of a relatively nonvolatile compound can be determined from the vapor pressure lowering of a solution of the compound in a volatile solvent. If sufficiently dilute solutions are employed, Raoult's Law is valid, and the molecular weight M of the solute can be calculated from the relation

$$M = \frac{M_s P}{P_0 - P} \cdot \frac{W}{W_s}$$

where P is the vapor pressure of the solution, P_0 is the vapor pressure of the pure solvent, M_s is the molecular weight of the solvent, W is the weight of the solute, and W_s is the weight of the solvent. It is best to determine the quantity $(P_0 - P)$ directly by means of a differential manometer, such as the one pictured in Fig. 8.33.[31] Vessel A contains the pure solvent and B contains the solution. The vessels are placed side-by-side in a constant-temperature bath. When filling and emptying the apparatus, the mercury at C is frozen temporarily. When applying the preceding method, one should choose a solvent having a high vapor pressure, for then $P_0 - P$ will be fairly large and can be measured with fair accuracy.[32]

The isopiestic method for determining molecular weights does not require pressure measurements. Solutions of two different solutes in the same solvent are equilibrated in a closed space; distillation of solvent occurs until the solutions have the same vapor pressure (i.e., they are isopiestic). If the solutions are contained in suitably calibrated glass vessels, it is possible to measure their volumes, V_1 and V_2. If one of the solutes is of known molecular weight M_1, then the molecular weight M_2 of the other solute can be calculated from the relation

$$M_2 = M_1 \frac{W_2}{W_1} \cdot \frac{V_1}{V_2}$$

where W_1 and W_2 are the weights of the solutes.

[30] The freezing-point depression method is discussed in Chap. 20.

[31] A. Stock, *Hydrides of Boron and Silicon*, Cornell University Press, Ithaca, N.Y., 1933, p. 194.

[32] R. W. Parry, G. Kodama, and D. R. Schultz [*J. Am. Chem. Soc.*, **80**, 24 (1958)] determined the molecular weights of various substances in liquid ammonia.

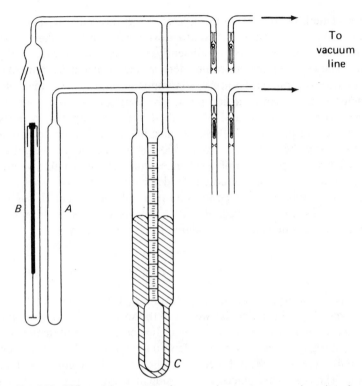

Fig. 8.33. Differential manometer for measuring vapor pressure lowering of a solvent. Vessel *B* is provided with a magnetically operated pulsating stirrer.

Analysis

"Synthetic Analysis." Nothing can be lost or overlooked when studying a reaction by properly conducted vacuum line procedures. Consequently, it is possible to obtain quantitative measures of the reagents consumed and products formed. If the reaction proceeds with very little or no side reaction, the quantitative data can be used to compute the yield and composition of any new compound that is formed. We shall give two examples of such "synthetic analysis."

Xenon hexafluoride (3.06 mmoles) and germanium tetrafluoride (0.666 mmole) were allowed to react at room temperature for 22 hours.[33] The reaction mixture was then held at 0° and subjected to pumping. Analysis of the material pumped away showed it to be pure XeF_6. After 6.5 hours of pumping, 0.4303 g of XeF_6 was pumped away, corresponding to the product composition $GeF_4 \cdot 1.97\ XeF_6$. A further 2.5 hours of pumping removed only 12 mg of XeF_6, showing that the excess XeF_6 had been removed and that a compound of composition $GeF_4 \cdot 2\ XeF_6$ had undoubtedly formed.

[33] K. E. Pullen and G. H. Cady, *Inorg. Chem.*, **6**, 1300 (1967).

A solution of 5.30 mmoles of sodium hydroaluminate ($NaAlH_4$) in diglyme was treated at room temperature with 53.6 mmoles of ammonia until hydrogen evolution ceased (6 hours).[34] A total of 20.8 mmoles of hydrogen was evolved; 32.9 mmoles of ammonia was recovered, corresponding to the reaction of 20.7 mmoles of ammonia. The data indicate that, for each mole of $NaAlH_4$, 4 moles of NH_3 were consumed and 4 moles of hydrogen were evolved. The following reaction is thus strongly supported:

$$NaAlH_4 + 4\,NH_3 \longrightarrow NaAl(NH_2)_4 + 4\,H_2$$

Analysis by Chemical Reaction. Unfortunately, not all reactions are so clean cut that the product composition can be determined by "synthetic analysis." More often, several concurrent reactions occur, the products of which must be separated and individually analyzed. Here we shall discuss several analytical methods that are facilitated by vacuum line techniques.

In Frazer's method[35] for carbon, hydrogen, and nitrogen, the compound is sealed into a quartz bomb with a mixture of copper oxide and copper. The bomb is heated for 1 hour at 950°, cooled, and then opened to the vacuum line. Carbon, hydrogen, and nitrogen in the original compound are thereby converted to CO_2, H_2O, and N_2, which are readily separable on the vacuum line. The CO_2 and N_2 can be measured directly, and the water can be reduced to H_2 with uranium turnings at 750°.

Sometimes the structure of a new compound can be inferred from the products of a simple hydrolysis. For example, potassium germyltrihydroborate, KH_3GeBH_3, was characterized by treatment with an excess of aqueous acid.[36] Each mole of compound gave 1 mole of germane and 3 moles of hydrogen, as expected for the structure $H_3GeBH_3^-$.

$$H_3GeBH_3^- + 3\,H_2O + H^+ \longrightarrow GeH_4 + 3\,H_2 + B(OH)_3$$

In another study, the diphosphoxane $(CF_3)_2POP(CF_3)_2$ was hydrolyzed by treatment with hot aqueous sodium hydroxide.[37] Each mole of compound gave, as expected for the indicated structure, 4 moles of trifluoromethane.

$$(CF_3)_2POP(CF_3)_2 + 4\,OH^- + H_2O \longrightarrow 4\,HCF_3 + 2\,HPO_3^{2-}$$

A pyrolysis can often be used for quantitative analysis. Thus, the boron hydrides decompose quantitatively to boron and hydrogen at temperatures above 500°.[38] Metal carbonyls decompose to carbon monoxide and the

[34] A. E. Finholt, C. Helling, V. Imhof, L. Nielsen, and E. Jacobson, *Inorg. Chem.*, **2**, 504 (1963).

[35] J. W. Frazer, University of California Radiation Laboratory Reports UCRL–5134 (February 1958) and UCRL–6764T (February 1962); also see C. W. Koch and E. E. Jones, *Mikrochim. Acta*, **4**, 734 (1963).

[36] D. S. Rustad and W. L. Jolly, *Inorg. Chem.*, **7**, 213 (1968).

[37] J. E. Griffiths and A. B. Burg, *J. Am. Chem. Soc.*, **84**, 3442 (1962).

[38] A. Stock, *Hydrides of Boron and Silicon*, Cornell University Press, Ithaca, N.Y., 1933, p. 107.

corresponding metals. Many hydrates, ammoniates, etherates, and so forth, liberate the coordinated solvent upon heating. Certain oxides, such as AgO and BaO_2, liberate part or all of their oxygen upon pyrolysis. Dozens of other examples could be cited.

Problems

1. A mixture of Xe (b.p. $= -109°$), AsH_3 (b.p. $= -62.5°$), and As_2H_4 (b.p. $\approx +100°$) are to be separated by distillation into a series of three successive traps. What slush baths or refrigerants would you use in each trap?

2. What cold baths (give temperatures and liquids used) would you use in the separation of NF_3 (b.p. $= -120°$), PF_5 (b.p. $= -85°$), GeF_4 (sub. p. at 1 atm $= -35°$), and BrF_5 (b.p. $= 41°$) by fractional condensation? Indicate the particular fluoride that would condense in each trap.

3. How would you separate a mixture of GeH_4 (b.p. $= -88°$), Ge_2H_6 (b.p. $= 30°$), and Ge_3H_8 (b.p. $= 111°$)?

4. A particular Toepler pump reduces the pressure in a system from P to $0.8P$ during each cycle of its operation. For how many cycles must the pump operate to remove 99.5 per cent of the gas in the system?

5. The molecular weight of a pure gas is determined to be 29 ± 1. What are the various possibilities and how would you uniquely identify each?

6. How would you separate (without destruction) a mixture of
   (a) helium and argon?
   (b) ammonia and water?
   (c) hydrogen chloride and water?

7. How would you analyze a mixture of
   (a) nitrogen and carbon monoxide?
   (b) hydrogen and helium?
   (c) methylamine and trimethylamine?

8. A student distilled a volatile compound into a trap connected to a manometer and cooled the trap first in a $-78.5°$ bath, then a $-63.5°$ bath, and finally a $-45.2°$ bath. The pressures corresponding to each of these temperatures were 40, 126, and 222 mm, respectively. He said: "The first two pressures agree fairly well with the known vapor pressures of ammonia at $-78.5°$ and $-63.5°$ (42 and 130 mm, respectively). However, the third pressure disagrees with the vapor pressure of ammonia at $-45.2°$ (405 mm), and so the compound cannot be ammonia." Criticize the student's conclusion.

9. A solution of sodium in liquid ammonia was treated with enough phosphine to discharge the blue color. A yellow solution remained. Enough methyl chloride was added to this solution to discharge the yellow color and precipitate NaCl (compound A formed). Sodium was then added until the solution was permanently blue (compound B formed). Methyl chloride was then added (compound C formed). Product C was separated by distillation and 30 cc (S.T.P.) was treated with 15 cc of diborane

with formation of a liquid, D, the vapor density of which indicated a molecular weight of 78. Compound D, when heated at 150° for 40 hours, gave 30.1 cc of H_2 and white crystalline products. The most volatile fraction of these products, E, was found to melt at 85 to 86° and had a molecular weight of 214 to 234. Identify compounds A, B, C, D and E.

10. After 2.3 mmoles of germane and 1.8 mmoles of iodine were allowed to react, a yellow liquid A and 2.4 mmoles of gas remained. The gas was separated by fractional condensation into two portions: 1.7 mmoles of material B, with a vapor pressure of about 70 mm at $-78.5°$; and 0.7 mmoles of material C, with a vapor pressure of 195 mm at $-112°$. One mmole of a compound A was allowed to react with 10 mmoles of HgO powder. Compound C (0.35 mmole) and water (0.86 mmole) were the only volatile products, except for a black substance, D, which distilled exceedingly slowly under high vacuum at room temperature. The nonvolatile residue was not further characterized. Identify compounds A, B, C, and D.

Gas
Chromatography

9

Gas chromatography is a technique for separating the components of a volatile mixture on the basis of their different vapor pressures when dissolved in a liquid of negligible vapor pressure or when adsorbed on a solid surface. The mixture is introduced at one end of a column containing a porous, high surface-area solid that is usually impregnated with an oil. A stream of gas (usually helium) is passed through the column and gradually carries the adsorbed mixture with it. The various components of the mixture emerge from the column separately and are detected as they emerge, usually by a device that measures the thermal conductivity of the gas. The components may be collected in separate cold traps.

THE CHROMATOGRAPH

Figure 9.1 is a schematic diagram of a gas chromatograph. The various parts are separately discussed in the following paragraphs.

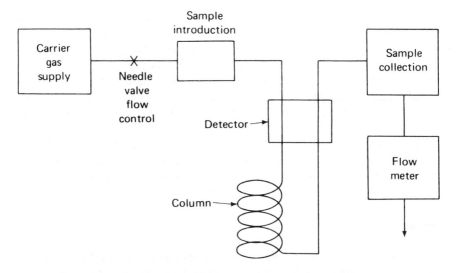

Fig. 9.1. Schematic diagram of a gas chromatograph.

Sample Introduction Point

Samples are introduced at a point immediately before the beginning of the column. Liquids are usually injected with a syringe and hypodermic needle through a rubber diaphragm into a chamber that is heated to ensure immediate volatilization of the entire sample. Condensible gases can be introduced by allowing the helium to pass through a liquid nitrogen-cooled trap containing the frozen sample and suddenly warming the trap up to room temperature. Noncondensible gases can be introduced by suddenly switching the flow of helium through a by-pass tube containing the sample gas.

For analytical purposes, sample sizes in the range of from 2 to 20 μl (1 $\mu l = 10^{-3}$ ml) for liquids and from 0.5 to 5 ml for gases are usually adequate. For preparative work, the sample size is limited principally by the dimensions of the apparatus, but generally single-batch samples are limited to 5 ml of liquid or less.

Column

The column is generally constructed of about 2 m of copper or stainless steel tubing in the form of a helix. For analytical work, $\frac{1}{4}$-in. O.D. tubing is commonly used; for preparative-scale work, $\frac{3}{4}$-in. O.D. tubing is often used. The column is packed with a high surface-area solid, such as molecular sieve or charcoal, or with an inert solid, such as crushed firebrick that has been impregnated with a high-boiling liquid. Liquids such as silicone oils,

Apiezon greases, polyethylene glycol, and dioctyl phthalate are often used. To prepare the column packing, the liquid is dissolved in a volatile solvent, such as chloroform or ether, and the solution is added to the powdered solid. The solvent is then evaporated, leaving a coating of the liquid on the solid. The coated solid is then packed into the column, usually before it has been coiled. During the packing, it is important to tap the column and to add only a little solid at a time to avoid channeling. Plugs of glass wool at each end hold the packing in place.

The column is usually placed in a thermostatted zone maintained at a temperature approximately equal to the boiling point of the highest-boiling fraction. Cold baths (see Chap. 5) are used for temperatures below 20°, and special ovens are used for temperatures above 20°.

Detector

Devices for detecting and measuring the separated fractions of the mixture as they leave the column make use of a difference in some physical property of the pure carrier gas and the carrier gas containing the eluted compound. The commonly used detectors are based on the measurement of thermal conductivity. A wire or semiconductor is heated by passing a current through it. The temperature (and hence the resistance) of the wire or semiconductor is dependent on the thermal conductivity of the gas that passes around it. Two such detectors—one in the entering gas stream and one in the exit gas stream—are made two of the arms of a Wheatstone bridge circuit. The imbalance of the bridge is fed to a strip chart recorder and a chromatogram is produced, showing a series of peaks corresponding to the separately eluted compounds. Figure 9.2 shows a typical gas chromatogram.[1]

Collection Apparatus

The eluted compounds can be collected in suitably cooled U traps, attached to the exit side of the detector. In the case of compounds insensitive to the air, these traps can be attached one at a time as the fractions emerge, but in the case of air-sensitive compounds, one must use a closed system of traps designed so that by stopcock manipulation the effluent gas can be directed through individual traps.

Flowmeter

The flow rate of the carrier gas can be determined with a rotameter, a differential manometer, or a soap-film meter. A rotameter is based on the

[1] C. S. G. Phillips and P. L. Timms, *Anal. Chem.,* **35**, 505 (1963).

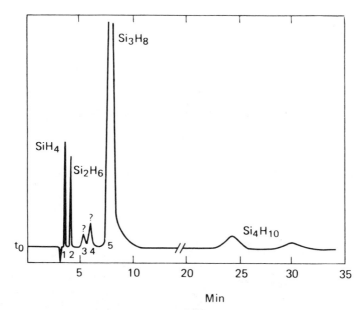

FIG. 9.2. Chromatogram of products obtained by passing an electric discharge through trisilane.[1] (Reproduced with permission of the American Chemical Society.)

displacement of a small sphere in a slightly tapered tube. A differential manometer is one which measures the pressure drop across a capillary tube through which the gas flows. The latter devices require calibration. A soap-film meter (see Fig. 9.3) consists of a buret or other calibrated tube through which the gas passes and through which thin films of soap solution are pushed. A measurement is made by momentarily squeezing the rubber syringe containing soap solution to produce bubbles and by then timing the movement of a soap film up the tube.

APPLICATIONS

Low-boiling Gases

Molecular sieves (Zeolites, manufactured by the Linde Company) without any liquid coating are commonly used as column packing for the separation of "noncondensible" gases, such as the rare gases, hydrogen, nitrogen, oxygen, carbon monoxide, and so forth.[2] For these separations, the column is generally held at room temperature or at $-78°$. Ortho- and parahydrogen, HD, and D_2 can be separated using a pair of columns of active alumina and

[2] E. W. Lard and R. C. Horn, *Anal. Chem.*, **32**, 878 (1960).

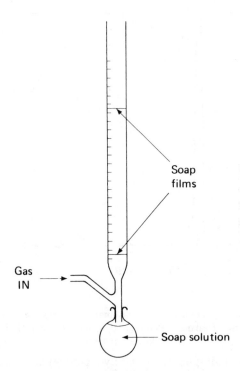

FIG. 9.3. A soap-film flow meter.

active alumina coated with ferric oxide at $-195°$.[3] Many other examples are cited by Brandt.[4]

Metal Halides and Chelates

Niobium and tantalum pentachlorides (b.p. = 254 and 239°, respectively) and tin and titanium tetrachlorides (b.p. = 114 and 136°, respectively) can be readily separated[5] using a column of Chromosorb (Johns-Manville) with a high molecular weight hydrocarbon as the liquid phase, operated at 100 to 200°. A large number of metal acetylacetonate mixtures have been separated gas chromatographically.[6] The rare earths, when chelated with 2,2,6,6-

[3] S. Furuyama and T. Kwan, *J. Phys. Chem.*, **65**, 190 (1961).

[4] W. W. Brandt, *Technique of Inorganic Chemistry*, vol. 3, H. B. Jonassen and A. Weissberger, eds., Interscience Publishers, 1963, p. 1.

[5] R. A. Keller and H. Freiser, *Gas Chromatography 1960*, R. P. W. Scott, ed., Butterworth & Co., London, 1960, p. 301.

[6] See references in E. W. Berg and F. R. Hartlage, Jr., *Anal. Chim. Acta*, **33**, 173 (1965); R. W. Moshier and R. E. Sievers, *Gas Chromatography of Metal Chelates*, Pergamon Press, Oxford, 1965.

tetramethyl-3,5-heptanedione, can be separated[7] using an Apiezon column at 157°. A chromatogram is shown in Fig. 9.4.

Volatile Hydrides

When magnesium silicide is hydrolyzed in aqueous acid, or when silane (SiH_4) is passed through a silent electric discharge, a large number of silicon hydrides analogous to the alkanes are formed.[8] These can be separated by gas chromatography. Figure 9.2 shows the chromatogram of the mixture obtained by subjecting trisilane (Si_3H_8) to an electric discharge. The two isomers of tetrasilane are cleanly separated in the chromatogram. Mixtures

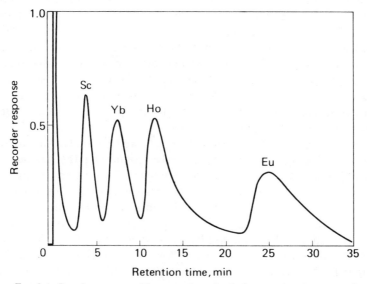

FIG. 9.4. Gas chromatographic separation of volatile rare earth complexes.[7] (Reproduced with permission of the American Chemical Society.)

of germanium hydrides can be similarly separated. Dutton and Onyszchuk[9] found that methyl(germyl)silanes that could not be separated by repeated vacuum line fractionation could be separated by chromatography on a silicone-on-firebrick column (17 mm I.D., 8 ft long). They found that the compounds were decomposed by thermal conductivity detectors; thus, they used an infrared spectrophotometer set at 2050 cm^{-1} (the Ge–H stretching frequency) as a detector.

[7] K. J. Eisentraut and R. E. Sievers, *J. Am. Chem. Soc.*, **87**, 5254 (1965).

[8] Phillips and Timms, *Anal. Chem.*, **35**, 505 (1963); S. D. Gokhale, J. E. Drake, and W. L. Jolly, *J. Inorg. Nucl. Chem.*, **27**, 1911 (1965).

[9] W. A. Dutton and M. Onyszchuk, *Inorg. Chem.*, **7**, 1735 (1968).

The boron hydrides are often obtained as mixtures that are relatively difficult to separate by conventional techniques. However, gas chromatography using paraffin oil-Celite columns at room temperature gives complete separation[10] of diborane, tetraborane and pentaborane(9). Unfortunately, pentaborane(11) undergoes decomposition to diborane and tetraborane on the column.

A SIMPLE CHROMATOGRAPH FOR
SEPARATING NITROGEN AND OXYGEN

The apparatus and wiring diagram for a very simple analytical gas chromatograph are pictured in Fig. 9.5. The column is a coiled 2-m length of 8 to 9-mm O.D. glass tubing, packed with Linde molecular sieve 5A. The detector consists of a thermistor[11] situated in the gas stream and made one arm of a Wheatstone bridge. A 0 to 10-mv strip chart recorder is used to record the chromatogram. The sample chamber is made of 7-mm O.D. glass tubing and has a volume of about 2 ml. Despite the crudeness of the apparatus, it is fairly sensitive and can easily detect as little as 1 per cent oxygen in nitrogen (or vice versa) in a 2-ml sample.

The following procedure should be followed to analyze a mixture of nitrogen and oxygen. The sample holder containing the gas sample is attached to the gas chromatography line with pinch clamps, using a little vacuum grease on the ball joints. At the beginning of the run, stopcocks B and C should be closed and stopcocks D and E should be open. The helium pressure is reduced to 5 to 10 lb per sq in. gauge by suitable adjustment of the reduction valve, and needle valve A is adjusted so that the flow rate, as measured with the soap-bubble flow meter, is 0.7 to 1.5 cc sec^{-1}. The system is flushed with helium for about 10 min. Meanwhile, an ice bath is placed around the detector, the constant voltage power supply is turned on, and the recorder power switch is turned on. The "sensitivity" knob should be turned to about the 5 per cent point, and the recorder pen should be checked for satisfactory writing. After the recorder has warmed up for several minutes, the pen drive motor is turned on. While momentarily short-circuiting the recorder input terminals, turn the "zero adjust" knob to adjust the pen to the 1-mv position on the chart (10 per cent of full scale). Then turn on the chart drive and bring the pen to the 1-mv position (the "zero" line) by suitable adjustment of the resistance box. Turn the sensitivity up to 100 per cent, and finally adjust the pen to the zero line with the "fine adjust" knob.

In rapid succession, open cocks B and C and close cock D. At this point mark the chart paper to indicate the point at which the sample was

[10] J. J. Kaufman, J. E. Todd, and W. S. Koski, *Anal. Chem.*, **29**, 1032 (1957).

[11] Victory Engineering Corporation, Springfield, N.J.; thermistor 32A1.

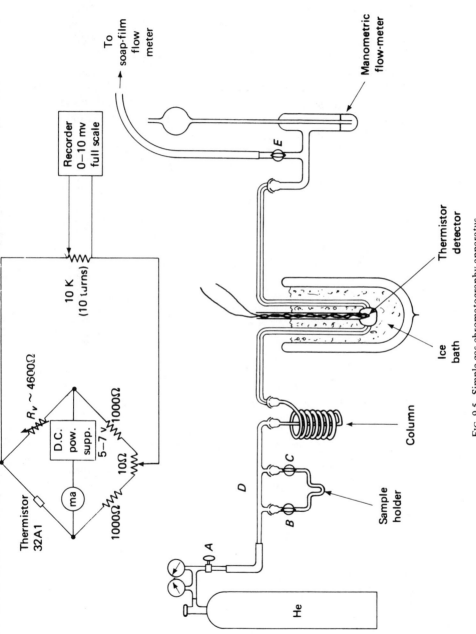

FIG. 9.5. Simple gas chromatography apparatus.

introduced. The oxygen is eluted before the nitrogen. After the gas sample has been completely eluted, it is wise to run a sample of ordinary air in order to calibrate the relative sensitivity of the detector toward N_2 and O_2. Remove the sample holder, open cocks B and C, flush the holder with air by attaching one end to a vacuum momentarily, and close cocks B and C. Open cock D and again attach the sample holder to the line. After flushing the apparatus for about 10 min, obtain the chromatogram for air as described above. Air is resolved into two peaks: the first corresponds to O_2 + Ar and the second corresponds to N_2.[12] Assume that O_2 and Ar are detected with equal sensitivities (in air, the abundance ratio is $N_2/(O_2 + Ar) = 3.57$).

[12] Lard and Horn, *Anal. Chem.*, **32**, 878 (1960).

Electrolytic
Synthesis

10

ADVANTAGES AND DISADVANTAGES OF
ELECTROLYTIC SYNTHESES

Electrochemical oxidation and reduction processes have several features that can make them superior to straight chemical processes. One feature frequently taken advantage of is the ability to apply very high potentials in electrolysis, and the ability thereby to achieve oxidizing or reducing power that cannot be equaled by any ordinary chemical reagent. The electrochemical syntheses of fluorine and ozone are of this type. Another advantage of electrochemical synthesis is that it is not necessary to contaminate a solution with a reducing agent (or oxidizing agent) and its corresponding oxidation product (or reduction product). A third advantage, found in many electrochemical processes, is the ability to control conveniently the electrode potential and thereby to carry out selective oxidations and reductions. A fourth advantage is that electrochemistry permits the preparation of many substances that cannot be prepared by other means. For example, the synthesis of chromium hydride ($CrH_{1.4}$) cannot be duplicated with chemical

191

reagents,[1] and many metal and alloy platings can only be formed by electro-plating.

The main disadvantages of electrochemical synthesis are that it is usually relatively slow and that special equipment is required.

CURRENT EFFICIENCY AND ENERGY EFFICIENCY

The theoretical yield of an electrolytic process can be readily calculated from the current (I) and the time (t) for which it flows.

$$\text{theoretical yield} = \frac{(\text{equivalent weight of product}) \cdot I \cdot t}{96,500}$$

where I is measured in amperes and t in seconds. In practice, yields fall below the theoretical value because of competing side reactions causing formation of unwanted by-products. The "percentage yield," or current efficiency of an electrolytic process is given by the relation

$$\text{current efficiency} = \frac{\text{yield}}{\text{theoretical yield}} \cdot 100$$

Inasmuch as electricity is sold in terms of watt-hours, not ampere-hours, a quantity of more commercial significance is the energy efficiency, or yield per kilowatt-hour.

$$\text{yield per kilowatt-hour} = \frac{\text{yield} \cdot 1000}{\text{voltage} \cdot \text{amperage} \cdot \text{hours}}$$

Current efficiency may be increased by reducing the extent of the side reactions. The yield per kilowatt-hour may be increased by increasing the current efficiency and by decreasing the voltage. Of course, the applied voltage cannot be decreased below the thermodynamically required value (the reversible potential); in fact, for reasons we shall discuss, the applied potential is almost always appreciably greater than the reversible potential. Actually, in laboratory electrolytic syntheses, one is very seldom concerned with energy efficiency (the costs of the experimenter's time and of the depreciation of the equipment far outweigh the cost of the electricity). However, the current efficiency is of considerable concern because it is directly related to the yield that can be achieved in a given time.

[1] C. A. Snavely, *Trans. Electrochem. Soc.*, **92**, 537 (1947); C. A. Snavely and D. A. Vaughan, *J. Am. Chem. Soc.*, **71**, 313 (1949).

THE CURRENT-VOLTAGE CURVE

The electrical properties of an electrolytic cell are much different from those of a simple ohmic resistance. Let us consider a cell having only one net reaction, that is, having electrode processes that occur with 100 per cent current efficiency. In general, the current varies with the applied voltage as curve B of Fig. 10.1. Practically no current flows, even when the applied voltage equals the reversible cell potential, E_R. If the electrode processes were completely reversible, increasing the potential beyond this point would cause the current to rise along line A, the slope of which equals the reciprocal of the cell resistance. But actually, opposing potentials (the magnitudes of which increase with increasing current) are established in the cell, and these prevent the current from increasing without bound.

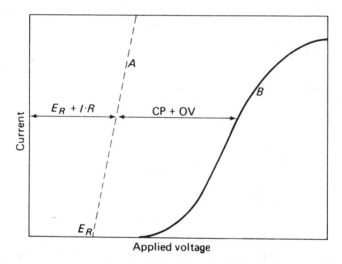

FIG. 10.1. A typical electrolytic current-voltage curve. (E_R is the reversible cell potential, $I \cdot R$ is the product of the current and the cell resistance, CP is the concentration polarization, and OV is the overvoltage.)

Inasmuch as ions are either formed or removed at the electrode surfaces, concentration gradients are built up in the vicinity of the electrodes. These concentration gradients cause a potential drop called concentration polarization. An electrode reaction, like any chemical reaction, possesses an activation energy. This activation energy appears in the form of an *overvoltage*. Some electrode reactions, such as metal electrodepositions, have very low or practically no overvoltages; others, such as those involving gas evolution, have high overvoltages. Overvoltage is briefly discussed in the following sections. It is a combination of concentration polarization and overvoltage that causes current-voltage curve B of Fig. 10.1 to be observed rather than line A.

At sufficiently high currents, the concentration of reacting species (the depolarizer) becomes practically zero at an electrode surface. Then the current becomes almost independent of the applied voltage and is controlled mainly by diffusion and migration of the ions. If the solution is well stirred and contains a high concentration of inert electrolyte, and if there is no appreciable overvoltage, the value of the limiting current is directly proportional to the concentration of the reacting species.

Now, in most electrolytic cells there are several electrode reactions that can take place, and for each of these there is a current-voltage curve, as indicated in Fig. 10.2. The resultant current-voltage curve, the solid curve, is

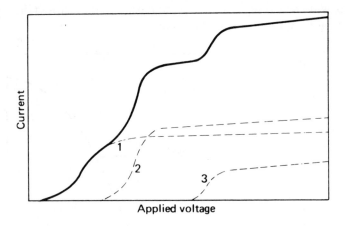

FIG. 10.2. A current-voltage curve for a cell involving three electrolytic processes.

the sum of the individual curves. Thus, at low applied voltages (and corresponding low currents), only process 1 occurs. At sufficiently high voltages and currents, processes 1, 2, and 3 occur simultaneously.

Obviously for any particular electrolytic cell containing a particular electrolytic solution, there is a one-to-one correspondence between the applied voltage and the current. However, if the electrode surface areas are increased, the value of the current at a particular voltage will increase proportionately.[2] Therefore, a more useful and fundamental quantity than the current is the *current density*, or the current divided by the electrode surface area. Descriptions of electrochemical processes almost always include the specification of the current density at the appropriate electrode.

---

[2] Actually, the current is not exactly proportional to the electrode area because the $I \cdot R$ drop across the cell changes, thus changing the effective electrode potential.

HYDROGEN AND OXYGEN OVERVOLTAGES

If a potential greater than 1.23 v is applied to inert electrodes immersed in an aqueous solution of some "inert" electrolyte (such as H_2SO_4, NaOH, or Na_2HPO_4), the decomposition of water into hydrogen and oxygen is thermodynamically possible. But even when concentration polarization effects are minimized, it is usually found that potentials much greater than 1.23 v must be applied before appreciable gas evolution occurs. These extra potentials are required because the overvoltages for evolution of hydrogen and oxygen are very high at most electrodes.

Table 10.1 Hydrogen Overvoltages in 1 *M* H_2SO_4 in Volts*

Cathode	Current Density (amp cm^{-2})		
	10^{-2}	10^{-1}	1
Pt (black)	0.03	0.04	0.05
Pt (smooth)	0.07	0.29	0.68
Au	0.39	0.59	0.80
Fe	0.56	0.82	1.29
Cu	0.58	0.80	1.25
C	0.70	0.89	1.17
Ni	0.74	1.05	1.24
Zn	0.75	1.06	1.23
Ag	0.76	0.98	1.10
Al	0.83	1.00	1.29
Hg	0.93	1.03	1.07
Pb	1.09	1.18	1.26
Cd	1.13	1.22	1.25

* See ref. 3.

Tables 10.1 and 10.2 give hydrogen and oxygen overvoltages, respectively, for various electrode surfaces.[3] Mercury, lead, and cadmium exhibit extremely high hydrogen overvoltages and platinum and gold exhibit high oxygen overvoltages. Because overvoltage is markedly affected by the condition of the electrode surface, the values are only approximate.

Most cathodic and anodic processes in aqueous solution are accompanied by the evolution of hydrogen and oxygen, respectively. There are various ways whereby the ratio of hydrogen to desired cathode product or the ratio of oxygen to desired anode product may be decreased.

[3] W. M. Latimer, *Oxidation Potentials*, 2nd ed., Prentice-Hall, Inc., Englewood Cliffs, N.J., 1952.

Table 10.2   Oxygen Overvoltages in $1M$ KOH
in Volts*

Cathode	Current Density ($amp\ cm^{-2}$)		
	10^{-2}	10^{-1}	1
Ni	0.35	0.73	0.87
Pt (black)	0.40	0.64	0.79
Cu	0.42	0.66	0.84
C	0.52	1.09	1.24
Ag	0.58	0.98	1.14
Au	0.67	1.24	1.68
Pt (smooth)	0.72	1.28	1.38

* See ref. 3.

1. One may use an electrode possessing a high overvoltage for hydrogen or oxygen evolution and then apply a voltage such that the ratio of hydrogen or oxygen to desired product is minimized.

2. The concentration of the species leading to the desired product may be increased.

3. The temperature may be decreased (inasmuch as hydrogen and oxygen overvoltages are higher than most other overvoltages, a given decrease in temperature will usually cause a greater decrease in the rate of formation of hydrogen or oxygen than in the rate of formation of the desired product).

Method 1 may be used to increase the current efficiency of the reduction of chromic ion to chromous ion in acidic solution. The thermodynamically favored cathode process is the reduction of hydrogen ion; it may be made the exclusive process if a "platinized" platinum electrode is used. But if a mercury cathode is used, hydrogen evolution does not occur until a relatively high voltage is applied. By choosing an applied voltage somewhat lower than that, chromic ion reduction is made the important process. The application of methods 2 and 3 to the decrease of oxygen evolution is discussed later in this chapter under Effect of Concentration and Effect of Temperature.

EFFECT OF THE ELECTRODE MATERIAL

Some electrolytic processes have overvoltages comparable in magnitude to hydrogen and oxygen overvoltages. The cathodic reduction of nitric acid is an interesting case in which changes in the cathode material profoundly influence the course of the reaction. Table 10.3 gives the current efficiencies for the production of hydroxylammonium ion and ammonium ion at various cathodes. In all cases, the current density was 0.24 amp cm^{-2}.

Table 10.3   Current Efficiencies in Reduction of
Nitric Acid (0.4 g HNO_3 in 20 ml 50 per cent H_2SO_4,
cooled with ice)*

Cathode	NH_3OH^+ (per cent)	NH_4^+ (per cent)
Amalgamated lead	69.7	16.9
Smooth zinc	45.8	38.3
Rough lead	26.8	57.6
Smooth copper	11.5	76.8
Spongy copper	1.5	93.8

* J. Tafel, *Z. anorg. Chem.*, **31**, 289 (1902).

Catalysts are often added to an electrolyte to change the current efficiency at an electrode. For example, salts of mercury, lead, or cadmium are often added to increase the hydrogen overvoltage at a cathode. Salts of titanium, vanadium, chromium, iron, or cerium are sometimes added to an electrolyte when the metal ions are capable of undergoing oxidation-reduction reactions both at the electrode and with the depolarizer.

The literature abounds with data on electrolytic catalysts; in many cases the mechanisms of the catalysts are unknown.

EFFECT OF THE ELECTRODE POTENTIAL

There are often several processes that can occur at an electrode. If these several processes have current-potential curves that are well separated, it is possible, by careful control of the applied electrode potential, to carry out the process having the most positive electrode potential to the exclusion of the others. When this electrode process has been carried to completion (say, when the current has dropped to less than 1 per cent of its initial value), the electrode potential can be adjusted to carry out the process with the next most positive potential. Thus, the several processes can be carried out in turn. The technique of *electrolysis with controlled cathode potential* is used quite commonly in the analysis of solutions containing several kinds of metal ions. For example, a solution containing copper, bismuth, and lead as tartrate complexes can first be electrolyzed with a cathode potential of -0.30 v (vs. the saturated calomel electrode) until practically all the silver has been plated out; then the bismuth may be plated out at a potential of -0.40 v, and, finally, the lead is plated out at -0.60 v.[4]

The reduction at a mercury electrode of the coenzyme of Vitamin B_{12} [a cobalt(III) metal alkyl] furnishes an example in which the reducible sub-

[4] J. J. Lingane, *Electroanalytical Chemistry*, 2nd ed., Interscience Publishers, New York, 1958, p. 408.

stance may be reduced to either of two substances, depending on the cathode potential.[5] With a cathode potential of -0.4 v (vs. the standard calomel electrode), the principal product is Vitamin B_{12r}, a cobalt(II) complex. With a cathode potential of -1.2 v, the principal product is Vitamin B_{12s}, which contains either cobalt(I) or a Co^{III}-H group.

Controlled-potential electrolyses of $B_{10}H_{10}{}^{2-}$ solutions using a rotating platinum gauze anode indicate that two oxidations take place.[6] At potentials of $+0.4$ and $+0.6$ v (vs. S.C.E.), the principal product is $B_{20}H_{19}{}^{3-}$, whereas exhaustive electrolysis at $+0.9$ v yields principally $B_{20}H_{18}{}^{2-}$.

An electrode potential cannot be calculated readily from the overall applied voltage; it must be measured with the aid of a suitable reference anode at essentially zero current flow. The simple apparatus pictured in Fig. 10.3 is adequate for measuring an electrode potential during an electrolysis.

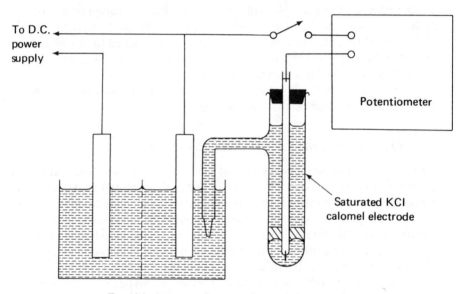

FIG. 10.3. Apparatus for measuring electrode potentials.

Even though the applied voltage is held constant during an electrolysis, the potential of the electrode at which the important reaction is taking place will not remain constant. If we assume, for simplicity, that the potential of the opposing electrode remains constant, we have the relation:

$$E_{\text{electrode}} + (I \cdot R) = \text{applied voltage} + \text{constant}$$

[5] H. A. O. Hill, J. M. Pratt, and R. J. P. Williams, *Chem. Ind. London*, 197 (1964); S. L. Tackett, J. W. Collat, and J. C. Abbott, *Biochemistry*, **2**, 919 (1963).

[6] R. L. Middaugh and F. Farha, Jr., *J. Am. Chem. Soc.*, **88**, 4147 (1966); also see R. J. Wiersema and R. L. Middaugh, *J. Am. Chem. Soc.*, **89**, 5078 (1967).

After the electrolysis has proceeded a finite amount, the concentration of depolarizer will have decreased, causing a decrease in the current I. The decrease in the $I \cdot R$ drop will result in a corresponding increase in the electrode potential. To hold the electrode potential constant, it is necessary to decrease continually the applied voltage throughout the electrolysis. This can be accomplished manually or by using an electronic potentiostat.[7]

EFFECT OF CONCENTRATION

Consider an electrode at which several competing reactions are taking place. A change in the electrolyte composition can change the relative rates of the electrode reactions, and hence change their current efficiencies. For example, when a cold solution of sulfuric acid is electrolyzed with platinum electrodes, three anodic processes can occur:

$$2\,H_2O = O_2 + 4\,H^+ + 4\,e^-$$
$$2\,SO_4^{2-} = S_2O_8^{2-} + 2\,e^-$$
$$3\,H_2O = O_3 + 6\,H^+ + 6\,e^-$$

At $0°$, the formation of oxygen and peroxydisulfate are the only important processes. From Table 10.4, we see that the relative production of oxygen and peroxydisulfate is quite sensitive to the concentration of sulfuric acid.

Table 10.4   Oxidation of Sulfuric Acid to Peroxydisulfuric Acid (at $0°$ with Pt Anode)*

Specific Gravity of H_2SO_4	Current Efficiency (per cent) Current Density (amp cm^{-2})		
	0.05	0.5	1.0
1.15			7.0
1.20		4.4	20.9
1.25		29.3	43.5
1.30	1.8	47.2	51.6
1.35	3.9	60.5	71.3
1.40	23.0	67.7	75.6
1.45	32.9	73.1	78.4
1.50	52.0	74.5	71.8
1.55	59.6	66.7	65.3
1.60	60.1	63.8	50.8
1.65	55.8	52.0	
1.70	40.0		

* K. Elbs and O. Schönherr, *Z. Elektrochem.*, **1**, 417 (1894).

[7] Ref. 4, pp. 308–39.

EFFECT OF TEMPERATURE

Increasing the temperature of the electrolyte reduces the resistance of the cell and increases the rate of diffusion of the depolarizer. For these reasons, the energy efficiency is increased. Electrode reactions, just like ordinary reactions, have activation energies. Hence, if more than one reaction is occurring at an electrode, a change in the temperature of the electrode usually changes the relative rates of the electrode reactions. We have already seen (Table 10.4) how the yield of peroxydisulfate in the electrolysis of aqueous sulfuric acid is affected by changes of concentration at 0°. In Table 10.5, data are presented that show the effect of temperature on the production of ozone by the electrolysis of sulfuric acid of specific gravity 1.200 with a platinum anode.

Table 10.5   Ozone by Electrolysis of Sulfuric Acid*

Approximate Anode Temperature (°C)	*Current Density* ($amp\ cm^{-2}$)	*Current Efficiency* (*per cent*)
0	0.29	0.78
	0.73	1.11
	2.93	1.27
−63	0.25	23.6
	0.75	32.4
	1.24	27.5

* J. D. Seader and C. W. Tobias, *Ind. Eng. Chem.*, **44**, 2207 (1952).

ELECTRODEPOSITION OF METALS

The nature of electrodeposited metals (the physical appearance, hardness, etc.) is affected by several experimental factors that will be considered in turn.[8]

Current Density

At low current densities, there is ample time for the growth of crystal nuclei without the formation of new nuclei. Consequently, under these conditions metals form coarsely crystalline deposits. At higher current densities, the growth of new nuclei is encouraged; consequently, smaller crystals form, and the deposits are more fine grained. However, at very high current densities,

[8] See S. Glasstone, *The Fundamentals of Electrochemistry and Electrodeposition*, 2nd ed., Franklin Publishing Co., Palisade, N. J., 1960, pp. 76–80.

crystals tend to grow out toward the bulk of the solution where the metal ions are more concentrated, resulting in "trees" or nodules. Also, high current densities can cause hydrogen evolution, resulting in spots on the plate and, because of the local increase in pH, precipitation of hydroxides or basic salts.

Electrolyte Concentration

By increasing the concentration of the reducible metal ions, the solution near the cathode does not become exhausted as readily. Consequently, the ill effects of high current density can be largely offset by the use of a high electrolyte concentration.

Temperature

The effect of temperature is not the same for all electrodepositions, and it cannot be readily predicted, probably because an increase in temperature produces effects which are opposed. Thus, a higher temperature favors diffusion to the cathode and smoother deposits, but it also favors a faster growth of crystal nuclei and coarser deposits. In addition, hydrogen overvoltage is decreased, and therefore hydrogen evolution and its consequences can also become important at elevated temperatures.

Additives

Often the addition to an electroplating bath of small amounts of organic substances—such as sugars, camphor, gelatine, glue, casein, and the like—causes the metal deposit to change from coarse to smooth and fine grained. The additives are probably effective because they are adsorbed by the surface and cover crystal nuclei. Thus, fresh nuclei growth is encouraged, with the result of a fine-grained deposit.

Metal Ion Complexation

Generally, unsatisfactory deposits are obtained when a solution of a simple metal salt is used as an electrolytic bath. For example, if silver is plated out from a silver nitrate solution, the deposit is usually nonadherent and composed of large crystals. However, if the electrolyte contains excess cyanide [the silver being complexed as $Ag(CN)_2^-$], the deposit is firm and smooth. In fact, cyanide-containing baths are frequently used for the deposition of gold, zinc, copper, and cadmium. Other complexing anions, such as fluoride, chloride, and tartrate, are used for other electrodepositions. Suitable electrolyte compositions for a large number of metals are given by Rogers.[9]

[9] T. M. Rogers, *Handbook of Practical Electroplating*, The Macmillan Company, New York, 1959.

ELECTROLYSIS APPARATUS

Some laboratories are wired with direct current from a dynamo. Such d.c. power is usually supplied at 110 v, and for most electrolyses it is necessary to reduce the voltage with a rheostat. When such a power supply is not available, one may rectify the ordinary 60-cycle alternating current. A simple apparatus, such as that used for charging storage batteries, is satisfactory if direct current with a great deal of "ripple" is not detrimental to the electrolysis. However, when a steady direct current with a voltage that is readily controlled is desired, an apparatus such as that diagrammed in Fig. 10.4 is useful.[10] This apparatus produces d.c. current with very little "ripple"; the applied voltage may be varied from 0 to 30 v, and currents up to 5 amp may be drawn (some simple laboratory electrolytic cells are described on pp. 447, 448, and 450).

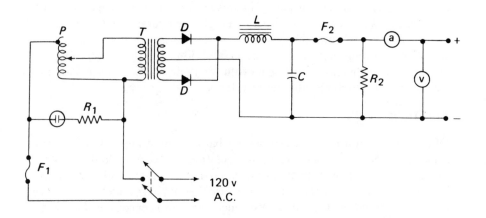

FIG. 10.4. A d.c. power supply, 0–30 v at 5 amp.

F_1 3 AG slow blow fuse, 2 amp
F_2 8 AG fuse, 5 amp
R_1 56 K resistor
R_2 250 Ω, 25 w resistor
P Powerstat, tube #10B
T Triad #F-49-U, paralleled secondaries
D 1 N 3210 diodes, 200 v, 10 amp
L Triad #C-48-U, parallel connected
C 6000 μf condenser, 50 v d.c.
a 0–5 amp ammeter
v 0–30 v voltmeter

[10] Even simpler devices are described by R. Barnard and R. Woodriff, *J. Chem. Educ.*, **38**, 521 (1961), and O. S. de Ita and M. Urquiza, *J. Chem. Educ.*, **41**, 655 (1964).

Problems

1. Potassium chlorate was prepared by the electrolytic oxidation of potassium chloride. A current of 1 amp was passed for 2 hours. Assuming a current efficiency of 50 per cent, how many grams of $KClO_3$ should have been obtained?

2. If germane may be prepared by the cathodic reduction of aqueous H_2GeO_3 with a current efficiency of 2 per cent, how long would it take to prepare 5 mmoles of germane with a current of 2 amp?

3. Explain why it is possible to reduce the sodium ion at a mercury cathode even though the reduction of water to hydrogen is the thermodynamically favored process. How would you expect changes in pH and temperature to affect the current efficiency?

4. What is the flow rate [l. (S.T.P.) per hour] of fluorine from a molten KHF_2 electrolytic generator that operates with a current of 10 amp and an efficiency of about 100 per cent?

High-Temperature Processes

11

TEMPERATURE MEASUREMENT

Mercury-in-glass thermometers are commonly used for measuring temperatures up to about 400°. When higher temperatures are to be measured, or when the use of a thermometer would be experimentally inconvenient, thermocouples can be used. Two wires of different metals joined at their ends to form a loop constitute a thermocouple. If the two junctions are at the same temperature, no current will flow through the loop; however, if the junctions are at different temperatures, there will usually be a current flow. If the circuit is broken and a potentiometer is connected across the break, one may measure an electromotive force that depends only on the metals used and the temperatures of the junctions. Thus, thermocouples may be used as temperature-measuring devices having definite advantages over the usual liquid-in-glass thermometers: they are less liable to break and have a much wider temperature range.

Thermocouples made of the following wire pairs are commonly used: Chromel P (90 per cent Ni, 10 per cent Cr) vs. Alumel (98 per cent Ni, 2 per

Table 11.1   Temperature-Electromotive Force Data for Thermocouples
(in mv for Fixed Junction at 0°)

Temperature (°C)	Chromel vs. Alumel	Pt vs. Pt-10 per cent Rh	Fe vs. Constantan	Cu vs. Constantan
0	0	0	0	0
20	0.80	0.11	1.02	0.79
40	1.61	0.24	2.06	1.61
60	2.43	0.36	3.11	2.47
80	3.26	0.50	4.19	3.36
100	4.10	0.64	5.27	4.28
120	4.92	0.79	6.36	5.23
140	5.73	0.95	7.45	6.20
160	6.53	1.11	8.56	7.21
180	7.33	1.27	9.67	8.23
200	8.13	1.44	10.78	9.29
220	8.94	1.61	11.89	10.36
240	9.75	1.78	13.01	11.46
260	10.57	1.96	14.12	12.57
280	11.39	2.14	15.22	13.71
300	12.21	2.32	16.33	14.86
320	13.04	2.50	17.43	16.03
340	13.88	2.69	18.54	17.22
360	14.71	2.87	19.64	18.42
380	15.55	3.06	20.75	19.64
400	16.40	3.25	21.85	20.87
420	17.24	3.44	22.95	
440	18.09	3.64	24.06	
460	18.94	3.83	25.16	
480	19.79	4.02	26.27	
500	20.65	4.22	27.39	
520	21.50	4.42	28.52	
540	22.35	4.62	29.65	
560	23.20	4.82	30.80	
580	24.06	5.02	31.95	
600	24.91	5.22	33.11	
620	25.76	5.43	34.29	
640	26.61	5.64	35.48	
660	27.45	5.84	36.69	
680	28.29	6.05	37.91	
700	29.14	6.26	39.15	
720	29.97	6.47		
740	30.81	6.68		
760	31.65	6.90		
780	32.48	7.11		
800	33.30	7.33		
820	34.12	7.55		
840	34.93	7.77		
860	35.75	7.99		

Table 11.1   Temperature-Electromotive Force Data for Thermocouples
(in mv for Fixed Junction at 0°) *(cont.)*

Temperature (°C)	Chromel vs. Alumel	Pt vs. Pt-10 per cent Rh
880	36.55	8.21
900	37.36	8.43
920	38.16	8.66
940	38.95	8.88
960	39.75	9.11
980	40.53	9.34
1000	41.31	9.57
1050	43.25	10.15
1100	45.16	10.74
1150	47.04	11.34
1200	48.89	11.94
1250		12.54
1300		13.14
1350		13.74
1400		14.34

cent Al), platinum vs. platinum-10 per cent rhodium, iron vs. Constantan, and copper vs. Constantan. For accurate work, it is necessary that a thermo-couple be calibrated by measuring the electromotive force corresponding to various temperatures. However, for most synthetic purposes, it is adequate to use the approximate data in Table 11.1, which gives the electromotive force as a function of the hot-junction temperature when the cold junction is held in an ice-water bath. Whenever it is unnecessary to know high tempera-tures to better than ±5°, one may dispense with the ice-water bath and simply allow the cold junction to remain at room temperature. The data of Table 11.1 may still be used in such cases if the observed voltages are increased by the amount corresponding to room temperature. In fact, it is most convenient simply to interpose the potentiometer between the two dis-similar metal wires (see Fig. 11.1). Some potentiometers, such as Leeds and Northrup's Model 8657, have devices for compensating for a cold junction that is at room temperature instead of 0°. With such compensation, the potentiometer readings may be used without correction.

When rather crude temperature measurement (±10°) will suffice, a simple high-resistance millivoltmeter may be used instead of a potentiometer. Such meters usually give slightly low temperature readings because of the finite $I \cdot R$ drop through the thermocouple.[1]

Care must be taken to prevent shorting the thermocouple wires. If the wires are allowed to touch at a place other than the hot junction, a spurious

[1] Special meters with temperature scales can be purchased.

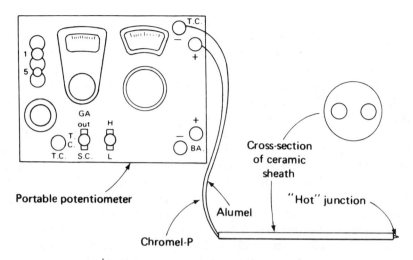

Fig. 11.1. Simple thermocouple circuit.

reading will be obtained, because the voltage will correspond to the temperature of the point of contact. Thermocouple wires are generally insulated and protected by a double-bore ceramic sheath for a distance of about 30 cm from the point at which the wires are welded. When it is necessary to make the sheath airtight, or when it is necessary to protect the thermocouple wires from chemical attack, a glass, ceramic, or metal tube (closed at one end) may be slipped over the ceramic sheath.

HEATING APPARATUS

We shall assume that the reader is already familiar with simple heating devices, such as Bunsen burners and hot plates. Round-bottomed flasks may be heated to temperatures up to about 400° by means of form-fitting hemispherical electric heating mantles.[2] These mantles consist of wire heating elements, coiled and anchored in glass cloth and insulated by glass wool. The temperature is controlled by a regulator based on adjustable periodic current interruption or by a variable transformer. Glass tubes and irregularly shaped objects can be heated by wrapping with heating tape (glass cloth-enclosed resistance wire) and application of an appropriate voltage with a variable transformer.

When it is important that a vessel be heated to a uniform temperature, it should be immersed in a liquid that can serve as a heat transfer medium. For

[2] Special mantles with quartz fiber fabric can be used up to 650°.

temperatures up to about 200°, mineral oil or molten wax can be used; for temperatures up to about 250°, silicone oil can be used. A bath of Wood's metal (a Bi-Pb-Sn-Cd alloy) can be used from its melting point (about 70°) up to about 400°. Various salts (e.g., $PbCl_2$, m.p. = 501°) or eutectic mixtures of salts (e.g., $KNO_3/NaNO_3$, m.p. = 230°) can also be used. The use of a refluxing liquid for maintaining a constant elevated temperature has been described in Chap. 4 in connection with the Abderhalden drier.

There are three main types of electric furnaces available for studying high-temperature reactions: muffle furnaces, "pot" furnaces, and tube furnaces. Muffle furnaces are simply high-temperature ovens; they are primarily used for heating materials in crucibles without a controlled atmosphere. Muffle furnaces are usually provided with their own temperature-controlling mechanisms and thermocouple pyrometers. Pot and tube furnaces are usually used for heating materials in controlled atmospheres. Typical setups for these types of furnaces are pictured in Fig. 11.2. These furnaces usually do not have their own temperature-regulating mechanisms; it is customary to control their power input (and consequently their temperature) with autotransformers.

While a reaction vessel is being warmed up in a pot furnace or tube furnace, its outer surface may be several hundred degrees warmer than its interior.

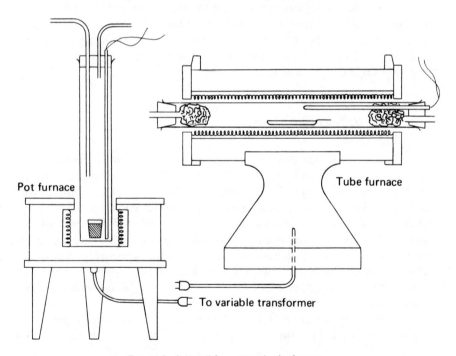

Pot furnace

Tube furnace

To variable transformer

Fig. 11.2. A "pot" furnace and tube furnace.

Thus, one might easily melt a glass reaction vessel during a warmup if one naïvely relied upon the readings of a thermocouple placed inside the vessel. Precautions must also be taken when the thermocouple is placed outside the reaction vessel. It must not be placed too close to the heating elements or the temperature reading will be higher than that inside the reaction vessel, even after the warmup period.

CHOOSING CONTAINERS

Consider the following laboratory mishaps.

1. An iron crucible was ruined when it was used as a reaction vessel for the dehydration of NaH_2PO_4 to $(NaPO_3)_x$. A platinum or porcelain crucible should have been used.

2. A platinum boat was ruined when aluminum was melted in it. A graphite container would have been satisfactory, inasmuch as the fusion was carried out in a nitrogen atmosphere.

3. A quartz tube was ruined when magnesium was heated in direct contact with it. The magnesium should have been contained in an alumina boat.

4. Numerous platinum dishes have been ruined because they were used for alkaline fusions. In most cases, porcelain or iron dishes could have been used.

Each of these mishaps could have been avoided by careful planning. When choosing a container for a high-temperature reaction, the following questions should be considered.

1. Are the reactants acidic? If so, will the container be affected?
2. Will the reactants alloy with the container?
3. Is the reactant a strong reducing agent? If so, will the container be affected? (Strong oxidizing agents should be considered similarly.)
4. Will the oxygen of the air react with the container?

CHEMICAL TRANSPORT REACTIONS

Consider the reaction, at a particular temperature, of a nonvolatile solid substance with a gas to form exclusively vapor-phase reaction products, which, in turn, undergo the reverse reaction in a part of the system maintained at a different temperature. The process appears to be a sublimation, but is really a *chemical transport reaction*. This transport technique has been extensively used as a method of purification and as a means for growing large, single crystals. The subject has been reviewed by Schäfer.[3] A few examples

[3] H. Schäfer, *Chemical Transport Reactions*, Academic Press Inc., New York, 1963; also see M. C. Ball, *J. Chem. Educ.*, **45**, 651 (1968).

of transport reactions are given (the arrow between the indicated temperatures refers to the transport direction in the thermal gradient).

$$Ni + 4\,CO \;\rightleftharpoons\; Ni(CO)_4 \qquad 80 \rightarrow 200° \tag{11.1}$$

$$Zr + 2\,I_2 \;\rightleftharpoons\; ZrI_4 \qquad 280 \rightarrow 1450° \tag{11.2}$$

$$Ir + \tfrac{3}{2}\,O_2 \;\rightleftharpoons\; IrO_3 \qquad 1325 \rightarrow 1130° \tag{11.3}$$

$$Au + \tfrac{1}{2}\,Cl_2 \;\rightleftharpoons\; \tfrac{1}{2}\,Au_2Cl_2 \qquad 1000 \rightarrow 700° \tag{11.4}$$

$$C + CO_2 \;\rightleftharpoons\; 2\,CO \qquad 1000 \rightarrow 600° \tag{11.5}$$

$$Ge + GeI_4 \;\rightleftharpoons\; 2\,GeI_2 \qquad 500 \rightarrow 350° \tag{11.6}$$

$$Fe_2O_3 + 6\,HCl \;\rightleftharpoons\; Fe_2Cl_6 + 3\,H_2O \qquad 1000 \rightarrow 750° \tag{11.7}$$

$$NbCl_3 + NbCl_5 \;\rightleftharpoons\; 2\,NbCl_4 \qquad 390 \rightarrow 355° \tag{11.8}$$

$$FeS + I_2 \;\rightleftharpoons\; FeI_2 + \tfrac{1}{2}\,S_2 \qquad 900 \rightarrow 700° \tag{11.9}$$

An experimental setup for carrying out the transport of Fe_2O_3 by reaction (11.7) is shown in Fig. 32.22.

In all the above cases, the starting material is chemically identical to the transported material; however, this feature is not a requirement for chemical transport. In fact, a number of syntheses can be facilitated by transport reactions.[3] For example, beautiful crystals of iron tungstate can be prepared by the reaction

$$FeO + WO_3 \;\longrightarrow\; FeWO_4 \tag{11.10}$$

using gaseous HCl as the transporting agent. In the absence of HCl, no reaction occurs because of the relative nonvolatility of FeO and WO_3; in the presence of HCl these materials are effectively brought together in the form of the vapors of $FeCl_2$ and H_2O and $WOCl_4$ (?) and H_2O.

Another example is the synthesis of Nb_5Si_3 by the reaction of niobium powder with quartz glass at 1000°:

$$11\,Nb + 3\,SiO_2 \;\longrightarrow\; Nb_5Si_3 + 6\,NbO \tag{11.11}$$

No reaction takes place in an evacuated system, but complete reaction takes place in the presence of a minute amount of hydrogen. The hydrogen serves to transport the silica to the niobium by the process

$$SiO_2 + H_2 \;\longrightarrow\; SiO(g) + H_2O(g) \tag{11.12}$$

The reader is referred to the previously noted monograph by Schäfer for many other examples.

THERMAL ANALYSIS

Ordinary thermal analysis[4] is based on the fact that a phase change usually produces an evolution or absorption of heat. When a pure compound cools

---

[4] See A. Findlay, *The Phase Rule*, 9th ed., rev. by A. N. Campbell and N. O. Smith, Dover Publications, Inc., New York, 1951, Appendix II, pp. 471–79.

without change of state, the plot of temperature vs. time is continuous and exponential in form, when the surroundings are at a constant temperature. However, in the case of a pure molten material, freezing will commence at a characteristic temperature. Freezing is an exothermic process, and the rate of freezing is determined by the rate of heat loss by radiation and conduction. Thus, while the sample is freezing the cooling curve is a horizontal straight line. The cooling curve for a pure substance is shown in Fig. 11.3A. If the crystallization is slow or the cooling is too fast, the horizontal line may become a point of inflection.

Now let us consider the cooling of a solution of two substances that are completely immiscible in the solid state and that do not form a compound. This system will cool until the initial crystallization of one component occurs (point X, Fig. 11.3B). Because a solution does not freeze at a constant temperature, the temperature continues to fall—but less rapidly—until the eutectic temperature is reached (point Y, Fig. 11.3B). This temperature is maintained until all is solid (point Z, Fig. 11.3B).

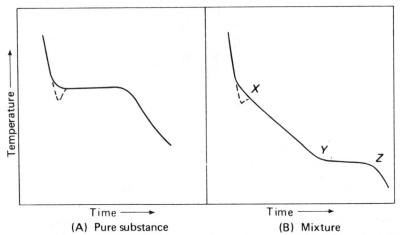

FIG. 11.3. Typical cooling curves. (Dashed portions indicate possible effects of supercooling.)

A simple apparatus for studying solid-liquid systems from room temperature to about 450° is illustrated in Fig. 11.4.[5] By replacing the brass and glass parts with appropriate ceramic material, the same cooling-curve technique can be used, in furnaces, to temperatures above 1000°.

Differential thermal analysis (DTA) is often used instead of the simple cooling-curve method.[6] The heat effects are detected by a *differential*

[5] F. Daniels, *et al.*, *Experimental Physical Chemistry*, 6th ed., McGraw-Hill Book Company, New York, 1962, pp. 116–21.

[6] See H. B. Jonassen and A. Weissberger, eds., *Technique of Inorganic Chemistry*, vol. 1, Interscience Publishers, New York, 1963, Chap. 6, pp. 209–57.

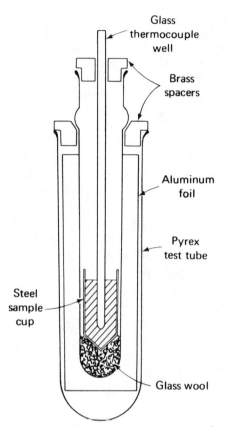

FIG. 11.4. Sample tube and cooling jacket for thermal analysis.[5] (Reproduced with permission of the McGraw-Hill Book Company.)

method. A furnace is used that contains two symmetrically located and identical chambers, each containing an identical thermocouple. The sample is placed in one chamber and an inert material, such as α alumina, is placed in the other. The furnace is heated at a uniform rate, and the temperature difference between the chambers (as measured by the difference in the electromotive force values of the thermocouples) is recorded as a function of time or of the furnace temperature. If an endothermic reaction takes place in the sample chamber, a peak will be obtained; an exothermic reaction will give a peak in the opposite direction. A typical DTA plot is shown in Fig. 11.5.

In the technique of *thermogravimetric analysis*,[7] the weight of a heated sample is followed as a function of temperature and time. The apparatus consists of a sample container supported or suspended in a vertical tube furnace, with suitable instrumentation for recording the weight of the sample container and its temperature as a function of time. Chemical reactions

[7] C. Duval, *Inorganic Thermogravimetric Analysis*, Elsevier, New York, 1953.

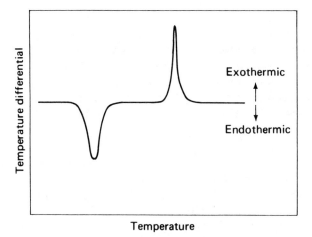

FIG. 11.5. Typical differential thermal analysis curve.

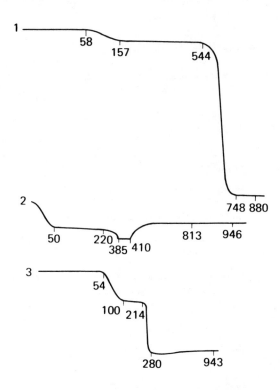

FIG. 11.6. Thermogravimetric pyrolysis curves. The numbers indicate temperatures, °C. 1. Strontium iodate. 2. Calcium sulfite. 3. Manganese(II) oxalate.

involving the evolution of volatile products can be recognized by weight loss, and those involving the absorption of gases (as oxidations) can be recognized by weight gain. Some typical pyrolysis curves are given in Fig. 11.6.[7] Curve 1 is that for freshly precipitated strontium iodate, which is a mixture of the anhydrous and monohydrated compounds. The curve shows that the product is anhydrous at 157°. At 600°, conversion to paraperiodate begins and is accompanied by the release of iodine vapor above 677°:

$$5\,Sr(IO_3)_2 \longrightarrow Sr_5(IO_6)_2 + 4\,I_2 + 9\,O_2 \tag{11.13}$$

Curve 2 is that for the pyrolysis of precipitated calcium sulfite. The curve indicates that at 50° the weight corresponds approximately to $CaSO_3 \cdot \frac{1}{2} H_2O$. This compound loses weight continuously, giving the anhydrous $CaSO_3$ between 385 and 410°. From 410 to 946°, oxygen is gradually absorbed, and at the latter temperature the weight corresponds quantitatively to $CaSO_4$. Curve 3 corresponds to precipitated manganese(II) oxalate. The dihydrate is stable to 54°; complete dehydration occurs at 100°. Beyond 214°, elimination of carbon monoxide and dioxide yields MnO, which gradually gains weight as it is converted into Mn_3O_4.

References for Further Reading

American Institute of Physics, *Temperature; Its Measurement and Control in Science and Industry*, vol. 1, Reinhold Publishing Corp., New York, 1941 (see also vol. 2, 1955).

Bautista, R. G. and J. L. Margrave, *Technique of Inorganic Chemistry*, vol. 4, H. B. Jonassen and A. Weissberger, eds., Interscience Publishers, New York, 1965, p. 65.

Brauer, G., *Handbook of Preparative Inorganic Chemistry*, vol. 1, 2nd ed., Academic Press Inc., New York, 1963, pp. 32–42.

Dike, P. H., *Thermoelectric Thermometry*, Leeds and Northrup Co., Philadelphia, 1954.

Dodd, R. E. and P. L. Robinson, *Experimental Inorganic Chemistry*, Elsevier, New York, 1954, pp. 69–72.

Duval, C., *Inorganic Thermogravimetric Analysis*, Elsevier, New York, 1953.

Lyon, R. N., ed., *Liquid Metals Handbook*, 2nd ed. (rev.), Office of Naval Research, 1954 (especially for choosing a container for a molten metal).

Electrical
Discharge
Synthesis

12

ELECTRICAL DISCHARGE REACTIONS

In a sufficiently high electrical field, a gas becomes conducting and an electrical discharge is established. The electrons that are formed may interact with the molecules by the following processes.[1]

1. Elastic impacts between electrons (e^-) and molecules[2] (AB), where there is no great energy transfer.

2. Electron additions in which negative ions are formed:

$$e^- + AB \rightarrow AB^-$$

The negative ion may dissociate as follows:

$$AB^- \rightarrow A + B^-$$

[1] F. K. McTaggart, *Plasma Chemistry in Electrical Discharges*, Elsevier, New York, 1967.
[2] The letters A and B represent molecular fragments (atoms or radicals) into which the molecule could conceivably dissociate.

215

3. Ionization impacts of the types

$$e^- \text{ (fast)} + AB \rightarrow A^+ + B^- + e^- \text{ (slow)}$$
$$e^- \text{ (fast)} + AB \rightarrow AB^+ + 2\,e^- \text{ (slow)}$$

The positive molecular ion may dissociate to give a positive ion and a radical or atom:

$$AB^+ \rightarrow A^+ + B$$

4. Excitation impacts of the type

$$e^- \text{ (fast)} + AB \rightarrow AB* + e^- \text{ (slow)}$$

Excited molecules often return to their normal state by collisional de-excitation, but they may also do so by emission of radiation,

$$AB* \rightarrow AB + h\nu$$

or they may dissociate into free radicals or atoms:

$$AB* \rightarrow A + B$$

Thus, in an electrical discharge it is possible to have chemical reactions initiated by ions, excited and metastable molecules, radicals, and atoms. In many respects the situation is similar to that existing in systems exposed to high-energy radiation (gamma rays, X rays, ultraviolet light, beta rays, etc.).

Some discharge-induced reactions yield thermodynamically stable products. These reactions seldom are of practical interest because thermodynamically stable materials generally can be made by other methods that are simpler and more economical. However, discharge reactions often yield thermodynamically unstable products, of unusual structure, that are difficult to prepare by other methods. Such reactions are of great interest to chemists who hope to discover new types of compounds. The systematics of electric discharge reactions, as applied to inorganic synthesis, has been reviewed.[3]

ELECTRODE DISCHARGES[1,4]

The classic glow discharge tube consists of a glass tube (say, 1 to 3 cm in diameter and 20 to 100 cm long) at each end of which is a metallic electrode with an electrical lead sealed through the wall of the tube. The electrodes are usually connected to the secondary coil of a step-up transformer. A typical discharge tube and circuit are pictured in Fig. 12.1. Usually, an applied potential of several kilovolts at 60 cycles is adequate to maintain a stable glow discharge in a gas at several millimeters pressure. When diatomic gases, such as H_2, O_2, N_2, and so on, are passed through a discharge tube such as that in Fig. 12.1, the gas leaving the discharge zone contains significant

[3] W. L. Jolly, in *Advan. Chem. Ser.*, **80**, 156 (1969).
[4] W. L. Jolly, *Technique of Inorganic Chemistry*, vol. 1, H. B. Jonassen and A. Weissberger, eds., Interscience Publishers, New York, 1963, Chap. 5, p. 179.

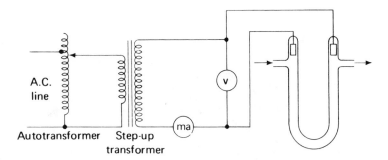

FIG. 12.1. Typical ac discharge circuit and electrode discharge tube.

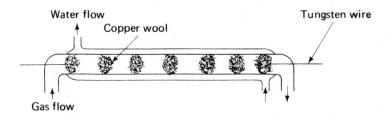

FIG. 12.2. Discharge tube used for preparing B_2Cl_4.[5] (Reproduced from ref. 4 with permission of Wiley–Interscience Publishers.)

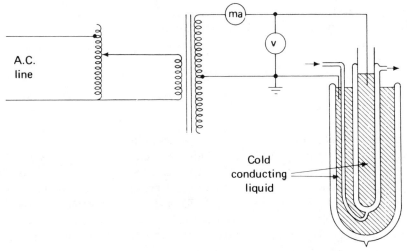

FIG. 12.3. Typical ozonizer setup.[4] (Reproduced with permission of Wiley–Interscience Publishers.)

Pyrex, Vycor, or quartz tube

Helical coil (copper tubing)

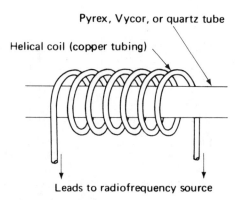

Leads to radiofrequency source

FIG. 12.4. Inductively coupled discharge tube.[4] (Reproduced with permission of Wiley–Interscience Publishers.)

concentrations of the atoms (H, O, N, etc.). By allowing this gas stream to come in contact with other compounds, we can study the chemical reactions of these atomic species.

The principal difficulty in using electrode discharges for chemical synthesis is that the electrodes usually undergo reaction with the species in the discharge zone. However, this problem can be turned into an asset by intentionally making the electrodes of a material that can react with one of the product species and thereby preventing back reaction to reform the starting materials. For example, the copper wool electrodes in the discharge tube of Fig. 12.2 serve as chlorine getters in the discharge synthesis of B_2Cl_4:[5]

$$2\,BCl_3 + 2\,Cu \longrightarrow B_2Cl_4 + 2\,CuCl$$

(In an electrodeless discharge, B_2Cl_4 and Cl_2 form, but, because of the back reaction, very poor yields of B_2Cl_4 are obtained.[6])

ELECTRODELESS DISCHARGES[1,4]

Ozonizers

If, instead of sealing electrodes through the glass walls of a discharge tube, the electrodes are simply placed on the outside wall, a discharge may still be established if alternating current is used and if the voltage is high enough. Placing the electrodes on the outside wall surfaces amounts to adding capacitors in series with the discharge circuit. Thus, it is advantageous to make the electrode areas large, to use thin glass walls, and to use a high-frequency current source. The usual discharge tube consists of two concentric glass

[5] T. Wartik, R. Rosenberg, and W. B. Fox, *Inorg. Syn.*, **10**, 118 (1967).

[6] J. W. Frazer and R. T. Holzmann, *J. Am. Chem. Soc.*, **80**, 2907 (1958).

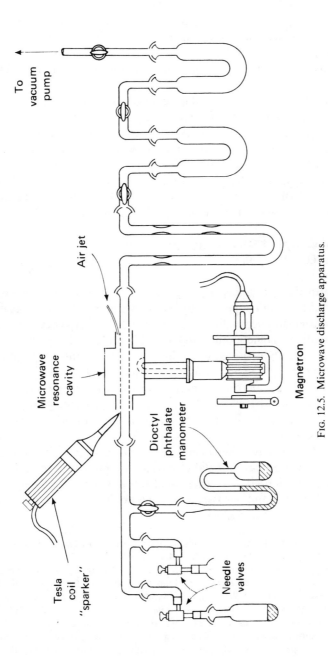

Fig. 12.5. Microwave discharge apparatus.

tubes between which the discharge is established. This type of device is called an ozonizer. A typical setup is pictured in Fig. 12.3. Applied voltages of 5 to 20 kv are usually employed. The principal application of the ozonizer, as implied by the name, is in the preparation of ozone from oxygen. Ozone concentrations of 2 to 8 per cent can be prepared in oxygen at atmospheric pressure.

The ozonizer has been very useful for converting simple volatile hydrides into more complicated hydrides; for example, germane can be converted into a mixture of higher germanes (Ge_2H_6, Ge_3H_8, etc.).[7] The ozonizer is mild in its action on molecules, and it does not cause drastic fragmentation and rearrangement. There is some evidence that, by the judicious choice of reagents, it can be used for the preparation of specific isomers.[8] Thus, when a mixture of SiH_3PH_2 and SiH_4 is passed through an ozonizer discharge, $(SiH_3)_2PH$ is formed, whereas a mixture of Si_2H_6 and PH_3 yields $Si_2H_5PH_2$.

Inductively Coupled Discharges and Microwave Discharges

A common form of electrodeless discharge tube consists of a glass tube passing through several turns of copper tubing, as shown in Fig. 12.4. The copper-tubing solenoid is supplied with power from a radio-frequency generator. Rapidly fluctuating magnetic and electric fields are established within the tube, and these maintain the glow discharge. A very similar type of glow discharge can be established by having a glass tube pass through a microwave resonance cavity.[9] The microwave energy is generated in a continuous-wave magnetron oscillator tube which is powered by a high-voltage d.c. power supply.[10] The radiation passes through a coaxial cable or other wave guide to the resonance cavity.

These discharges have the advantages that they can produce quite energetic plasmas and that no metal electrodes are present in the discharge zone. They have been used in a variety of chemical syntheses. A diagram of a microwave discharge and the associated vacuum line for preparing Ge_2Cl_6 from $GeCl_4$ is given in Fig. 12.5.[11]

[7] S. D. Gokhale, J. E. Drake, and W. L. Jolly, *J. Inorg. Nucl. Chem.*, **27**, 1911 (1965).

[8] S. D. Gokhale and W. L. Jolly, *Inorg. Chem.*, **4**, 596 (1965).

[9] The design of microwave resonance cavities is discussed by E. C. Pollard and J. M. Sturtevant, *Microwaves and Radar Electronics*, John Wiley & Sons, Inc., New York, 1948, pp. 44–54; and by M. Zelikoff, P. H. Wyckoff, L. M. Aschenbrand, and R. S. Loomis, *J. Opt. Soc. Am.*, **42**, 818 (1952). Also see McTaggart, *Plasma Chemistry in Electrical Discharges*, 1967.

[10] Commercial diathermy units are useful laboratory microwave generators. The units manufactured by Baird Associates, Inc., of Cambridge, Mass., and by Raytheon Manufacturing Co., Waltham, Mass., operate at a wavelength of 12.2 cm with a power output of 125 w. See ref. 1 for a discussion of microwave equipment.

[11] D. Shriver and W. L. Jolly, *J. Am. Chem. Soc.*, **80**, 6692 (1958).

Ion Exchange

13

ION EXCHANGERS

An ion exchanger is a porous, insoluble polymer containing immobile, electrically charged groups that bind an equivalent number of mobile, oppositely charged ions. A cation exchanger contains groups—such as sulfonate, phosphate, or carboxylate—the negative charges of which must be neutralized by an equivalent number of cations. The usual anion exchanger contains quaternary ammonium groups having positive charges that must be neutralized by an equivalent number of anions. Some commercially available ion exchangers are listed in Table 13.1.

The *capacity* of an ion exchanger is the number of available ionic sites per gram of dry exchanger—usually expressed as milliequivalents per gram (abbreviated meq g^{-1}). The exchangers in Table 13.1 vary in capacity from about 1 to 10 meq g^{-1}. Dowex 50, a strong acid type resin, has a capacity of about 5.0 meq g^{-1}; Dowex 1 and 2, strong base type resins, have capacities of about 3.5 meq g^{-1}.

Table 13.1   Some Commercially Available Ion Exchangers

Trade Name	Type
Cation Exchangers:	
Amberlite IR–120	sulfonated polystyrene
Amberlite IRC–50	carboxylic acid
Bio–Rad ZP–1	zirconium phosphate
Bio–Rad ZT–1	zirconium tungstate
Bio–Rad ZM–1	zirconium molybdate
Bio–Rad AMP–1	ammonium molybdophosphate
Dowex 50	sulfonated polystyrene
Duolite C–3, –10, –20, –25	sulfonic acid
Duolite C–62, –63	phosphonic acid
Duolite C–65	phosphoric acid
Duolite CS–101	carboxylic acid
Permutit Q	sulfonic acid
Anion Exchangers:	
Amberlite IRA–400, –401, –410	polystyrene quaternary amine
Amberlite IR–45	polystyrene polyamine
Dowex 1, 2, 3	polystyrene-base amines
Duolite A–40, –42, –101, –102	quaternary ammonium
Duolite A–30	polyalkylene amines
Duolite A–41, –43, –6, –14, –114	primary, secondary, tertiary, and/or quaternary amines

ION EXCHANGER SELECTIVITY

Usually divalent ions are more tightly held by an exchanger than monovalent ions, and trivalent ions are more tightly held than divalent ions. However, even when the ions undergoing exchange are of the same charge, the exchanger shows selectivity. For Dowex 50, the following two series of ions show qualitatively the order of decreasing affinity[1]:

$$Ba^{2+} > Sr^{2+} > Ca^{2+} > Mg^{2+} > Be^{2+}$$

$$Ag^+ > Tl^+ > Cs^+ > Rb^+ > NH_4^+ > K^+ > Na^+ > H^+ > Li^+$$

For Dowex 1,

$$SCN^- > I^- > NO_3^- > Br^- > CN^- > HSO_4^- \approx HSO_3^- > NO_2^- >$$

$$Cl^- > HCO_3^- > CH_3CO_2^- > OH^- > F^-$$

For Dowex 2, OH^- lies between Cl^- and HCO_3^-. For any two cations or any two anions, the relative selectivity may be expressed quantitatively by means

[1] *Dowex Ion Exchange*, Dow Chemical Co., Midland, Mich., 1958.

of an equilibrium concentration quotient. Thus, for the general reaction,

$$bA^{a\pm}_{(exch.)} + aB^{b\pm}_{(soln.)} \rightleftharpoons bA^{a\pm}_{(soln.)} + aB^{b\pm}_{(exch.)}$$

$$Q_A^B = \frac{(A_S^{a\pm})^b}{(A_E^{a\pm})^b} \cdot \frac{(B_E^{b\pm})^a}{(B_S^{b\pm})^a}$$

where $(A_S^{a\pm})$ and $(B_S^{b\pm})$ are the concentrations in solution, and where $(A_E^{a\pm})$ and $(B_E^{b\pm})$ are the amounts or concentrations in the exchanger. The aqueous concentrations are expressed as molarity and the exchanger concentrations are expressed as fractions of the exchanger capacity or as milliequivalents per unit weight of exchanger. For ions of like charge, the concentration units are immaterial. Although aqueous activity coefficients are sometimes available, the aqueous concentrations are seldom corrected to activities because the corresponding activity coefficient data are unavailable for the ions in the exchanger. Thus, even an equilibrium quotient in which activities are used for the aqueous ions $(K_A^B = Q_A^B \cdot \gamma_{A_S}^b / \gamma_{B_S}^a)$ is not constant with varying exchanger composition. In Fig. 13.1, the logarithm of K_H^{Na} has been plotted vs. the mole

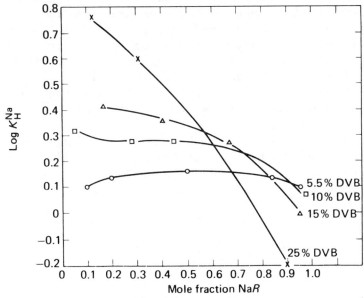

FIG. 13.1. Variation of K_H^{Na} with mole fraction of sodium in exchanger.[2] (Reproduced with permission of John Wiley & Sons, Inc.)

fraction of sodium in the exchanger—$(Na_E^+)/[(Na_E^+ + (H_E^+)]$,—for sulfonated polystyrene resins of different degrees of cross linking.[2] Here, K_H^{Na} refers to the reaction,

$$Na^+ + H^+_{(exch.)} = Na^+_{(exch.)} + H^+$$

[2] R. Kunin, *Ion Exchange Resins*, 2nd ed., John Wiley & Sons, Inc., New York, 1958, p. 24.

BATCH OPERATION VS. COLUMN
OPERATION

Ion exchangers have two principal functions in synthetic chemistry: the *replacement* of one ion by another, and the *separation* of different kinds of ions from one another. Let us consider the possibility of applying batch operation to each of these functions.

As an example of the first function, let us suppose that we have an aqueous solution of sodium chloride that we wish to convert quantitatively to hydrochloric acid. We might dump into the solution a strongly acidic cation exchanger in the hydrogen form in order to carry out the reaction previously shown, but unless either an enormous excess of exchanger were used or the quotient Q_H^{Na} were extremely high, appreciable amounts of sodium ion would remain in the solution at equilibrium. Thus, batch operation is seldom employed in ionic replacement processes.

As an example of the second use, let us suppose that we have a solution containing both sodium ions and calcium ions, which are to be separated. We might add a hydrogen-form cation exchanger to the solution in the hope that the calcium ions would be preferentially exchanged. Although a higher proportion of the calcium ions would probably be exchanged, there would always be a finite concentration of aqueous calcium ions in equilibrium with the exchanger. If a sufficiently large excess of the exchanger were employed to reduce this concentration to a negligible value, a considerable fraction of the sodium ions would also undergo exchange. Thus, no clean-cut separation would be possible.

There are a few isolated examples in which ion-exchange reactions are carried to completion by the formation of insoluble precipitates or weakly ionized species. In such cases, batch operation is successful. For example, a hydrogen-form cation exchanger will react quantitatively with an NaOH solution,

$$H_E^+ + Na^+ + OH^- \longrightarrow Na_E^+ + H_2O$$

and a silver-form exchanger will desalt seawater

$$Ag_E^+ + Na^+ + Cl^- \longrightarrow Na_E^+ + AgCl$$

However, in most situations, ion exchangers are used in columns, and the remainder of this chapter will describe column operation.

Column operation may be looked upon as a series of batch operations; the column acts as a long series of solution-exchanger equilibration stages, or "theoretical plates." Thus, the process is analogous to fractional distillation, countercurrent extraction, and other processes that involve repetitive equilibria (the reader is referred to Appendix 8 for a discussion of the theory of theoretical plates).

COLUMN PREPARATION

A buret can serve very adequately as the supporting tube for an ion exchange column (see Fig. 13.2). Columns of greater capacity may be readily constructed; two types are shown in Fig. 13.3. A plug of glass wool is generally placed in the bottom of the tube, the tube is filled with water, and the stopcock or screw clamp is adjusted so that the water level falls at the rate of about 0.5 cm sec^{-1}. A water slurry of the ion exchanger is added at a rate so as to maintain the water level near the top of the tube. In this way, a very evenly packed column of ion exchanger particles will be built up. It is very important that the liquid level is never allowed to fall below the top of the ion exchanger

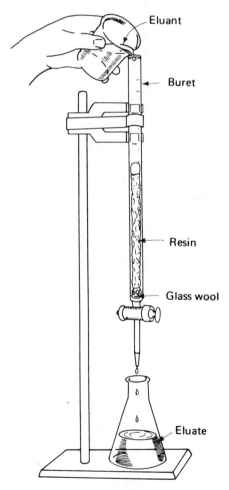

Eluant

Buret

Resin

Glass wool

Eluate

FIG. 13.2. An ion-exchange column made from a buret.

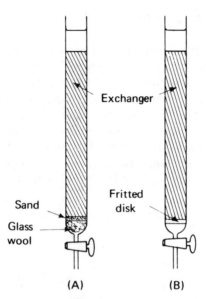

Fig. 13.3. Other types of ion-exchange columns.

bed; otherwise, channeling will occur and it will be necessary to remake the column. When a sufficiently high column of ion exchanger has been deposited, the water flow is stopped.

If the ion exchanger had previously been converted quantitatively to the ionic form in which it is to be used (as by thoroughly washing the exchanger with an appropriate aqueous solution), the column is now ready for use. However, if there is any doubt about the ions in the exchanger, the exchanger should now be washed with an appropriate reagent. Thus, if one has a cation exchanger that is to be in the hydrogen form, several column volumes of 3 M HCl are passed through the column, followed by sufficient distilled water to remove completely all the free chloride ions from the column. If one has an anion exchanger that is to be completely in the hydroxide form, a similar treatment is carried out using a concentrated solution of sodium hydroxide.

In certain types of column operation (such as displacement chromatography, to be discussed later), it is important that both the top and bottom of the exchanger column be flat. A flat top is relatively easily maintained as long as solutions are carefully added to the top of the column so as not to disturb the top layer of exchanger particles. A flat bottom may be achieved by placing a flat-topped layer of clean sand immediately over the glass wool plug, as shown in Fig. 13.3A. The exchanger column is then built up on the sand base. A flat bottom is achieved most elegantly with a fritted-glass disk (Fig. 13.3B).

REPLACEMENT OF ONE ION BY ANOTHER

Suppose that one wishes to prepare a solution of tetraethylammonium hydroxide from a solution of tetraethylammonium fluoride. One prepares a column containing about twice as many equivalents of an anion exchanger in the hydroxide form as there are equivalents of fluoride in the solution. The solution of tetraethylammonium fluoride is allowed to pass slowly through the column. The effluent, when combined with the solution still held in the column (which may be rinsed out with distilled water), contains a quantitative yield of fluoride-free tetraethylammonium hydroxide.

Figure 13.4 shows how the composition of the ion exchange column varies with the depth of the column at the end of the process. Whenever the displacing ion is held more strongly than the ion originally on the column, a self-sharpening boundary is obtained. When the displacing ion is held less strongly than the ion originally on the column, a diffuse boundary is obtained, as shown in Fig. 13.4. When quantitative exchange is essential, an adequate excess of ion exchanger is required to prevent premature breakthrough of the displacing ion.

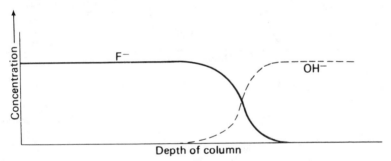

FIG. 13.4. Composition of hydroxide-form anion exchange column through which a fluoride solution has been passed.

SEPARATION OF IONS

In ion exchange chromatography, a mixture of ions is absorbed at the top of a column and these ions are then successively desorbed. Desorption is accomplished in either of two ways: by displacement chromatography or by elution chromatography.

Displacement Chromatography

The solution containing the mixture of ions to be separated is first added to a column of exchanger in which the exchangeable ions are less strongly held than any of the ions in the mixture. If a sufficient excess of exchanger is used, the mixture of ions will be absorbed in a sharp band at the top 5 to 10 per cent of the column. Next, a solution containing ions that are more strongly held than any of the ions in the mixture is allowed to pass down the column. As the ions in the mixture are displaced down the column, they accumulate in sharply defined touching bands, which appear in the order of increasing exchanger affinity.

Consider the following separation of a mixture of sodium and calcium chlorides. The cations are absorbed at the top of a column of cation exchanger in the hydrogen form. They are displaced by a solution of aluminum chloride. The effluent first consists of pure water (corresponding to the column volume beneath the initially absorbed ions), followed successively by hydrochloric acid, sodium chloride, calcium chloride, and finally aluminum chloride. A plot of solution composition vs. effluent volume is given in Fig. 13.5A.

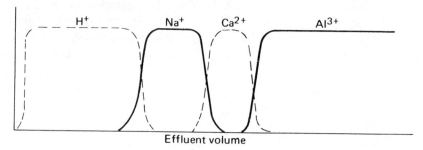

Effluent volume

(A) Normality of effluent solution

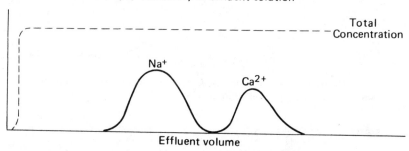

Effluent volume

(B) Normality of effluent solution

Fig. 13.5. Composition vs. volume for effluent. (A) Displacement chromatography, (B) Elution chromatography.

Elution Chromatography

Elution chromatography is very similar to displacement chromatography, except that the displacing ion (now called the eluting ion) is held less strongly than any ion in the mixture. The eluting ion may be the same as the exchangeable ion originally on the column. In the process of elution, the eluted ions accumulate in bands which, although they eventually are well separated from one another, have more or less diffuse leading and trailing edges. The eluate concentration of any ion in the mixture is always less than that of the initial eluting agent. If the eluting ions are held very weakly by the exchanger, large volumes of eluant solution are required.

A mixture of sodium and calcium ions may be separated by elution from a cation exchange column with hydrochloric acid. A plot of eluate composition vs. eluate volume for this separation is given in Fig. 13.5B.

A common method of separating cations on a column is to use an eluant containing some substance that will complex the cations. The various cations will form complexes of various degrees of stability and will appear in the eluate in the order of decreasing complex stability. Rare earths, for example, may be fractionally eluted from a cation exchange resin with a solution of tartrate or citrate.

Problems

For each pair of compounds listed, give both a classical chemical method and an ion exchange method for quantitatively converting one compound into the other (and vice versa).

1. KBr and KOH

2. NH_4NO_3 and HNO_3

3. $(NH_3OH)HSO_4$ and $(NH_3OH)Cl$

References

General references

Dowex Ion Exchange, Dow Chemical Co., Midland, Mich., 1958.

Hiester, N. K. and R. C. Phillips, *Chem. Eng.*, **61**, 161 (1954).

Ion Exchange and Its Applications, Society of Chemical Industry, 1955.

Kunin, R., "Ion Exchange Techniques," in *Technique of Inorganic Chemistry*, vol. 4, H. B. Jonassen and A. Weissberger, eds., Interscience Publishers, 1965.

Salmon, J. E. and D. K. Hale, *Ion Exchange*, Academic Press Inc., New York, 1959.

Samuelson, O., *Ion Exchangers in Analytical Chemistry*, John Wiley & Sons, Inc., New York, 1953.

Specific references

The Separation of Chromium(III) Thiocyanate Cationic Complexes. E. L. King and E. B. Dismukes, *J. Am. Chem. Soc.*, **74**, 1674 (1952).

The Separation of *Cis-* and *Trans*-Isomers of Complex Ions. E. L. King and R. R. Walters, *J. Am. Chem. Soc.*, **74**, 4471 (1952). J. T. Hougen, K. Schug, and E. L. King, *J. Am. Chem. Soc.*, **79**, 519 (1957).

Preparation of Polyphosphoric Acids and Their Tetramethylammonium Salts, J. Van Wazer, E. Griffith, and J. McCullough, *J. Am. Chem. Soc.*, **77**, 287 (1955).

The Preparation of Tetra-, Penta-, Hexa-, Hepta- and Octaphosphates. E. J. Griffith and R. L. Buxton, *J. Am. Chem. Soc.*, **89**, 2884 (1967).

Preparation of Monosilicic Acid. G. B. Alexander, *J. Am. Chem. Soc.*, **75**, 2887 (1953).

Preparation of Heteropolyacids. L. C. W. Baker, B. Loev, and T. P. McCutcheon, *J. Am. Chem. Soc.*, **72**, 2374 (1950).

Preparation of Complex Acids of Cobalt and Chromium. T. P. McCutcheon and W. J. Schuele, *J. Am. Chem. Soc.*, **75**, 1845 (1953).

Preparation of Monosulfatopentaaquochromium(III) Chloride. J. E. Finholt, R. W. Anderson, J. A. Fyfe, and K. G. Caulton, *Inorg. Chem.*, **4**, 43 (1965).

Preparation of Ligandpentaaquochromium(III) Sulfate Monohydrate Compounds. P. Moore and F. Basolo, *Inorg. Chem.*, **4**, 1670 (1965).

High-pressure
Apparatus

14

An autoclave (or high-pressure "bomb") is used for carrying out reactions with gases under high pressure and for handling solutions at temperatures above their boiling points. A common type of laboratory bomb is pictured in Fig. 14.1. When this type of bomb is used, moderate agitation of the reaction mixture can be accomplished by rocking the bomb to and fro. If vigorous stirring is required, special bombs fitted with stirring shafts and propellers can be used. In some bombs of this type, the stirrer shaft passes through a packing gland and is connected directly to a pulley, which in turn is connected by a belt to a motor. In others, the stirrer shaft is completely enclosed and is magnetically coupled to an outer pulley.

The apparatus shown in Fig. 14.2 is similar to that used in most high-pressure laboratories for charging a rocking type bomb with gas, for heating the bomb, and for rocking it. In the following paragraphs we give directions for operating the apparatus. These directions can easily be adapted to modifications of this apparatus.

231

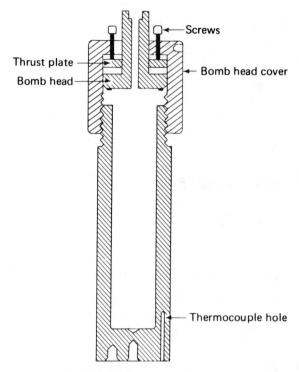

FIG. 14.1. High-pressure bomb.

INSTRUCTIONS FOR CHARGING THE BOMB WITH GAS

The bomb is always stored coated with oil in order to prevent rusting. Therefore immediately before use, the bomb is rinsed with several small portions of benzene and once with acetone; it is finally dried with a stream of air.

The clean bomb is set into a bomb vise. When placing the bomb in the vise, the position of the three holes in its bottom should be noted. The large center hole in the bottom of the bomb is for a holding screw, and the two small holes on either side are for a thermocouple and a heating jacket insert. The hole for the thermocouple is the smaller of the two.

The reactants are placed in the bomb. (The total volume of reactants, including solvent, should never exceed one-half the total capacity of the bomb.) The bomb head is placed on the top of the bomb and is moved about gradually until the gasket slides into the recess provided for it. The thrust plate is then placed on the bomb head so that the side which has been previously exposed to the screws is on top. The bomb-head cover is screwed down gently until it almost touches the thrust plate and is then reversed

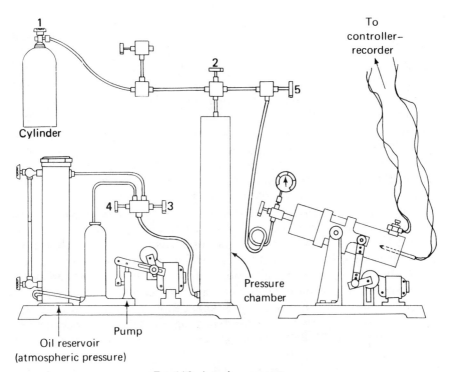

1

To
controller–
recorder

Cylinder

2

5

4   3

Pressure
chamber

Oil reservoir
(atmospheric pressure)

Pump

Fig. 14.2. Autoclave apparatus.

one-half turn. (It is important that all the screws are slightly recessed into the bomb head cover before the cover is placed on the bomb.) The head should not turn as the cover is screwed down. The screws are then turned down, with a small hex wrench, until they just reach the thrust plate; they are tightened with a torsion wrench. (Generally, a torque of 20 ft-lb should be used with a copper gasket, and a torque of 40 ft-lb should be used with a stainless steel gasket.) A torsion wrench should be grasped at both ends so that a firm fit is had. The wrench should always be kept parallel to the table top in order not to strip the screw or the fitting of the wrench. The screws should be gradually tightened according to the following pattern:

```
                3
          6         8
     1                   5
          4         2
                7
```

The tightening cycle is repeated until each of the screws has been tightened to the recommended torque, which should not be exceeded. During this entire operation, caution should be exercised to prevent movement of the bomb head.

The pressure gauge is now attached to the bomb[1] by holding the gauge in one hand and screwing the pressure fitting with the other. A wrench should not be employed in this operation. The gauge and the bomb are now carefully lifted from the vise, holding the gauge in one hand so it will not rotate freely. Do *not* support one end of the bomb by the gauge. The bomb is placed into the heating jacket. When properly fitted into the jacket, the top of the bomb head cover should be flush with the end of the heating jacket. The back of the gauge is loosely attached by means of a thumb screw to a metal arm extending from the jacket. Now, a small stiff metal rod, the size of the thermocouple, is inserted into the thermocouple hole in the bottom of the heating jacket to line up the bomb with the thermocouple hole. Next, the large screw is placed in the center hole in the rear of the heating jacket and it is tightened with a wrench, using gentle pressure, to hold the bomb in the heating jacket. The gauge is then tightened to the bomb, most safely done by placing one wrench on the hex nut entering the bomb and one wrench (fixed wrench, *not* an adjustable one, such as a crescent wrench) on the hex nut entering the gauge. By gently applying pressure in opposite directions, one prevents excessive strain on the gauge connection. All fittings are of the high-pressure type and function by a good fit rather than by the application of excessive pressure. All that excessive pressure will do is spread the precision fitting and render it more susceptible to leaks. Finally, the thumb screw on the rear side of the gauge is tightened.

The metal rod is removed from the thermocouple hole and the thermocouple carefully inserted. It should be placed so that it just touches the end of the well, and then the end holder is firmly screwed in place.

Gas Addition[2]

The coiled high-pressure tubing is put in place by screwing one end into needle valve 5 (see Fig. 14.2) and the other end into the bottom of the gauge valve. The needle valve on the gauge and needle valve 5 are then closed. A needle valve should only be turned gently, and excessive pressure should never be applied.

Use of the Booster

Before any gas is added, it should be ascertained that the oil in the glass side arm is approximately at the upper etched line. If not, valve 3 is carefully

---

[1] Usually high pressure is measured with a Bourdon gauge, which has a pointer and a calibrated dial. However, it is also possible to use an electrical strain gauge or transistor type pressure gauge. With such a device, the pressure can be measured or recorded in a safe location.

[2] In the following discussion, it has been assumed that the gas cylinder pressure is lower than the desired starting pressure and that it will be necessary to use the booster pump. It has also been assumed that the compression cylinder has been previously flushed with the gas from the gas cylinder. When boosting is not required, the same directions apply, except that it is then never necessary to open valve 2 or to operate the pump.

opened and the oil is allowed to return slowly from the pressure chamber. It is imperative that this operation be done slowly, because the presence of gas bubbles in the oil will stop the booster from working and the system will need to be primed. Valve 3 is then closed.

With all the other valves closed, valve 1 (the valve at the top of the gas cylinder) is opened. Next, valve 2 (on top of the compression cylinder) is slowly opened, which fills the compression cylinder to tank pressure. Valve 5 and the gauge valve are then opened to allow the gas to fill the bomb. At this point, the bomb may be purged of air by closing valve 5 and loosening the connection between the coiled high-pressure tubing and valve 5. Retighten the connection and reopen valve 5.

Valve 1 is closed so that the booster and bomb system is isolated from the gas cylinder. *Next,* valve 4 is opened and *then* the pump is turned on. *If the motor is turned on before valve 4 is open,* it will *break the shaft* of the hydraulic pump, because it will be pumping against a closed system. The compressing of the gas is allowed to continue until the desired pressure is reached or until the oil level in the glass side arm has reached the lower etched line. The motor should be shut off at this time, for excessive pumping at this point will fill the hydraulic system with air and the pump will not function. Valve 4 is closed.

If after one boosting the pressure still is not sufficiently high, the process must be repeated. Valve 5 is closed. Valve 3 is carefully opened, the oil is returned to the atmospheric reservoir, and then the valve is closed. Next, valve 1 is opened, the chamber is filled with hydrogen from the tank, and then valve 1 is closed. *Valve 4 is then opened* and the pump is turned on. After a short while, valve 5 is *very slowly* opened and the pressure increase carefully watched. When the desired pressure is reached, first turn off the valve in the gauge. Immediately turn off the motor of the booster, and then shut valves 4, 5, and 2. Next, return the oil to the atmospheric chamber through valve 3 and then close valve 3. Disconnect the coiled high-pressure tubing so that the rocking mechanism will be able to work.

Use of the
Temperature Controller

After these operations have been completed, the reactants are ready to be heated. The rocking motor and the recorder-controller are turned on. The pointer on the recorder-controller is set at the desired temperature, and the switch on the heating jacket is turned to HIGH. When a temperature about 30° below the desired temperature has been reached, it may be necessary to turn the switch to MEDIUM or LOW to obtain the smallest fluctuation in temperature. The apparatus should be watched until the temperature controller is working properly.

Opening Bomb

The bomb is allowed to cool to room temperature (about two hours), the gas is vented (in a hood if poisonous), and the bomb is opened by reversing the procedure for closing. The bomb and head are lightly but thoroughly oiled after the product has been removed and the bomb has been cleaned.

General Precautions

Do not smoke when working with inflammable gases such as H_2 or CO. Keep the autoclave room well ventilated, and vent poisonous gases such as CO in a hood.

If anything gives way, it is likely to be the gauge. Do not put your face near it; especially do not put your face *behind* the gauge, for blowout would most likely be to the rear.

Do not carry out a large-scale reaction that has not been tried on a small scale. (There may be a violent reaction in which the temperature and pressure increase uncontrollably.)

If anything is noticeably wrong with the equipment, it should be considered unsafe and unusable. Especially, if the pressure screws in the head closure cannot be turned by hand, they should either be replaced or the holes rethreaded.

INSTRUCTIONS FOR SEALED-TUBE REACTIONS

Occasionally it is desired to heat a solution above its boiling point when it would be undesirable for the reactants or products to come in contact with the metal parts of the bomb. In such cases, the solution is sealed in a glass ampoule, the ampoule is carefully packed into the bomb with the aid of glass wool, and about 50 ml of the pure solvent is placed in the bomb. The bomb head is attached as previously described, but instead of the gauge, a closed plug is attached to the top of the bomb. The bomb may now be heated without having the glass ampoule burst. After cooling the bomb to room temperature, the ampoule may be removed and opened for recovery of the products.

References

Adkins, H., *Reactions of Hydrogen*, University of Wisconsin Press, Madison, Wisc., 1937, Chap. 3.

Komarewsky, V. I., C. H. Riesy, and F. L. Morritz, in *Technique of Organic Chemistry*, vol. 2, 2nd ed., A. Weissberger, ed., Interscience Publishers, New York, 1956, pp. 1–255.

Wiberg, K. B., *Laboratory Technique in Organic Chemistry*, McGraw-Hill Book Company, New York, 1960, pp. 231–36.

Photochemical
Synthesis

15

PHOTOCHEMICAL PRINCIPLES

Sometimes a chemical reaction will proceed only when the reactants are illuminated. In such a reaction, the rate-determining process is the transfer of the light energy to the reactants. The initial effect of the energy transfer is usually to excite certain of the reactant molecules to a higher energy level of greater reactivity or to dissociate them into reactive fragments. Obviously, the photons must possess sufficient energy to carry out the desired process. Therefore, it is important to recognize the relationship between photon energy and wavelength, as expressed in the following equation:

$$\text{kcal mole}^{-1} = 2.86 \times 10^5/\text{Å}$$

Table 15.1 gives the energies (in kcal mole^{-1}) corresponding to several wavelengths (in Ångstrom units). Obviously, a reaction in which the initial step is a dissociation requiring 100 kcal could not be effected by irradiation with visible light; ultraviolet light of wavelength shorter than 3000 Å would be required.

238

Table 15.1   Relationship Between
Wavelength and Energy

Å	$kcal\,mole^{-1}$
2000	143
3000	96
4000	72
5000	57
7000	41
10,000	29

For light to be effective, it must be absorbed by the reacting species (or by some species that can transfer the energy to the reacting species). Many reactions of fluorine (e.g., its reaction with xenon to form XeF_2) require the initial dissociation of fluorine molecules into atomic fluorine. The dissociation energy of F_2 is 36.7 kcal, corresponding to 7800 Å. However, fluorine has no strong absorption bands in the region of 7800 Å, and thus the photochemical synthesis of XeF_2 cannot proceed in red light. However, fluorine does have an intense ultraviolet absorption band, the tail of which extends into the blue region of the visible spectrum, and XeF_2 can be made by irradiating $Xe + F_2$ with unfiltered sunlight or ultraviolet light.[1]

Many photochemical reactions that proceed at a negligible rate because of the absence of absorption bands in the reactant molecules can be sensitized (catalyzed) by the presence of mercury vapor (particularly if a mercury vapor lamp is used as a light source). Thus, hydrogen, with a dissociation energy of $104\,kcal\,mole^{-1}$, is not dissociated by light in the wavelength range from 2000 to 3000 Å because it is transparent in this spectral region. However, in the presence of mercury vapor, hydrogen molecules can be dissociated by impact with mercury atoms that have absorbed 2537 Å photons [corresponding to the excitation $Hg(^1S_0) \rightarrow Hg(^3P_1)$].

LIGHT SOURCES

Mercury Low-pressure Lamps.   Mercury low-pressure lamps are electrical discharges through mercury vapor and several millimeters pressure of a rare gas, such as neon or argon. Most of the radiation in these lamps is concentrated at the wavelengths 1849 and 2537 Å, the so-called resonance lines. The lamps are useful when very short wavelength light is required and for mercury-sensitized reactions.

Mercury Medium-pressure Lamps.   These lamps are electrical discharges through mercury vapor and 1 atm or more pressure of a rare gas. The emission spectrum consists of many lines of high intensity, with about 30 per

[1] S. M. Williamson, *Inorg. Syn.*, **11**, 147 (1968).

cent of the total energy in the ultraviolet portion of the spectrum, 18 per cent in the visible, and the remainder in the infrared. A typical lamp of this type, with accessory equipment, is pictured in Figs. 32.10 and 32.15.

Great care must be taken when working with mercury vapor lamps. It is imperative that the apparatus be completely shielded so that none of the light could possibly fall on anybody's eyes. Even a fleeting glance at an operating mercury vapor lamp can cause irreparable damage. Albumins in the eye may be coagulated, and the effect is cumulative and irreversible. The skin can be badly burned by short exposures to ultraviolet radiation; the effects usually are not noticed until several hours after exposure. Ozone, a very poisonous gas, is formed in the air around ultraviolet lamps; consequently, the apparatus should be set up in a hood. Care should be taken to avoid touching the high-voltage leads of the lamp.

Tungsten Lamps. A 200- to 1000-w tungsten-filament lamp is often satisfactory for syntheses effected by visible or near ultraviolet light. Photoflood lamps are very useful. Such lamps emit large amounts of infrared radiation, and the reaction systems generally require cooling.

Sunlight. Few laboratory sources can match the intensity of bright summer sunlight at wavelengths above 3500 Å. Many photochemical reactions can be carried out by simply allowing the reaction vessel to stand in sunlight.

EXAMPLES OF PHOTOSYNTHESIS

Carbonyls and Carbonyl Derivatives

Iron pentacarbonyl may be readily converted to diiron enneacarbonyl by exposure of a solution of the pentacarbonyl in acetic acid to sunlight[2] or an ultraviolet lamp[3]:

$$2\,Fe(CO)_5 \longrightarrow Fe_2(CO)_9 + CO$$

Apparently the irradiation causes a labilization of the CO groups in $Fe(CO)_5$, because ultraviolet irradiation of mixtures of $Fe(CO)_5$ with nitrogen bases (L) in a nonpolar solvent yields complexes of the type $Fe(CO)_4L$.[4] Similarly, when solutions of the metal hexacarbonyls $Cr(CO)_6$, $Mo(CO)_6$, and $W(CO)_6$ in acetonitrile are exposed to ultraviolet radiation, species of the type $M(CO)_5CH_3CN$, $M(CO)_4(CH_3CN)_2$, and, in the cases of Mo and W, $M(CO)_3(CH_3CN)_3$ are formed.[5] By irradiation of the same hexacarbonyls in

---

[2] E. Speyer and H. Wolf, *Ber.*, **60**, 1424 (1927).

[3] E. H. Braye and W. Hübel, *Inorg. Syn.*, **8**, 178 (1966).

[4] E. H. Schubert and R. K. Sheline, *Inorg. Chem.*, **5**, 1071 (1966).

[5] G. R. Dobson, M. F. Amr El Sayed, I. W. Stolz, and R. K. Sheline, *Inorg. Chem.*, **1**, 526 (1962); W. Strohmeier and G. Schonauer, *Ber.*, **94**, 1346 (1961).

the presence of either $SnCl_3^-$ or $GeCl_3^-$ ions, complexes containing metal–metal bonds have been obtained.[6] For example,

$$\left[(C_6H_5)_4As\right]\left[GeCl_3\right] + Mo(CO)_6 \xrightarrow[CH_2Cl_2]{hv} \left[(C_6H_5)_4As\right]\left[Mo(CO)_5GeCl_3\right] + CO$$

Boron Compounds

Some interesting derivatives of the benzenelike molecule borazine $(B_3N_3H_6)$ have been prepared by the vapor-phase irradiation of mixtures of borazine with oxygen, water, and ammonia.[7] B-hydroxyborazine $(B_3N_3H_6O)$ was formed in the reaction with oxygen, diborazinyl ether $(B_6N_6H_{10}O)$ was formed in the reactions with oxygen and water, and B-aminoborazine $(B_3N_4H_7)$ was formed in the reaction with ammonia.

The photochemical reaction of boron trichloride with oxygen yields trichloroboroxine $(B_3O_3Cl_3)$.[8] The following mechanism was tentatively suggested:

$$BCl_3 \xrightarrow{hv} BCl + Cl_2$$

$$BCl + O_2 \longrightarrow BClO_2$$

$$BClO_2 + BCl \longrightarrow 2\,BOCl$$

$$3\,BOCl \longrightarrow (BOCl)_3$$

The polyhalogenated clovoborane anions, $B_{10}Cl_{10}^{2-}$ and $B_{12}Cl_{12}^{2-}$, are inert toward nucleophiles. However, substitution reactions are readily effected by irradiation of aqueous solutions of the ions in the presence of nucleophiles.[9] For example, irradiation of $B_{12}Cl_{12}^{2-}$ in the presence of cyanide ion gives $B_{12}Cl_5(CN)_7^{2-}$, and similar treatment of $B_{10}Cl_{10}^{2-}$ gives a mixture of $B_{10}Cl_9CN^{2-}$, $B_{10}Cl_8(CN)_2^{2-}$, and $B_{10}Cl_7(CN)_3^{2-}$. A wide variety of such nucleophilic substitution reactions has been studied.[9]

Mercury-Sensitized Photosyntheses

The effect of mercury vapor on the photochemical reaction of silane with ethylene can be seen in the data of White and Rochow,[10] presented in Table 15.2. It was assumed that the primary step was $SiH_4 + Hg(^3P_1) \rightarrow SiH_3 + H + Hg(^1S_0)$, and that the products were formed by reactions such as the following:

$$SiH_3 + C_2H_4 \longrightarrow C_2H_4SiH_3$$

$$C_2H_4SiH_3 + SiH_4 \longrightarrow C_2H_5SiH_3 + SiH_3$$

[6] J. K. Ruff, *Inorg. Chem.*, **6**, 1502 (1967).

[7] G. H. Lee and R. F. Porter, *Inorg. Chem.*, **6**, 648 (1967).

[8] D. J. Knowles and A. S. Buchanan, *Inorg. Chem.*, **4**, 1799 (1965).

[9] S. Trofimenko, *J. Am. Chem. Soc.*, **88**, 1899 (1966).

[10] D. G. White and E. G. Rochow, *J. Am. Chem. Soc.*, **76**, 3897 (1954).

**Table 15.2   Effect of Mercury Vapor on Reaction of Silane
with Ethylene (Mercury Vapor Lamp Irradiation)**

	Yields in Millimoles	
Product	No Hg Added	Hg Added
$H_3SiC_4H_8SiH_3$	0.09	0.47
$n\text{-}C_4H_9SiH_3$	0.15	0.71
$EtSiH_3 + Si_2H_6$	0.14	0.47

It has been observed that, in the mercury-sensitized photolyses of organosilanes, the initial step was always loss of an H atom from an Si–H bond, if available; otherwise, C–H scission occurred.[11] Thus, the major products of the photolysis of dimethylsilane were 2,3-dimethyl-2,3-disilabutane, $(Me_2SiH)_2$, and hydrogen. The only products found in the photolysis of tetramethylsilane were 2,2,5,5-tetramethyl-2,5-disilahexane, $(Me_3SiCH_2)_2$, and hydrogen. In an extension of this work to alkylchlorosilanes, it was shown that 1,1,2,2-tetrachloro-1,2-dimethyldisilane could be obtained in high purity from the mercury-sensitized photolysis of methyldichlorosilane[12] (the usual, nonphotochemical, synthetic method for this compound yields an impure, difficult-to-purify product).

Various volatile hydride derivatives have been prepared by the mercury-sensitized photolyses of mixtures of simple compounds.[13] In Table 15.3, the products identified in these photolyses are listed.

**Table 15.3   Photosynthesis of Volatile Hydride
Derivatives**

Reactants	Products
$CH_3I + SiH_4$	CH_3SiH_3, Si_2H_6
$GeH_4 + SiH_4$	GeH_3SiH_3, Ge_2H_6, Si_2H_6
$CH_3OH + GeH_4$	CH_3OGeH_3, Ge_2H_6, $(GeH_3)_2O$

Oxygen atoms (^3P) may be prepared by the mercury-sensitized photolysis of nitrous oxide:

$$N_2O + Hg(^3P_1) \longrightarrow N_2 + O + Hg(^1S_0)$$

Thus, by irradiation of mixtures of nitrous oxide, mercury vapor, and various compounds, it is possible to study the effect of atomic oxygen on the compounds. In this way, the reaction of oxygen atoms with tetrafluoroethylene

---

[11] M. A. Nay, G. N. C. Woodall, O. P. Strausz, and H. E. Gunning, *J. Am. Chem. Soc.*, **87**, 179 (1965).

[12] D. Reedy and G. Urry, *Inorg. Chem.*, **6**, 2117 (1967).

[13] G. A. Gibbon, Y. Rousseau, C. H. Van Dyke, and G. J. Mains, *Inorg. Chem.*, **5**, 114 (1966).

has been shown to yield CF_2O and cyclo-C_3F_6.[14] These products were believed to form by the mechanism

$$O + C_2F_4 \longrightarrow CF_2O + CF_2$$
$$CF_2 + C_2F_4 \longrightarrow C_3F_6$$

Coordination Compounds

Although extensive application has been made of the photochemical synthesis of organometallic compounds,[15] and although the mechanisms and products of photochemical reactions of metal complexes have been studied,[16] essentially no practical photosyntheses of Werner complexes are known. Bauer and Basolo[17] have made a start in this direction, however. They have reported a useful photochemical route for the preparation of mixed diacidobis(ethylenediamine) complexes of the type *trans*-$[M(en)_2XY]^+$ [M = Rh(III) or Ir(III)]. Their general method is given by the reaction scheme

$$trans\text{-}[M(en)_2X_2]^+ \xrightarrow[H_2O]{hv} trans\text{-}[M(en)_2(H_2O)X]^{2+} + X^-$$

$$\downarrow Ag^+$$

$$trans\text{-}[M(en)_2XY]^+ \xleftarrow{Y^-} trans\text{-}[M(en)_2(H_2O)X]^{2+} + AgX\downarrow$$

Products of this type are difficult to prepare by other methods because of the differing *trans*-labilizing effects of the halides. Thus, the reaction of *trans*-$[M(en)_2Cl_2]^+$ with bromide or iodide yields the disubstituted complex exclusively.

Salts of the *nitro*pentaamminecobalt(III) ion, $Co(NH_3)_5NO_2{}^{2+}$, are converted to the corresponding salts of the *nitrito*pentaamminecobalt(III) ion, $Co(NH_3)_5ONO^{2+}$, by exposure to sunlight[18] or ultraviolet light.[19] This photoisomerization has been studied only in the solid state; if future studies show that it can take place in aqueous solutions, it will constitute a useful synthetic method.

It has been reported that, in acidic aqueous solutions, *cis*-$Co(en)_2(H_2O)_2{}^{3+}$ is converted to *trans*-$Co(en)_2(H_2O)_2{}^{3+}$ by strong illumination.[20] Possibly photoisomerizations of this type are common and can be developed into a general synthetic process.

[14] D. Saunders and J. Heicklen, *J. Am. Chem. Soc.*, **87**, 2088 (1965).

[15] E. O. Fischer, J. P. Kögler, and P. Kuzel, *Chem. Ber.*, **93**, 3006 (1960); W. Strohmeier and K. Gerlach, *Z. Naturforsch.*, **15b**, 413 (1960); R. S. Nyholm, S. S. Sandhu, and M. H. B. Stiddard, *J. Chem. Soc.*, 5916 (1963).

[16] A. W. Adamson, *J. Phys. Chem.*, **71**, 798 (1967), and references therein.

[17] R. A. Bauer and F. Basolo, *J. Am. Chem. Soc.*, **90**, 2437 (1968).

[18] B. Adell, *Z. anorg. allgem. Chem.*, **279**, 219 (1955).

[19] W. W. Wendlandt and J. H. Woodlock, *J. Inorg. Nucl. Chem.*, **27**, 259 (1965).

[20] Footnote 9 in W. Kruse and H. Taube, *J. Am. Chem. Soc.*, **83**, 1280 (1961).

Suggestions for Further Reading

Calvert, J. G. and J. N. Pitts, Jr., *Photochemistry*, John Wiley & Sons, Inc., New York, 1966.

Noyes, W. A., Jr., G. S. Hammond, and J. N. Pitts, Jr., *Advances in Photochemistry*, series of volumes, John Wiley & Sons, Inc., New York, 1963.

Turro, N. J., "Molecular Photochemistry," *Chem. Eng. News*, May 8, 1967, p. 84.

Reviews on the photochemistry of coordination compounds:
A. W. Adamson *et al.*, *Chem. Revs.*, **68**, 541 (1968).
E. L. Wehry, *Quart. Rev. (London)*, **21**, 213 (1967).
D. Valentine, *Advan. Photochem.*, **6**, 124 (1968).
A. W. Adamson, *Coord. Chem. Rev.*, **3**, 169 (1968).

Growing Crystals

from

Aqueous Solutions[1]

16

There are two general methods for growing large single crystals from aqueous solutions. In one method, crystal growth occurs as a saturated solution is gradually cooled to a temperature at which the solution is appreciably supersaturated. In the other case, crystal growth occurs as a saturated solution is allowed to gradually evaporate at a constant temperature. In both methods, it is first necessary to prepare a saturated solution at the temperature at which crystal growth is to occur (usually near room temperature) and to prepare some "seed crystals," one of which will be suspended by a thread in the saturated solution.

A saturated solution may be conveniently prepared by the following procedure. In about 500 ml of water at 50°, dissolve somewhat more of the salt whose crystals are to be grown than will dissolve at the expected growing temperature. While stirring vigorously, cool the solution to the expected

---

[1] For further details and for delightful reading, the reader is referred to A. Holden and P. Singer, *Crystals and Crystal Growing*, Doubleday & Company, Inc., Garden City, N.Y., 1960. A more complete discussion is found in Chap. 8 and 11 in J. J. Gilman, ed., *The Art and Science of Growing Crystals*, John Wiley & Sons, Inc., New York, 1963.

growing temperature (do not cool below this temperature). If crystallization does not take place, add a small crystal to induce it. Stir the suspension of crystals for about 15 min and then let the solution stand in contact with the crystals in a covered beaker or stoppered flask in the crystal-growing room for a day or longer. Finally decant the solution from the precipitated crystals. These crystals may be spread out on a piece of filter paper to dry, and among them may be found a suitable seed crystal. A seed crystal must be a single crystal, and, so that it may be easily suspended by a thread, it should be at least 3 mm long. Smaller crystals are not only difficult to attach to the thread, but they also may be buoyed up to the surface of the solution by the thread. Save all good seed crystals, for your initial attempt at crystal growing may not be successful.

Crystal Growth by Cooling

A solution, just saturated at the temperature of the crystal-growing room, is heated to about 15° above this temperature; a small additional amount of the salt[2] is dissolved, and the solution is filtered. The solution (which would now be supersaturated at the growing temperature) is carefully poured into a clean 600-ml beaker. When the temperature is about 3° above growing temperature, the seed crystal is suspended in the middle of the solution by a thin piece of sewing thread (a monofilament thread, such as nylon, is best) the upper end of which is attached to a piece of wood that completely covers the beaker (see Fig. 16.1). The thread is most conveniently attached to the seed by means of a slip knot. The free end of the thread should be cut off as close as possible to the knot. The beaker should be allowed to stand in a room in which the temperature fluctuates less than 2° throughout the entire day. To avoid rapid temperature fluctuations, the beaker may be covered with several cardboard boxes or a large crock.

After an hour or two, examine the beaker to see whether or not the seed crystal has dissolved. If it has dissolved, it will be necessary to begin again, using a solution containing a little more salt. Actually, a little dissolution of the seed crystal at the beginning is desirable. A seed crystal generally has several other tiny crystals adhering to it. Thus, if the seed crystal is placed in an undersaturated solution, these tiny crystals will dissolve away, leaving only one crystal. Of course, one hopes that the solution cools rapidly enough so that the solution becomes supersaturated before the seed crystal dissolves completely.

The crystal should grow to a good size in about 3 to 6 days. Remove the fully grown crystal from the solution, and carefully dry it with filter paper or absorbent tissue.

---

[2] From 2 to 20 per cent of the amount required to saturate the solution at room temperature, depending on the ease of supersaturation.

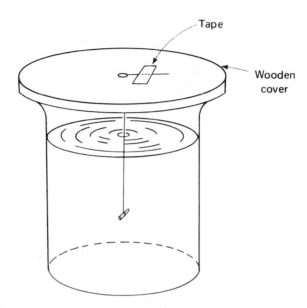

FIG. 16.1. Seed crystal suspended by thread in covered beaker.

Crystal Growth by Evaporation

A solution, saturated at the growing temperature, is heated to about 10° above this temperature and carefully filtered into a clean 600-ml beaker. When the temperature is 1 to 2° above growing temperature, introduce the seed crystal suspended by a glass support. A suspension of this type is pictured in Fig. 16.2. The seed crystal is hung by a monofilament thread from the gallows-shaped support, the top of which must always be below the surface of the solution.

Cover the beaker with a cloth, and hold it in place with a string or rubber band. The beaker should stand in a room in which the temperature fluctuates

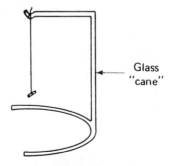

FIG. 16.2. Seed crystal suspended from glass support.

less than 5° throughout the entire day. The rate of crystal growth depends on the rate at which water evaporates from the solution. When the crystal has reached a satisfactory size, or when the top of the suspension is about to protrude through the surface of the solution, remove the crystal and dry it with filter paper or absorbent tissue.

Liquid-Liquid
Extraction

17

A simple solute dissolved in one phase in equilibrium with another, immiscible, phase will distribute itself between the two phases so that the ratio of the activities in the two phases is a constant at a fixed temperature. If we are concerned with dilute solutions, we may make the approximation that the activity of a solute is proportional to its concentration, and we write

$$K = \frac{C_1}{C_2}$$

The constant K is the *distribution constant*, and C_1 and C_2 are the concentrations in the two phases. The distribution constant is readily determined by shaking the solute with the two liquids in a separatory funnel, followed by analyzing the separated phases.

In most simple extraction processes, a solute is transferred from one liquid (the extrahend) to another liquid (the extractant). The first step in the procedure consists of pouring the liquids into the separatory funnel, stoppering the funnel, and shaking gently for a few seconds. The funnel is then inverted while the stopper is pressed in with the forefinger, and the stopcock is momentarily opened to release the internal pressure developed by the vaporization of the liquids. These operations are repeated until the internal pressure equals atmospheric pressure. Then more vigorous shaking can be used. Finally the funnel is allowed to stand upright in a ring until the phases have completely separated. The stopper is removed, and the denser phase is drawn off through the stopcock into another vessel (or another separatory funnel, if further extraction of the denser phase is required). The lighter phase may be collected in a separate vessel, or, if further extraction of it is required, more of the denser liquid can be added and the extraction continued.

SINGLE-STAGE CONTINUOUS EXTRACTION

Unless the distribution constant is very large, a single extraction with a separatory funnel will not remove sufficient solute from the extrahend (the solvent from which the solute is extracted) if a reasonably quantitative

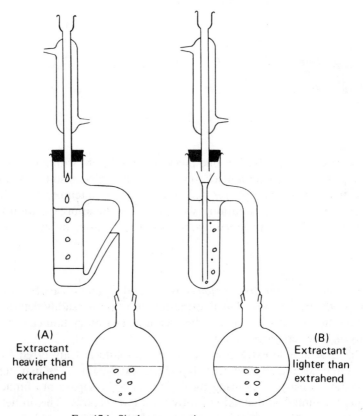

(A)
Extractant
heavier than
extrahend

(B)
Extractant
lighter than
extrahend

FIG. 17.1. Single-stage continuous extractors.

separation is desired. Several successive extractions with fresh portions of extractant can be employed, but such a procedure is often tedious or involves excessive volumes of the extractant.

A *continuous extractor* often solves the problem. In this type of extractor, the extractant is continuously furnished to the extrahend by distillation and passes through it.

There are two main types of continuous extractors: those in which the extractant must be lighter than the extrahend, and vice versa. These two types are pictured in Fig. 17.1.

BATCHWISE COUNTERCURRENT EXTRACTION

Two solutes may be separated from each other by selective extraction if their distribution coefficients are sufficiently different. The efficiency of the extraction depends on the ratio of the distribution coefficients, or the *separation factor, S*:

$$S = \frac{K_X}{K_Y}$$

If the separation factor is very large or very small (e.g., 10^4 or 10^{-4}), the two solutes may practically be separated quantitatively by simple methods such as those previously discussed. But if the separation factor is near unity, special schemes must be devised for fractionally extracting the solutes. One of the most effective procedures is that known as countercurrent extraction. The procedure now described is more accurately labeled "batchwise countercurrent extraction with center feed."[1] A discussion of the concept of theoretical plates as applied to separation schemes is given in Appendix 6.

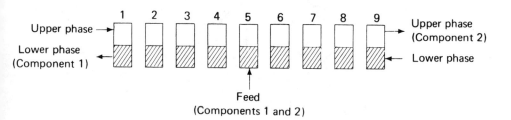

FIG. 17.2. Countercurrent extraction with center feed (nine stages).

[1] For an extensive description of other fractional extraction schemes, the reader is referred to the article by L. C. Craig and D. Craig, in *Technique of Organic Chemistry*, vol. 3, 2nd ed., A. Weissberger, ed., Interscience Publishers, New York, 1956, Part 1, p. 149.

Imagine a series of nine equilibration stages, as pictured in Fig. 17.2. First, a mixture of the two solutes is distributed between the two phases in stage 5. Then the upper phase is placed in stage 6 with fresh lower phase and the lower phase is placed in stage 4 with fresh upper phase. Again, a mixture of the two solutes is placed in stage 5 with fresh phases. Now the three stages

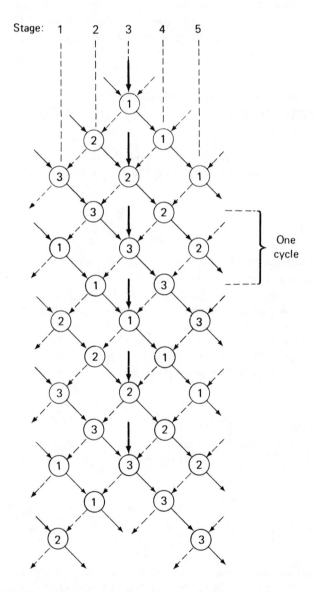

FIG. 17.3. Countercurrent extraction with center feed (five stages).

are equilibrated. The upper phase moves from stage 6 to stage 7, from stage 5 to stage 6, and from stage 4 to stage 5. The lower phase moves from stage 4 to stage 3, from stage 5 to stage 4, and from stage 6 to stage 5. Fresh upper phase is put in stages 3 and 4, and fresh lower phase is put in stages 6 and 7. Now all five stages are equilibrated and the cycle is repeated over and over. After four cycles, lower phase is ejected from stage 1 and upper phase is ejected from stage 9. If the relative volumes of the two countercurrent phases have been properly chosen the effluent lower phases will contain much more of one solute than the other and the effluent upper phase will contain much more of the other than the one.

Many cycles must be performed before the concentrations of solutes in the effluents reach their steady-state values. The closer the separation factor is to unity the more stages are required to effect a "quantitative" separation of the two solutes. Usually the volumes of the two phases are adjusted so that the following relation holds:

$$\frac{V_1}{V_2} = \frac{1}{\sqrt{K_X K_Y}}$$

Figure 17.3 gives another way of schematically representing a counter-current extraction. In this figure, five stages are involved, requiring only three separatory funnels. Suppose we are separating two substances, X and Y, which are preferentially extracted by ether and water, respectively. In separatory funnel 1 we mix and equilibrate the mixture of X and Y (feed) with portions of ether and water. The aqueous phase is drained into funnel 2; ether is added to funnel 2, and then funnel 2 is shaken. While funnel 2 is settling, water is added to funnel 1 and funnel 1 is shaken. The aqueous phase in funnel 2 is drained into funnel 3; ether is added to funnel 3, and funnel 3 is shaken. The aqueous phase in funnel 1 is drained into funnel 2; another feed sample is introduced into funnel 2, and funnel 2 is shaken. Water is added to funnel 1 and funnel 1 is shaken. The aqueous phase is drained from funnel 3 into a beaker. This product is the first emerging from the fractiona-tion, and it should be relatively rich in component Y. The aqueous phase is drained from funnel 2 into funnel 3 and funnel 3 is shaken. The aqueous phase from funnel 1 is drained into funnel 2 and funnel 2 is shaken. The ether phase remaining in funnel 1 may now be transferred to a beaker; this sample is the first X-enriched product. Fresh ether is now put in funnel 1 and funnel 1 re-enters the scheme at stage 1.

Problem

1. How many separatory funnels would be required to carry out a countercurrent extraction with center feed with nine stages?

Suggestions for Further Reading

Alders, L., ed., *Liquid-liquid Extraction. Theory and Laboratory Practice*, 2nd ed., D. Van Nostrand Co., Inc., Princeton, N.J., 1959.

Craig, L. C., and D. Craig, *Technique of Organic Chemistry*, vol. 3, 2nd ed., A. Weissberger, ed., Interscience Publishers, New York, 1956, Part 1, p. 149.

Morrison, G. H., and H. Freiser, *Solvent Extraction in Analytical Chemistry*, John Wiley & Sons, Inc., New York, 1957.

Peppard, D. F., "Liquid-liquid Extraction of Metal Ions," *Adv. Inorg. Chem. Radiochem.*, **9**, 1 (1966).

Liquid
Moving-Phase
Chromatography

18

We shall now discuss several chromatographic separation procedures in which the moving phase is a liquid. In this general type of chromatography, the stationary phase can be a solid (adsorption or liquid-solid chromatography) or a liquid (partition or liquid-liquid chromatography). The stationary phase is commonly supported in a column, but exceptions are found in thin-layer chromatography (a kind of adsorption chromatography) and paper chromatography (a special case of partition chromatography). The fundamental principles of theoretical plates (described in Appendix 6) can be applied to these processes just as they can to fractional distillation, gas chromatography, fractional liquid-liquid extraction, and so on.

Adsorption Column Chromatography

This method is based on the difference in adsorption of the components of a solution on a finely divided solid. The adsorbent is placed in a cylindrical column and the solution is passed through it. The solutes to be separated are thereby adsorbed in a narrow zone at the top of the column. Fresh solvent, or

255

a suitable mixture of solvents, is then passed through the column, and by successive desorption and adsorption steps, the solutes are eluted down the column. The individual components of the mixture are eluted in bands moving at rates inversely proportional to the degree of adsorption. If the components are of different colors, varicolored bands will develop in the column, explaining the origin of the word chromatography. If the components are colorless, the composition of the eluate can be monitored by spectrophotometry, chemical tests on eluate fractions, or by simple examination of the residue formed upon evaporation of eluate fractions. Uniform packing of the adsorbent is essential in order to obtain regular bands. It is achieved by adding a thin slurry of the adsorbent and the solvent to the column while allowing the solvent to drain slowly from the column. The glass columns described in Chap. 13 (Ion Exchange) and pictured in Figs. 13.2 and 13.3 are quite satisfactory for column chromatography; however, it is important that Teflon stopcocks be used, inasmuch as organic solvents are usually used as eluants. The following adsorbents are commonly used: activated alumina, silica gel, charcoal, molecular sieve, sucrose, starch, and powdered refractories, such as magnesium silicates.

Adsorption column chromatography has been used for the separation of the *cis* and *trans* isomers of coordination compounds, such as $Pt(PBu_3)_2Cl_2$, $Pt(SEt_2)_2Cl_2$,[1]

$$\left[\begin{array}{c} \overset{H}{\underset{CH_3}{\overset{|}{\underset{|}{C-O}}}} \\ HC \bigcirc Cr, \\ \overset{|}{\underset{CH_3}{C-O}} \end{array} \right]_3^2 \quad \text{and} \quad \left[\begin{array}{c} \overset{CH_3}{\overset{\backslash}{C-O}} \\ HC \bigcirc M. \\ \overset{|}{\underset{C_6H_5}{C-O}} \end{array} \right]_3^3$$

In each of these cases, the *cis* isomer was more strongly adsorbed than the *trans* isomer. Adsorption column chromatography has been used to separate mixtures of the following eight-membered ring compounds: S_8, S_7NH, and the positional isomers of $S_6(NH)_2$ and $S_5(NH)_3$.[4] The separation of S_7NH and S_8 is described in Synthesis 28.

Partition Column Chromatography

This technique is based on the difference in partition coefficients of the components of a mixture distributed between two liquid phases; thus, it is akin to countercurrent liquid-liquid extraction. One phase is fixed by adsorption on

[1] G. B. Kauffman, R. P. Pinnell, and L. T. Takahashi, *Inorg. Chem.*, **1**, 544 (1962).

[2] J. P. Collman, E. T. Kittleman, W. S. Hart, and N. A. Moore, *Inorg. Syn.*, **8**, 144 (1966).

[3] R. C. Fay and T. S. Piper, *J. Am. Chem. Soc.*, **84**, 2303 (1962); M = Cr, Co, and Rh.

[4] H. G. Heal and J. Kane, *Inorg. Syn.*, **11**, 184 (1968).

an inert solid substance; the mobile phase flows through the column. Some typical immiscible solvent pairs are nitromethane-*iso*octane, dimethylformamide-*iso*octane, ethylene dichloride-ethylene glycol, *n*-butanol-aqueous medium, capryl alcohol-aqueous medium, ethylene dichloride-formamide, and chloroform-propylene glycol. The more polar phase wets the solids best and should be used as the stationary phase.

Although partition column chromatography gives sharper separations and is less likely to induce chemical changes than adsorption column chromatography, the former does not seem to have achieved much popularity among inorganic chemists. The reason is probably that there is somewhat more effort required to set up a partition chromatographic column than to set up an adsorption column. However, we may cite the experiments of Buckingham *et al.*[5] who used a cellulose column (essentially a stationary aqueous phase) with *n*-butanol as eluant for the separation of ions such as $[Co(trien)OH(H_2O)]^{2+}$ and $[Co(trien)H_2NCHRCO_2]^{2+}$.

Thin-layer Chromatography[6]

Thin-layer chromatography is a kind of adsorption chromatography in which thin and precisely uniform layers (100 to 2000 μ) of adsorbents, such as silica and alumina, are applied to a flat surface of inert material, such as glass or polyester film. The mixture to be separated is placed a short distance from one end of the layer and is separated by a solvent ascending through the layer by capillary action. Only minute amounts (0.1 to 500 μg) of material can be separated by this technique; therefore, it is essentially an analytical method for checking the purity of a compound or for giving a preliminary indication of how larger quantities of material can be separated by column chromatography. When colored mixtures are separated, the separate spots on the thin layer are readily distinguished. Colorless species can sometimes be made visible by fluorescence, using ultraviolet light; in other cases, the spots can be sprayed with some reagent that reacts with the compounds to give colored products. The relative rates of displacement of the components of a mixture are usually expressed by R_f values, defined as follows:

$$R_f = \frac{\text{distance traveled by substance}}{\text{distance traveled by solvent front}}$$

[5] D. A. Buckingham, J. P. Collman, D. A. R. Happer, and L. G. Marzilli, *J. Am. Chem. Soc.*, **89**, 1082 (1967).

[6] For comprehensive discussions of this topic see J. M. Bobbit, *Thin-Layer Chromatography*, Reinhold Publishing Corp., New York, 1963; K. Randerath, *Thin-Layer Chromatography*, Academic Press Inc., New York, 1963; E. Stahl and Associates, *Thin-Layer Chromatography—A Laboratory Handbook*, Academic Press Inc., New York, 1965; E. V. Truter, *Thin-Layer Chromatography*, Interscience Publishers, New York, 1963; F. L. J. Sixma and H. Wynberg, *A Manual of Physical Methods in Organic Chemistry*, John Wiley & Sons, Inc., New York, 1964, pp. 43–47.

Thin-layer chromatography has been used to monitor the eluate from the column chromatography of mixtures of the sulfur imides,[4] to check the purity of sulfur-containing chelates of nickel(II),[7] and to verify the presence of single stereoisomers in samples of complexes of the type $[IrClICF_3(CO)(P(C_6H_5)_2CH_3)_2]$.[8]

Paper Chromatography

Paper chromatography is a kind of partition chromatography in which the moisture of a strip of filter paper is the stationary phase. Either a water-immiscible solvent, such as butanol, or a water-miscible phase, such as acetone, can serve as the moving phase. The solvent can move either up or down the paper strip by capillary action. The individual spots of the developed chromatogram can be distinguished by the same methods as used in thin-layer chromatography, and R_f values are calculated in the same way.

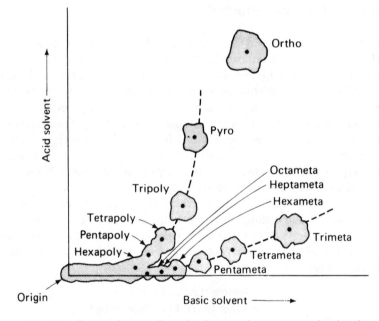

FIG. 18.1. Corner of a two-dimensional paper chromatogram showing the positions of the pentameta—through octametaphosphate rings in relation to the positions of the well-known ring and chain phosphates. The basic solvent moved 9 inches in 24 hours, whereas the acid solvent moved 4.5 inches in 5.5 hours. [From J. Van Wazer, *Phosphorus and Its Compounds*, Vol. 1, Fig. 11-7 (New York: John Wiley & Sons, Inc., Interscience, 1958).]

[7] D. Coucouvanis and J. P. Fackler, Jr., *Inorg. Chem.*, **6**, 2047 (1967).
[8] J. P. Collman and C. T. Sears, Jr., *Inorg. Chem.*, **7**, 27 (1968).

A wide variety of inorganic qualitative analytical separations can be effected on a small scale by paper chromatography.[9] As with thin-layer chromatography, paper chromatography is used in synthetic work principally as a means for checking purity or for purposes of identification. One recent application was the separation and identification of the complex ions formed by treating $[Co(en)_2OH(H_2O)]^{2+}$ with amino acid esters, amino acid amides, and dipeptides.[10]

An interesting variant is two-dimensional paper chromatography, in which, after a chromatogram is developed in one direction along one side of a large sheet of paper, the paper is turned at right angles and it is developed with a different solvent system. This technique is useful for the separation of complex mixtures of closely related substances; for example, a mixture of linear and cyclic polyphosphates can be separated, as shown in the two-dimensional chromatogram of Fig. 18.1.[11]

[9] A. I. Vogel, *A Textbook of Quantitative Inorganic Analysis*, 3rd. ed., John Wiley & Sons, Inc., New York, 1961, pp. 724–30.

[10] D. A. Buckingham and J. P. Collman, *Inorg. Chem.*, **6**, 1803 (1967).

[11] J. Van Wazer, *Phosphorus and Its Compounds*, vol. 1, Interscience Publishers, New York, 1958.

COMPOUND
CHARACTERIZATION

Part

Structure

from

Chemical Data

19

A chemist is very seldom satisfied with simply preparing a compound for the first time. The question immediately arises: What is the structure of this new compound? So the chemist then proceeds to collect sufficient data to determine the structure. Once the structure of a new compound is known, it is usually possible to rationalize the synthesis in terms of a mechanism, and then to postulate the synthesis of further unknown compounds of related structures. Such activity is the essence of synthetic chemistry, and it would be impossible without structure determination.

Consider, for example, the cage compounds containing the trirhenium(III) cluster. The $Re_3Cl_{12}^{3-}$ ion has the structure shown in Fig. 19.1.[1] The Re-Re distance in this cluster is 2.5 Å. It has been shown[2] that, of the 12 halogen atoms (6 axial, 3 equatorial bridging, and 3 equatorial nonbridging), only three—presumably the equatorial bridging ones—are not subject to exchange

[1] (a) W. T. Robinson, J. E. Fergusson, and B. R. Penfold, *Proc. Chem. Soc.*, 116 (1963); (b) J. A. Bertrand, F. A. Cotton, and W. A. Dollase, *J. Am. Chem. Soc.*, **85**, 1349 (1963).

[2] B. H. Robinson and J. E. Fergusson, *J. Chem. Soc.*, 5683 (1964).

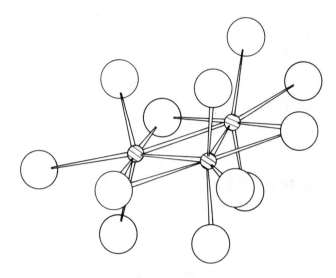

FIG. 19.1. Structure of $Re_3Cl_{12}^{3-}$ ion. [1b] (Reproduced with permission of the American Chemical Society.)

with other halide ions. Cotton and Lippard[3] reasoned that anions such as AsO_4^{3-}, in which three oxygen atoms define an equilateral triangle of approximate edge 2.5 Å, would be ideally suited for forming neutral adducts of the $Re_3X_3^{6+}$ ion of the types $Re_3X_3(AsO_4)_2$. Thus, they began an investigation of the reactions of Re_3Cl_9 and Re_3Br_9 with MO_3^{3-} and MO_4^{3-} anions, where M = As and P. The proposed structure of one member of the series of products, $Re_3Br_3(AsO_4)_2(DMSO)_3$, is shown in Fig. 19.2.

The chemist has a wide variety of methods at his disposal for determining structure. In the earlier days of inorganic chemistry, relatively few structural methods were available, and, in general, synthetic work progressed slowly and unsystematically. Hugh Taylor tells the story[4] of a compound that he synthesized[5] in 1911 and for which the structure was not determined until 50 years later.[6]

Structural methods may be roughly classified into three categories: *spectroscopic methods*, *macroscopic physical methods*, and *chemical methods*.

Spectroscopic methods generally give detailed, specific, structural information. The important spectroscopic methods are listed. For those methods which are discussed at some length in this book, the appropriate chapters are indicated. The other methods are described briefly in Chap. 29.

[3] F. A. Cotton and S. J. Lippard, *J. Am. Chem. Soc.*, **88**, 1882 (1966).
[4] H. S. Taylor, *Am. Scientist*, **51**, 371 (1963).
[5] H. Bassett and H. S. Taylor, *J. Chem. Soc.*, **99**, 1402 (1911); *Z. Anorg. Chem.*, **73**, 75 (1912).
[6] J. Danielsen and S. E. Rasmussen, *Acta Chem. Scand.*, **15**, 1398 (1961).

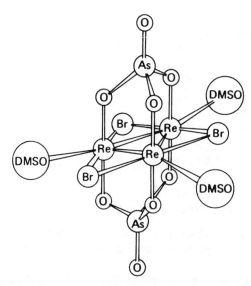

FIG. 19.2. Perspective view of the proposed structure of the molecule Re_3Br_3-$(AsO_4)_2(DMSO)_3$.[3] (Reproduced with permission of the American Chemical Society.)

Microwave spectrometry
Infrared spectrometry (see Chap. 21)
Raman spectrometry
Visible-ultraviolet spectrometry (see Chap. 22)
Nuclear magnetic resonance (see Chap. 24)
Nuclear quadrupole spectroscopy
Mössbauer spectroscopy
Electron spin resonance (see Chap. 26)
X-ray diffraction (see Chap. 28)
Neutron diffraction
Electron diffraction
Mass spectrometry (see Chap. 27)
Photoelectron spectroscopy
X-ray spectroscopy (see Chap. 28)

Macroscopic physical methods are those which cannot determine structural details but which can establish the general structure type. Measurements of gross physical properties, such as the following, fit this category:

Melting point (see Chap. 20)
Boiling point (see Chap. 4)
Vapor pressure (see Chap. 8)
Viscosity

Density:
  of solids (see Chap. 20)
  of liquids (see Chap. 20)
Dielectric constant
Surface tension
Refractive index
Electrical conductivity
Optical rotation[7] (see Chap. 23)
Molecular weight determination:
  cryoscopic (see Chap. 20)
  ebulliometric
  vapor density (see Chap. 8)
  vapor-pressure depression (see Chap. 8)

Again, the appropriate chapters are indicated for those methods discussed in this book. The other methods are discussed in various physical chemistry laboratory manuals.[8]

By *chemical methods*, we mean the study of chemical reactions involving the compound whose structure is sought. Such information, if carefully interpreted, may sometimes be used to choose one structure from several alternative structures.

In this chapter, we shall discuss the application of chemical data to various structural problems.

STOICHIOMETRY

Elemental analysis is often the most important step in the characterization of a new compound. As any beginning chemistry student should know, the data can be used to establish the empirical formula of the compound, if that formula is simple enough. It is important to be aware of the accuracy of typical analytical data and to recognize the consequent limitations of the usefulness of such data. The average percentage deviation between the calculated and found percentages for the elemental analyses reported in one issue of *Inorganic Chemistry*[9] was found to be ±3.7 per cent. Probably in many cases the discrepancies resulted from poor analytical techniques rather than impure compounds. Such inaccurate analytical data are of little use in determining the composition of a sample. For example, consider the com-

---

[7] Optical rotatory dispersion is properly classified as a spectroscopic technique.

[8] See, for example, F. Daniels, *et al.*, *Experimental Physical Chemistry*, 6th ed., McGraw-Hill Book Company, New York, 1962, and D. P. Shoemaker and C. W. Garland, *Experiments in Physical Chemistry*, McGraw-Hill Book Company, New York, 1962.

[9] *Inorg. Chem.*, 7, 2471–2675 (1968). The average deviation for carbon analyses is considerably lower than the overall average deviation.

pound $SiI_4 \cdot 4\,C_5H_5N$, which has been reported by several workers[10,11] and which is a true adduct of silicon tetraiodide and pyridine. The observed and theoretical analytical data in Table 19.1 show how difficult it is to differentiate the adduct from its hydrolysis products by elemental analysis.[12]

Table 19.1   Analytical Data for the Adduct $SiI_4 \cdot 4\,C_5H_5N$ and Its Hydrolysis
Product (weight per cent)*

Substance	C	H	N	I	Si
$SiI_4 \cdot 4\,C_5H_5N$ (theory)	28.2	2.4	6.6	59.6	3.3
$SiI_4 \cdot 4\,C_5H_5N$ (found)†	27.4	2.5	6.6	60.0	3.2
$SiI_4 \cdot 4\,C_5H_5N$ (found)**	27.9	3.2	5.9	57.2	—
$4\,C_5H_5N \cdot HI + SiO_2$ (theory)	27.0	2.7	6.3	57.1	3.2

* See ref. 12.   † See ref. 10.   ** See ref. 11.

When a chemist makes a compound for the first time, he naturally has a preconceived notion of its constitution. Because of this idea, he may be tempted to postulate prematurely a formula (and structure) for the compound on the basis of too few analytical and physical data. Consider, for example, the reaction of S_4N_4 with nickel chloride or nickel carbonyl. On the basis of satisfactory analyses for nickel, sulfur and nitrogen, the product of this reaction was first formulated as NiS_4N_4.[13] However, as discussed on p. 296, an infrared study showed the presence of N-H bonds in the compound, and further study led to its reformulation as $NiS_4N_4H_2$.[14] Presumably, the hydrogen was introduced into the compound during workup with organic solvents.

A more spectacular example of this type is found in the reaction of *trans*-$[(C_2H_5)_3P]_2PtHCl$ with tetrafluoroethylene at 120° in a sealed Pyrex tube. In the first study of this reaction,[15] one of the products was characterized as $[(C_2H_5)_3P]_2PtHCl(C_2F_4)$ [calculated analysis (in per cent): F, 13.4; P, 10.9; Pt, 34.4; found (in per cent): F, 12.7; P, 10.72; Pt, 34.4]. Soon thereafter, a single-crystal X-ray diffraction study[16] of the compound showed that it contained the cation $[(C_2H_5)_3P]_2PtCl(CO)^+$ and the anion BF_4^-. The reaction is remarkable in that BF_4^- and CO formed by reaction of C_2F_4 with borosilicate glass under relatively mild conditions.

[10] U. Wannagat, R. Schwarz, H. Voss, and K. G. Knauff, *Z. Anorg. Chem.*, **277**, 73 (1954).

[11] E. L. Muetterties, *J. Inorg. Nucl. Chem.*, **15**, 182 (1960).

[12] I. R. Beattie, *Quart. Rev. London*, **17**, 382 (1963).

[13] M. Goehring and A. Debo, *Z. anorg. allgem. Chem.*, **273**, 319 (1953).

[14] T. S. Piper, *J. Am. Chem. Soc.*, **80**, 30 (1958).

[15] H. C. Clark and W. S. Tsang, *J. Am. Chem. Soc.*, **89**, 529 (1967).

[16] H. C. Clark, P. W. R. Corfield, K. R. Dixon, and J. A. Ibers, *J. Am. Chem. Soc.*, **89**, 3360 (1967).

Even when the composition of a compound has been definitely established, the structure cannot be deduced reliably. Thus, the empirical formula $CsReCl_4$ gives no hint of the structure shown in Fig. 19.1, with three 7-coordinate rhenium atoms. Likewise, the formula SbF_5 gives no indication of this compound's polymeric structure, in which 6-coordinate antimony atoms are linked by bridging fluorine atoms.[17]

DISTINGUISHING NONEQUIVALENT ATOMS

Many molecules and ions contain more than one atom of a particular element. In determining the structure of such a species, it is very helpful to know whether all the atoms of the element are equivalent. For example, all the hydrogens in ethane (C_2H_6) are equivalent. On the other hand, there are two kinds of hydrogen atoms—four of one kind and two of another—in diborane (B_2H_6). Hence, the "hydrogen bridge" structure for diborane is quite reasonable:

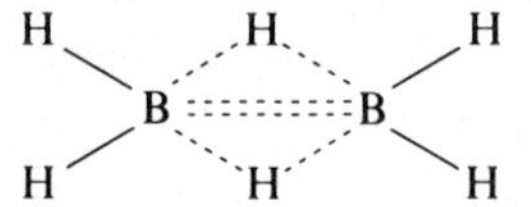

In the following paragraphs, we shall examine some of the types of chemical data used to distinguish two or more kinds of atoms of the same element in compounds.

Discrimination by Means of Thermodynamic Properties

The structures of many phosphorus oxyacids have been deduced from a knowledge of the relative acidities of the protons. For example, only two of the hydrogens in phosphorous acid (H_3PO_3) may be titrated with alkali. Hence, the following structure has been proposed for the acid:

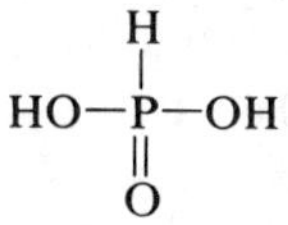

$$\begin{array}{c} H \\ | \\ HO-P-OH \\ \| \\ O \end{array}$$

The chain polyphosphoric acids consist of chains of $-O-PO(OH)-$ groups:

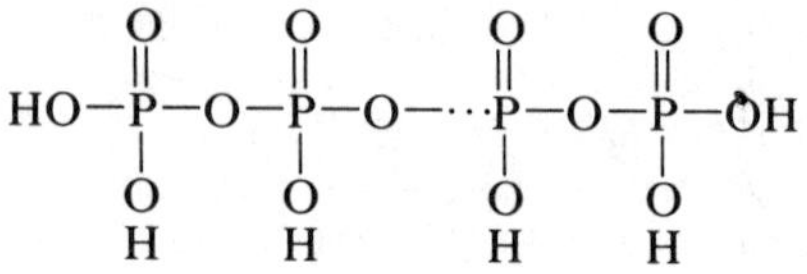

[17] C. A. Hoffman, B. E. Holder, and W. L. Jolly, *J. Phys. Chem.*, **62**, 364 (1958).

Theoretically, the chain length could be ascertained from a knowledge of the ratio of total hydrogen to total phosphorus, but for chains with n greater than 4 or 5, ordinary analytical precision leads to very uncertain values for n. A simpler, more accurate procedure takes advantage of the fact that after one of the two identical protons on each terminal $-PO(OH)_2$ group dissociates, the remaining proton dissociates with much greater difficulty. Thus, there are apparently n strongly acidic protons per chain (one per phosphorus atom) and two weakly acidic protons per chain (one at each end). The "strong" hydrogens have ionization constants around 10^{-2}–10^{-3}, and the "weak" hydrogens have ionization constants around 10^{-7}–10^{-8}. Thus, they may be distinguished readily by a pH titration.[18] Triphosphoric acid, for example, gives a titration curve with two main inflections, spaced so as to indicate that the ratio of "weak" to "strong" hydrogens is $2:3$.

Discrimination by Means of Kinetic Properties

When silver nitrate is added to a freshly prepared solution of anhydrous chromium(III) chloride in water, only one-third of the chlorine is precipitated as silver chloride. Therefore, it is believed that two of the chloride ions are coordinated directly to the chromium atom and that the other is ionized. The formula of the complex ion is $Cr(H_2O)_4Cl_2^+$, and the salt $[Cr(H_2O)_4Cl_2]$-$Cl \cdot 2H_2O$ may be crystallized from a solution of $CrCl_3$ in water.

Discrimination by Means of Isotopic Labeling

Tracer techniques offer two simple methods for determining the equivalence or nonequivalence of atoms in a molecule.

1. The compound may be synthesized in such a way that only part of the atoms in question are labeled. The compound is then degraded to two different species containing these atoms. If the concentrations of labeled atoms are markedly different in these two species, it may be concluded that the atoms in the original compound were nonequivalent.

2. If part of the atoms in a compound exchange with another species more rapidly than the other part, it may be concluded that the atoms in the compound are nonequivalent.

Examples of each of these methods will be discussed.

Synthesis-decomposition Method. The thiosulfate ion, $S_2O_3^{2-}$, might be formulated as either

$$O-S-O-S-O \quad \text{or} \quad O-\overset{\overset{\displaystyle O}{|}}{\underset{\underset{\displaystyle O}{|}}{S}}-S$$

[18] E. J. Griffith, *J. Am. Chem. Soc.*, **79**, 509 (1957).

In the first structure, the two sulfur atoms are equivalent, whereas in the second structure they are nonequivalent. The second structure is compatible with the observation that when thiosulfate is prepared from sulfite and radioactive sulfur and then decomposed to sulfide and sulfate (by decomposition of the silver salt), the radioactivity appears only in the sulfide.[19]

$$S^* + SO_3^{2-} \longrightarrow S^*SO_3^{2-} \xrightarrow{\ Ag^+\ } Ag_2S^*SO_3 \qquad (19.1)$$

$$Ag_2S^*SO_3 + H_2O \xrightarrow{\ \Delta\ } Ag_2S^* + 2H^+ + SO_4^{2-} \qquad (19.2)$$

The equivalence of the lead atoms in lead sesquioxide, Pb_2O_3, was investigated by preparing Pb_2O_3 with radioactive plumbite and ordinary plumbate.[20]

$$HPb^*O_2^- + PbO_3^{2-} + H_2O = Pb^*PbO_3 + 3OH^- \qquad (19.3)$$

The precipitate was allowed to stand for 3 hours and was then decomposed with 12 M KOH:

$$Pb^*PbO_3 + 4OH^- = PbO_3^{2-} + Pb^*O_2^{2-} + 2H_2O \qquad (19.4)$$

The plumbate was separated by the precipitation of barium plumbate. Because practically all the activity was found in the plumbite fraction, it was concluded that Pb_2O_3 may be regarded as lead(II) metaplumbate(IV), $Pb(PbO_3)$.

Exchange-rate Method. The equivalence of the iodine atoms in diphenyliodonium iodide was studied by allowing the normal compound to exchange with radioactive iodide ion.[21] The material was then treated with silver oxide and the organic product was found to be inactive:

$$2(C_6H_5)_2I_2^* + Ag_2O + H_2O = 2(C_6H_5)_2IOH + 2AgI^* \qquad (19.5)$$

It is apparent that the iodide exchanged with only one of the iodine atoms of diphenyliodonium iodide. These results are in agreement with the structure $[C_6H_5{-}I^+{-}C_6H_5]I^-$.

ISOMERISM

Many physical methods, discussed in the following chapters, can be used to distinguish isomers. Occasionally a chemical method can be used. For

[19] E. B. Andersen, *Z. physik. Chem.*. **B32**. 237 (1936).
[20] E. Zintl and A. Rauch, *Ber.*, **57**, 1739 (1924).
[21] F. Juliusberger, B. Topley, and J. Weiss, *J. Chem. Soc.*, 1295 (1935).

example, there are two isomers of the thiosulfatopentaamminecobalt(III) ion, $[Co(NH_3)_5S_2O_3]^+$. In one, the thiosulfate ion is bonded to the cobalt through an oxygen atom; in the other, the bonding is through a sulfur atom. Both isomers are reduced by chromium(II):

$$Co(NH_3)_5S_2O_3{}^+ + Cr^{2+} + 5\,H^+ \longrightarrow 5\,NH_4{}^+ + Co^{2+} + CrS_2O_3{}^+$$

One isomer is reduced 70 times more rapidly than the other.[22] The fast isomer is believed to be the oxygen-bonded isomer because the rate of reduction is similar to that observed for other pentaamminecobalt(III) complexes (such as the sulfato- and sulfito-complexes), in which the electron is transferred via the path $Cr-O-S-O-Co$. The slow isomer is believed to be the sulfur-bonded isomer, in which the electron is transferred via the path $Cr-O-S-S-Co$. Other examples of kinetic identification of isomers are given in Chap. 3.

Sometimes structural information can be deduced from the number of substituted derivatives and positional isomers known for a compound. Thus, no more than four of the six hydrogen atoms of diborane have been replaced with methyl groups.[23] Only five methyl derivatives are known: monomethyl diborane, trimethyl diborane, tetramethyl diborane, and two isomers of dimethyl diborane. These facts suggest that two of the six hydrogen atoms in diborane are bound differently from the others, and they support the hydrogen-bridge model for diborane. If diborane had an ethane-like structure and if there were no steric effects preventing complete substitution, one would expect nine different methyl derivatives.

PRINCIPLE OF "RETENTION OF CONFIGURATION"

Chemists have long made use of the fact that when an organic compound reacts, most of the bonds between the atoms are unaffected. In fact, in most organic reactions, the minimum structural change consonant with the empirical change occurs. The latter generalization has been of great help in structure determinations. We shall consider several examples from the field of inorganic chemistry in which this generalization has been useful in the assignment of structure.

Two isomers of dichlorobis(ethylenediamine)chromium(III) chloride (a violet form and a green form) are known. In Fig. 19.3, the reactions used to assign the *cis* configuration to the violet form and the *trans* configuration to the green form are represented schematically.[24] The chloride ions in the *trans*

---

[22] D. E. Peters and R. T. M. Fraser, *J. Am. Chem. Soc.*, **87**, 2758 (1965).

[23] H. I. Schlesinger and A. B. Burg, *Chem. Revs.*, **31**, 1 (1942).

[24] P. Pfeiffer, *Z. anorg. Chem.*, **56**, 261 (1908).

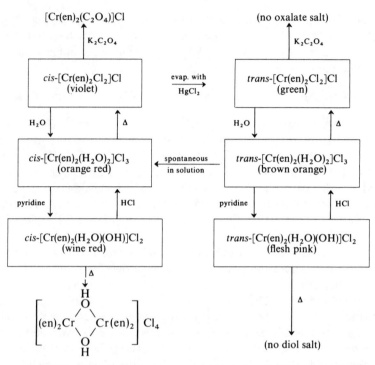

Fig. 19.3.   Transformations   among   bis(ethylenediamine)chromium(III) complexes.

complex cannot be replaced by the oxalate ion because the oxalate ion cannot span the *trans* positions. Similarly, the *trans* form of $[Cr(en)_2(H_2O)(OH)]Cl_2$ will not yield the binuclear complex $[Cr_2(en)_4(OH)_2]Cl_4$ because it is geometrically impossible to link two octahedra through two groups that are *trans* in each octahedron. The reader will appreciate that, in view of the many rearrangements which coordination compounds are known to undergo, structure assignments of this type are very tentative.

It is possible to prepare at least two isomers of trimethylborazine, $B_3N_3H_3(CH_3)_3$. One isomer, with a boiling point of 133° and a melting point of $-7.5°$, may be prepared by heating a mixture of methylammonium chloride and lithium borohydride.[25] Another isomer, with a boiling point of 129° and a melting point of 31.5°, may be prepared by heating a mixture of boron trimethyl and borazine.[26] If we assume that in each of these reactions a minimum number of bonds is broken, then we may assign structures to the isomers. The first compound is N-trimethylborazine and the second is B-trimethylborazine.

[25] G. W. Schaeffer and E. R. Anderson, *J. Am. Chem. Soc.*, **71**, 2143 (1949).
[26] H. I. Schlesinger, D. M. Ritter, and A. B. Burg, *J. Am. Chem. Soc.*, **60**, 1296 (1938).

N-trimethylborazine

B-trimethylborazine

These deductions have been corroborated by the differences in the products of hydrolysis of the compounds. The first compound gives 3 moles of hydrogen upon hydrolysis, whereas the second gives essentially none.

The silicate structures in the silicate minerals are integral parts of complex, highly cross-linked inorganic polymers. However, some of these react with acidic aqueous solutions of hexamethyldisiloxane, $[(CH_3)_3Si]_2O$, to yield specific trimethylsilyl silicates having the same silicate structure as the mineral from which they are derived.[27] Thus, olivine (an orthosilicate mineral in which magnesium and iron atoms separate the monomeric SiO_4 groups) gives a 70 per cent yield of the orthosilicate derivative $[(CH_3)_3Si]_4SiO_4$. Hemimorphite (a zinc disilicate) gives predominantly the disilicate derivatives $[(CH_3)_3SiO]_3SiOSi[OSi(CH_3)_3]_3$ and $[(CH_3)_3SiO]_3SiOSi(OH)[OSi(CH_3)_3]_2$. Natrolite (an aluminosilicate containing trisilicate chains) gives principally a trimethylsilyl-substituted trisilicate. When the technique was tried on a mineral of unknown structure, laumontite $(CaAl_2Si_4O_{12}\cdot4\,H_2O)$, an 81 per cent yield of the cyclic compound $\{[(CH_3)_3SiO]_2SiO\}_4$ was obtained. It may be tentatively concluded that the crystal structure of this mineral contains cyclic tetrasilicate units. The technique has also been used to analyze sodium silicate solutions.

The principle of "retention of configuration" is only a rough guide. It seems to hold reasonably well with compounds of elements in the first row of the periodic table, but one should be skeptical of its application to other compounds, particularly those involving d orbitals. For example, when the hypophosphate ion, $P_2O_6^{4-}$, is oxidized by bromine in aqueous bicarbonate solution, the diphosphate ion, $^{2-}O_3P{-}O{-}PO_3^{2-}$, is formed. This observation was once interpreted as an indication of the asymmetric structure, $^{2-}O_3P{-}O{-}PO_2^{2-}$, for the hypophosphate ion.[28] It was considered unlikely that the species contained a P—P bond, as in $^{2-}O_3P{-}PO_3^{2-}$, because it was thought that the oxidizing species would break the P—P bond and give orthophosphate as the sole product. However, several physical techniques have definitely shown that the last structure is correct.[29] If we

[27] C. W. Lentz, *Inorg. Chem.*, **3**, 574 (1964).

[28] B. Blaser and P. Halpern, *Z. anorg. allgem. Chem.*, **215**, 33 (1933).

[29] J. R. Van Wazer, *Phosphorus and its Compounds*, Interscience Publishers, New York, 1958.

assume that, in an intermediate step of the oxidation, the phosphorus atoms are pentacovalent (by use of d orbitals), there is no difficulty in explaining the formation of diphosphate.

hypothetical intermediate
in oxidation of hypophosphate

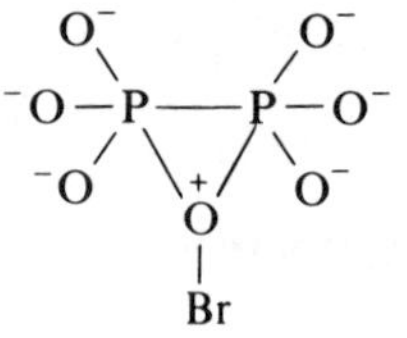

The reaction of phenylmagnesium bromide with the phosphonitrilic chloride $P_4N_4Cl_8$ (an eight-membered ring compound) gives two isomeric tetraphenyl derivatives, $P_4N_4(C_6H_5)_4Cl_4$, with melting points of 181 and 212.5°. Hydrolytic degradation gave entirely different results for the two compounds[30]:

$$P_4N_4(C_6H_5)_4Cl_4 \text{ (m.p.} = 181°) \longrightarrow 1\,(C_6H_5)_3PO + 1\,C_6H_5PO(OH)_2 + 2\,H_3PO_4$$

$$P_4N_4(C_6H_5)_4Cl_4 \text{ (m.p.} = 212.5°) \longrightarrow 2\,(C_6H_5)_2POOH + 2\,H_3PO_4$$

The results are most easily reconciled by assuming that the lower-melting isomer contains a six-membered ring:

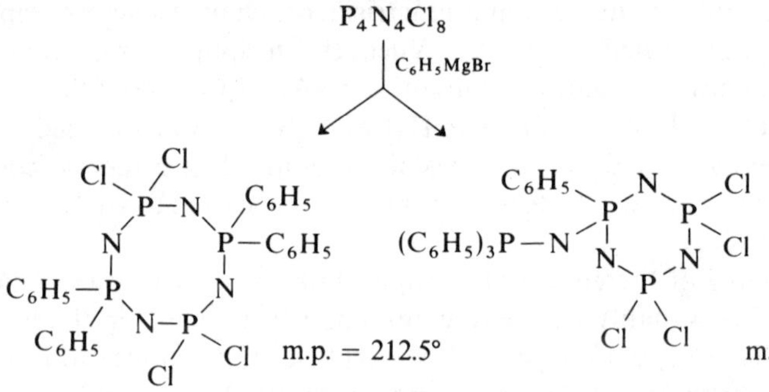

A mechanism involving initial ring cleavage followed by cyclization to a six-membered ring structure is compatible with the ring contraction reaction. Infrared and nmr spectral data confirm the assigned structures.

HOW STRUCTURES "CHANGE WITH TIME"

It is interesting to trace the histories of certain compounds whose structures were elucidated only very slowly. The following examples illustrate the pitfalls awaiting those who rely on scanty data.

[30] M. Biddlestone and R. A. Shaw, *Chem. Commun.*, 205 (1965).

Diborane Diammoniate

When diborane is allowed to react with excess ammonia at low temperatures, a diammoniate of composition $B_2H_6 \cdot 2\,NH_3$ forms. A number of structures have been proposed for this compound, some of which are represented in Table 19.2. Structure I is certainly not correct, inasmuch as a material with

Table 19.2   Proposed Structures for "Diborane Diammoniate"

I	$H_3B:NH_3$	ammonia-borane adduct
II	$(NH_4)_2B_2H_4$	ammonium tetrahydrodiborate $(2-)$
III	$NH_4(H_3B:NH_2:BH_3)$	ammonium μ-amidohexahydrodiborate $(1-)$
IV	$(NH_4)_2NH(BH_3)_3$	ammonium *tris*(borane)hydronitrate $(2-)$
V	$NH_4 \cdot BH_2NH_2 \cdot BH_4$	"ammonium hydroborate + aminoborane"
VIA	$(H_3N:BH_2:NH_3)BH_4$	diamminedihydroboron $(1+)$ hydroborate
VIB	$HB(NH_3)_3(BH_4)_2$	triamminehydroboron $(2+)$ hydroborate

this structure has been prepared by another method and has been found to possess properties entirely different from those of "diborane diammoniate"[31] (e.g., ammonia-borane is quite volatile, whereas "diborane diammoniate" is nonvolatile). Structure II was proposed by Wiberg[32] to explain the observed acid character of liquid-ammonia solutions of the diammoniate, but quantitative studies have shown that when a fresh cold solution of the diammoniate is allowed to react with excess sodium in ammonia, only one equivalent of hydrogen is evolved[33]; therefore, structure III was proposed. However, from solutions of the diammoniate that have previously been kept near the boiling point, a total of about 1.33 equivalents of hydrogen are evolved per mole of ammoniate, and it has been suggested that structure III undergoes rearrangement to structure IV.[34] When the liquid ammonia is evaporated after allowing 1 g-atom of sodium to react with 1 mole of diammoniate in ammonia, sodium hydroborate, $NaBH_4$, and aminoborane, $(NH_2BH_2)_x$, are left.[35] Thus, the diammoniate reacts somewhat as if it were a mixture of ammonium hydroborate and aminoborane (structure V). In fact, when the diammoniate is treated with excess trimethyl amine, the products are exactly what would be expected on this basis: hydrogen, ammonia, trimethylamine-borane, and aminoborane.

$$B_2H_6 \cdot 2\,NH_3 + (CH_3)_3N \longrightarrow NH_3 + H_2 + (CH_3)_3N:BH_3 + \tfrac{1}{3}(NH_2BH_2)_3$$

Vapor-pressure studies in liquid ammonia showed that the remaining aminoborane is trimeric, and reactions with sodium showed the trimer to be

[31] S. G. Shore and R. W. Parry, *J. Am. Chem. Soc.*, **80**, 8 (1958).
[32] E. Wiberg, *Ber.*, **69B**, 2816 (1936).
[33] H. I. Schlesinger and A. B. Burg, *J. Am. Chem. Soc.*, **60**, 290 (1938).
[34] W. L. Jolly, University of California Radiation Laboratory Report No. 4505, May, 1955.
[35] G. W. Schaeffer, M. D. Adams, and F. J. Koenig, *J. Am. Chem. Soc.*, **78**, 725 (1956).

a monobasic acid[36]:

$$N_3B_3H_{12} + Na \longrightarrow Na^+ + N_3B_3H_{11}^- + \tfrac{1}{2}H_2$$

Thus, the evolution of 1.33 equivalents of hydrogen in the reaction of the diammoniate with excess sodium was presumably explained.

After an exhaustive study of the diammoniate, Parry[37] proposed structures VIA and VIB, in which the cations are B^{3+} ions complexed by hydride ions and ammonia molecules. Evidence for the $H_2B(NH_3)_2^+$ cation was obtained in the reactions of ammonium halides with the diammoniate:

$$[H_2B(NH_3)_2]BH_4 + 2NH_4X \longrightarrow 2[H_2B(NH_3)_2]X + 2H_2$$

and evidence for the BH_4^- ion was obtained by the precipitation of $[Mg(NH_3)_6](BH_4)_2$ upon addition of magnesium thiocyanate to a solution of the diammoniate in ammonia.[37] Structure VIA was presumed to represent the species in fresh, cold ammonia solutions; the reaction with sodium is explained as follows:

$$(H_3N:BH_2:NH_3)BH_4 + Na \longrightarrow NaBH_4 + \frac{1}{x}(BH_2NH_2)_x + \tfrac{1}{2}H_2 + NH_3$$

Structure VIA was thought to rearrange in warm ammonia solutions to structure VIB, which reacts with sodium as follows:

$$HB(NH_3)_3(BH_4)_2 + 2Na \longrightarrow 2NaBH_4 + HB(NH_2)_2 + H_2 + NH_3$$

Conclusive evidence for the species $H_2B(NH_3)_2^+$ and BH_4^- in ammonia solutions of the diammoniate has been found in the boron-11 nuclear magnetic resonance spectrum.[38] In fact, it has been shown that $H_2B(NH_3)_2^+$ is but one member of a large family of borane cations of formula $H_2B(base)_2^+$.[39] However, the nature of the species in ammonia solutions of the diammoniate that have been allowed to become warm has not yet been firmly established.

Polyaryl Chromium Compounds

Hein and his co-workers[40] prepared an extensive series of polyaryl chromium compounds starting with the crude product obtained from the reaction of phenyl magnesium bromide with chromium(III) chloride. The scheme for preparing some of the more important members of the series is outlined in

[36] H. Hornig, W. Jolly, and G. Schaeffer, Paper Presented at Miami A.C.S. Meeting, April, 1957, Division of Inorganic Chemistry. However, see work of K. W. Boeddeker, S. G. Shore, and R. K. Bunting, *J. Am. Chem. Soc.*, **88**, 4396 (1966).

[37] R. W. Parry, *et al.*, *J. Am. Chem. Soc.*, **80**, 4–30 (1958).

[38] T. P. Onak and I. Shapiro, *J. Chem. Phys.*, **32**, 952 (1960).

[39] N. E. Miller and E. L. Muetterties, *J. Am. Chem. Soc.*, **86**, 1033 (1964); Nöth, Beyer and Vetter, *Chem. Ber.*, **97**, 110 (1964).

[40] See H. Zeiss, *Organometallic Chemistry*, H. Zeiss, ed., Reinhold Publishing Corp., New York, 1960, Chap. 8, pp. 380–425.

Fig. 19.4. The structures proposed by Hein for these compounds are given in the first column of Fig. 19.5. These structures are unreasonable for at least two reasons. First, the magnetic susceptibilities of compounds *A*, *B*, and *C* correspond to the presence of one free electron per chromium.[41] Second, when 1 mole of compound *B* is reduced with lithium hydroaluminate, 2 moles of biphenyl (and no benzene) is formed.[42] These observations may be accounted for by the structures (due to Klemm and Neuber[41]) given in the second column of Fig. 19.5, where each chromium is in the +5 oxidation state.

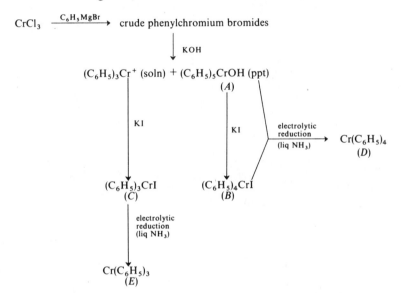

FIG. 19.4. The preparation of aromatic chromium compounds.

However, phenyl magnesium bromide is recognized as a reducing agent but not as an oxidizing agent. Thus, it is difficult to account for the formation of tetra-, penta-, and hexavalent chromium compounds. One would also expect that hydrogen atoms that are directly bonded to chromium atoms (as in the Klemm and Neuber structures) would be hydridic in character; however, these compounds are quite stable in aqueous solution. In addition, when compound *B* is reduced with lithium deuteroaluminate, the biphenyl produced contains only 5*D* per cent instead of the 10*D* per cent expected from the Klemm and Neuber structure. A similar reduction of compound *C* gives biphenyl containing 6.7*D* per cent instead of 10*D* per cent. Zeiss and Tsutsui[42] have explained these data in terms of the sandwich structures given in the third column of Fig. 19.5. Presumably, when a deuteride ion attacks

[41] W. Klemm and A. Neuber, *Z. anorg. Chem.*, **227**, 261 (1936).

[42] H. Zeiss and M. Tsutsui, *J. Am. Chem. Soc.*, **79**, 3062 (1957). (In this reference L. Onsager is credited with proposing the structures in the third column of Fig. 19.5.)

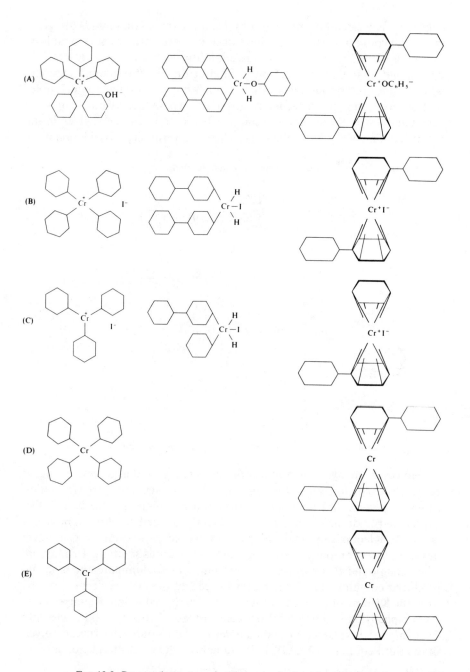

FIG. 19.5. Proposed structures for the aromatic chromium compounds.

one of the four phenyl groups of compound B [bis(biphenyl)chromium(I) iodide], the entire complex collapses with expulsion of a nondeuterated biphenyl molecule and the formation of one monodeuterobiphenyl molecule and evolution of hydrogen. Indiscriminate deuteride reduction at either benzene or biphenyl sites in compound C [benzenebiphenylchromium(I) iodide] would explain the $6.7D$ per cent in this case.

In the bis(arene)chromium(0) compounds (D and E), each chromium atom accepts 12π electrons to attain a rare gas (krypton) electronic configuration. The same effective electronic configuration is attained by the iron atoms in the analogous compounds bis(cyclopentadienyl)iron(II) [$Fe(C_5H_5)_2$] and cyclobutadieneiron(0) tricarbonyl [$C_4H_4Fe(CO)_3$].

Problems

1. One of the products of the pyrolysis of beryllium diethyl is a white solid. When this solid is treated with excess water, both methane and hydrogen are evolved in the ratio $CH_4/H_2 = \frac{1}{4}$. If the white solid is a pure compound, what can be said about its structure?

2. There exist two isomers (A and B) of dimethyl diborane. When 1 mole of isomer A is hydrolyzed, 1 mole of boric acid and $\frac{1}{2}$ mole of bis(dimethylboryl)oxide, $(CH_3)_2BOB$-$(CH_3)_2$, are formed. When 1 mole of isomer B is hydrolyzed, $\frac{2}{3}$ mole of trimethyl-boroxin is formed:

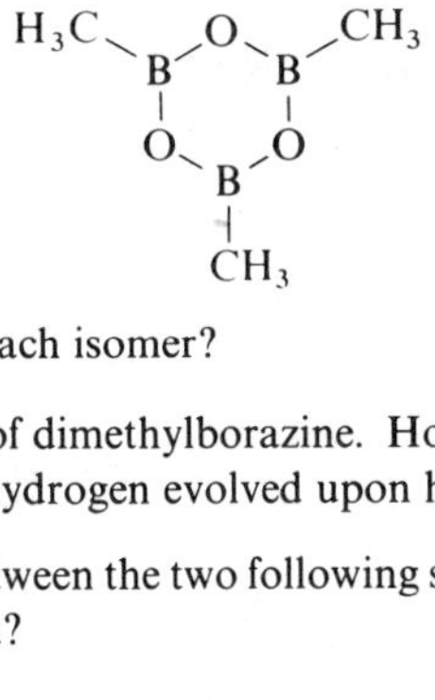

What is the structure of each isomer?

3. There exist four isomers of dimethylborazine. How would you partially distinguish them by the amounts of hydrogen evolved upon hydrolysis?

4. How might one decide between the two following structures for the pyrosulfite ion on the basis of chemical data?

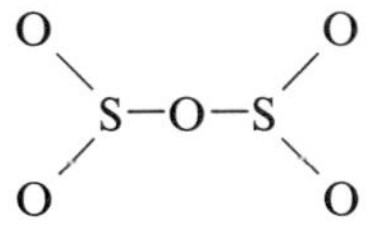

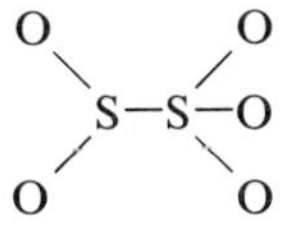

5. A sodium polyphosphate solution is passed through a cation exchange resin column in the hydrogen form (thus converting all the sodium ions to protons). The resulting solution is titrated with $0.1\ M$ NaOH. Two endpoints are found (at pH 4 and 9) at 42 ml and 50 ml. If the polyphosphate consists of linear chains, what is the average number of phosphorus atoms per chain?

6. The equilibrium constant for the dissociation of a chloride ion from $AgCl_{(aq)}$ is much smaller than that for the dissociation of a chloride ion from $AgCl_2{}^-$.

$$AgCl_{(aq)} = Ag^+ + Cl^- \qquad K = 4.9 \times 10^{-4}$$
$$AgCl_2{}^- = AgCl_{(aq)} + Cl^- \qquad K = 1.2 \times 10^{-2}$$

May we conclude that the two chlorine atoms in $AgCl_2{}^-$ are structurally different?

7. The compound N_2O_3 is always in equilibrium with appreciable amounts of NO and NO_2. When $^{15}NO_2$ is introduced into such a mixture, the ^{15}N is rapidly exchanged among the ^{14}N atoms in the N_2O_3 and NO. From this information, what, if anything, can be said about the structure of N_2O_3?

8. When liquid SO_2 is saturated with BF_3, and 70 per cent H_2SO_4 is then added, the compound $S_3O_8F_2$ separates out. This compound is a colorless liquid that hydrolyzes slowly in KOH solution, according to the equation:

$$S_3O_8F_2 + 4\,OH^- \longrightarrow 2\,SO_3F^- + SO_4{}^{2-} + 2\,H_2O$$

Suggest a structure in accord with these facts.

Chemical Analysis
and
Elementary
Physical Methods

20

QUANTITATIVE CHEMICAL ANALYSIS

There are two main approaches to the chemical analysis of a substance. In one approach, the analytical techniques are designed to determine or check the composition of the principal constituent. In the other approach, the analytical techniques are designed to determine the amounts of likely impurities. The techniques used in these two approaches often are quite different.

Techniques commonly used in the analysis of a principal constituent are titrimetry, gravimetric analysis, electrolytic analysis, and gas analysis. Titrimetric methods include acid-base titrations, precipitation titrations, oxidation-reduction titrations, and complexometric titrations. The endpoints in these titrations can be determined in a variety of ways, including the use of indicators, potentiometry, coulometry, conductometry, and so forth. Titrimetric analyses are generally preferred over other methods because they are usually reasonably accurate and simple to carry out. Two titrations are particularly important because of their wide applicability: the titration of an acid solution with standard 0.1 M sodium hydroxide, using an indicator for endpoint determination; and the titration of a triiodide solution with standard 0.1 N sodium thiosulfate, using starch as the endpoint indicator.

When the aim of an analysis is to estimate the purity of a compound rather than to establish its stoichiometric composition, it is important to know the most likely impurity present. Very little is learned by analyzing for an element or group that is present to about the same percentage in both the compound and the impurity. For example, elemental sulfur is usually the principal contaminant in heptasulfur imide, S_7NH; therefore, an analysis of S_7NH for nitrogen or hydrogen would be much more revealing than one for sulfur.

Often, the best indication of purity can be obtained from an analysis for the most likely impurity. Analytical techniques used for such purposes must be applicable to the determination of small amounts, yet they need not be as accurate as techniques designed for the principal constituent. The techniques include colorimetry, spectrophotometry, emission spectrographic analysis, flame photometry, gas chromatography, and other techniques discussed in the following chapters.

MELTING POINT DETERMINATION

The most common and simple method for the determination of the melting point of a solid consists in placing a small sample in a glass capillary tube, gradually heating the capillary, and visually noting the temperature at which melting occurs. The capillary must be thin walled, of small diameter (about 1 mm), and clean. Samples unaffected by air can be held in capillaries open at one end; otherwise, completely sealed capillaries should be used. Evacuated sealed capillaries are convenient for air-sensitive materials, but solids that are appreciably volatile near their melting points must be sealed in capillaries containing an inert gas at 0.5 to 1.0 atm pressure.

The usual procedure for packing a capillary with solid consists in pushing the open end of the capillary into a small heap of the solid and knocking the material to the bottom of the capillary by tapping it on the table top. This procedure is repeated until the capillary is filled to a depth of 0.5 to 1.0 cm. When an air-sensitive material is being studied, these operations are carried out in an inert atmosphere box or bag, and the open end of the capillary is sealed with a piece of Apiezon Q sealing compound or Duck Seal. The capillary is then brought out into the atmosphere, and the upper end is sealed off with a flame.[1]

The melting-point capillary is placed in a stirred liquid bath or an electrically heated metal block. Mineral oil is commonly used as a bath liquid

[1] A capillary sealed off at atmospheric pressure should only be about one-third immersed in the melting-point bath if the melting point is above 100° because of the danger of bursting the capillary.

(up to about 200°), but silicone oil can be heated to about 350°. By using special thermometers or thermocouples and a metal heating block, melting points as high as 500° can be observed. It is important to have a heating rate of less than $1° \text{ min}^{-1}$ in the neighborhood of the melting point, but much faster heating is desirable when the temperature is more than 10° below the melting point. The capillary is placed in the bath so that the end containing the solid touches the bulb of the thermometer. If the solid is absolutely pure, the melting and freezing points are identical, and the melting-point range (the temperature interval in which observable amounts of both solid and liquid are present) is zero. Impurities cause a slight lowering of the freezing point and a marked lowering of the melting point. Consequently, the melting range is a commonly used measure of the purity of a substance.

The thermometer may be calibrated by using the melting points of known pure substances. A better method is to compare the thermometer directly against a standardized thermometer in a well-stirred bath at a series of temperatures. The working thermometer should be immersed to the same depth as it is in the melting-point apparatus, and, if necessary, emergent stem corrections should be applied to the readings of the standardized thermometer. The amount that should be added to the reading of a mercury thermometer designed for total immersion when it is only partially immersed is

$$\Delta T = 0.000154(t - n)l$$

where t is the observed temperature, n is the average temperature of the emergent stem, and l is the length in degrees of the mercury column above the bath.

A technique for determining melting points below room temperature is described on pp. 174–5, and the cooling-curve method of determining high-temperature melting points is described on p. 211.

CRYOSCOPIC DETERMINATION OF MOLECULAR WEIGHT

The freezing point depression of a solution is commonly used to determine the molecular weight of a solute. The molecular weight M may be calculated from the relation

$$M = K_c \frac{w}{\Delta T}$$

where w is the weight (in grams) of solute per 1000 g of solvent, ΔT is the observed depression, and K_c is the so-called cryoscopic constant of the solvent, defined as

$$K_c = \frac{RT_f^2 M_s}{1000 \Delta H_f}$$

where R is the ideal gas constant, T_f is the freezing point of the solvent (°K), M_s is the molecular weight of the solvent, and ΔH_f is the heat of fusion of the solvent. The freezing points and cryoscopic constants for some solvents commonly used for cryoscopic molecular weight determinations are given in Table 20.1. Apparatus of the type shown in Fig. 20.1 can be used for determining freezing-point depressions in benzene solutions. The inner test tube,

Table 20.1 **Some Cryoscopic Constants**

Solvent	Freezing Point (°C)	K_c
Ammonia	−77.7	0.957
Water	0	1.86
Benzene	5.51	5.12
Sulfuric acid	10.37	6.12
Acetic acid	16.7	3.9
Naphthalene	80.2	6.9
Camphor	179	47.5

containing the solution, the thermometer, and a ring-shaped stirrer, is insulated from the surrounding ice bath by an air gap. The thermometer is a Beckmann thermometer with a 5 or 6° range adjusted to span the temperature range expected. The thermometer should be graduated in units of 0.01 or 0.02° and should be read (by interpolation) to 0.001 or 0.002°. The observed cooling curve will be similar to that of Fig. 11.3B, but the curve beyond the freezing point, X, will not be as steep as indicated. If a small amount of supercooling occurs, the true freezing point may be estimated by extrapolating back that part of the curve that corresponds to freezing out of the solvent until it intersects the cooling curve of the liquid solution.

Cryoscopic molecular weights are only valid for solutions which are sufficiently dilute that their properties are those of ideal solutions. It is advisable to determine cryoscopic molecular weights as a function of concentration in order to detect any possible variation with concentration. If such variation is found, the data should be extrapolated to infinite dilution. Other methods for determining molecular weights have been discussed in Chap. 8.

DENSITY DETERMINATION

Liquids

A variety of methods are available for determining the density of a liquid. The more accurate methods involve weighing a known volume of the liquid. When an accuracy of ±0.5 per cent is adequate, and the liquid is not highly

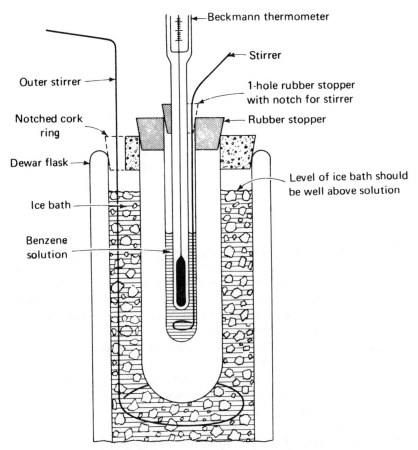

FIG. 20.1. Apparatus for cryoscopic determination of molecular weight. (Adapted from D. P. Shoemaker and C. W. Garland, "Experiments in Physical Chemistry," McGraw-Hill Book Company, New York, 1962, p. 137.)

volatile, the known volume can be measured out from a buret into a weighing bottle. For the highest accuracy, the liquid is contained in a pycnometer, which is a vessel with a marked capillary top. The volume is determined by filling the tared pycnometer with pure water to the mark, weighing, and noting the temperature. From a knowledge of the density of water (see Table 20.2), the volume is readily calculated. The pycnometer is then dried, filled with the liquid of unknown density, and weighed.

When a relatively large amount of liquid is available, buoyancy methods can be used. The simplest buoyancy device is the hydrometer—a glass float, weighted at the bottom, with an upright calibrated stem at the top. One floats the hydrometer in the unknown liquid in a cylindrical vessel, sights across the surface of the liquid, and reads the density on the stem of the hydrometer. A more accurate buoyancy device is the Westphal balance,

Table 20.2   The Density of Water at Various Temperatures

Temperature (°C)	Density
14	0.9993
16	0.9990
18	0.9986
20	0.9982
22	0.9978
24	0.9973
25	0.9971
26	0.9968
28	0.9963
30	0.9957
32	0.9950
34	0.9944

which is based on the Principle of Archimedes. The apparatus consists of a glass body, weighted with mercury and containing a thermometer, which is suspended by a thin platinum wire from one arm of a balance. The glass body (the volume of which is accurately known) is weighed both in air and while suspended in the unknown liquid. The difference in weight divided by the known volume gives the density of the liquid.

A very simple but relatively inaccurate device for determining density is based on the fact that the hydrostatic pressure of a column of liquid is proportional to both the height of the liquid and the density of the liquid. One end of a U tube containing a liquid of known density is connected to one end of another U tube containing a liquid of unknown density; then the pressure in the connected region is either increased or decreased. The ratio of the height differences in the U tubes is inversely proportional to the ratio of the liquid densities.

Solids

For determining the density of a solid, one can use a pycnometer with a wide mouth and a ground-glass stopper through which a capillary is bored. Following the pycnometric procedure described above, the density is determined for a liquid of density lower than that of the solid and in which the solid has no detectable solubility or reactivity. The weight of the dry pycnometer containing the granulated solid is determined. The pycnometer is then filled with the liquid (plus the solid) and reweighed. The greatest source of error in this technique is the trapping of air by the solid. This problem may be avoided by applying a partial vacuum until air bubbles have ceased rising from the solid and then by refilling the pycnometer with liquid.

The density of a solid sometimes can be determined by finding an inert liquid in which the solid neither sinks nor floats (and in which it does not dissolve). Fine adjustments in the density of the supporting liquid can be made by mixing two liquids of different densities. The density of the final liquid can then be determined by one of the methods previously described.

Probably the most accurate way to determine the density of a solid is by determining the dimensions of the unit cell by X-ray diffraction (see Chap. 28). From a very crude (± 5 per cent) density determination (say, by flotation) and a knowledge of the unit cell volume, the number of "molecules" per unit cell can be calculated. Then the latter two quantities can be used to calculate a very accurate density.

Infrared

Spectrometry

21

Infrared spectrometry is used by chemists for three different purposes: the identification of compounds; the quantitative analysis of mixtures; and structure determination.

IDENTIFICATION OF COMPOUNDS

Practically every compound has a unique infrared spectrum. Thus, an infrared spectrum can be used to characterize a compound in much the same way that the melting point, boiling point, or other physical property is used. The infrared spectrum of a compound is very easy to obtain, even when only a small amount of material is available. The main features of the spectrum are not markedly affected by the presence of other substances; therefore, even an impure sample can be identified if the spectrum of the pure material is known.

The infrared spectrum of a compound may be used as a "fingerprint" of the compound. To identify a compound, its spectrum is compared with the spectra of a limited number of possible compounds suggested by other properties and the source of the compound. When a matching spectrum is found, identification is complete.

Unfortunately, infrared spectra are not as commonly tabulated as melting points, boiling points, specific gravities, refractive indexes, and so on; consequently, one is often forced to do considerable library research to find comparison spectra. However, there are a few collections of spectra and bibliographies of spectral studies that are helpful in such searches. These are American Petroleum Institute Research Project 44, Catalog of Infrared Spectral Data (an extensive catalog of spectra, principally for hydrocarbons). R. H. Pierson, A. N. Fletcher, and E. S. Gantz, *Anal. Chem.*, **28**, 1218 (1956) (a catalog of the gas spectra of 65 volatile compounds, both organic and inorganic). F. A. Miller and C. H. Wilkins, *Anal. Chem.*, **24**, 1253 (1952) (a catalog of the Nujol-mull spectra of 159 inorganic compounds, mostly salts containing polyatomic ions). J. M. Hunt, M. P. Wisherd, and L. D. Bonham, *Anal. Chem.*, **22**, 1478 (1950) (a catalog of the spectra of 64 minerals and related compounds). F. A. Miller, G. L. Carlson, F. F. Bentley, and W. H. Jones, *Spectrochim. Acta*, **16**, 135 (1960) [infrared spectra of inorganic ions in the cesium bromide region (300 to 700 cm^{-1})]. *An Index of Published Infrared Spectra*, vol. 2, Ministry of Aviation, London, 1960, pp. 747–805 (an extensive bibliography, arranged alphabetically according to compounds). K. Nakamoto, *Infrared Spectra of Inorganic and Coordination Compounds*, John Wiley & Sons, Inc., New York, 1963 (a very useful source book for the spectra of a wide variety of inorganic compounds). *The DMS (Documentation of Molecular Spectroscopy) Index*, Butterworth & Co., Ltd., London (a collection of infrared spectra and a literature service that can be consulted by a punched-card index). "High Resolution Spectra of Inorganic and Related Compounds," Sadtler Research Labs., Inc., 1965 (a collection of 600 spectra of solids on cards).

As an example of the use of infrared spectroscopy in the identification of a compound, consider the compound heptasulfur imide, S_7NH. It is a very pale yellow, almost colorless solid that melts at 113.5°. It is difficult to distinguish from elementary sulfur (m.p. = 113 to 119°) either by its appearance or by its melting point. However, sulfur has no absorption peaks in the rock salt region, whereas heptasulfur imide has the spectrum shown in Fig. 21.1. The peak at 3300 cm^{-1} corresponds to the N-H stretching vibration.

The infrared spectrum of a mixture of compounds can be used to identify the components of the mixture if the absorption bands of the various components do not overlap too much. However, unknown mixtures containing more than three major components usually have spectra that are so complicated as to be undecipherable, and it is usually advisable in such cases to separate the components before attempting spectral identification.

Infrared spectra are very useful criteria of purity. If the infrared spectrum of a compound shows a band that does not exist in the spectrum of the pure

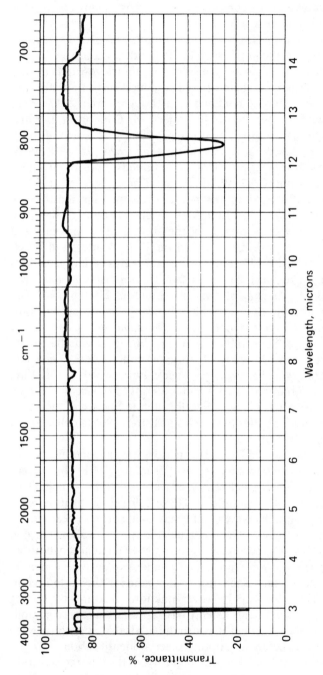

FIG. 21.1. Infrared spectrum of S$_7$NH in benzene against a reference cell containing benzene.

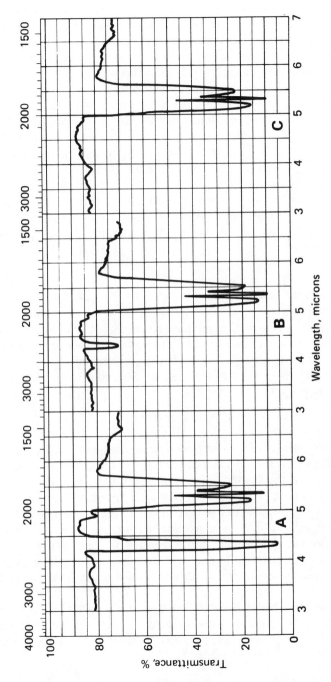

Fig. 21.2. Infrared spectra of SnH_4-CO_2 mixtures at various stages of CO_2 removal.

material, it is obvious that the compound is contaminated. One usually has some ideas as to possible impurities, so even impurities present in low concentration (and hence giving weak absorption bands) can often be identified.

For an example of the use of infrared spectroscopy in establishing the purity of a compound, consider the purification of stannane. When stannane is prepared by the addition of a stannite-hydroborate solution to acid, the principal impurities are water and carbon dioxide.[1] The water is easily removed by fractional condensation, but the carbon dioxide cannot be separated in this way. Therefore, an investigation was made to determine whether stannane could be freed of carbon dioxide by passing it through a trap packed with a molecular sieve. The infrared spectrum of a sample of stannane contaminated with carbon dioxide is presented in Fig. 21.2A. The peak at 2300 cm^{-1} is due to CO_2; the other features are due to SnH_4. After the gas passed twice through a molecular sieve, the infrared spectrum was as shown in Fig. 21.2B. Two more passes yielded stannane with the spectrum shown in Fig. 21.2C.

QUANTITATIVE ANALYSIS

In the quantitative analysis of a mixture containing several compounds exhibiting absorption spectra, it is necessary to choose, for each compound to be determined, an absorption peak such that there is minimum interference by the peaks of the other compounds. Then, for each such absorption peak, one may apply Beer's Law:

$$A = \log\left(\frac{T_0}{T}\right) = \varepsilon l c$$

where A is the absorbance of the compound being analyzed, ε is the decadic molar extinction coefficient of the compound, l is the cell thickness, c is the concentration of the compound in moles per liter, T is the fraction of the incident light transmitted by the mixture (the transmittance), and T_0 is the fraction of the light transmitted when $c = 0$. This relation is applicable to compounds in liquid solutions, in potassium bromide pellets, and in the gas phase. Most infrared spectrometers record spectra on scales that are linear in transmittance. The quantity T_0 may be determined by simply recording the base line, that is, the spectrum recorded in the absence of the substance to be analyzed.

The compound $[Co(NH_3)_5NO_2]Cl_2$ exists in two isomeric forms. In the "nitro" isomer, the NO_2 group is coordinated through its nitrogen atom: $Co-NO_2$. In the "nitrito" isomer, the NO_2 group is coordinated through one of its oxygen atoms: $Co-O-N=O$.

---

[1] W. L. Jolly and J. E. Drake, *Inorg. Syn.*, **7**, 34 (1963).

The infrared spectra[2] of both isomers show absorption peaks at 850, 1315, and 1595 cm^{-1}. These peaks correspond to those observed in the spectrum of $[Co(NH_3)_6]Cl_3$, and they are attributable to NH_3 vibrations.[3] In addition to these peaks, the nitrito isomer has peaks at 1065 and 1460 cm^{-1}, and the nitro isomer has peaks at 825 and 1430 cm^{-1}.[3] From Fig. 21.3 we see that

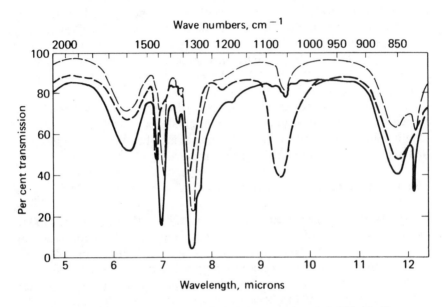

FIG. 21.3. Infrared spectra of nitro- and nitritopentaminecobalt (III) chloride complexes.[2] (Reproduced with permission of the American Chemical Society.)

——————— $[Co(NH_3)_5NO_2]Cl_2$
– – – – – – $[Co(NH_3)_5ONO]Cl_2$, freshly prepared
· · · · · · · · $[Co(NH_3)_5ONO]Cl_2$, aged

the nitrito peak at 1065 cm^{-1} is free from interference by other bands. Beattie and Satchell[4] used this absorption peak, in conjunction with those at 1430 and 1460, to determine the relative amounts of the two isomers present in mixtures. Thus, they were able to measure rates of isomerization and equilibrium constants for isomerization at a series of temperatures. Similar measurements have been made for the analogous complexes of rhodium(III), iridium(III), and platinum(IV).[5]

[2] R. B. Penland, T. J. Lane, and J. V. Quagliano, *J. Am. Chem. Soc.*, **78**, 887 (1956).

[3] The peak at 1315 cm^{-1} accidentally corresponds to both an NH_3 symmetric deformation and an NO_2 symmetric stretching.

[4] I. R. Beattie and D. P. N. Satchell, *Trans. Faraday Soc.*, **52**, 1590 (1956).

[5] F. Basolo and G. S. Hammaker, *Inorg. Chem.*, **1**, 1 (1962).

STRUCTURE DETERMINATION[6]

Group Frequencies

It is usually very difficult to assign vibrations to all the absorption bands of a molecule containing more than about six atoms. Nevertheless, infrared spectra of complicated molecules are often extremely useful in structure determinations, because certain groups of atoms always absorb at roughly the same frequency, regardless of the rest of the molecule in which they are located. For example, the group frequencies of the $-NH_2$ group are 3000 to 3500, 1550 to 1700, 1050 to 1200, and 700 to 900 cm^{-1}. A chart giving the group frequency ranges for 31 inorganic anions is given by Ferraro.[7] Charts for phosphorus and sulfur compounds, hydrogen stretching frequencies, oxygen stretching and bending frequencies, halogen stretching frequencies, 22 inorganic anions, and 11 simple ligands in metal complexes are given by Nakamoto.[8] A chart of organic groups is given by Colthup.[9]

Hydrogen Stretching Frequencies

Practically all hydrogen stretching frequencies lie in the range from 1900 to 4000 cm^{-1}. This is rather a wide frequency range, and this information by itself is not of much value in identifying hydrogen atoms in molecules. However, the range of frequencies is greatly reduced if one specifies the atom to which the hydrogen is bound (e.g., Si–H frequencies fall in the range from 2100 to 2250 cm^{-1}). It has been observed that the average frequency for X–H groups varies in a regular way with the electronegativity of X.[10] In Fig. 21.4, X–H stretching frequencies are plotted against the Pauling electronegativities of the X atoms. The points fall on a series of straight lines, each line characteristic of a particular periodic table family. For a series of compounds containing the same X–H group, the stretching frequency is affected by the electronegativity of the other atom or atoms attached to the X atom. Thus, it has been shown that the Si–H frequency varies linearly with the sum of either the electronegativities of the three atoms attached to the Si or the Taft inductive factors (σ^*) of the three groups attached to the Si. A plot of Si–H frequencies against $\Sigma\sigma^*$ is given in Fig. 21.5.[11] It is clear that one

[6] See E. A. V. Ebsworth, "Inorganic Applications of Infra-Red Spectroscopy," in *Infra-Red Spectroscopy and Molecular Structure*, M. Davies, ed., Elsevier, New York, 1963, Chap. 9.

[7] J. R. Ferraro, *J. Chem. Educ.*, **38**, 201 (1961).

[8] K. Nakamoto, *Infrared Spectra of Inorganic and Coordination Compounds*, John Wiley & Sons, Inc., New York, 1963.

[9] N. B. Colthup, *J. Opt. Soc. Am.*, **40**, 397 (1950).

[10] L. J. Bellamy, *The Infra-red Spectra of Complex Molecules*, 2nd ed., John Wiley & Sons, Inc., New York, 1958, p. 392.

[11] H. W. Thompson, *Spectrochim. Acta*, **16**, 238 (1960).

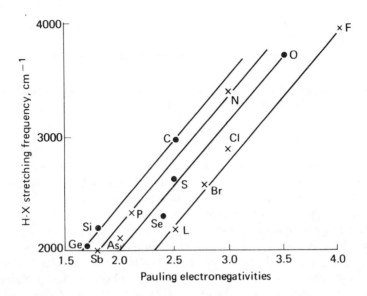

FIG. 21.4. X–H stretching frequency against electronegativity of X.[10] (Reproduced with permission of John Wiley & Sons, Inc.)

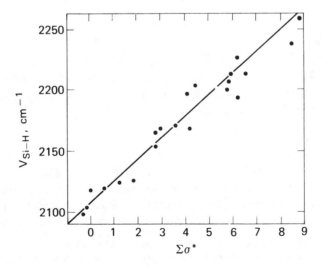

FIG. 21.5. Si–H stretching frequency against $\Sigma\sigma^*$ for groups attached to Si atom.[11] (Reproduced with permission of Pergamon Press.)

can predict a particular Si–H frequency in a compound with an uncertainty of about $\pm 10 \, \text{cm}^{-1}$.

Whenever there is any doubt that a particular absorption band corresponds to a motion of hydrogen atoms, the corresponding deuterated compound can be prepared, and the spectrum compared. A vibration frequency may be approximately represented by the function

$$v = \frac{1}{2\pi c} \sqrt{\frac{k}{\mu}}$$

where k is the force constant of the vibration and μ is the reduced mass. Because deuterium substitution causes the reduced mass to roughly double without appreciably changing the force constant, deuteration causes hydrogen stretching frequencies to change by roughly 1/1.4.

An interesting example of the use of infrared spectroscopy is the work of Piper[12] on the compound $NiS_4N_4H_2$. This compound, prepared by the reaction of S_4N_4 with anhydrous $NiCl_2$ in boiling methanol, had been formulated earlier as $Ni(SN)_4$.[13] However, the infrared spectrum of a Nujol mull of the compound contained bands at 3235 and 3095 cm^{-1} that were suspected of being due to N–H stretching vibrations. This assignment was confirmed by showing that these bands were shifted to 2435 and 2360 cm^{-1} upon deuteration. The compound probably has a structure similar to the following:

Later work[14] showed that the hydrogen atoms may be replaced by methyl groups by first forming a silver salt by treatment with silver nitrate, followed by reaction with methyl iodide.

Metal Carbonyls

The stretching frequency of carbon monoxide is 2168 cm^{-1}. The C—O bond order is slightly lower[15] when the carbon monoxide molecule is coordinated to a metal atom; consequently, the stretching frequency for such a CO group lies in a slightly lower frequency range, 2000 to 2100 cm^{-1}. A bridging carbon

[12] T. S. Piper, *Chem. Ind. London*, 1101 (1957); *J. Am. Chem. Soc.*, **80**, 30 (1958).

[13] M. Goehring, *Ergebnisse und Probleme der Chemie der Schwefelstickstoffverbindungen*, Akademie-Verlag, Berlin, 1957, pp. 151–52.

[14] J. Weiss and M. Ziegler, *Z. anorg. allgem. Chem.*, **322**, 184 (1963).

[15] Because of π bonding between the metal and carbon atoms.

monoxide molecule—that is, one which is bonded to two different metal atoms—is analogous to a keto group. Because the C—O bond order in a keto group is less than that in carbon monoxide, it is not unexpected that ketonic frequencies lie in a lower frequency range, 1700 to 1800 cm^{-1}. Sheline and Pitzer[16] observed a band at 1828 cm^{-1} (in addition to the usual metal carbonyl frequencies near 2000 cm^{-1}) in the infrared spectrum of $Fe_2(CO)_9$, which was known to contain bridging CO groups. Hence, they assigned the 1828 cm^{-1} band to the bridging CO groups. Whenever bands are observed in the vicinity of 1800 to 1900 cm^{-1} in other metal carbonyls, they usually may be taken as evidence of bridging CO groups.

The infrared spectrum of $Co_2(CO)_8$ shows only terminal (about 2050 cm^{-1}) and bridging (approximately 1860 cm^{-1}) CO stretching bands[17] (see Fig. 21.6). This result is consistent with the X-ray-determined structure shown in Fig. 21.7. The infrared spectrum of $Mn_2(CO)_{10}$ shows no absorption in the region of 1800 to 1900 cm^{-1} (see Fig. 21.8); consequently, it was

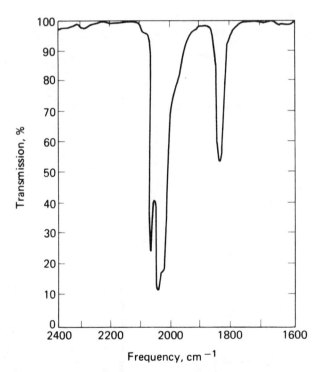

Fig. 21.6. Infrared spectrum of $Co_2(CO)_8$.[17] (Reproduced with permission of the American Chemical Society.)

[16] R. K. Sheline and K. S. Pitzer, *J. Am. Chem. Soc.*, **72**, 1107 (1950).

[17] J. W. Cable, R. S. Nyholm, and R. K. Sheline, *J. Am. Chem. Soc.*, **76**, 3373 (1954).

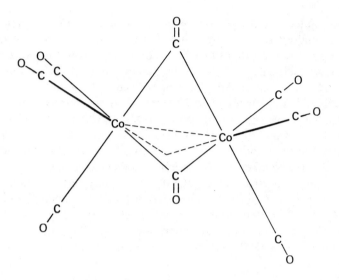

FIG. 21.7. Structure of $Co_2(CO)_8$.

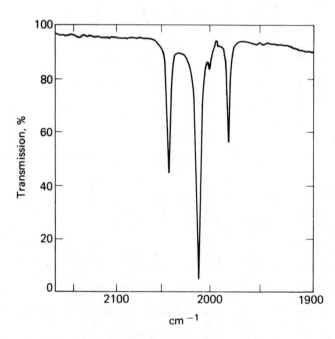

FIG. 21.8. Infrared spectrum of $Mn_2(CO)_{10}$.[19] (Reproduced with permission of the American Chemical Society.)

correctly predicted[18,19] that in this compound there are no bridging CO groups and that the molecule consists of two $Mn(CO)_5$ groups joined by an Mn—Mn bond (Fig. 21.9).

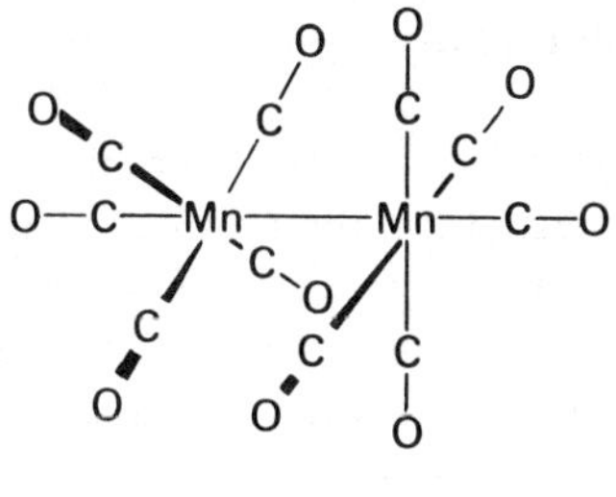

FIG. 21.9. Structure of $Mn_2(CO)_{10}$.

Dinitrogen Trioxide

The nitro group, in compounds of the type $R-O-NO_2$, has a characteristic N—O asymmetric stretching frequency near $1600 \, cm^{-1}$ and an N—O symmetric stretching frequency near $1300 \, cm^{-1}$. The —N=O group in nitrites $(RO-N=O)$ has a stretching frequency in the range 1600 to 1700 cm^{-1}. Most nitrites show two bands in this region because of the existence of both *cis* and *trans* isomers. However, when the —N=O group is attached to an electronegative group, such as chlorine or bromine, the N=O frequency falls near $1800 \, cm^{-1}$. The apparently increased N—O bond order in the latter compounds is attributable to contributions from resonance forms of the type $O \equiv N^+ X^-$.

Gaseous N_2O_3 shows bands at 1830, 1615, and $1309 \, cm^{-1}$. The first band has been assigned to the —N=O group and the last two bands to the —NO_2 group.[20] Thus, infrared spectroscopy supports the $O=N-NO_2$ structure for N_2O_3, as opposed to the $O=N-O-N=O$ structure.

Normal Modes and Infrared Activity

The complex vibrational motions of a molecule are the result of the superposition of a number of simple vibrational motions called the *normal modes* of vibration of the molecule. An N-atom molecule has $3N - 6$ normal modes of vibration. By a very simple application of group theory, it is possible to determine the symmetry types of these normal modes—that is, the irreducible representations for which the normal modes are bases. Then, using the character table for the molecular point group, it is an easy matter to determine

[18] E. O. Brimm, M. A. Lynch, Jr., and W. J. Sesny, *J. Am. Chem. Soc.*, **76**, 3831 (1954).

[19] N. Flitcroft, D. K. Huggins, and H. D. Kaesz, *Inorg. Chem.*, **3**, 1123 (1964).

[20] L. D'Or and P. Tarte, *Bull. Soc. Roy. Sci. Liege*, **22**, 276 (1953).

which modes are infrared active (i.e., observable as infrared absorption bands). From the symmetry types it is also obvious which vibrational modes are degenerate (i.e., which have identical vibrational frequencies). It is generally true that the more symmetrical a molecule is, the fewer are the number of observed infrared absorptions—that is, the simpler the infrared spectrum is.

The number of absorption bands observed in an infrared spectrum is usually considerably greater than the number of active normal modes. The extra bands, which are usually relatively weak, are attributable to the presence of *overtones, combination bands*, and *difference bands*. For example, consider the spectrum of sulfur dioxide, given in Table 21.1. The three expected fundamentals are observed as intense bands at 519, 1151, and 1361 cm^{-1}. The band at 2305 cm^{-1} is the overtone of v_1, corresponding to excitation to the second vibrational level. The bands at 1871 and 2499 cm^{-1} are combination bands, in which two normal modes are simultaneously excited. The band at 606 cm^{-1} is a difference band, in which there is a transition from one excited mode to another excited mode.

Table 21.1   Infrared Spectral Data for SO_2

$v(cm^{-1})$	Intensity	Assignment
519	strong	v_2
606	weak	$v_1 - v_2$
1151	very strong	v_1
1361	very strong	v_3
1871	very weak	$v_2 + v_3$
2305	weak	$2v_1$
2499	medium	$v_1 + v_3$

A free molecule will exhibit not only vibrational motions but also translational and rotational motions. The combination of all these motions in an N-atom molecule at any instant can be represented by the $3N$ Cartesian coordinate vectors of the atoms. These vectors can serve as the basis for a reducible representation of the molecular point group. When this reducible representation is reduced to its irreducible representations, those irreducible representations corresponding to translation and rotation can be readily identified and deleted. Those remaining are the labels of the vibrational modes.

The following procedure is followed:

1. Determine the point group of the molecule.
2. For each class of symmetry operation R, determine $n_u(R)$, the number of atoms having positions unchanged by carrying out the operation.
3. Multiply each value of $n_u(R)$ by the appropriate factor $f(R)$, found in

**Table 21.2   Values of $f(R)$ for
Various Operations**

R	$f(R)$
E	3
σ	1
i	-3
C_2	-1
$C_3^{\ 1}, C_3^{\ 2}$	0
$C_4^{\ 1}, C_4^{\ 3}$	1
$C_6^{\ 1}, C_6^{\ 5}$	2
$S_3^{\ 1}, S_3^{\ 2}$	-2
$S_4^{\ 1}, S_4^{\ 3}$	-1
$S_6^{\ 1}, S_6^{\ 5}$	0

Table 21.2[21] (for the meaning of the symbols under R in Table 21.2, see Appendix 11, including footnote 8 on p. 554). The resulting products $n_u(R)f(R)$, are the characters of the reducible representation of the $3N$ Cartesian coordinate vectors.

4. Reduce the reducible representation into its irreducible representations.

5. From the right side of the character table, we can determine which irreducible representations correspond to the translations and rotations (translations are indicated by x, y, and z, and rotations by R_x, R_y, and R_z). These irreducible representations are deleted from the list determined in step 4. Remember that the total degeneracy of the deleted irreducible representations must be six (e.g., $2T$, $A + E + T$, etc.). The remaining irreducible representations are the symmetry types of all the genuine normal vibrations of the molecule. The total degeneracy of the irreducible representations equals the number of normal vibrational modes. Those modes of doubly degenerate symmetry types (e.g., E) occur in pairs having identical frequencies, and those of triply degenerate symmetry types (e.g., T) occur in groups of three having identical frequencies.

6. For a molecular vibration to be excited to a higher energy level by the absorption of light of the appropriate frequency, it is necessary that some component of the dipole moment change during the vibration. Inasmuch as the dipole moment vectors can be represented by the coordinates x, y, and z, a vibration with an irreducible representation for which x, y, or z forms the basis is infrared active. All such vibrations are readily identified from the character table.

As an example of the application of the preceding rules, let us calculate the symmetry types of the normal modes of vibration for the three conceivable configurations of SF_4 shown in Fig. 21.10. Let us first consider the structure

[21] The $f(R)$ values may be calculated from the expression $2\cos(2\pi/n) + 1$ for C_n and from $2\cos(2\pi/n) - 1$ for S_n. Note that $E = C_1$, $\sigma = S_1$, and $i = S_2$.

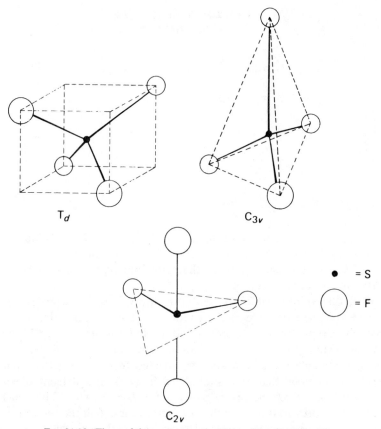

Fig. 21.10. Three of the many conceivable configurations for SF_4.

with C_{2v} symmetry. For each of the operations listed in the C_{2v} character table, we calculate the number of atoms with unchanged positions. These numbers appear in the following table:

C_{2v}	E	C_2	$\sigma_v(xz)$	$\sigma'_v(yz)$
$n_u(R)$	5	1	3	3

Beneath each number we write the appropriate factor $f(R)$ (from Table 21.2), and by multiplying the corresponding values of $n_u(R)$ and $f(R)$, we obtain the characters of the reducible representation.

C_{2v}	E	C_2	$\sigma_v(xz)$	$\sigma'_v(yz)$
$n_u(R)$	5	1	3	3
$f(R)$	3	-1	1	1
Γ	15	-1	3	3

The representation is readily reduced to $\Gamma = 5A_1 + 2A_2 + 4B_1 + 4B_2$. From the right side of the character table we see that, of the six translational and rotational vectors, one is of type A_1, one of type A_2, two of type B_1, and two of type B_2. By deleting these representations from Γ, we are left with

$$4A_1 + A_2 + 2B_1 + 2B_2$$

as the irreducible representations of the genuine normal vibrations. We see that there are nine such vibrations, in agreement with the formula $3N - 6$ for a five-atom molecule. From the character table, we determine that eight of the fundamental vibrations $(4A_1 + 2B_1 + 2B_2)$ are infrared active and that all nine are Raman active.

For the SF_4 structure with C_{3v} symmetry, we calculate the reducible representation as follows:

C_{3v}	E	$2C_3$	$3\sigma_v$
$n_u(R)$	5	2	3
$f(R)$	3	0	1
Γ	15	0	3

This representation reduces to $\Gamma = 4A_1 + A_2 + 5E$. The translations and rotations are $A_1 + A_2 + 2E$. Two of these symmetry species are doubly degenerate, and the total degeneracy of the translations and rotations is six. After deleting these representations, we are left with the symmetry types of the genuine vibrations, $3A_1 + 3E$. All six of these fundamental vibrational frequencies are both infrared and Raman active.

Finally, for the SF_4 structure with T_d symmetry, we calculate

T_d	E	$8C_3$	$3C_2$	$6S_4$	$6\sigma_d$
$n_u(R)$	5	2	1	1	3
$f(R)$	3	0	-1	-1	1
Γ	15	0	-1	-1	3

The reducible representations are $\Gamma = A_1 + E + T_1 + 3T_2$. The translations and rotations are $T_1 + T_2$, with a total degeneracy of six. After deleting these, we are left with $A_1 + E + 2T_2$ for the normal vibrations. The two triply degenerate vibrations are infrared active; all four vibrational frequencies are Raman active.

The infrared spectrum of SF_4 shows at least five bands, which, because of their intensity and occurrence in overtones or combinations, qualify as fundamentals.[22] This information is sufficient to rule out the T_d structure, but it does not permit one to choose between the C_{3v} and C_{2v} structures. It will be shown on p. 307 that, from a consideration of the band contours, it is

[22] R. E. Dodd, L. A. Woodward, and H. L. Roberts, *Trans. Faraday Soc.*, **52**, 1052 (1956).

possible to rule out all but the C_{2v} structure. Nuclear magnetic resonance and microwave spectroscopy have confirmed this structure assignment.[23]

The free sulfate ion, $SO_4{}^{2-}$, belongs to the high symmetry point group T_d, in which only two fundamentals, v_3 and v_4, are infrared active. In the complex $[Co(NH_3)_6]_2(SO_4)_3 \cdot 5\,H_2O$, v_3 and v_4 appear as strong bands (at 1130 to 1140 and 617 cm^{-1}, respectively) and v_1 appears as a very weak band at 973 cm^{-1}.[24] We may conclude that T_d symmetry is approximately maintained for the $SO_4{}^{2-}$ ion; the appearance of v_1 is probably due to a perturbation of the crystal field. In the complex $[Co(NH_3)_5SO_4]Br$, one of the sulfate oxygen atoms is coordinated to the cobalt atom. Hence, the local symmetry of the sulfate ion is C_{3v}, and one predicts six infrared-active bands. Indeed, six bands are observed: v_1, v_2, v_3 (split into a doublet), and v_4 (split into a doublet). In the complex $\left[(NH_3)_4Co \diagup\!\!{NH_2}\!\!\diagdown_{SO_4}\!\!\diagup\!\!\diagdown Co(NH_3)_4 \right]_2 (NO_3)_3$, the local symmetry of the bridging sulfate ion is C_{2v}, and one predicts eight infrared-active bands for the $SO_4{}^{2-}$ group, which are observed: v_1, v_2, v_3 (split into a triplet), and v_4 (split into a triplet). The infrared spectra of these three compounds in the region of v_1 and v_3 are shown in Fig. 21.11, where it is obvious that the reduction in local symmetry is accompanied by an increase in spectral complexity.[24]

Bond Stretching and Deformation[25]

Every fundamental vibration may be considered to be a bond stretching, a bond angle bending, or a combination of these motions. We shall now discuss the method for determining which fundamentals are made up, at least in part, of bond stretchings, and which fundamentals are made up, at least in part, of bond angle bendings.

First, let us consider the SF_4 molecule, which, as we have already stated, has C_{2v} symmetry. The four S—F bonds serve as the basis for the reducible representation

C_{2v}	E	C_2	$\sigma_v(xz)$	$\sigma_v'(yz)$
Γ	4	0	2	2

[23] F. A. Cotton, J. W. George, and J. S. Waugh, *J. Chem. Phys.*, **28**, 994 (1958); W. M. Tolles and W. D. Gwinn, *J. Chem. Phys.*, **36**, 1119 (1962).

[24] K. Nakamoto, *Infrared Spectra of Inorganic and Coordination Compounds*, John Wiley & Sons, Inc., New York, 1963, pp. 161–65.

[25] For more complete discussions, see F. A. Cotton, *Chemical Applications of Group Theory*, Interscience Publishers, New York, 1963, and E. Wilson, J. Decius, and P. Cross, *Molecular Vibrations*, McGraw-Hill Book Company, New York, 1955.

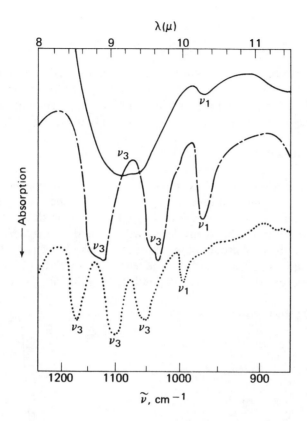

FIG. 21.11. Infrared spectra of $[Co(NH_3)_6]_2(SO_4)_3 \cdot 5H_2O$ (solid line);

$$\left[(NH_3)_4 Co \underset{SO_4}{\overset{NH_2}{\diagdown\diagup}} Co(NH_3)_4 \right] (NO_3)_3$$

$[Co(NH_3)_5SO_4]Br$ (dot-dash line);

(dotted line).[24] (Reproduced with permission of John Wiley & Sons, Inc.)

This reduces to $\Gamma = 2A_1 + B_1 + B_2$, showing that all the fundamental vibrations of symmetry types A_1, B_1, or B_2 involve some degree of S—F bond stretching. For each of these symmetry types, there are fewer terms in the reducible representation than there are vibrational fundamentals. Consequently, we may state that all the fundamentals of these symmetry types also involve some F—S—F bending. Only one fundamental vibration, of symmetry A_2, involves no bond stretching. This Raman-active vibration involves only bond angle deformations.

Second, let us consider the carbonate ion, which has D_{3h} symmetry. By following the procedure already outlined, we determine that there are four fundamental vibrational frequencies,

$$A_1' + 2E' + A_2''$$

The three C—O bonds serve as the basis for the reducible representation 3 0 1 3 0 1, which reduces to $A_1' + E'$. We conclude that the A_1' vibration is a pure stretching mode, that the E' modes combine both stretchings and bond angle bendings, and that the A_2'' mode involves only bond angle bendings.

Band Contours[26]

Vibrational transitions of gaseous molecules can be accompanied by relatively small changes in rotational energy, which give fine structure to the vibrational bands. This structure is usually unresolved for all but very light molecules. However, the band envelopes are observed, and these have contours related to the magnitudes of the three moments of inertia of the molecule, I_A, I_B, and I_C (given in order of increasing magnitude). Four typical contours are observed: *PQR*, **PQR**, *PR*, and *PQQR* (illustrated in Fig. 21.12).

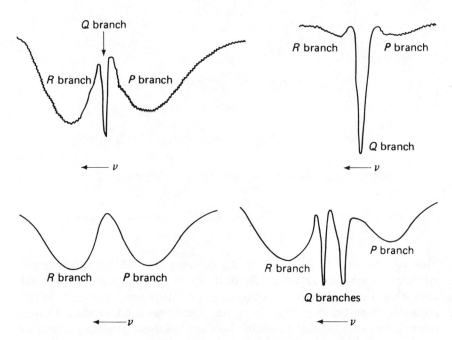

FIG. 21.12. Typical infrared band contours: *PQR*, **PQR**, *PR*, and *PQQR*.

[26] See M. J. Ware, in *Physical Methods in Advanced Inorganic Chemistry*, H. A. O. Hill and P. Day, eds., Interscience Publishers, New York, 1968, Chap. 6, p. 214, and R. M. Badger and L. R. Zumwalt, *J. Chem. Phys.*, **6**, 711 (1938).

The *PQR* band shows three branches of similar intensity. The *PQR* band shows a very prominent middle (*Q*) branch. The *PR* band has no *Q* branch, and the *PQQR* band has a doublet *Q* branch.

Linear molecules ($I_A = 0$, $I_B = I_C$), such as CO_2, show *PR* and *PQR* bands.[27] Spherical tops ($I_A = I_B = I_C$), such as SiH_4, show only *PQR* bands. Symmetric tops ($I_A < I_B = I_C$, or $I_A = I_B < I_C$), such as BCl_3 and GeH_3I, show *PQR* and *PQR* bands. Asymmetric tops ($I_A < I_B < I_C$), such as CSF_2, show *PQR*, *PQR*, and either *PR* or *PQQR* bands.

A consideration of band contours in the vapor infrared spectrum of SF_4 was decisive in establishing the structure of this molecule.[21] One of the principal bands (that at 728 cm^{-1}) was observed to have the *PQQR* type contour. This observation eliminates all the structures indicated in Fig. 21.10 except the C_{2v} structure.

PROCEDURES[28]

Spectrophotometers

A large number of relatively inexpensive (less than \$7000) infrared spectrophotometers are available for use in the spectral region from about 650 to about 4000 cm^{-1}. Both NaCl prism and diffraction-grating instruments are available. Most of the spectrophotometers are double-beam instruments and record the spectra on scales linear in transmittance. They are all easy to use, and, despite their simplicity, are very satisfactory for the qualitative and semiquantitative identification of compounds.

The spectral region below 650 cm^{-1} can be covered using spectrophotometers that employ diffraction gratings or prisms made of KBr or CsBr.

Gaseous Samples

A convenient type of gas absorption cell is pictured in Fig. 21.13. The body of the cell is a glass cylinder with a groove at each end for a rubber O ring. A sodium chloride plate is pressed against the O ring at each end so as to make a vacuum-tight seal. Gases are introduced through a greaseless stopcock, and, if desired, they may be condensed in a small side arm protruding from the body of the cell. The metal frame holding the NaCl plates onto the cell has a rectangular plate at one end that permits the cell to be easily mounted in the sample beam of the spectrophotometer.

[27] Diatomic molecules have only *PR* bands.

[28] See A. E. Martin, "Instrumentation and General Experimental Methods," *Infra-Red Spectroscopy and Molecular Structure*, M. Davies, ed., Elsevier Publishing Co., New York, Chap. 2, 1963.

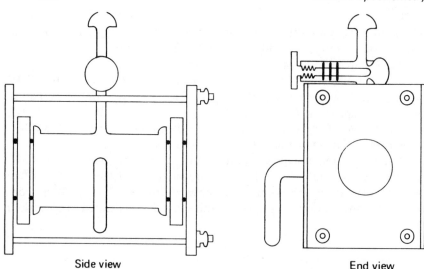

Side view   End view

FIG. 21.13. Infrared cell for gases.

The proper pressure to use for a satisfactory spectrum differs from compound to compound and is usually determined by trial and error. Thus, for a cell with a 10-cm path length, pressures of 700, 70, and 7 mm give good spectra for the compounds CO, NH_3, and Si_3H_8, respectively.

Liquid Samples

If a liquid compound has a sufficiently high vapor pressure at room temperature, it may be distilled into a gas cell and the spectrum of the vapor can be measured. Alternatively, the spectrum of the pure liquid or a solution of the liquid in some solvent can be measured.

The Barnes Engineering Company[29] manufactures "cavity cells" made of sodium chloride, which serve as simple and inexpensive infrared cells for liquids and solutions. These cavity cells are made from single blocks of salt and are filled with either a hypodermic syringe fitted with a number 26 needle or a medicine dropper. Path lengths from 0.1 to 5 mm are available. The Limit Research Corporation[30] manufactures inexpensive "throwaway" cells made of silver chloride. Path lengths from 0.025 to 1.0 mm are available. Silver chloride is transparent to about $22\,\mu$ and can be used with many aqueous solutions.

Demountable cells with path lengths that can be varied are more versatile than those just described. In this type of cell, the liquid sample is sandwiched between two salt plates held together in a metal frame. The liquid is prevented

[29] Located at Stamford, Conn.
[30] Located at Darien, Conn.

from squeezing out of the cell by a thin ring of Teflon or amalgamated metal serving as a spacer. By changing the spacer, the path length can be changed. In the more elegant cells of this type, one of the salt plates has two holes drilled through it, and provision is made for filling the cell with a hypodermic syringe without the need of demounting the cell. Such a cell must be demounted for studying very viscous samples or mulls (see section entitled Solid Samples).

The main problem in determining the spectrum of a sample in solution is finding a solvent that does not absorb in the same region as the sample. Two commonly used solvents are carbon disulfide and carbon tetrachloride, the spectra of which are presented in Figs. 21.14 and 21.15. Note that CS_2 is relatively free of interfering bands between 7.5 and 15 μ, and that CCl_4 is practically free of interfering bands between 2.5 and 12 μ. Thus, if the spectrum of a sample is run in each of these solvents, the entire NaCl spectral region will be covered.

Liquids that dissolve sodium chloride (aqueous solutions, low molecular weight alcohols, acids, etc.) cannot be used in the usual cells with sodium chloride windows. Calcium fluoride windows are usable at wavelengths less than 10 μ, barium fluoride windows at wavelengths less than 13 μ, and Irtran-2 windows at wavelengths less than 14 μ.

After a cell is used to determine the spectrum of a liquid, it should be flushed with a clean, thoroughly dry solvent, such as benzene, and then dried by blowing a stream of absolutely dry air through it. Care should be taken when handling cells not to let the skin come near the salt surfaces or the surfaces may become fogged. Rubber gloves or finger cots should be used whenever it is necessary to touch the salt surfaces.

Solid Samples

In many cases, solid samples will dissolve to form solutions the spectra of which can be determined as described in the section entitled Liquid Samples. If it is not possible to find a suitable solvent for a solid, there are two common ways of determining the spectrum: by determining the spectrum of a mull; and by determining the spectrum of a pellet made from a mixture of the material with KBr.

In preparing a mull, the sample is ground with a weakly absorbing, nonvolatile liquid with a mortar and pestle. The resulting paste is spread on one salt plate and is covered with another; the sample thickness is adjusted by rotating and pressing the plates together to squeeze out excess material. The mulling liquid, having an index of refraction much closer to that of the sample than that of air, minimizes the light scattering of the powdered sample. It is important that the sample be ground to a very fine particle size to reduce light scattering and salt plate scratching. The most common mulling agent is

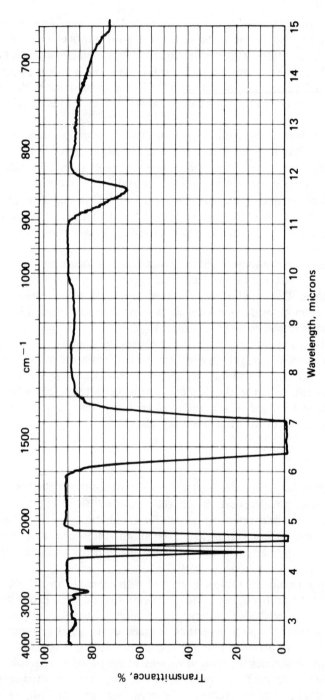

Fig. 21.14. Infrared spectrum of carbon disulfide.

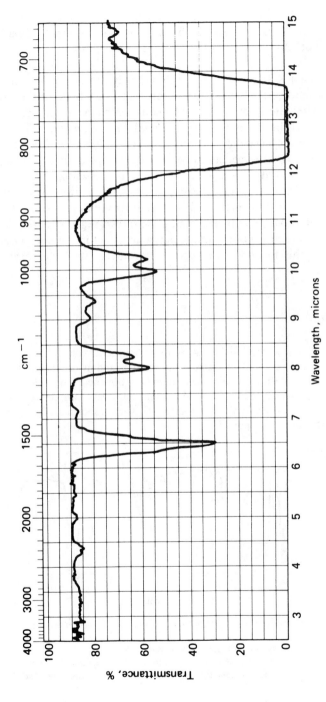

FIG. 21.15. Infrared spectrum of carbon tetrachloride.

mineral oil (Nujol), which is reasonably transparent in the infrared except for narrow bands at 2900, 1450, and 1375 cm^{-1}. An alternative mulling liquid, which does not absorb in these regions, is a perfluorokerosene, such as Fluorolube S.[31]

In preparing a KBr pellet, the sample is finely pulverized with pure dry potassium bromide, and the mixture is pressed in an hydraulic press to a transparent pellet, the spectrum of which can then be measured. It is important that the solids be extremely finely divided and well mixed. The pellet is usually pressed in a special die that can be evacuated in order to avoid entrapped air, which causes the pellets to be cloudy. However, fairly satisfactory pellets may be prepared in a simple device consisting essentially of a large nut or steel cylinder into which two flat-ended bolts are screwed from opposite ends.[32] The powder to be pelletized is placed between the ends of the bolts and is squeezed into a pellet by tightening the screws together with the aid of a wrench. Upon removal of the bolts, the pellet is left in the nut. The pellet is mounted in this form on the spectrophotometer, and the spectrum is obtained.

Problems

1. Iron pentacarbonyl was refluxed for 40 hours with an equal volume of dicyclopentadiene. The volatile material was removed in vacuum, and by recrystallization of the residue from chloroform a crystalline compound was obtained. The molecular weight in benzene was determined to be 368, and the following analytical data were obtained: C, 47.55; H, 2.85; and Fe, 31.53. The infrared spectrum is shown in Fig. 21.16. What is the structure of the compound?

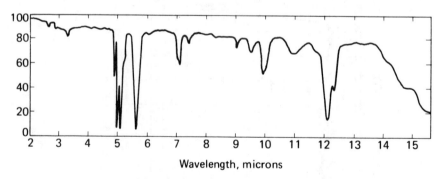

Wavelength, microns

FIG. 21.16. Spectrum of Fe(CO)$_5$, dicyclopentadiene reaction product.

[31] Available from Hooker Electrochemical Company.

[32] Potassium bromide pellet presses of this type are manufactured by Wilk's Scientific Corporation, South Norwalk, Conn. 06854, and Research Analysis Associates, Inc., Stratford, Conn. 06497.

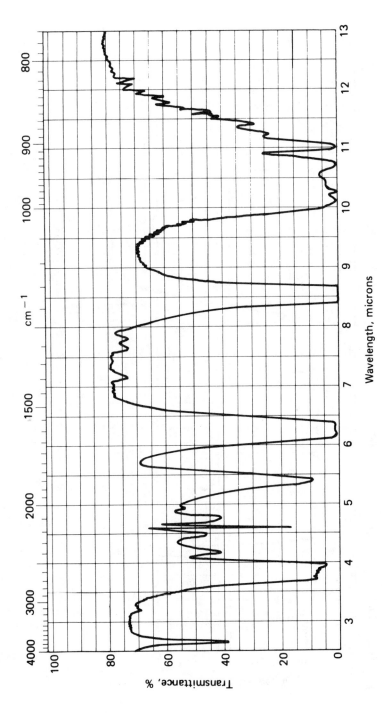

FIG. 21.17. Spectrum of the material collected in the −196° trap. (Problem 2.)

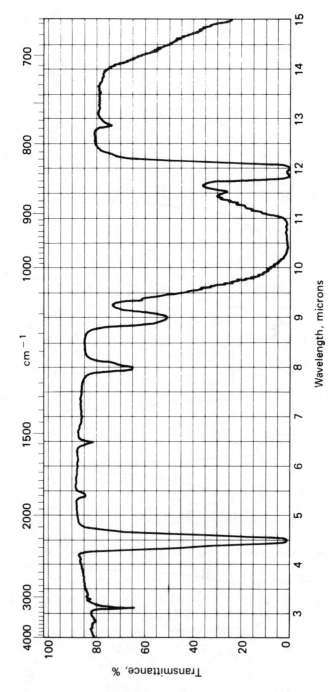

Wavelength, microns

FIG. 21.18. Spectrum of the material collected in the −130° trap. (Problem 2.)

2. Boron tribromide (0.2 mmole) and silane (1.0 mmole) were allowed to react together at 0° for 6 hours. The resulting mixture was distilled on the vacuum line through a $-130°$ trap into a $-196°$ trap, and the infrared spectra of the contents of the traps were determined (see Figs. 21.17 and 21.18). Identify the contents of the traps.

3. When iron pentacarbonyl is treated with aqueous alkali, it dissolves to give a solution which on acidification gives a gas having an infrared absorption around $1900 \, cm^{-1}$. This absorption band is believed to be due to a hydrogen stretching. Explain how you might prove this by using D_2O. Do you think the hydrogen is attached to the oxygen of a carbonyl group, to the carbon of a carbonyl group, or to an iron atom? Explain.

4. The compound $CH_3Co(CO)_4$ reacts with $P(C_6H_5)_3$ to form a material of empirical formula $CH_3Co(CO)_4P(C_6H_5)_3$. Suggest structures for, and explain the bonding in each of these cobalt compounds. Explain how you might use infrared spectra to verify your suggested structures.

5. Give the number and symmetry species of the infrared-active normal modes of vibration of $POCl_3$. This molecule belongs to the C_{3v} point group.

6. What are the symmetry species of the infrared-active vibrations of the water molecule?

Literature

For more detailed discussion of the experimental procedures of infrared spectroscopy, the reader is referred to the following sources of information:

Bauman, R. P., *Absorption Spectroscopy*, John Wiley & Sons, Inc., New York, 1962.

Cotton, F. A., in *Modern Coordination Chemistry*, J. Lewis and R. G. Wilkins, eds., Interscience Publishers, New York, 1960, Chap. 5, pp. 301–99.

Infrared Sampling and Techniques, vol. 2, Instruction Manual for Perkin Elmer Infrared Equipment.

Instruction Manual for Infrared Sampling Accessories, Connecticut Instrument Company, Wilton, Conn.

Potts, W. J., Jr., *Chemical Infrared Spectroscopy*, vol. 1, John Wiley & Sons, Inc., New York, 1963.

Electronic
Spectra

22

ULTRAVIOLET AND VISIBLE SPECTROMETRY

Most compounds absorb light somewhere in the spectral region between 200 and 1000 nm. These transitions correspond to the excitation of the molecules to higher *electronic* states; therefore, spectra in this region are often called electronic absorption spectra. The "visible" spectral region ranges from about 400 nm (violet) to about 700 nm (red), and the "ultraviolet" spectral region corresponds to wavelengths less than about 400 nm. Both of these regions, as well as part of the "near infrared" region (700 to 2500 nm), can be studied with quartz prism spectrophotometers.

The energy of a spectral transition is related to the frequency of the light by the relation $\Delta E = hv$, where h is Planck's constant and v is the frequency. In practice, the energy of light is usually specified in terms of its wavelength, λ, or its wave number, $\bar{v}$. These quantities are related to the frequency as follows: $\bar{v} = 1/\lambda = v/c$. The commonly used units and symbols for wavelength and wave number are given in Table 22.1. The intensity of light transmitted by a solution of a sample may be expressed by the relation

$$I = I_0 \cdot 10^{-\varepsilon l c} \quad \text{or} \quad \log\left(\frac{I_0}{I}\right) = \varepsilon l c = A$$

where I_0 is the intensity of the incident light, ε is the molar absorptivity (or decadic molar extinction coefficient), l is the cell length in centimeters, c is the concentration in moles per liter, and A is the absorbance or optical density.

Table 22.1   Units of Wavelength and Wave Number

Unit	Symbol	Definition
Wavelength (λ)		
micron	μ	10^{-4} cm
nanometer	nm	10^{-7} cm
millimicron	mμ	10^{-7} cm
ångström	Å	10^{-8} cm
Wave number ($\bar{v}$)		
kayser	K	cm^{-1}
kilokayser	kK	1000 cm^{-1}

Most spectrophotometers have two optical beams that are compared. A cell containing the solution is placed in one beam, and a reference cell containing the solvent is placed in the other. The instrument usually records graphically the difference in absorbance (i.e., the absorbance of the solute) against wavelength. It is important that the solvent does not absorb strongly in the spectral region of interest. Most solvents are completely transparent in the visible region and begin to absorb strongly at wavelengths below a characteristic cutoff wavelength. The cutoff wavelengths for some common solvents are given in Table 22.2.

Table 22.2   Ultraviolet Cutoff Wavelengths for Solvents

Solvent	Ultraviolet Cutoff (nm)
Water	< 200
Diethyl ether	205
Ethanol	205
Methanol	210
Cyclohexane	210
Isooctane	215
Acetonitrile	220
Dichloromethane	234
Chloroform	245
Carbon tetrachloride	265
Benzene	280
Pyridine	305
Acetone	330
Carbon disulfide	375

THE THEORY OF ELECTRONIC SPECTRA

Molecular Energy Levels and Their Labels

The electrons of a molecule in its ground state usually fill the lowest available molecular orbitals. If the highest-energy occupied molecular orbitals are degenerate, the electrons occupy as many of these degenerate orbitals as possible, with the maximum possible number of unpaired electrons. Sometimes two different orbitals, or two sets of degenerate orbitals, differ in energy by an amount less than the energy required to pair electrons. In such a case, the electrons occupy the orbitals as if the entire group were degenerate.

Whenever the electrons are distributed among the molecular orbitals in a way other than the ground state configuration, we say that the molecule is in an excited state. We label the various states of a molecule by means of their symmetry species. When there are no partially filled molecular orbitals, the electronic state is totally symmetric—that is, of symmetry species such as $^1A_1$ or ^1A_g.[1] Thus, the ground state of the nitrate ion, with the electronic configuration $\cdots (\sigma^B e')^4 (\pi^B a_2'')^2 (\pi_0 e')^4 (\pi_0 e'')^4 (\pi_0 a_2')^2$, is $^1A_1'$. When there is only one unpaired electron, the state is the symmetry species of the orbital containing the odd electron. Thus, the ground state of the NO_3 radical, with the electronic configuration $\cdots (\sigma^B e')^4 (\pi^B a_2'')^2 (\pi_0 e')^4 (\pi_0 e'')^4 (\pi_0 a_2')^1$, is $^2A_2'$. When there are two or more singly occupied orbitals, the state may be any of the terms of the direct product of the symmetry species of the singly occupied orbitals. Consider the excited state of the nitrate ion, having the electronic configuration $\cdots (\sigma^B e')^4 (\pi^B a_2'')^2 (\pi_0 e')^4 (\pi_0 e'')^3 (\pi_0 a_2')^2 (\pi^* a_2'')^1$. This configuration corresponds to the terms $^1E'$ and $^3E'$ ($E'' \times A_2'' = E'$). Or consider an octahedral d^2 transitional metal complex, with the outer electronic configuration $(a_{1g})^2 (t_{1u})^6 (t_{2g})^2$. This configuration leads to the terms[2] $^1A_{1g}$, ^1E_g, $^1T_{2g}$, and $^3T_{1g}$ ($T_{2g} \times T_{2g} = A_{1g} + E_g + T_{1g} + T_{2g}$).

The lowest lying of a group of terms arising from the same electronic configuration is the term of highest multiplicity; when there is more than one term of highest multiplicity, the term having both the highest multiplicity and the highest orbital degeneracy ($T > E > A$) lies lowest. Thus, the $^3E'$ state lies lower than the $^1E'$ state for the excited nitrate ion just described, and the $^3T_{1g}$ state is the ground state for the d^2 octahedral complex previously described.

Selection Rules for Electronic Transitions

When a molecule absorbs a photon, the molecule is excited from its ground state to an excited state. However, the various conceivable transitions do not

---

[1] The left-hand superscript refers to the "multiplicity" (one more than the number of unpaired electrons).

[2] In this case, the assignment of the multiplicities is fairly complicated, and will not be described.

occur with equal probability: some transitions have high extinction coefficients, and others have low (sometimes negligible) extinction coefficients. It is possible to predict very approximately the intensities of electronic absorption bands by a consideration of three theoretical rules called selection rules.

1. Transitions between states of different multiplicity are forbidden. Thus, according to this rule, a molecule in a singlet state (e.g., $^1A_1{}'$) can, by the absorption of a light quantum of the appropriate frequency, be excited to a different singlet state (e.g., $^1A_1{}''$), but not to a triplet state (e.g., $^3A_1{}''$).

2. Transitions involving the excitation of more than one electron are forbidden. According to this rule, an octahedral complex with the ground state $\cdots (a_{1g})^2(t_{1u})^6(e_g)^4(t_{2g})^2[^3T_{1g}]$ cannot undergo a transition to the excited state $\cdots (a_{1g})^2(t_{1u})^6(e_g)^4(e_g{}^*)^2[^3A_2]$.

3. Where ψ_i and ψ_f are the wave functions of the initial and final states, respectively, and M_x, M_y, and M_z are the components of the dipole moment vector in the x, y, and z directions, the transition is allowed only if the product $\psi_i M_x \psi_f$, $\psi_i M_y \psi_f$, or $\psi_i M_z \psi_f$ belongs to or contains the totally symmetric symmetry species (e.g., A_1 or A_{1g}). This rule means that the direct product of the irreducible representations of ψ_i, ψ_f, and either M_x, M_y, or M_z must be totally symmetric (i.e., all the characters of the irreducible representation must be $+1$).

For an application of the third selection rule, consider an $A_1 \rightarrow A_2$ transition in a molecule of point group C_{2v}. The x, y, and z components of the dipole moment vector transform as the corresponding Cartesian coordinates x, y, and z. For the point group C_{2v}, these vectors transform as B_1, B_2, and A_1, respectively. The pertinent direct products are

$$\psi_i M_x \psi_f \qquad A_1 \times B_1 \times A_2 = B_2$$
$$\psi_i M_y \psi_f \qquad A_1 \times B_2 \times A_2 = B_1$$
$$\psi_i M_z \psi_f \qquad A_1 \times A_1 \times A_2 = A_2$$

Inasmuch as none of the direct products belongs to species A_1, the transition is symmetry-forbidden.

As further examples, consider the transitions $A_2 \rightarrow T_1$ and $A_2 \rightarrow T_2$ in a tetrahedral (T_d) molecule. The coordinates x, y, and z transform according to the T_2 representation. For the transition $A_2 \rightarrow T_1$, the intensity product is

$$A_2 \times T_2 \times T_1 = A_1 + E + T_1 + T_2$$

Because A_1 appears in the product, the transition is allowed. For the transition $A_2 \rightarrow T_2$, the intensity product is

$$A_2 \times T_2 \times T_2 = A_2 + E + T_1 + T_2$$

Because A_1 does not appear in the product, the transition is forbidden.

There is a corollary of the third selection rule, called the parity rule. Molecules with a center of symmetry have symmetry species of two types: gerade and ungerade. The product of two gerade functions, or of two ungerade functions, is a gerade function, and the product of an ungerade function and a gerade function is an ungerade function. Because of these facts, and because the Cartesian coordinates always transform as ungerade species, the product $\psi_i M \psi_f$ can be of species A_{1g} only when ψ_i and ψ_f are of opposite parity. That is, $u \to g$ and $g \to u$ transitions are allowed, but $g \to g$ and $u \to u$ transitions are forbidden.

Transitions that, according to one or more of the three selection rules, are forbidden actually do take place, because real molecules do not behave in the ideal manner assumed in the theoretical derivation of the rules. For example, molecules do not have firm geometries; asymmetric vibrations lower the symmetry and permit violation of the third selection rule. However, "forbidden" transitions can be distinguished from "allowed" transitions because the former are of much lower intensity than the latter. Completely allowed electronic transitions generally have extinction coefficients of the order of magnitude 10^3 to 10^5 l mole^{-1} cm^{-1}. Symmetry-forbidden transitions have extinction coefficients around 1 to 100, and spin forbidden transitions (rule 1 or 2) have extinction coefficients around 10^{-3} to 1.

The Nitrite Ion

The ultraviolet absorption spectrum[3] of the nitrite ion in water and acetonitrile is shown in Fig. 22.1. In water, a medium-intensity band ($\varepsilon = 22.5$) is observed at 28,200 cm^{-1}, a very weak band ($\varepsilon = 9.4$) at 34,800 cm^{-1}, and a strong band ($\varepsilon = 5380$) at 47,620 cm^{-1}. To interpret these electronic transitions, one must have some understanding of the symmetries and relative energies of the molecular orbitals of the nitrite ion. The latter topics are discussed in some detail in Appendix 12 (pp. 561 to 569). The molecular orbital energy level diagram is given in Fig. 22.2, where three possible electronic transitions are indicated by vertical arrows. The $^1A_1 \to {}^1A_2$ transition is symmetry forbidden and is believed to correspond to the very weak band in the aqueous spectrum at 34,800 cm^{-1}. In non-hydrogen-bonding solvents, such as dimethylformamide and acetonitrile, this band is not observed. Probably its appearance in water results from lowering of the molecular symmetry by hydrogen bonding, and the consequent relaxation of the symmetry selection rule. Both the $^1A_1 \to {}^1B_1$ and $^1A_1 \to {}^1B_2$ transitions are allowed by all the selection rules. The former may be described as an $n \to \pi^*$ transition, and the latter as a $\pi \to \pi^*$ transition.[4] Transitions of the

---

[3] S. J. Strickler and M. Kasha, *J. Am. Chem. Soc.*, **85**, 2899 (1963).

[4] See the discussion of these transitions in R. S. Drago, *Physical Methods in Inorganic Chemistry*, Reinhold Publishing Corp., New York, 1965, pp. 144–47.

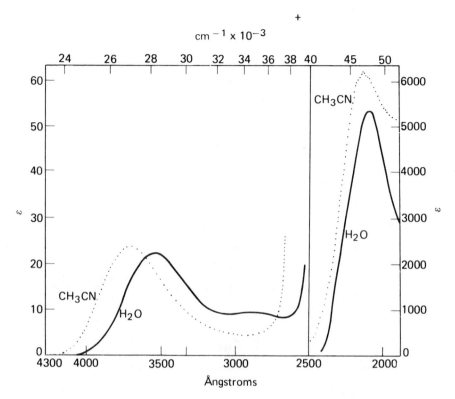

FIG. 22.1. Spectra of tetramethylammonium nitrite in acetonitrile and water at room temperature.[3] (Reproduced with permission of the American Chemical Society.)

$n \rightarrow \pi^*$ type generally have extinction coefficients less than 2000 and undergo shifts to lower energy on going from high dielectric constant or hydrogen-bonding solvents to low dielectric constant or non-hydrogen-bonding solvents. Such transitions usually have the lowest energy of the singlet-singlet transitions of a molecule. The band at 28,200 cm^{-1} has all of these characteristics, and therefore it is assigned to the $^1A_1 \rightarrow \, ^1B_1$ transition. The solvent shift may be understood when it is realized that the transition corresponds to the excitation of a nonbonding electron on an oxygen atom to a delocalized π antibonding orbital. When the oxygen atoms are hydrogen bonded to solvent molecules, the presence of the positive hydrogen atoms makes it more difficult to remove the electron from the nonbonding orbital.

Transitions of the $\pi \rightarrow \pi^*$ type generally have very high intensity, and, accordingly, the intense band at 47,620 cm^{-1} is assigned to the $^1A_1 \rightarrow \, ^1B_2$ transition. Usually $\pi \rightarrow \pi^*$ transitions undergo a slight shift to higher energy on going from a high dielectric to low dielectric solvent, whereas a slight shift to *lower* energy is observed for the 47,620 cm^{-1} band, as in the case of the

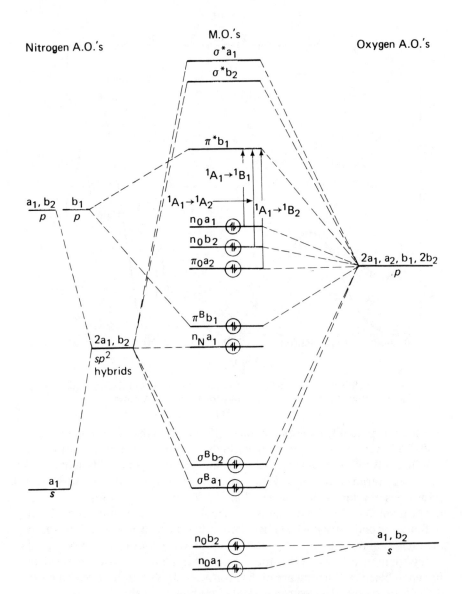

FIG. 22.2. Molecular orbital energy level diagram of the nitrite ion, showing the observed electronic transitions.

$n \to \pi^*$ band at 28,200 cm^{-1}. This behavior is probably attributable to the fact that in nitrite ion the $\pi_0 a_2$ orbital has the nature of a nonbonding orbital, because it is completely localized on the oxygen atoms.[3]

Ligand Field Spectra

The molecular orbital diagram for σ bonding in an octahedral complex is given in Fig. 22.3.[5] The six ligands contribute 12 electrons to the molecular

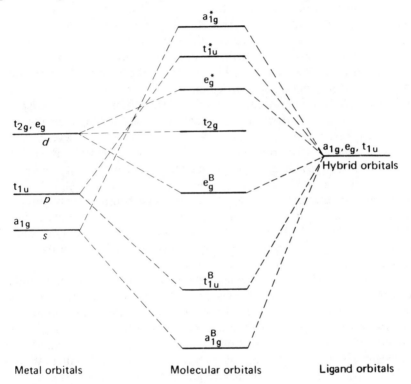

FIG. 22.3. Molecular orbital diagram for σ-bonding in an octahedral complex.

orbitals, thus filling the $a_{1g}{}^B$, $t_{1u}{}^B$, and $e_g{}^B$ orbitals. If the metal ion has no valence electrons (as in complexes of Al^{3+} and Ti^{4+}), then there are no more electrons to be added to the set of molecular orbitals, and the ground state of the complex is $^1A_{1g}$. However, in many complexes of transition metals the metal ion has valence electrons (usually d electrons) that enter the t_{2g} and $e_g{}^*$ molecular orbitals. The molecular orbital diagram for tetrahedral complexes is similar to that for octahedral complexes in that again the valence

[5] These molecular orbitals are discussed in Appendix 12 (pp. 567–569).

electrons of the metal enter t_2 and e orbitals.[6] However, in this case, the t_2 level lies above the e level.

In general, there are several possible electronic states corresponding to the possible distributions of d electrons in the two levels (t_{2g} and $e_g{}^*$, or e and t_2). Transitions among these states account for the visible and many of the ultraviolet absorption bands of transition metal octahedral and tetrahedral complexes. In this section we shall discuss the systematics of these spectra.

Electronic transitions in octahedral complexes from the t_{2g} to the e_g levels are forbidden by the parity selection rule, but they are observed as low-intensity bands, generally in the visible region of the spectrum. Because these bands in both the octahedral and tetrahedral complexes involve transitions between orbitals that have been perturbed by the field of the ligands, they are often referred to as *ligand field bands*. When there is only one electron in the lower level, there is only one conceivable ligand field transition and only one ligand field band is observed. However, the situation is considerably more complicated when there is more than one electron. Then, in general, several states of different energy arise from each of the possible configurations, such as $(t_{2g})^n(e_g{}^*)^m$ or $(e)^n(t_2)^m$. On p. 318 we have already seen how a $(t_{2g})^2$ configuration yields the terms $^1A_{1g}$, ^1E_g, $^1T_{2g}$, and $^3T_{1g}$. Similarly, a $(t_{2g})^1(e_g{}^*)^1$ configuration yields the terms $^1T_{1g}$, $^1T_{2g}$, $^3T_{1g}$, and $^3T_{2g}$, and an $(e_g{}^*)^2$ configuration yields the terms $^1A_{1g}$, ^1E_g, and $^3A_{2g}$. The reason why all these states have different energies, and why transitions can be observed involving the excitation of more than one electron [e.g., $(t_{2g})^2 \rightarrow (e_g{}^*)^2$], can be seen by taking account of the interaction of the electrons with one another.

If the ligand field of the complex were so very strong as to make inter-electronic interactions negligible by comparison, then the various terms arising from each molecular orbital configuration would be degenerate, and only one transition would be allowed for a d^2 complex: $(t_{2g})^2 \rightarrow (t_{2g})^1(e_g{}^*)^1$. Actually, of course, the interaction of the electrons with one another is significant, and the energy level for each molecular orbital configuration is split into various terms. If we were able gradually to decrease the ligand field strength to zero, we would find that each term would gradually approach an energy corresponding to one of the states of the free ion. That is, the states of the free ion may be correlated with those of the complex. In the following paragraph, we shall discuss the states of the free ion and how they are split by fields of various symmetry.

In the absence of an electric field, or in a spherically symmetric field, a transition metal ion can exist in one or more electronic states, each state corresponding to a different energy level. The number and nature of the states depend on the number of d electrons in the ion. The terms, or spectro-

---

[6] The g subscripts are missing because tetrahedral complexes lack a center of symmetry.

scopic symbols, for the 11 possible d^n systems are[7]

$$d^1 \text{ or } d^9: {}^2D$$

$$d^2 \text{ or } d^8: {}^1S, {}^1D, {}^1G, {}^3P, {}^3F$$

$$d^3 \text{ or } d^7: {}^2P, 2{}^2D, {}^2F, {}^2G, {}^2H, {}^4P, {}^4F$$

$$d^4 \text{ or } d^6: 2{}^1S, 2{}^1D, {}^1F, 2{}^1G, {}^1I, 2{}^3P, {}^3D, 2{}^3F, {}^3G, {}^3H, {}^5D$$

$$d^5: {}^2S, {}^2P, 3{}^3D, 2{}^2F, 2{}^2G, {}^2H, {}^2I, {}^4P, {}^4D, {}^4F, {}^4G, {}^6S$$

$$d^0 \text{ or } d^{10}: {}^1S$$

The last term in each series is the ground state for that system. If the ion is placed in an electric field having less than spherical symmetry, most of the listed terms will split into two or more terms. The representations of the various states that arise from S, P, D, F, and G terms in fields of O_h, T_d, and D_{4h} symmetry are given in Table 22.3.

Table 22.3   States Arising in Fields of Various Symmetry

Free Ion	States in Point Groups		
Term	O_h	T_d	D_{4h}
S	A_1	A_1	A_1
P	T_1	T_2	A_2, E
D	E, T_2	E, T_2	A_1, B_1, B_2, E
F	A_2, T_1, T_2	A_2, T_1, T_2	$A_2, B_1, B_2, 2E$
G	A_1, E, T_1, T_2	A_1, E, T_1, T_2	$2A_1, A_2, B_1, B_2, 2E$
H	$E, 2T_1, T_2$	$E, T_1, 2T_2$	$A_1, 2A_2, B_1, B_2, 3E$
I	$A_1, A_2, E, T_1, 2T_2$	$A_1, A_2, E, T_1, 2T_2$	$2A_1, A_2, 2B_1, 2B_2, 3E$

Let us consider a d^1 ion, which exists in the ^2D state in the absence of a field. In a field of O_h symmetry (as when surrounded by an octahedron of water molecules), two states are possible for the ion: a doubly degenerate E_g state, and a triply degenerate T_{2g} state. Now, an atomic orbital (such as p_x, p_z, $d_{x^2-y^2}$, d_{xy}, etc.) transforms in the same way as its subscript. Therefore, the transformation properties of an orbital on an atom at the origin of the coordinate system can be determined by looking up the subscript in the appropriate character table. Under O_h symmetry, the d orbitals fall into two inseparable groups—one consisting of d_{z^2} (really $d_{2z^2-x^2-y^2}$) and $d_{x^2-y^2}$; and the other consisting of d_{xy}, d_{xz}, and d_{yz}. We see that d_{z^2} and $d_{x^2-y^2}$ transform as E_g, and d_{xy}, d_{xz}, and d_{yz} transform as T_{2g}. In a metal ion surrounded by six negative ligands at the corners of a regular octahedron, the d_{z^2} and $d_{x^2-y^2}$ orbitals point toward the ligands, and the d_{xy}, d_{xz}, and d_{yz} orbitals point between the ligands. Therefore, a complex in the E_g state (electron in

---

[7] The method for calculating the terms for a particular electronic configuration is given by H. B. Gray, *Electrons and Chemical Bonding*, W. A. Benjamin, Inc., New York, 1964, pp. 22–28.

a d_{z^2} or $d_{x^2-y^2}$ orbital) will have a higher energy than a complex in the T_{2g} state (electron in a d_{xy}, d_{xz}, or d_{yz} orbital). Thus, it is clear that, for a d^1 metal ion in an octahedral environment, the ground state is T_{2g}.

An octahedrally coordinated d^1 ion has an energy level diagram as shown in Fig. 22.4. We predict only one *d-d* transition in the absorption spectrum:

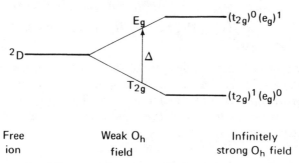

FIG. 22.4. Correlation diagram for an octahedral d^1 complex.

$T_{2g} \rightarrow E_g$. This prediction is fairly well borne out by experiment. The violet $Ti(H_2O)_6^{3+}$ ion has only one ligand field absorption band at $20,300\ cm^{-1}$ with a slight shoulder at $17,400\ cm^{-1}$. The shoulder has been interpreted in terms of a slight distortion of the octahedron of water molecules. The value of Δ for this complex is simply the energy of the main transition, $20,300\ cm^{-1}$.

We shall use the notation Δ_o and Δ_t to indicate Δ for octahedral and tetrahedral complexes, respectively. A number of generalizations regarding the magnitudes of Δ_o and Δ_t can be of help in interpreting spectra.

1. Values of Δ_o for complexes of the first transition series fall in the range 7500 to $12,500\ cm^{-1}$ for $+2$ ions and in the range 14,000 to $25,000\ cm^{-1}$ for $+3$ ions.

2. For metals in the same group and having the same oxidation state, Δ_o increases by 30 to 50 per cent on going from the first transition series to the second, or from the second to the third.

3. Values of Δ_t are about $\frac{4}{9}$ of Δ_o, other things being equal.

4. Ligands may be arranged in a series according to their ability to cause *d*-orbital splitting—that is, according to the relative magnitudes of the Δ values. For some common ligands, this series (called the *spectrochemical series*) is $CN^- > NO_2^- > o\text{-phen} > dipy > en > NH_3 \sim py > -NCS^- > H_2O \sim C_2O_4^{2-} > OH^- > F^- > Cl^- > Br^- > I^-$. Many exceptions to this series are known.

It has been suggested that Δ_o can be written as the product of two factors, f and g, where f is characteristic of the ligand and g that of the metal.[8] This

[8] C. K. Jorgensen, reported by L. E. Orgel, in *An Introduction to Transition-Metal Chemistry*, 2nd ed., John Wiley & Sons, Inc., New York, 1966, p. 48.

rule must be applied with considerable caution. Some of the f and g values are given in Table 22.4.

Table 22.4   Values of f and g for Estimating Δ_o*

Ligand	f	Metal Ion	$g \times 10^{-3} (cm^{-1})$
Br^-	0.76	V^{2+}	12.3
Cl^-	0.80	Cr^{3+}	17.4
F^-	0.9	Mn^{2+}	8.0
H_2O	1.00	Mn^{4+}	23.0
NH_3	1.25	Fe^{3+}	14.0
CN^-	1.7	Co^{2+}	9.3
		Co^{3+}	19.0
		Ni^{2+}	8.9
		Rh^{3+}	27.0
		Ir^{3+}	32.0

* See ref. 8.

We are now ready to consider the correlation diagram for a d^2 octahedral complex, shown in Fig. 22.5. The detailed methods of construction[9] are beyond the scope of this book, but certain features of the diagram are noteworthy. First, there is a one-to-one correspondence between states on opposite sides of the diagram. (Thus, a $^3T_1$ state on one side correlates with a $^3T_1$ state on the other side, etc.) Second, as the field strength increases from the free ion to the infinitely strong case, states of the same spin and symmetry do not cross. (Thus, the $^3T_1$ state arising from the ^3P free ion state cannot correlate with the $^3T_1$ state arising from the $t_2{}^2$ configuration.)

From the diagram we may glean the facts that no matter how strong or weak the field, the ground state is always $^3T_{1g}$, and that the only spin-allowed transitions are $^3T_{1g}(F) \rightarrow {}^3T_{2g}$, $^3T_{1g}(F) \rightarrow {}^3A_{2g}$, and $^3T_{1g}(F) \rightarrow {}^3T_{1g}(P)$.

It is possible to construct a correlation diagram for each electronic configuration in the series d^1 to d^9 for both octahedral and tetrahedral ions.[10] However, when such diagrams are to be used only for qualitative predictions of spectra, some consolidation and simplification can be effected. The configurations corresponding to infinitely strong field can be omitted, inasmuch as they are never achieved in practice. Likewise, all terms of multiplicity other than that of the ground state can be omitted, inasmuch as transitions involving these states are forbidden. The energy level order of the states arising from the splitting of a term state for a particular ion in an octahedral field is the opposite of that for the ion in a tetrahedral field. In addition, the

[9] The reader is referred to F. A. Cotton, *Chemical Applications of Group Theory*, Interscience Publishers, New York, 1963, pp. 196–205.

[10] See, for example, B. N. Figgis, *Introduction to Ligand Fields*, Interscience Publishers, New York, 1966, pp. 152–59.

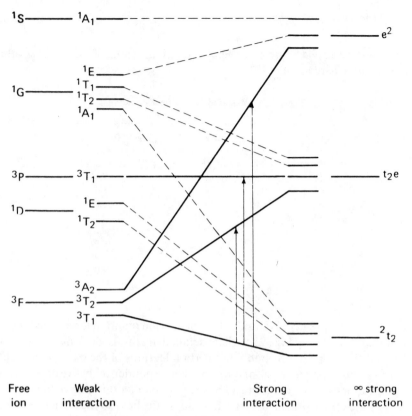

FIG. 22.5. Correlation diagram for an octahedral d^2 complex.

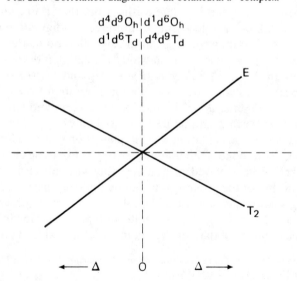

FIG. 22.6. Qualitative energy level diagram for octahedral and tetrahedral complexes.

energy level order of the states arising from the splitting of a term state of a d^n ion in a particular field is the opposite of that for a d^{10-n} ion in the same field. These simplifications and rules have been used in constructing the diagrams in Figs. 22.6 through 22.9.

In the case of certain octahedral configurations (d^4, d^5, d^6, and d^7), the multiplicity and symmetry of the ground state change on going from a weak field to a strong field. This situation is a consequence of the spin pairing that occurs when the $t_{2g} - e_g{}^*$ energy gap becomes greater than the pairing energy.

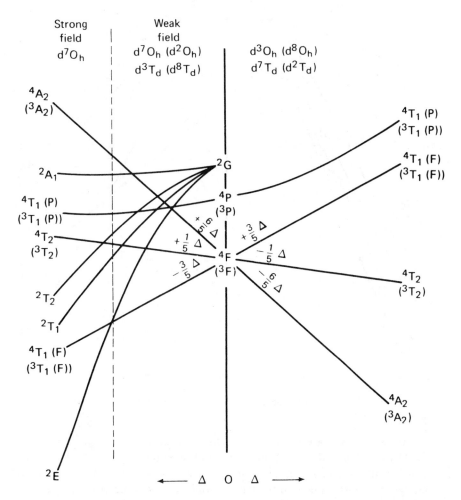

FIG. 22.7. Qualitative energy level diagram for octahedral and tetrahedral complexes.

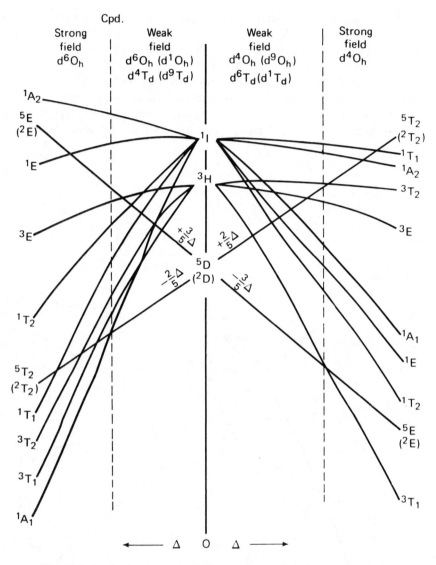

Fig. 22.8. Qualitative energy level diagram for octahedral and tetranedral complexes.

We shall illustrate the use of the energy diagrams in Figs. 22.6 through 22.9 by discussing the electronic spectra of several complexes.[11] First, let us consider the case of a d^2 ion in an octahedral environment. The appropriate energy levels are shown on the left side of Fig. 22.7. One of the best studied

[11] See C. J. Ballhausen, *Introduction to Ligand Field Theory*, McGraw-Hill Book Company, New York, 1962, pp. 226–92.

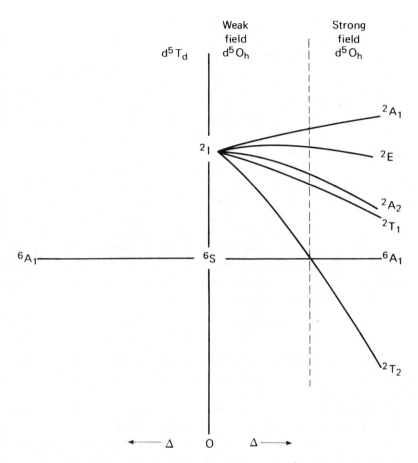

Fig. 22.9. Qualitative energy level diagram for d^5 complexes.

examples of this system is a dilute solid solution of V_2O_3 in Al_2O_3. The absorption spectrum shows three strong bands at 17,400, 25,200, and 34,500 cm^{-1}, which have been interpreted as the transitions $^3T_{1g}(F) \rightarrow {}^3T_{2g}$, $^3T_{1g}(F) \rightarrow {}^3T_{1g}(P)$ and $^3T_{1g}(F) \rightarrow {}^3A_{2g}$, respectively. The $V(H_2O)_6{}^{3+}$ ion (in vanadium alum) shows two bands at 12,400 and 26,200 cm^{-1}, corresponding to $^3T_{1g}(F) \rightarrow {}^3T_{2g}$ and $^3T_{1g}(F) \rightarrow {}^3T_{1g}(P)$. The third band $(^3T_{1g}(F) \rightarrow {}^3A_{2g})$, which should appear around 34,000 cm^{-1}, is not seen because of a strong charge-transfer band[12] in this region of the spectrum.

Next, let us consider a cobalt(II) ion in an octahedral environment. We look at the energy level diagram for a d^7 ion in O_h symmetry (the left side of Fig. 22.7) and find that in the case of weak fields there are three possible transitions from the $^4T_{1g}(F)$ ground state, to the states $^4T_{2g}$, $^4A_{2g}$, and $^4T_{1g}(P)$.

[12] See p. 334 for a discussion of charge-transfer bands.

The aqueous $Co(H_2O)_6^{2+}$ ion has bands at 8350 and 19,000 cm^{-1}, which have been assigned to the transitions $^4T_{1g}(F) \rightarrow {}^4T_{2g}$ and $^4T_{1g}(F) \rightarrow {}^4T_{1g}(P)$, respectively. Using the simple weak-field approximation that the $^4T_{1g}(F) \rightarrow {}^4T_{2g}$ energy is $\frac{4}{5}\Delta_o$ and that the $^4T_{1g}(F) \rightarrow {}^4A_{2g}$ energy is $\frac{9}{5}\Delta_o$, we calculate that the latter transition should occur at 19,000 cm^{-1}. Now, this transition would be expected to be much weaker than the other transitions because it corresponds to a two-electron excitation. It is believed that the band is so weak that it is completely obscured by the $^4T_{1g}(F) \rightarrow {}^4T_{1g}(P)$ band occurring at practically the same frequency.

Next, let us consider the cobalt(II) ion in a tetrahedral environment, as in $CoCl_4^{2-}$. From the right side of Fig. 22.7, we see that three transitions are expected. From the data for $Co(H_2O)_6^{2+}$, we estimate $\Delta_o = 10,000$ cm^{-1}. Because chloride generally acts about 0.80 as strong a ligand as water (see Table 22.4), and because $\Delta_t \approx \frac{4}{9}\Delta_o$, we would expect Δ_t for $CoCl_4^{2-}$ to be about 3700 cm^{-1}. We therefore estimate 3700 and 6600 cm^{-1} for $^4A_2 \rightarrow {}^4T_2$ and $^4A_2 \rightarrow {}^4T_1(F)$, respectively. In the gaseous Co^{2+} ion, the ^4P state lies 14,500 cm^{-1} above the ^4F ground state. Thus, we estimate that the $^4A_2 \rightarrow {}^4T_1(P)$ transition should occur at

$$\tfrac{6}{5}\Delta + 14{,}500 \text{ cm}^{-1} = 18{,}900 \text{ cm}^{-1}$$

Experimentally, bands have been observed at 6300 and about 15,000 cm^{-1}. It seems reasonable to assign these to the $^4A_2 \rightarrow {}^4T_1(F)$ and $^4A_2 \rightarrow {}^4T_1(P)$ transitions, respectively.

Practically all cobalt(III) complexes are strong-field octahedral complexes. Thus, the ground state is $^1A_{1g}$, and transitions to the $^1T_{1g}$ and $^1T_{2g}$ states should be expected. (The ^1E_g and $^1A_{2g}$ states are of extremely high energy and correspond to the excitation of more than one electron.) The $^1A_{1g} \rightarrow {}^1T_{1g}$ and $^1A_{1g} \rightarrow {}^1T_{2g}$ transitions are observed in the bands at 21,400 and 29,600 cm^{-1} in $Co(en)_3^{3+}$. This complex also shows a very broad spin-prohibited band at approximately 14,000 cm^{-1}, which has been assigned to the $^1A_{1g} \rightarrow {}^3T_{1g}$ transition.

In a cobalt(III) complex of type CoA_4B_2, the symmetry is lowered, and the $^1A_{1g} \rightarrow {}^1T_{1g}$ band is split.[13] The splitting in the *trans* complex is about twice that in the *cis* complex. Because the *trans* complex has a center of symmetry and the *cis* complex does not, the bands of the *trans* complex are considerably weaker than those of the *cis* complex. For complexes in which neither geometrical isomer possesses a center of symmetry (e.g., CoA_4BC and CoA_3B_3), the *cis* and *trans* isomers have approximately equal areas under the absorption curves. Examples of spectra[13] illustrating these effects are given in Fig. 22.10.

[13] F. Basolo, C. J. Ballhausen, and J. Bjerrum, *Acta Chem. Scand.*, **9**, 810 (1955).

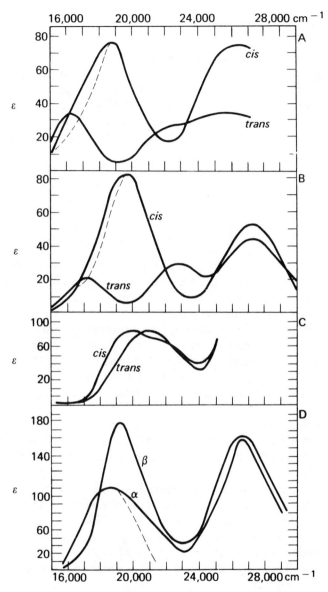

FIG. 22.10. Molar extinction coefficient ε vs. wavenumber in cm^{-1} [13]

(A) cis-[CO en$_2$Cl$_2$]Cl 0.02015 M
    tr.-[Co en$_2$Cl$_2$]Cl 0.01672 M
(B) cis-[Co en$_2$F$_2$]NO$_3$ 0.0177 M
    tr.-[Co en$_2$F$_2$]NO$_3$ 0.0184 M
(C) cis-[Co en$_2$(NO$_2$)Cl]Cl 0.01617 M
    tr.-[Co en$_2$(NO$_2$)Cl]NO$_3$ 0.00771 M
(D) α-[Co(NH$_2$CH$_2$COO)$_3$], 2H$_2$O
                       0.00188 M
    β-[Co(NH$_2$CH$_2$COO)$_3$], H$_2$O
                       0.000511 M

(Reproduced with permission of *Acta Chemica Scandinavica*.)

Charge Transfer Spectra

An electronic transition between molecular orbitals that are essentially centered on different atoms is called a *charge transfer* transition. The absorption band is usually very intense, with ε around 10^4 or greater.

The ion pair N-methylpyridinium iodide undergoes a transition of this type in which an electron is essentially transferred from the iodide ion to an antibonding orbital of the pyridine ring. Molecular iodine and various Lewis

bases form adducts exhibiting intense charge transfer absorptions. The latter transitions correspond to the excitation of an electron from an orbital that is principally nonbonding on the Lewis base to an orbital that is principally antibonding on the iodine molecule.

Charge transfer bands are often very prominent in the spectra of metal complexes in which there are electrons in π orbitals on the ligands. An energy level diagram for such a complex is shown in Fig. 22.11. The spectra of $RuCl_6^{2-}$ and $IrBr_6^{2-}$ (d^4 and d^5, respectively) show two sets of bands that have been tentatively assigned to the transitions $(t_{1u}, t_{2u}) \rightarrow t_{2g}^*$ and $(t_{1u}, t_{2u}) \rightarrow e_g^*$.

In $IrBr_6^{3-}$ (d^6), the t_{2g} orbitals are filled, and therefore only the $(t_{1u}, t_{2u}) \rightarrow e_g^*$ bands can be observed. All these bands[14] are parity-allowed transitions, and they are much more intense than the usual "ligand-field" or *d-d* bands.

In the series of complexes $Co(NH_3)_5X^{2+}$ (where X is a halogen), strong charge transfer bands are observed in the ultraviolet.[15] On going from the chloro- to the bromo- to the iodo-complex, these bands appear at progressively longer wavelengths, as shown in Fig. 22.12. In fact, in $Co(NH_3)_5I^{2+}$, these bands largely obscure the weaker *d-d* transitions. These charge transfer bands have been assigned to transitions in which an electron moves from the halide to the metal.

The intensely colored permanganate ion, MnO_4^-, is an example of a d^0 metal ion in a tetrahedral environment. The molecular orbital energy level diagram of Fig. 22.13 has been proposed by Viste and Gray.[16] The three

[14] Figgis, *Introduction to Ligand Fields*, 1966, pp. 245–47.

[15] M. Linhard and M. Weigel, *Z. anorg. Chem.*, **266**, 49 (1951).

[16] A. Viste and H. B. Gray, *Inorg. Chem.*, **3**, 1113 (1964); C. J. Ballhausen and H. B. Gray, *Molecular Orbital Theory*, W. A. Benjamin, New York, 1964, p. 127.

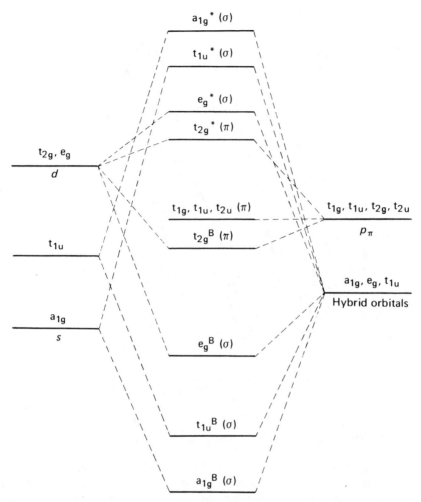

FIG. 22.11. Molecular orbital diagram for an octahedral complex with both σ and π bonding.

principal absorption bands at 18,300, 32,200, and 44,000 cm^{-1} have been assigned to the charge transfer transitions $^1A_1 \rightarrow {}^1T_2$ $(t_1 \rightarrow 2e)$, $^1A_1 \rightarrow {}^1T_2$ $(3t_2 \rightarrow 2e)$, and $^1A_1 \rightarrow {}^1T_2$ $(t_1 \rightarrow 4t_2)$. In each of these transitions, an electron is promoted from an essentially nonbonding ligand orbital to one that is largely on the manganese atom.

Charge transfer to solvent. The absorption of light in the ultraviolet region by aqueous halide ions is believed to cause a transition to a state in which a valence electron has been promoted to an expanded orbital, symmetrically centered on the parent atom, and involving the first hydration

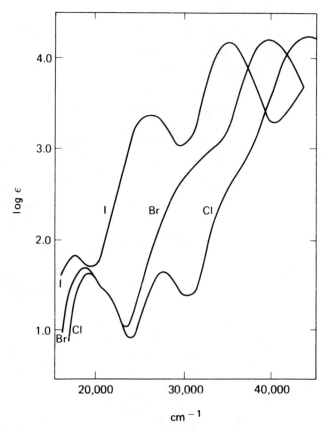

FIG. 22.12. The spectra of the $[Co(NH_3)_5X]^{2+}$ ions, where X is a halide ion.

layer. The transition is referred to as *charge transfer to solvent*. This excited state is very short lived, and goes over into a halogen atom and a hydrated electron. The hydrated electron can then recombine with the halogen atom or lead to hydrogen evolution by the recombination

$$2\,e_{aq}^- + 2\,H_2O \longrightarrow H_2 + 2\,OH^-$$

or via

$$e_{aq}^- + H_2O \longrightarrow H + OH^-$$

As would be expected for electronic transitions of this type, the transition energy can be correlated with the oxidation potential of the halide ion—the more strongly reducing the halide, the lower energy the transition. A wide variety of aqueous anions, including azide, phenolate, and ferrocyanide, show spectra of this type. The subject has been reviewed by Stein.[17]

[17] G. Stein, in *Solvated Electron*, Advances in Chemistry Series No. 50, A.C.S., Washington, D.C., 1965, p. 230.

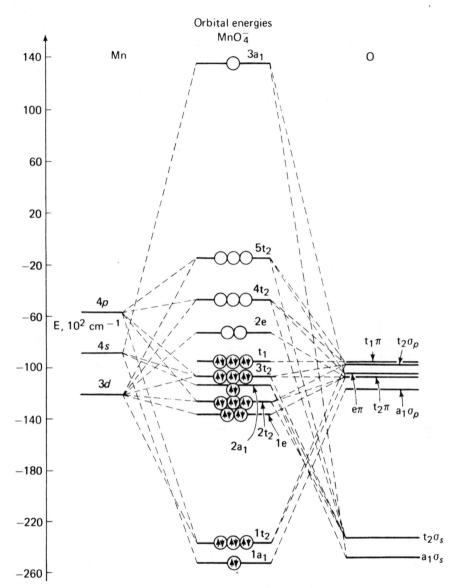

FIG. 22.13. Molecular-orbital energy-level diagram for MnO_4^-. (Adapted from ref. 16.)

Problems

1. The absorption spectrum of $Ni(NH_3)_6^{2+}$ has bands at 10,750, 17,500, and 28,200 cm^{-1}. Evaluate Δ and assign transitions to the bands. Using the value of Δ and the fact that $\Delta E(^3P - ^3F) = 15,800 \ cm^{-1}$ for Ni^{2+}, estimate the frequency of the third band and compare the estimated value with the experimental value $28,200 \ cm^{-1}$.

2. The bond angle in NO_2 is 132°. The ground-state MO electronic configuration is $\cdots (a_2)^2(b_1)^2(a_1)^1(b_2)^0$. Label the transitions to the excited states $\cdots (a_2)^2(b_1)^1(a_1)^2(b_2)^0$ and $\cdots (a_2)^2(b_1)^1(a_1)^1(b_2)^1$. Indicate whether or not each transition is allowed.

Optical Activity

23

INTRODUCTION

A species is optically active if its structure cannot be superimposed on its mirror image, which is usually the case whenever a species has neither a plane nor a center of symmetry. However, a more rigorous criterion[1] of optical activity is the nonexistence of a rotation-reflection axis (S_n).

Optically active compounds can exist in either of two isomeric forms (mirror images), which are often designated dextro (d) and levo (l). These appellations indicate the direction (right and left) in which polarized light at a particular wavelength (usually the sodium D line) is rotated by the isomers. A better way of indicating isomers is to use the notation $(+)$ for an isomer that is dextrorotatory at the sodium D line and $(-)$ for its levorotatory mirror-image *enantiomer* (isomer of opposite configuration). When the rotation is measured at a wavelength other than the sodium D line, the wavelength is indicated by a subscript, thus: $(+)_{5461}$. An equimolar mixture of the $(+)$ and

[1] See the discussion on pp. 23–24 in H. H. Jaffe and M. Orchin, *Symmetry in Chemistry*, John Wiley & Sons, Inc., New York, 1965.

339

(−) enantiomers of an optically active compound does not rotate polarized light and is called a *racemic mixture* or *racemate*. Often a sample of one enantiomer of an optically active compound will spontaneously lose its activity because of conversion to the corresponding racemate (a process that is always thermodynamically favored if there is no phase change). This process is referred to as *racemization*. A process for the separation of the enantiomers in a racemate is called a *resolution*.

POLARIMETRY

The *specific rotation* [α] of a dissolved substance is given by the expression [α] = α/lc, where α is the observed rotation in degrees, *l* is the path length of the sample in *decimeters*, and *c* is the concentration in *grams per milliliter*. The angle α is considered positive if the electric vector turns as a left-hand screw as the light passes through the solution. The *molecular rotation* [α_M] of a compound is given by the expression [α_M] = M [α]/100, where M is the molecular weight.

Many types of polarimeters (instruments for measuring the rotation of plane-polarized light by substances) are commercially available.[2] In the following paragraph, we very briefly describe the construction and operation of simple "visual" polarimeters.

Optical rotation is generally measured using light from a sodium-vapor lamp, which gives essentially monochromatic radiation (the yellow sodium *D* line is a doublet at 5890 and 5896 Å). A beam of light is polarized by passage through a nicol prism (the polarizer), which consists of two calcite prisms cemented together so that only one of the two rays formed by double refraction is transmitted. The beam of polarized light passes through the solution and then through a second nicol prism (the analyzer; see Fig. 23.1). When no optically active material is placed between the prisms (0° rotation), the prisms are positioned at right angles so that no light is transmitted. When an optically active material is placed between the prisms, the analyzer must be turned in order to maintain darkness in the field of view. The optical rotation is the angle by which the analyzer is turned (clockwise with respect to the observer) in order to reach darkness. It is very difficult to determine by eye the setting for complete darkness, because positions near the completely dark position are very dark. Therefore, many instruments are constructed such that the field of view is divided into two equal parts, and the analyzer is adjusted so as to equalize the light intensity in each half of the field.

---

[2] See W. Heller and D. Fitts, *Physical Methods of Organic Chemistry*, 3rd ed., in *Technique of Organic Chemistry*, vol. I, A. Weissberger, ed., Interscience Publishers, New York, 1960, Part 3, pp. 2184–2297.

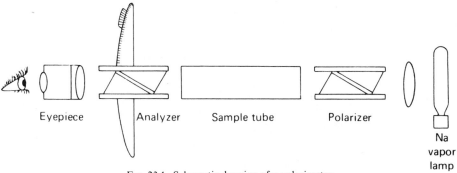

Eyepiece   Analyzer   Sample tube   Polarizer   Na vapor lamp

FIG. 23.1. Schematic drawing of a polarimeter.

PREPARATION OF OPTICALLY ACTIVE COMPOUNDS

Fractional Separation of Diastereoisomers[3]

This method is the commonest and most useful for the resolution of racemic ionic species. A typical procedure consists of treating a solution containing the racemic ions with a "resolving agent"—for example, a salt containing optically active ions of charge opposite to that of the racemic ions. By taking advantage of the fact that the two diastereoisomers[3] differ in solubility, we may separate them by fractional crystallization. The optically active ions of the resolving agent may be removed by a subsequent reaction (such as precipitation of an insoluble salt), leaving the separated enantiomers of the original racemic compound free of optically active contaminants. For example, racemic cis-$Co(en)_2(NO_2)_2^+$ can be resolved using potassium antimonyl-$(+)$-tartrate:

$$(+)\text{-}cis\text{-}Co(en)_2(NO_2)_2^+ + (-)\text{-}cis\text{-}Co(en)_2(NO_2)_2^+ + 2(+)\text{-}SbOC_4H_4O_6^- \longrightarrow$$

$$[(+)\text{-}cis\text{-}Co(en)_2(NO_2)_2][(-)\text{-}SbOC_4H_4O_6] +$$

$$[(-)\text{-}cis\text{-}Co(en)_2(NO_2)_2][(+)\text{-}SbOC_4H_4O_6]$$

The diastereoisomer $[(-)\text{-}cis\text{-}Co(en)_2(NO_2)_2][(+)\text{-}SbOC_4H_4O_6]$ is only slightly soluble and precipitates as yellow crystals, leaving the other diastereoisomer in solution. By treatment with sodium iodide, this precipitate may be converted to the corresponding iodide $[(-)\text{-}cis\text{-}Co(en)_2(NO_2)_2]I$. The dextro enantiomer may be isolated as $[(+)\text{-}cis\text{-}Co(en)_2(NO_2)_2]Br$ by addition of ammonium bromide to the mother liquor of the first precipitation. Some commonly used resolving agents are listed in Table 23.1.

[3] The diastereoisomers are the two combinations of the resolving agent ion and the ions of opposite charge that are to be resolved (separated).

Table 23.1   Some Resolving Agents

Cations	*Anions*
Strychnine*	Tartrate
Brucine*	Antimonyl tartrate
Chinchonine*	α-Bromocamphor-π-sulfonate
Quinine*	Camphor-π-sulfonate
α-Phenylethylamine*	Nitrocamphoronate
Quinidine*	Dibenzoyl-(+)-tartrate
Morphine*	Diacetyltartrate
Nor-4-ephedrine*	Co(edta)$^-$
Co(en)$_3$$^{3+}$	
cis-Co(en)$_2$(NO$_2$)$_2$$^+$	
Ni(phen)$_3$$^{2+}$	

* As the protonated species.

In principle, the resolution of a racemic species should not be able to give more than a 50 per cent yield of either one of the enantiomers; however, such yields can be achieved if racemization of the enantiomer left in solution occurs readily. Thus, if a solution of racemic iron(II) *o*-phenanthroline complex is allowed to stand for a long time, the whole of the iron complex is precipitated as the (−)-complex (+)-antimonyl tartrate, leaving a colorless solution behind.[4] In Synthesis 14 (p. 469), the precipitate is collected soon after the reagents are mixed, so as to allow the separate precipitation of the (+)-complex before significant racemization has occurred. On p. 75, a high-yield resolution of Co(en)$_3$$^{3+}$, effected by a Co(II)-catalyzed racemization of the unprecipitated enantiomer, is described.

Methods other than fractional crystallization have been used to separate diastereoisomers—for example, zone melting, chromatography, and distillation. However, these techniques are not as popular as the method based on solubility differences, principally because they usually result in only partial resolution.

Resolution by Adsorption

Unchanged optically active compounds can be resolved by preferential adsorption on optically active adsorbents. In the initial use of the technique by Tsuchida *et al.*[5] (±)-[Co(dmg)$_2$(NH$_3$)Cl] (dmg is the abbreviation for dimethylglyoximate) was partially resolved by shaking a solution of the substance with powdered levoquartz. By passing racemates through chromatographic columns containing optically active adsorbents or ion-exchangers, complete resolution has been achieved in a few cases. Thus,

[4] F. P. Dwyer and E. C. Gyarfas, *J. Proc. Roy. Soc. N. S. Wales*, **83**, 263 (1950).

[5] R. Tsuchida, M. Kobayashi, and A. Nakamura, *J. Chem. Soc. Japan*, **56**, 1339 (1935); *Bull. Chem. Soc. Japan*, **11**, 38 (1936).

$[Co(en)_n((-)-pn)_{3-n}]^{3+}$ (pn is the abbreviation for propylenediamine) complexes have been resolved on cellulose,[6] various metal acetylacetonates have been resolved on $(+)$-lactose hydrate,[7] and the hexakis (2-aminoethanethiolo) tricobalt(III) cation has been resolved using cation exchange cellulose.[8]

Displacement Reactions

When the displacement of ligands from an optically active complex by other ligands occurs with retention or inversion of configuration, the reaction can be used for the synthesis of optical isomers. Thus, $(+)$-$[Co(en)_3]^{3+}$ is formed by the treatment of $(+)$-$[Co\,EDTA]^-$ with ethylenediamine,[9] and $(-)$-$[Co((-)-pn)_3]^{3+}$ is formed by the reaction of $(+)$-$[Co\,EDTA]^-$ with $(-)$-propylenediamine.[10] It is particularly interesting that whereas treatment of $(-)$-$[Co$-$(en)_2Cl_2]^+$ with potassium carbonate gives $(+)$-$[Co(en)_2CO_3]^+$, treatment with silver carbonate gives $(-)$-$[Co(en)_2CO_3]^+$, showing that retention of configuration occurs in one case and inversion occurs in the other.[11]

Dissymmetric Synthesis

A dissymmetric synthesis is one in which the presence of one enantiomer of an auxiliary agent causes the preferential formation of one of the enantiomers of the compound synthesized. For example, oxidation of the product of the reaction between aqueous Co^{2+} and $(+)$-*trans*-1,2-diaminocyclopentane $[(+)$-cptn] yields almost exclusively $(-)$-$\{Co[(+)-cptn]_3\}^{3+}$.[12]

STRUCTURE DETERMINATION

Information from Resolvability

Sometimes the simple knowledge that a compound is resolvable into optically active enantiomers is very significant, because it permits one to eliminate certain conceivable configurations that could not possibly be optically active. The classic example of this sort of application of optical activity data is the

[6] F. P. Dwyer, T. E. MacDermott, and A. M. Sargeson, *J. Am. Chem. Soc.*, **85**, 2913 (1963).

[7] T. Moeller and E. Guylas, *J. Inorg. Nucl. Chem.*, **5**, 245 (1958); J. P. Collman, *et al.*, *Inorg. Chem.*, **2**, 576 (1963).

[8] G. R. Brubaker, J. I. Legg, and B. E. Douglas, *J. Am. Chem. Soc.*, **88**, 3446 (1966).

[9] F. P. Dwyer, E. C. Gyarfas, and D. P. Mellor, *J. Phys. Chem.*, **59**, 296 (1955).

[10] S. Kirschner, Y. K. Wei, and J. C. Bailar, Jr., *J. Am. Chem. Soc.*, **79**, 5877 (1957).

[11] J. C. Bailar, Jr., and R. W. Auten, *J. Am. Chem. Soc.*, **56**, 774 (1934).

[12] F. M. Jaeger and H. B. Blumendal, *Z. Anorg. Allgem. Chem.*, **175**, 161 (1928); *Proc. Acad. Sci. Amsterdam*, **40**, 108 (1937).

work of Mills and Quibell,[13] who resolved the (*meso*-stilbenediamine)-(*iso*-butylenediamine)platinum(II) cation and thus showed that the complex does not have a tetrahedral configuration of nitrogens around the platinum. As can be seen from Fig. 23.2, the tetrahedral configuration would have a plane of symmetry and thus be unable to show optical activity. On the other hand, a square-planar configuration (Fig. 23.3) is not superimposable on its mirror image and is consistent with the demonstration of optical activity.

Another application of resolvability data is in distinguishing *cis* and *trans* isomers of certain octahedral complexes. For example, the complex $Co(en)_2Cl_2^+$ has two geometric forms, *cis* and *trans*, illustrated in Fig. 23.4. The *trans* form has several planes of symmetry and therefore is inactive; the *cis* form has no plane of symmetry and is not superimposable on its mirror image. Now, $Co(en)_2Cl_2^+$ is known in two forms, purple and green. It has been shown that the purple form can be resolved into optically active enantiomers; therefore, it is clear that the purple form is the *cis* isomer and the green form is the *trans* isomer.

It should be obvious that the inability of an investigator to resolve a compound should not be taken as evidence of a plane or center of symmetry in the compound. Only a successful resolution should be taken as positive evidence for the *lack* of such symmetry.

Absolute Configuration

When any one of the various resolution techniques that we have discussed is successful, it is clear that the racemate has been separated into its two mirror-image components. However, in general, one does not know which of the separated enantiomers corresponds to a particular mirror image. A method based on "anomalous X-ray diffraction"[14] has been used to determine the absolute configurations of a number of optical isomers, and it is at present the only completely reliable method for absolute configuration determinations. However, the technique is relatively complicated, and chemists have attempted to devise simpler ways for such determinations. Three such techniques are discussed in the following paragraphs.

"*Active Racemates.*" When aqueous solutions of $(-)$-[Rh(en)$_3$]Br$_3$ and $(-)$-[Co(en)$_3$]Br$_3$ are mixed, a precipitate of composition [Co(en)$_3$][Rh(en)$_3$] Br$_6$·5 H$_2$O is formed. When a similar experiment is performed with $(+)$-[Rh(en)$_3$]Br$_3$ and $(-)$-[Co(en)$_3$]Br$_3$, no such precipitate is formed, and, upon evaporation, mixed crystals separate.[15] Delépine *assumed* that, in the one-to-one compound formed in the first case, the $(-)$-Co(en)$_3^{3+}$ ions were

[13] W. H. Mills and T. H. H. Quibell, *J. Chem. Soc.*, 839 (1935).

[14] J. M. Bijvoet, *Endeavor*, **14**, 71 (1955).

[15] M. Delépine, *Bull. Soc. Mineral France*, **53**, 73 (1930).

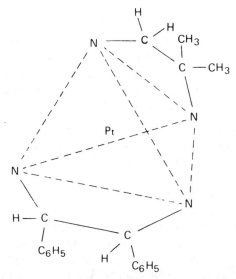

FIG. 23.2. Imaginary tetrahedral configuration of (meso-stilbenediamine)-(isobutylenediamine)platinum(II) cation. A plane of symmetry is perpendicular to the C—C bond of the stilbenediamine and contains the C—C bond of the isobutylenediamine.

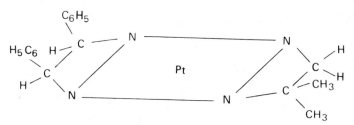

FIG. 23.3. The disymmetric square-planar configuration of (meso-stilbene-diamine) (isobutylenediamine)platinum(II) cation.

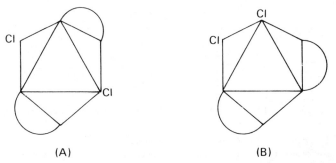

FIG. 23.4. (A) *trans*-Co(en)$_2$Cl$_2$$^+$. (B) One of the optical isomers of *cis*-Co(en)$_2$Cl$_2$$^+$.

of opposite configuration to the $(-)$-Rh(en)$_3^{3+}$ ions, and he called the compound an "active racemate"[16] (the optical activity would be zero only by accident). A number of such active racemates can be formed by the interaction of appropriate optical isomers. If it is *assumed* that each of these is formed by ions of opposite absolute configuration, and if the absolute configuration of one of the ions is known (say, from X-ray data), then obviously the absolute configuration of the other ion is determined.

The Cotton Effect. Optical rotation varies with the wavelength of the light, especially in the region of an electronic absorption band of the atom situated at the center of dissymmetry. A plot of optical rotation vs. wavelength is called an optical rotatory dispersion (ORD) curve. An idealized ORD curve in the region of an "optically active" absorption band is shown in Fig. 23.5. This curve is idealized in the sense that it is assumed that there are no other nearby absorption bands. Actual ORD curves for some Co(III) chelates are shown in Fig. 23.6. It should be understood that the value of $[\alpha]$ for one enantiomer is equal and opposite to that of the other enantiomer at all wavelengths.[17] The behavior of an ORD curve in the vicinity of an active absorption band is usually called the *Cotton effect*. A positive Cotton effect corresponds to an ORD curve in which $[\alpha]$ goes from negative to positive values with increasing wavelength in the neighborhood of the absorption

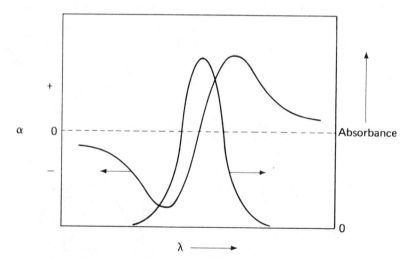

FIG. 23.5. Idealized plots of optical rotation and absorbance vs. wavelength.

[16] M. Delépine, *Bull. Soc. Chim. France*, **29**, 656 (1921).
[17] The difference in extinction coefficient for left- and right-circularly polarized light ($\varepsilon_l - \varepsilon_r$) for one enantiomer is equal and opposite to that of the other enantiomer. A plot of $\varepsilon_l - \varepsilon_r$ vs. wavelength is called a *circular dichroism* plot.

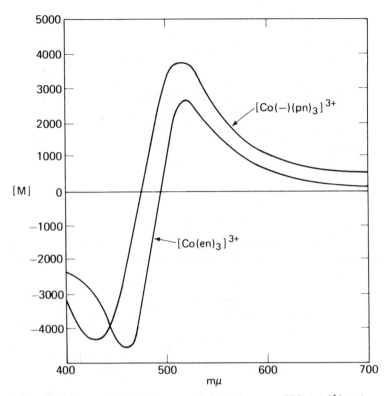

FIG. 23.6. Rotatory dispersion curves of the (+) isomers of $[Co(en)_3]^{3+}$ and $[Co(-)(pn)_3]^{3+}$.

band; a negative Cotton effect corresponds to an ORD curve with the opposite shape.

It is generally assumed that if two different optically active isomers of similar structure {e.g., (+)-$[Co(en)_3]^{3+}$ and (+)-$[Co(pn)_3]^{3+}$} have Cotton effects of the same sign at corresponding electronic absorption bands, then the isomers have the analogous absolute configurations. Thus, one concludes that the isomers having the ORD curves shown in Fig. 23.6 have essentially the same absolute configuration. When this generalization is used, it is most important to identify the electronic transitions to make sure that analogous Cotton effects are being compared.

Piper has suggested that optical isomers, the absolute configurations of which are like left-handed screws (as in the structure in Fig. 23.4B), should be referred to as Λ configurations, and that their mirror images should be referred to as Δ configurations.[18] Another terminology, called the octant sign, has been proposed. The octant sign is determined by straightforward

[18] T. S. Piper, *J. Am. Chem. Soc.*, **83**, 3908 (1961).

geometrical considerations and is applicable even when a configuration cannot be visualized as an imaginary screw. Some success has been had in correlating the octant sign with the sign of the Cotton effect.[19]

Stereospecificity. Stereospecific effects are observed in many reactions of complexes containing optically active chelating ligands. These are believed to arise because of nonbonded interactions between chelate rings causing certain conformations to be more stable than others.[20] For example, consider the ligand (+)-propylenediamine, which is known to have the absolute configuration,

$$CH_2NH_2$$
$$H_2N \blacktriangleright C \blacktriangleleft H$$
$$CH_3$$

The puckered chelate ring formed by a (+)-propylenediamine molecule and a Co^{3+} ion can have either of the two configurations,

∂-(+)-pn λ-(+)-pn

In the left-hand configuration, the methyl group is "equatorially" oriented; in the right-hand configuration, the methyl group is "axially" oriented.[21] Corey and Bailar[20] point out that nonbonded interactions are minimized when the methyl group is equatorial; therefore, we can reasonably assume that all three rings in $Co[(+)-pn]_3{}^{3+}$ have the first configuration.

Now, there are two principal ways of orienting the three rings around the Co^{3+} ion:

$\Lambda\partial\partial\partial$ $\Delta\partial\partial\partial$

[19] C. J. Hawkins and E. Larsen, *Acta Chem. Scand.*, **19**, 185 (1965); Anonymous, *Chem. Eng. News*, Oct. 2, 1967, pp. 48–49.

[20] E. J. Corey and J. C. Bailar, Jr., *J. Am. Chem. Soc.*, **81**, 2620 (1959).

[21] The nomenclature *equatorial* and *axial* is analogous to that used in discussing the substituents of the chair form of cyclohexane. It has become conventional to use the symbols ∂ and λ for chelate conformations that are shaped like right-handed and left-handed helices, respectively.

In the $\Lambda\partial\partial\partial$ configuration, the C—C bonds in the rings are almost parallel to the three-fold axis of the molecule, whereas in the $\Delta\partial\partial\partial$ configuration, the C—C bonds in the rings are more nearly perpendicular to the three-fold axis.[22] Corey and Bailar[20] have shown that nonbonded interactions are minimized in the former type of configuration. Therefore, stereochemical considerations indicate that $Co[(+)\text{-pn}]_3{}^{3+}$ should be most stable in the $\Lambda\partial\partial\partial$ configuration. In fact, the most stable complexes formed by Co^{3+} with racemic propylene-diamine are $(+)\text{-}\{Co[(+)\text{-pn}]_3\}^{3+}$ and $(-)\text{-}\{Co[(-)\text{-pn}]_3\}^{3+}$, and it has been shown that the first of these has the predicted absolute configuration. By similar arguments, one can show that $(-)$-propylenediamine should form principally the enantiomer of opposite configuration around the Co(III),

$\Delta\partial\partial\partial$

Suggestions for Further Reading

Cotton, F. A., and G. Wilkinson, *Advanced Inorganic Chemistry*, 2nd ed., Interscience Publishers, New York, 1966, pp. 646–51.

Gillard, R. D., "The Cotton Effect in Coordination Compounds," in *Progr. Inorg. Chem.*, **7**, 215 (1966).

Jones, M. M., *Elementary Coordination Theory*, Prentice-Hall, Inc., Englewood Cliffs, N.J., 1964, pp. 173–85.

Kirschner, S., "Optically Active Coordination Compounds," in *Preparative Inorganic Reactions*, vol. 1, W. L. Jolly, ed., Interscience Publishers, New York, 1964, p. 29.

Woldbye, F., "Technique of Optical Rotatory Dispersion and Circular Dichroism," in *Technique of Inorganic Chemistry*, vol. 4, H. B. Jonassen and A. Weissberger, eds., Interscience Publishers, New York, 1965, p. 249.

[22] It is helpful to examine models of the ions in order to be convinced of these differences.

Nuclear

Magnetic

Resonance

24

Nuclear magnetic resonance (nmr) is probably the chemist's most valuable tool for determining the structures of compounds in the liquid and dissolved states. In contrast to techniques such as X-ray diffraction and microwave spectroscopy (which are often capable of furnishing much more quantitative structural information for solids and gases, respectively), nmr spectroscopy is easily learned and yields spectra that are easily interpreted. In this chapter, we present a brief, heuristic description of the theory and interpretation of nmr spectra.

THE ENERGY LEVELS OF MAGNETIC NUCLEI

Every nucleus having a spin number I other than zero possesses angular momentum and, consequently, a magnetic moment. Some important magnetic nuclei and their spin numbers are listed in Table 24.1.[1] In a uniform

[1] When the atomic mass number, A, and the atomic number, Z, are both even numbers, I is zero, as in the cases of the ^{12}C and ^{16}O nuclei. Nuclear magnetic resonance is impossible with

Table 24.1 Properties of Some Important Nuclei*

Nucleus	Atomic Number (A)	Natural Abundance (per cent)	Spin Number (I)	Quadrupole Moment (Q)	nmr Frequency (in Mc for $H = 10^4$ G. v)
^{1}H	1	99.984	$\frac{1}{2}$	0	42.577
^{2}H	1	0.0156	1	0.0028	6.536
^{10}B	5	18.83	3	0.111	4.575
^{11}B	5	81.17	$\frac{3}{2}$	0.0355	13.660
^{13}C	6	1.108	$\frac{1}{2}$	0	10.705
^{14}N	7	99.635	1	0.02	3.076
^{15}N	7	0.365	$\frac{1}{2}$	0	4.315
^{17}O	8	0.037	$\frac{5}{2}$	−0.004	5.772
^{19}F	9	100	$\frac{1}{2}$	0	40.055
^{29}Si	14	4.70	$\frac{1}{2}$	0	8.460
^{31}P	15	100	$\frac{1}{2}$	0	17.235
^{33}S	16	0.74	$\frac{3}{2}$	−0.064	3.266
^{117}Sn	50	7.67	$\frac{1}{2}$	0	15.77
^{119}Sn	50	8.68	$\frac{1}{2}$	0	15.87

* See note 1.

applied magnetic field, H, a given nucleus can assume any one of the $2I + 1$ possible orientations relative to the applied field and will precess about the applied field at an angular velocity γH (where γ is the "magnetogyric" ratio— a characteristic constant for each nucleus). Each orientation corresponds to an energy

$$E = -\gamma h H \frac{M}{2\pi} \tag{24.1}$$

where h is Planck's constant and M has one of the following values: $I, I - 1$, $I - 2, \cdots -I$. Thus, the energy level diagram for a proton ($I = \frac{1}{2}$) is as follows:

$$M = -\frac{1}{2} \underline{\hspace{3cm}}$$
$$\text{- - - - - - - - -}$$
$$+\frac{1}{2} \underline{\hspace{3cm}}$$

and that for a deuteron ($I = 1$) is as follows:

$$M = -1 \underline{\hspace{3cm}}$$
$$0 \underline{\hspace{3cm}}$$
$$+1 \underline{\hspace{3cm}}$$

these nuclei. When A is even and Z is odd, I is integral, as in the cases of ^{2}H and ^{14}N (both with $I = 1$), and ^{10}B ($I = 3$). When A is odd, I is half-integral, as in the cases of ^{1}H, ^{31}P, ^{29}Si, and ^{19}F (all having $I = \frac{1}{2}$), and ^{11}B, ^{33}S, and ^{35}Cl ($I = \frac{3}{2}$). When $I = 0$ or $\frac{1}{2}$, the nuclear quadrupole moment is zero, and when $I \geqslant 1$, the nuclear quadrupole moment is finite. The data of Table 24.1 are taken from a chart of Varian Associates, Palo Alto, Calif.

The separation between adjacent energy levels is given by the expression $\Delta E = \gamma h (H/2\pi)$. At room temperature, $kT \gg \Delta E$, and consequently the energy levels are nearly equally occupied. Nuclear magnetic resonance spectrometry consists of detecting, and measuring the energies of, transitions between such energy levels. An nmr spectrometer consists of a magnet, a radio-frequency transmitter, and a radio-frequency receiver. In a typical nmr experiment, the precessional frequency of the nuclei is changed by varying the applied magnetic field while holding the transmitted radio-frequency constant. When the nuclear precessional frequency becomes equal to the transmitted radio frequency, a resonance absorption of energy takes place. In this process, some of the nuclei in each of the energy levels (except the highest) are promoted to the next higher energy level. Remembering that $\Delta E = h\nu$, we calculate for the resonance radio frequency

$$\nu = \frac{\gamma H}{2\pi}$$

CHEMICAL SHIFT

In an ordinary chemical system, the effective magnetic field H_{eff} at a nucleus is not the same as the applied field H_0 because of the shielding effect of the electrons in the vicinity of the nucleus. Where σ is the shielding constant for a nucleus in a particular environment, we write

$$H_{eff} = H_0(1 - \sigma) \tag{24.2}$$

Thus, in fixed-frequency nuclear magnetic resonance, the applied field must be greater to achieve resonance when a nucleus is shielded.

The difference in resonance field strength for nuclei in two different environments is called the *chemical shift*, δ, and is generally expressed in parts per million relative to some standard nucleus:

$$\delta = \frac{H_{sample} - H_{reference}}{H_{reference}} \times 10^6 \tag{24.3}$$

In the case of proton magnetic resonance of organic compounds, tetramethylsilane (TMS) is commonly used as a reference compound. Some chemical shifts for compounds of hydrogen,[2] nitrogen,[3] and fluorine[4] are given in Table 24.2. Each of the compounds in Table 24.2 has only one

[2] Tau, τ, values are used to avoid the use of negative values for δ. Data are from J. W. Emsley, J. Feeney, and L. H. Sutcliffe, *High Resolution Nuclear Magnetic Resonance Spectroscopy*, vol. 2, Pergamon Press, Inc., New York, 1966, Appendix B.

[3] Y. Masuda and T. Kand, *J. Phys. Soc. Japan*, **8**, 432 (1953); B. E. Holder and M. P. Klein, *J. Chem. Phys.*, **23**, 1956 (1955).

[4] R. S. Drago, *Physical Methods in Inorganic Chemistry*, Reinhold Publishing Corp., New York, 1965, p. 252.

Table 24.2 Some Chemical Shifts for ^1H, ^{14}N, and ^{19}F*

$\tau = 10 + \delta$ for ^1H†		δ for ^{14}N**		δ for ^{19}F§	
		NH_4^+	602	CH_3F	278
$(CH_3)_4Si$	10.000				
		$(C_2H_5)_3N$	575	HF	203
$(CH_3)_4Sn$	9.94				
		N_2H_4	566	SiF_4	177
$(CH_3)_4C$	9.06				
		$(CH_3)_4NBr$	552	BF_4^-	149
CH_3Br	7.32				
		NH_3OHCl	520	BF_3	133
$(CH_3)_2O$	6.76				
		SCN^-	406	$F^-(aq)$	120
CH_2I_2	6.10				
		CH_3CN	385	CF_3COOH	77
CH_3F	5.74				
		CN^-	380	CF_4	70
CH_2ClI	5.01				
		$C(NO_2)_4$	300	CCl_3F	0
CH_2Cl_2	4.67				
		C_5H_5N	276	SF_6	-42
$CHBr_3$	3.18				
		N_2	268	N_2F_4	-60
$CHCl_3$	2.75				
		NO_3^-	254	XeF_6	-78
C_6H_6	2.73				
		CH_3NO_2	254	NF_3	-140
		$C_6H_5NO_2$	252	OF_2	-250
		NO_2^-	0	F_2	-422
				UF_6	-746

* All values are in parts per million. † See ref. 2. ** See ref. 3. § See ref. 4.

structural kind of hydrogen atom, nitrogen atom, or fluorine atom; consequently, the spectra are very simple, generally consisting of single resonances. However, many compounds contain more than one structurally distinguishable magnetic nucleus of a given element, and the nmr spectra of such compounds usually consist of as many main resonances as there are types of magnetic nuclei. This fact, together with the fact that the signal intensities (measured as areas under the absorption curves) are proportional to the number of nuclei present, makes nmr extremely useful for determining the structures of compounds. Nuclear magnetic resonance spectra illustrating these points are presented in Fig. 24.1. Most of the resonance signals show fine structure, the cause of which will be discussed; in measuring the areas of

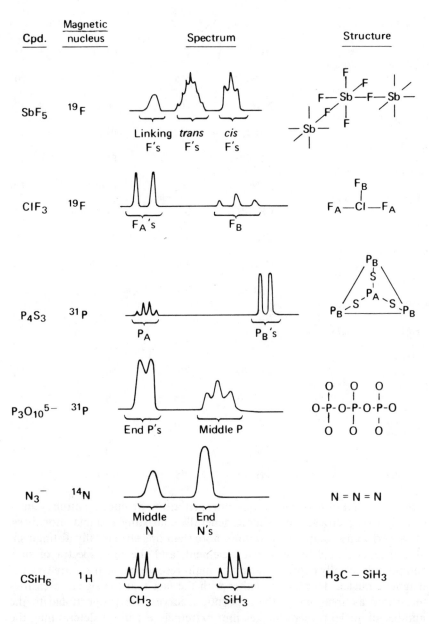

Fig. 24.1. nmr spectra of compounds containing non-equivalent magnetic nuclei.

such signals, one must add the areas of all the components of the signals. When this is done, the areas of the signals are in the same proportion as the numbers of structurally different nuclei.

Inasmuch as nmr signal intensities are proportional to the concentrations of the nuclei, nmr can be used as a method of quantitative analysis. For such purposes, it is necessary to include in the solution a known concentration of some standard substance.

FIRST-ORDER COUPLING

Energy Level Diagrams

The nmr spectra of molecules containing more than one magnetic nucleus of the same kind, all of which are structurally equivalent, consist of single-line resonances. The proton spectra of H_2 and CH_3Br are of this type. However, let us consider the situation in molecules containing more than one structural type of magnetic nucleus. For simplicity, let us first consider molecules such as HCOOH and HBr_2CCCl_2H, which contain just two different kinds of protons, A and X. The energy levels of a molecule of this type in an applied magnetic field are shown in Fig. 24.2. On the left side of the figure, we show a possible spacing of energy levels for the situation in which the two protons are essentially independent of one another—that is, one proton does not feel the magnetic moment of the other. This lack of "coupling" between the protons is expressed by stating that the coupling constant J_{AX} is zero. On the right side of Fig. 24.2, we show the energy level diagram for the situation in which the protons are affected by each other's magnetic moments; that is, J_{AX} is finite. Notice that when the two protons are aligned in the same direction, the energy level is raised when coupling takes place, and that when the two protons have their moments opposed, the energy level is lowered by coupling.[5] The only allowed transitions are those in which protons flip one at a time. These transitions are indicated by vertical lines in the energy level diagrams. In the absence of coupling, the spectrum consists of two lines— one corresponding to proton A, and one corresponding to proton X. With coupling of the spins, each of these lines is split into a doublet. The separation of the doublets is J_{AX}, usually expressed in frequency units (cycles per second, cps).

A somewhat more complicated energy level diagram—that for NH_3— appears in Fig. 24.3. In this molecule, the ^{14}N nucleus ($I = 1$) is coupled to the three equivalent protons ($I = \frac{1}{2}$). The ^{14}N magnetic resonance is split

---

[5] This is the situation when J_{AX} is positive; the opposite effect on the energy levels is obtained when J_{AX} is negative. The spectra are identical, however, and the sign of J_{AX} cannot be ascertained from the spectrum.

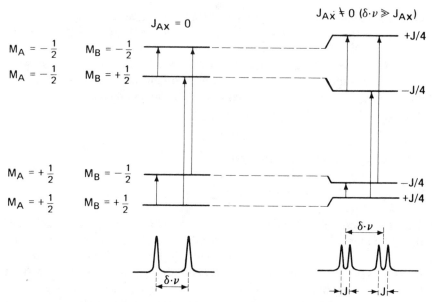

Fig. 24.2. Energy level diagram for molecule of type AX.

into a quartet by the protons, and the proton magnetic resonance is split into a triplet by the ^{14}N nucleus.

It may be stated as a general rule that the signal of a nucleus (or group of equivalent nuclei) that interacts with another group of n equivalent nuclei of spin I is split into $2nI + 1$ lines.

Intensity Ratios in Multiplets

The relative intensities of the lines in an nmr multiplet may be calculated by the method illustrated in the following examples. Consider the proton magnetic resonance of CH_3D. Although the three protons are equivalent, the spectrum is a triplet because of coupling to the deuteron ($I = 1$). The intensity ratio is $1:1:1$, because the deuterons in the sample can have their spins oriented according to any of three equally probable M values: $-1, 0$, or $+1$. Now consider the proton spectrum of CH_2D_2. We may first consider the effect of one deuteron, and then consider the effect of the second deuteron. As before, one deuteron will yield a $1:1:1$ triplet. The second deuteron will split each of these lines into a triplet (with the same coupling constant as in the first triplet). Obviously, some of the lines will be superimposed on each other. When two or more lines fall at the same point, the intensity of the resultant line is equal to the sum of the intensities of the individual lines. Thus, in the case of CH_2D_2, we may make a diagram as follows and conclude that the resultant signal will be a $1:2:3:2:1$ quintet.

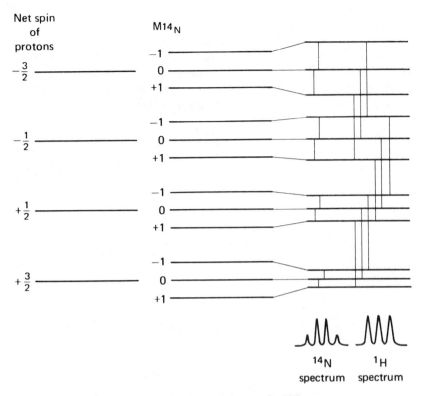

FIG. 24.3. Energy level diagram for NH_3.

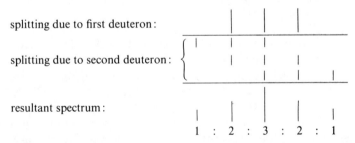

The reader should verify that the proton magnetic resonance spectrum of CHD_3 is predicted to be a septet with intensities in the ratio $1:3:6:7:6:3:1$.

SECOND-ORDER SPECTRA

Consequences of $\delta \cdot \nu \sim J$

The discussion of spin-spin coupling in the preceding section is based on the assumption that the chemical shift between the interacting nuclei is large compared to the coupling constant, when both are expressed in the same units (e.g., cps; $\delta \cdot v/J > \sim 6$). Whenever this condition is not fulfilled, the nmr spectrum is more complicated than that predicted on the basis of the simple theory, and it is generally difficult to analyze without the aid of a computer. For the relatively simple case of two nuclei, the spectrum changes, as shown in Fig. 24.4, as $\delta \cdot v/J$ varies from very large to zero.

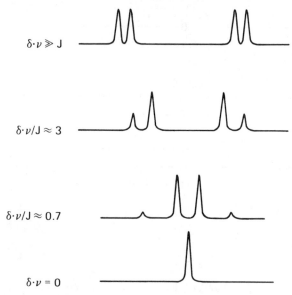

FIG. 24.4. Spectrum of a system of two nuclei as the chemical shift is changed relative to the coupling constant.

Spin-spin coupling does not depend on the presence of the magnetic field; consequently, J is independent of H_0. However $\delta \cdot v$ is proportional to H_0. Thus, the ratio $\delta \cdot v/J$ increases as H_0 increases, and complicated second-order spectra can be transformed into simple first-order spectra by suitably increasing the magnetic field strength. For example, consider the relatively complicated spectrum of tetraethylstannane at 60 Mc sec^{-1} (14 kG) in Fig. 24.5 (top) and the simplified spectrum of the same material at 200 Mc sec^{-1} (47 kG) in Fig. 24.5 (bottom).[6]

Magnetic Nonequivalence

The protons in CH_2F_2 are structurally indistinguishable, and the four H-F coupling constants are identical. Such nuclei are said to be both *chemically* and *magnetically equivalent*, and, when $\delta \cdot v \gg J$, they yield first-order spectra.

[6] Spectra of Drs. H. A. Weaver and Pier, Varian Associates, Inc. Taken from F. A. Bovey, *Chem. Eng. News*, Aug. 30, 1965, pp. 98–121.

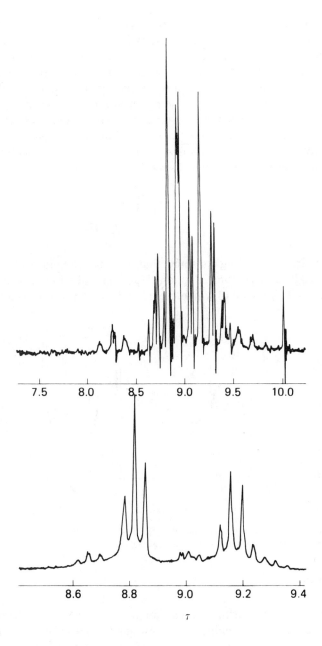

FIG. 24.5. These nmr spectra of tetraethyltin were obtained at 60 Mc (top) and 200 Mc (bottom).[6] (Reproduced with permission of the American Chemical Society.)

However, in $H_2C{=}CF_2$ there are two types of H-F coupling constants (*cis* and *trans*), and, although the protons are structurally equivalent (they have the same chemical shift), they are not *magnetically* equivalent. The nmr spectrum is very complicated. Often, nuclei that would otherwise be magnetically nonequivalent are magnetically equivalent because of rapid internal rotation. Thus, in CH_3CH_2I, the methyl group rotates so rapidly[7] that the three CH_3 protons are effectively magnetically equivalent, and the chemical shift is the weighted average of the chemical shifts for any one rotamer.

The proton spectrum of diphosphine, P_2H_4, is consistent with free interconversion between the two skew forms of the molecule,

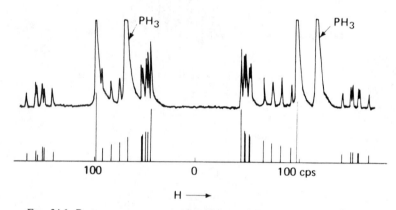

Thus, there are six coupling constants that must be taken into account: J_{PP}, $J_{HH}(gem)$, $J_{HH}(cis)$, $J_{HH}(trans)$, J_{PH}, and J_{PH}'. The observed spectrum (Fig. 24.6) contains 32 lines, all of which theoretically can be accounted for.[8]

FIG. 24.6. Proton nmr spectrum of P_2H_4 containing PH_3 impurity.[8] (Reproduced with permission of the Faraday Society.)

CONTACT SHIFTS

Normally, the nmr spectrum of a paramagnetic system cannot be observed because of extreme broadening of the resonances. However, in some paramagnetic systems, conditions are such that broadening is minimized, and the

---

[7] The rotational frequency is greater than the frequency difference corresponding to the chemical shift between the two environments of a given rotamer.

[8] R. M. Lynden-Bell, *Trans. Faraday Soc.*, **57**, 888 (1961).

principal effect of the unpaired electrons is to change the magnetic field experienced by the magnetic nuclei. This change causes very large temperature-dependent chemical shifts (sometimes approximately 6000 cps), which are referred to as nmr contact shifts. The contact shifts may be quantitatively related to spin densities (upfield shifts indicate positive spin densities and downfield shifts indicate negative spin densities). The proton spectrum[9] of the nickel(II) chelate of 1-(2-naphthylamino)-7-(2-naphthylimino)-1,3,5-cycloheptatriene is shown in Fig. 24.7. The spin densities[9] calculated from the observed contact shifts are indicated in Fig. 24.8. Obviously, spectra of this type can be valuable because they furnish detailed information regarding the nature of the bonding in chelates and organometallic compounds.

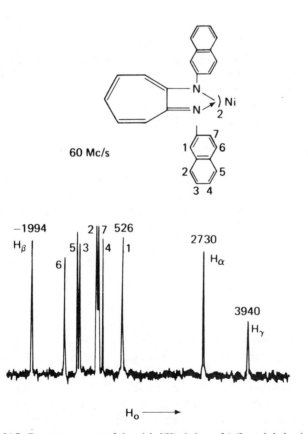

Fig. 24.7. Proton spectrum of the nickel(II) chelate of 1-(2-naphthylamino)-7-(2-naphthylimino)-1,3,5-cycloheptatriene in $CDCl_3$.[9] (Reproduced with permission of the American Chemical Society.)

[9] R. E. Benson, D. R. Eaton, A. Josey, and W. D. Phillips, *J. Am. Chem. Soc.*, **83**, 3714 (1961).

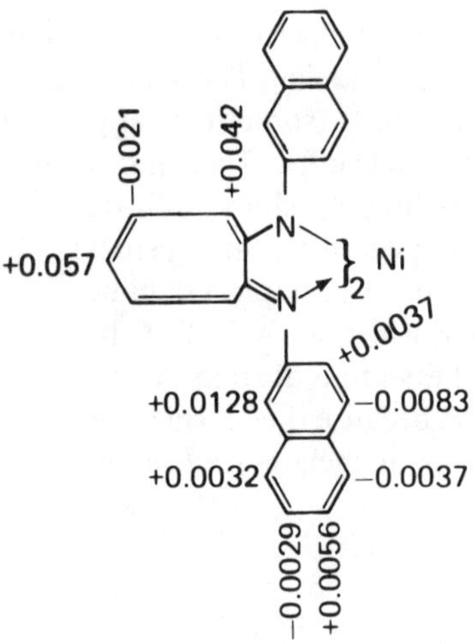

FIG. 24.8. Spin density distribution in nickel(II) chelate, calculated from spectrum in Fig. 24.7.[9] (Reproduced with permission of the American Chemical Society.)

LINE SHAPE

Quadrupolar Broadening

The resonances of nuclei that possess quadrupole moments (or that are adjacent to nuclei with quadrupole moments) are usually broadened when the quadrupolar nuclei are in unsymmetrical environments. For example, the ^{14}N spectrum of the $S_4N_3^+$ ion

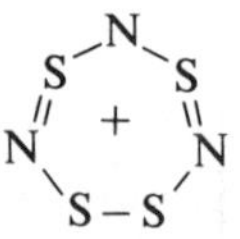

is so broadened as to make the signal unobservable (^{14}N has a spin of 1 and a quadrupole moment of 0.02). However, by preparing a sample highly enriched in ^{15}N (having no quadrupole moment), it was possible to obtain the sharp ^{15}N spectrum[10] shown in Fig. 24.9. Less extreme broadening is shown by trimethylamine, $N(CH_3)_3$, for which the ^{14}N signal line width (measured at half the signal amplitude) is 250 mG. In the more symmetrical $N(CH_3)_4^+$ ion, the ^{14}N line width is only 40 mG.

The effect of quadrupolar nuclei on the spectra of adjacent magnetic nuclei can be seen in the spectra of boron–hydrogen compounds. The proton

[10] N. Logan and W. L. Jolly, *Inorg. Chem.*, **4**, 1508 (1965).

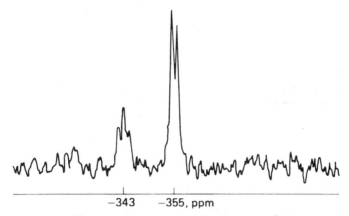

FIG. 24.9. The ^{15}N nmr spectrum of $S_4N_3^+$ in 70% HNO_3.[10] Chemical shifts relative to $^{15}NH_4^+$. (Reproduced with permission of the American Chemical Society.)

spectrum of the tetrahedral BH_4^- ion is a sharp quartet (coupling with ^{11}B) superimposed on a weaker sharp septet (coupling with ^{10}B); however, the proton spectrum of the relatively unsymmetrical $BH_2(OH_2)_2^+$ ion is so broad as to be unobservable.[11]

Chemical Exchange

The frequency separations between nuclei in different environments are comparable to the equilibrium rates of many chemical reactions. If the rate at which a nucleus changes its environment is slow compared to the frequency difference for the two environments, two separate distinct resonance signals are observed for the two environments, with an intensity ratio equal to the equilibrium abundance ratio of the nuclei in the two environments. If the exchange rate for the nucleus is fast compared to the frequency difference, a single sharp resonance is observed, with a chemical shift equal to the weighted average of those of the two environments. When the exchange rate has a magnitude comparable to the frequency difference, a broad resonance, between the positions of the two separate resonances, is observed.

Consider the exchange of fluorine atoms between the two nonequivalent sites in the SF_4 molecule.[12] At temperatures below $-85°$ (at 40 Mc sec^{-1}), the exchange is slow, and the ^{19}F spectrum shows two triplets (Fig. 24.10A). At $-58°$, the triplet structures have disappeared, and the separate resonances have begun to merge (Fig. 24.10B). At $-45°$, only one broad resonance is observed (Fig. 24.10C). At room temperature, the exchange is rapid, and one sharp line is observed (Fig. 24.10D).

---

[11] W. L. Jolly and T. Schmitt, *Inorg. Chem.*, **6**, 344 (1967).
[12] E. L. Muetterties and W. D. Phillips, *J. Am. Chem. Soc.*, **81**, 1084 (1959).

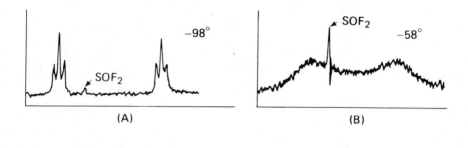

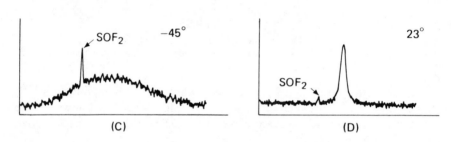

FIG. 24.10. Temperature dependence of the ^{19}F spectrum of SF_4.[12] (Reproduced with permission of the American Chemical Society.)

Problems

1. Explain the following spectra and describe the structures of the compounds.

(a) ^1H spectrum of $GeAsH_5$

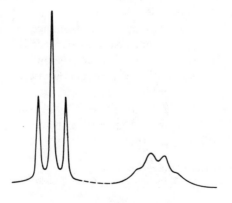

(b) ^{1}H spectrum of GePH$_5$

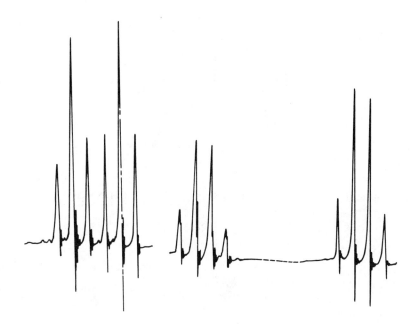

(c) ^{1}H spectrum of Si$_2$PH$_7$

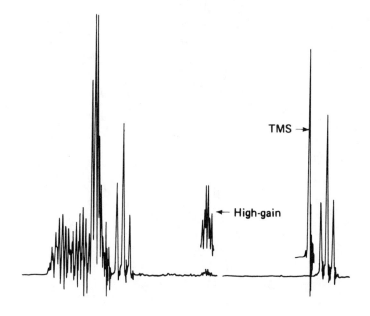

TMS →

← High-gain

(d) ^1H spectrum of $P_3N_3Cl_2[N(CH_3)_2]_4$

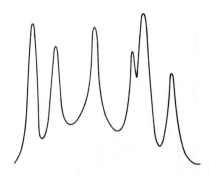

(e) ^1H spectrum of $[Co(en)_2Cl_2]Cl$ in D_2O (en $\equiv NH_2CH_2CH_2NH_2$)

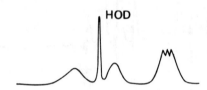

(f) ^{11}B spectrum of B_6H_{12}

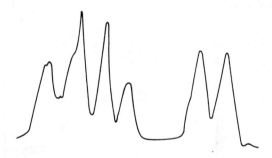

(g) ^{11}B spectrum of B_4H_{10}

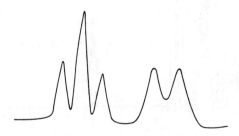

(h) ^{31}P spectrum of aqueous $Na_3HP_2O_6$

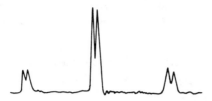

(i) ^{19}F spectrum of $(CH_3)_2PF_3$ (schematic[13])

A   B   A   B

2. Sketch the expected proton nmr spectrum of the *cis*-[Co(en)$_2$Cl NO$_2$]$^+$ ion.

3. Predict the proton magnetic resonance spectrum of the $^{11}BH_4^-$ ion.

4. Sketch the ^{19}F nmr spectrum of SF_4. Sulfur has no magnetic isotope in appreciable abundance.

5. The thallium nmr spectrum of crystalline Tl_2Cl_3 shows two components with intensities in the ratio of $3:1$. Suggest a structure for the compound.

6. Predict the proton magnetic resonance spectrum of *cis*-[Co(NH$_3$)$_4$(NO$_2$)Cl]$^+$. Neglect all possible spin-spin coupling.

7. A compound of empirical formula $Si_2C_6H_{16}$ was isolated from the pyrolysis products of $Si(CH_3)_4$. The structure first suggested was

$$CH_3 \diagdown \quad\quad \overset{H}{\underset{|}{C}}\quad CH_3 \diagup$$
$$CH_3 - \underset{\diagup}{\overset{\diagdown}{Si}} - C = Si$$
$$CH_3 \diagup \quad\quad\quad CH_3 \diagdown$$

However, the proton nmr spectrum showed two peaks, with intensities approximately in a $1:3$ ratio. Suggest a more plausible structure.

8. Explain how you might use ^{15}N or ^{14}N magnetic resonance to tell which of the following structures is correct for the $Ru(NH_3)_5N_2^{2+}$ ion.

$$\begin{array}{cc}
NH_3 & NH_3 \\
NH_3 \diagup\!\!+\!\!\diagdown NH_3 & NH_3\diagup\!\!+\!\!\diagdown NH_3 \\
NH_3 \diagdown\!\!|\!\!\diagup NH_3 & NH_3\diagdown\!\!|\!\!\diagup NH_3 \\
N\equiv N & N \\
& ||| \\
& N
\end{array}$$

Sketch the expected spectrum for each possibility.

[13] The overall integrated intensity of the A multiplets is twice that of the B multiplets.

9. One compound in each of the following pairs gives a simple first-order nmr spectrum; the other gives a relatively complicated second-order spectrum. Sketch and identify the first-order spectra. In which compounds are there magnetically nonequivalent nuclei?

^{1}H spectra: Si_2H_5Br; Si_3H_8 ^{31}P spectra: $P_4O_{13}{}^{6-}$;

^{19}F spectra: NF_3; *cis*-N_2F_2
        SF_5Cl; ClF_3

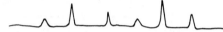

10. The ^{31}P spectrum of $P_2O_3Cl_2F_2$ is a doublet, the individual peaks of which show asymmetric triplet fine structure under very high resolution. After the sample had been heated at 50° for several days, the nmr spectrum was as shown. Explain the spectra.[14]

Suggested References for Further Reading

Introductory

Roberts, J. D., *Nuclear Magnetic Resonance*, McGraw-Hill Book Company, New York, 1959 (a good elementary introduction to the subject, at a level of sophistication comparable to that of this chapter).

Slightly more advanced nonmathematical treatments

Becker, E. D., "NMR Spectra: Appearance of Patterns from Small Spin Systems," in *J. Chem. Educ.*, **42**, 591 (1965).

Drago, R. S., *Physical Methods in Inorganic Chemistry*, Reinhold Publishing Corp., New York, 1965, Chap. 8, pp. 239–314.

Fluck, E., *Die kernmagnetische Resonanz und ihre Anwendung in der anorganischen Chemie*, Springer-Verlag, Berlin, 1963.

Jackman, L. M., *Applications of Nuclear Magnetic Resonance Spectroscopy in Organic Chemistry*, Pergamon Press, Inc., New York, 1959.

Muetterties, E. L. and W. D. Phillips, "The Use of Nuclear Magnetic Resonance in Inorganic Chemistry," in *Adv. Inorg. Chem. Radiochem.*, **4**, 231 (1962).

Relatively advanced but readable

Emsley, J. W., J. Feeney, and L. H. Sutcliffe, *High Resolution Nuclear Magnetic Resonance Spectroscopy*, vols. 1–2, Pergamon Press, Inc., New York, 1965–1966.

Pople, J. A., W. G. Schneider, and H. J. Bernstein, *High Resolution Nuclear Magnetic Resonance*, McGraw-Hill Book Company, New York, 1959.

Roberts, J. D., *An Introduction to the Analysis of Spin-Spin Splitting in High-Resolution Nuclear Magnetic Resonance Spectra*, W. A. Benjamin, Inc., New York, 1961.

[14] See M. M. Crutchfield, C. F. Callis, and J. R. Van Wazer, *Inorg. Chem.*, **3**, 280 (1964).

Magnetic
Susceptibility

25

THE CHEMICAL SIGNIFICANCE OF
SUSCEPTIBILITY DATA

Introduction

The ratio of the intensity of magnetization induced in a substance (I) to the magnetic field intensity (H) is called the *volume susceptibility*, κ.

$$\frac{I}{H} = \kappa \qquad (25.1)$$

The volume susceptibility is simply related to both the *gram susceptibility*, χ, and the *molar susceptibility*, χ_M:

$$\chi = \frac{\kappa}{d} \qquad (25.2)$$

$$\chi_M = \chi M = \frac{\kappa M}{d} \qquad (25.3)$$

where d and M are the density and the molecular weight of the substance, respectively.

369

For most substances, χ is negative and of a magnitude around 10^{-6} cgs units. Such substances are *diamagnetic*.

Substances having unpaired electrons that do not strongly interact with one another have positive values of χ around 100×10^{-6} cgs units. These substances are *paramagnetic*. In coordination compounds of transition metals with unpaired electrons, the diamagnetic ligands keep the metal atoms fairly well separated, and consequently such compounds are generally paramagnetic. Manganese(II) phthalocyanine (Synthesis 11) is an example of such a compound.

Substances in which the electrons of adjacent paramagnetic sites interact magnetically with each other are *ferromagnetic* or *antiferromagnetic*. A ferromagnetic substance is one in which the adjacent magnetic dipoles are oriented in the same direction. The gram susceptibility typically has positive values around 100 cgs units. Metallic iron is an example of a ferromagnetic material. An antiferromagnetic material is one in which adjacent magnetic dipoles are oppositely oriented. The gram susceptibility has positive values around 10^{-6} cgs units. Zinc ferrite, $ZnFe_2O_4$, (Synthesis 35) is an example of an antiferromagnetic material containing two interpenetrating lattices of ferric ions, each with its spins aligned, but with opposite orientations in the two lattices. Chromous acetate, $[Cr_2(OAc)_4] \cdot 2\,H_2O$ (Synthesis 1) is an example of a molecular antiferromagnetic compound in which the electronic coupling occurs between adjacent chromium atoms in each molecule.

Diamagnetism

Diamagnetism is attributable to the interaction of closed-shell electrons with an applied magnetic field. All substances, even paramagnetic substances, contain some closed shells of electrons. Consequently, paramagnetic substances have a negative (diamagnetic) contribution to their net susceptibility. In most cases, this diamagnetic contribution is only a small fraction of the total susceptibility, but in accurate work it is necessary to correct the measured susceptibility for the diamagnetic contribution. We use the relation

$$\chi_M{}^{\text{corr}} = \chi_M - \chi_{\text{dia}} \tag{25.4}$$

Because the diamagnetic contribution, χ_{dia}, is always negative, the corrected molar susceptibility is always greater than the uncorrected value.[1]

To a fair approximation, χ_{dia} may be considered an additive function of the closed shell atoms and ions in a molecule. The diamagnetic contributions

---

[1] In some compounds, a correction should be made for temperature-independent paramagnetism (T.I.P.) as well as the diamagnetic contribution. In such cases, $\chi_M{}^{\text{corr}}$ may be lower than the uncorrected value. Temperature-independent paramagnetism is discussed on pp. 377–378.

of a few ions and atoms (for covalent molecules) are given in Table 25.1. For a more extensive compilation, the reader is referred to Selwood.[2]

As an example of the use of Table 25.1, we shall estimate the diamagnetic contribution of the acetylacetonate ion. We shall assume that the diamagnetism of the ion is the same as that of the enol form of acetylacetone,

$$
\begin{array}{ccc}
\text{O} & & \text{OH} \\
\| & & | \\
\text{CH}_3-\text{C}-\text{CH}=\text{C}-\text{CH}_3
\end{array}
$$

$$
\begin{array}{rrl}
8\,\text{H}: & -2.93 \times 10^{-6} \times 8 = & -23.44 \times 10^{-6} \\
5\,\text{C}: & -6.00 \times 10^{-6} \times 5 = & -30.00 \times 10^{-6} \\
\text{O}: & -4.61 \times 10^{-6} \times 1 = & -4.61 \times 10^{-6} \\
\text{carbonyl} \quad \text{O}: & +1.72 \times 10^{-6} \times 1 = & +1.72 \times 10^{-6} \\
& \chi_{\text{dia}} = & -56.33 \times 10^{-6}
\end{array}
$$

The actual molar susceptibility of acetylacetone is 54.88×10^{-6} cgs units.[3]

**Table 25.1 Diamagnetic Contributions of Ions and Atoms $(-\chi \times 10^6)$*

Ag^+	24	F^-	11	Tl^+	34
Al^{3+}	2	Fe^{2+}	13	Zn^{2+}	10
BF_4^-	39	Fe^{3+}	10	H	2.93
Br^-	36	H^+	0	C	6.00
BrO_3^-	40	H_2O	13	N(open chain)	5.55
CN^-	18	I^-	52	N(ring)	4.61
SCN^-	35	K^+	13	N(amide)	1.54
CO_3^{2-}	34	Li^+	1	N(imide)	2.11
Ca^{2+}	8	Mg^{2+}	3	O	4.61
Cl^-	26	NH_4^+	12	O(carbonyl)	-1.72
ClO_3^-	32	NO_3^-	20	O(carboxyl)	3.36
ClO_4^-	34	Na^+	5	F	6.3
Co^{2+}	12	Ni^{2+}	12	Cl	20.1
Co^{3+}	10	O^{2-}	12	Br	30.6
Cr^{2+}	15	OH^-	12	I	44.6
Cr^{3+}	11	Rb^+	20	P	10
Cs^+	31	S^{2-}	38?	S	15
Cu^{2+}	11	SO_4^{2-}	40	B	-7
		Sn^{2+}	20	Si	13

* See ref. 2.

The Magnetic Moment

For an assemblage of magnetic dipoles having an alignment energy much larger than kT, the corrected susceptibility[4] may be represented by the Curie

---

[2] P. W. Selwood, *Magnetochemistry*, 2nd ed., Interscience Publishers, New York, 1956.

[3] *Handbook of Chemistry and Physics*, 44th ed., Chemical Rubber Publishing Co., Cleveland, Ohio, 1962.

[4] Corrected for diamagnetism and, if present, temperature-independent paramagnetism.

Law,

$$\chi_M{}^{corr} = \frac{C}{T} \tag{25.5}$$

The Curie constant, C, has the value

$$C = \frac{N(\mu_{eff})^2\beta^2}{3k} \tag{25.6}$$

where N is Avogadro's number, k is Boltzmann's constant, μ_{eff} is "the effective magnetic moment," and β is the Bohr magneton, 0.927×10^{-20} erg Oe^{-1}. The effective magnetic moment, μ_{eff}, is a quantity of considerable interest to chemists, which, after the values of the constants are substituted, can be expressed as

$$\mu_{eff} = 2.83(\chi_M{}^{corr})^{1/2} \, T^{1/2} \tag{25.7}$$

Many paramagnetic substances do not exactly obey the Curie Law, but rather a modification, called the Curie-Weiss Law,

$$\chi_M{}^{corr} = \frac{C}{T + \theta} \tag{25.8}$$

where θ is the Weiss constant that must be evaluated by determining the susceptibility as a function of temperature. When this law is followed, it is sometimes acceptable to calculate μ_{eff} by means of the formula

$$\mu_{eff} = 2.83(\chi_M{}^{corr})^{1/2}(T + \theta)^{1/2} \tag{25.9}$$

The importance of μ_{eff} to chemists lies in the fact that, for many compounds, it can be calculated theoretically from a knowledge of the structure and type of bonding. Thus, from a comparison of the experimental μ_{eff} with values calculated for different types of structure and bonding, it is often possible to choose one structure out of several possibilities, or at least to eliminate certain possibilities.

For a free, paramagnetic ion, in which spin-orbit coupling causes a splitting of the ground state that is much larger than kT, the effective magnetic moment may be calculated from the formulas

$$\mu_{eff} = g[J(J + 1)]^{1/2} \tag{25.10}$$

$$g = 1 + \frac{J(J + 1) + S(S + 1) - L(L + 1)}{2J(J + 1)} \tag{25.11}$$

where g is the electron gyromagnetic ratio (usually referred to as the "g factor" or "g value"), and J, S, and L refer to the ground-state spectroscopic term. Remember that when a shell is less than half-filled with electrons, the lowest state is specified by $J = |L - S|$; when it is more than half-filled, the lowest state has $J = |L + S|$. Equations (25.10) and (25.11) are applicable to many compounds of the lanthanide elements. For example, Pr^{3+} ($4f^2$) has the ground state $^3H_4$($J = 4$, $L = 5$, $S = 1$), and we calculate $g = 0.8$ and $\mu_{eff} = 3.58$. For the compound $Pr_2(SO_4)_3 \cdot 8 H_2O$, the experimental value at $300°K$ is $\mu_{eff} = 3.40$.

In compounds of the first transition series, the orbital contribution to the magnetic moment is almost completely quenched by the ligand fields. Hence, we may substitute $L = 0$ and $J = S$ in Eqs. (25.10) and (25.11), thereby obtaining $g = 2$ and the following "spin-only" formula, which may be used for approximate calculation of μ_{eff}.

$$\mu_{eff} = 2[S(S + 1)]^{1/2} \tag{25.12}$$

Because S is simply one-half the number of unpaired electrons, n, we may write

$$\mu_{eff} = [n(n + 2)]^{1/2} \tag{25.13}$$

Equation (25.13) is significant in the study of transition metal complexes because the number of unpaired electrons may be correlated with the type of bonding or structure of the complexes. For example, consider octahedral complexes with the configurations d^4, d^5, d^6, and d^7. In each of these cases, there are two ways of distributing the electrons between the t_{2g} and e_g orbitals: a high-spin state and a low-spin state. These distributions are indicated in Fig. 25.1. The spin state of a particular ion depends upon whether the ligand field strength, Δ_o, is greater or less than the mean energy required to cause pairing of two electrons in the same orbital. Thus, the $Fe(H_2O)_6^{2+}$ ion is paramagnetic because the field produced by the water molecules is relatively weak, whereas the $Fe(CN)_6^{4-}$ ion is diamagnetic

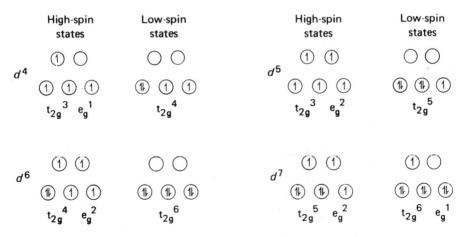

FIG. 25.1. Electron-occupancy diagrams for possible ground states of d^4, d^5, d^6, and d^7 ions in octahedral fields.

because the cyanide ions produce a very strong field. For ions of the types d^1, d^2, d^3, d^8, d^9, and d^{10}, the number of unpaired electrons is the same as in the free ion regardless of the magnitude of Δ_o. In tetrahedral complexes, high-spin and low-spin states are, in principle, possible for the configurations d^3, d^4, d^5, and d^6, but because Δ_t values are much smaller than Δ_o values, there are no known examples of low-spin states.

Let us consider the effect on a d^8 octahedral complex of a tetragonal distortion, created by moving the two z-axis ligands out to a greater distance than the ligands in the xy plane. The conceivable distributions are shown in Fig. 25.2. In the case of weak tetragonal distortion, there would be two unpaired electrons, just as in the case of a strictly octahedral complex; however, in the case of strong distortion or a square planar complex, a low-spin state results. Indeed, all the square planar complexes of Ni(II), Pd(II), Pt(II), Rh(I), Ir(I), and Au(III) are diamagnetic (unless the ligands have unpaired electrons).

In Table 25.2, the experimental and calculated effective magnetic moments for some octahedral complexes of the first transition series are tabulated. In most cases, the calculated and experimental values of μ_{eff} differ by less than 0.2. With only one exception, the complexes with discrepancies greater than 0.2 are those having T ground terms. For these complexes, there are theoretical reasons for expecting a considerable contribution to the magnetic moment from spin-orbit coupling. This subject is discussed in the next section.

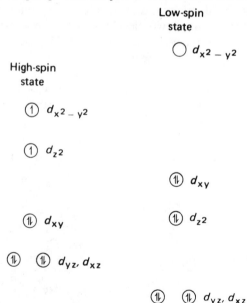

FIG. 25.2. Electron-occupancy diagrams for ground states of tetragonally distorted d^8 ions.

Table 25.2   Experimental and Calculated Values of the Effective Magnetic Moment

n	Compound	Ground State	Experimental μ_{eff}	$[n(n+2)]^{1/2}$
1	$CsTi(SO_4)_2 \cdot 12\,H_2O$	$^2T_{2g}$	1.8	1.73
	$K_4Mn(CN)_6 \cdot 3\,H_2O*$	$^2T_{2g}$	2.2	1.73
	$K_2PbCo(NO_2)_6*$	^2E_g	1.8	1.73
	$(NH_4)_2Cu(SO_4)_2 \cdot 6\,H_2O$	^2E_g	1.9	1.73
2	$(NH_4)V(SO_4)_2 \cdot 12\,H_2O$	$^3T_{1g}$	2.7	2.83
	$Cr(dipy)_3Br_2 \cdot 4\,H_2O*$	$^3T_{1g}$	3.3	2.83
	$(NH_4)_2Ni(SO_4)_2 \cdot 6\,H_2O$	$^3A_{2g}$	3.2	2.83
3	$KCr(SO_4)_2 \cdot 12\,H_2O$	$^4A_{2g}$	3.8	3.88
	$(NH_4)_2Co(SO_4)_2 \cdot 6\,H_2O$	$^4T_{1g}$	5.1	3.88
4	$CrSO_4 \cdot 6\,H_2O$	^5E_g	4.8	4.90
	$(NH_4)_2Fe(SO_4)_2 \cdot 6\,H_2O$	$^5T_{2g}$	5.5	4.90
5	$K_2Mn(SO_4)_2 \cdot 6\,H_2O$	$^6A_{1g}$	5.9	5.92

* High-field (low-spin) complexes.

Spin-Orbit Coupling

For an electron to have orbital angular momentum, it must be able to circulate about an axis. This fact requires the availability of an orbital, in addition to the orbital containing the electron, which has the following properties. It must have the same shape as, and be degenerate with, the orbital containing the electron. It must be superimposable with the orbital containing the electron by rotation about an axis. Finally, it cannot contain an electron having the same spin as the first electron. For example, in a free ion, the d_{xy} and $d_{x^2-y^2}$ orbitals are degenerate and can be rotated into each other about the z axis. In an octahedral or tetrahedral ligand field, these orbitals are no longer degenerate, and any orbital angular momentum associated with them is quenched. However, the d_{xz} and d_{yz} orbitals (which can be superimposed by rotation about the z axis) are degenerate in both the free ion and in the ligand field. Consequently, any orbital angular momentum associated with these orbitals is not quenched by the ligand field. In fact, unquenched orbital angular momentum will remain whenever any two of the t_2 orbitals (d_{xy}, d_{xz}, and d_{yz}) contain one or three electrons. This situation is found for T_1 and T_2 ground terms, which include the configurations $t_2{}^1$, $t_2{}^2$, $t_2{}^4$, and $t_2{}^5$.

Let us consider the splittings associated with a $^3T_{1g}$ ground term, as in an octahedral complex of V^{3+}. The ground term of the free ion, ^3F, is split by the ligand field into three states, the lowest of which is $^3T_{1g}$. Now, a T state is analogous to a P state ($L=1$) because both are triply degenerate, and the same rules apply to spin-orbit coupling of a T state as to ordinary LS coupling of a P state. Thus, J can have the values $|L + S|, |L + S - 1|, \cdots$ $|L - S|$; in this case, $J = 2$, 1, and 0, with $J = 2$ as the ground state. The splittings are indicated in Fig. 25.3. Typically, the splitting between the states of different J values is of the order of magnitude of kT at room temperature.

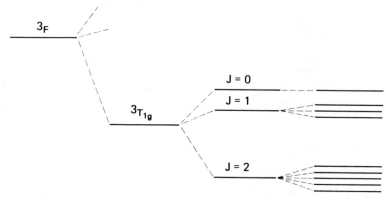

FIG. 25.3.

Thus, the magnetic moments of complexes with T ground terms not only differ from the spin-only values but also depend on temperature. The magnetic moment at a particular temperature may be obtained by properly summing over the states that arise from spin-orbit coupling. Calculations of this type have been carried out using spin-orbit coupling constants for free ions.[5] The results are given in Table 25.3; these data may be used as estimates of μ_{eff} for complexes containing any of the tabulated metal ions. Some second- and third-row transition metal ions are included.

Table 25.3   Calculated Values of μ_{eff} at 300°K for Octahedral and Tetrahedral Stereochemistries, Assuming Free-Ion Values of Spin-Orbit Coupling*

$^2T_{2g}$ States		$^3T_{1g}$ States		$^4T_{1g}$ States	
Ti^{3+}	1.9	V^{3+}	2.7	Cr^{3+}	3.4
Mo^{5+}	1.0	Cr^{2+}	3.5	Co^{2+}	5.1
Mn^{2+}	2.55	Mo^{4+}	1.9		
Fe^{3+}	2.45	W^{4+}	1.5	$^5T_{2g}$ States	
Ru^{3+}	2.1	Mn^{3+}	3.65	Cr^{2+}	4.7
Os^{3+}	1.85	Re^{3+}	2.4	Mn^{3+}	4.5
Ir^{4+}	1.8	Fe^{4+}	3.55	Fe^{2+}	5.65
Cu^{2+}	2.2	Ru^{4+}	3.2	Co^{3+}	5.75
		Os^{4+}	1.9		
		Ni^{2+}	4.0		

* See ref. 5.

[5] B. N. Figgis, *Introduction to Ligand Fields*, John Wiley & Sons, Inc., New York, 1966.

In view of the crudeness of using free-ion values of the spin-orbit coupling constants, one should not expect agreement to better than ± 0.2 between experimental and calculated values of μ_{eff}. Obviously, it is not necessary to use corrected χ_M values when calculating the experimental μ_{eff} values.

Complexes with *A* and *E* Ground States

Although the greatest discrepancies between experimental values of μ_{eff} and the calculated spin-only values are found in complexes with T ground states, smaller but significant discrepancies are found in complexes having A and E ground states. These smaller discrepancies arise because the quenching of spin-orbit coupling by the ligand field is not complete, and there exists an interaction between the ground state and excited states. These effects lead to g values different from the spin-only value and to a temperature-independent paramagnetism.

First, we shall discuss the effect of incomplete quenching of spin-orbit coupling. The electron gyromagnetic ratio, g, instead of being 2, as for a state lacking angular momentum [see Eq. (25.12)], takes the values $2(1 - 4\lambda/\Delta)$ for A_2 and A_{2g} states and $2(1 - 2\lambda/\Delta)$ for E and E_g states.[5] (The symbol λ refers to the spin-orbit coupling constant and Δ refers to the ligand-field splitting.) Thus, we may write for A_2 and A_{2g} states,

$$\mu_{\text{eff}} = (1 - 4\lambda/\Delta)[\mu_{\text{eff}}(\text{spin only})] \tag{25.14}$$

and for E and E_g states,

$$\mu_{\text{eff}} = (1 - 2\lambda/\Delta)[\mu_{\text{eff}}(\text{spin only})] \tag{25.15}$$

For A_1 and A_{1g} states, $g = 2.00$; hence, $\mu_{\text{eff}} = \mu_{\text{eff}}$ (spin-only). Selected values of spin-orbit coupling constants, λ, for various metal ions in octahedral and tetrahedral environments, follow[5] (values are in cm^{-1}): $V(IV)T_d[^2E]$, 250; $V(II)O_h[^4A_{2g}]$, 57; $Cr(III)O_h[^4A_{2g}]$, 92; $Cr(II)O_h[^5E_g]$, 58; $Mn(IV)O_h[^4A_{2g}]$, 138; $Mn(III)O_h[^5E_g]$, 89; $Fe(II)T_d[^5E]$, -100; $Co(II)T_d[^4A_2]$, -172; $Co(II)O_h[^2E_g]$, -515; $Ni(II)O_h[^3A_{2g}]$, -315; $Cu(II)O_h[^2E_g]$, -830. Values of λ for many other metal ions can be calculated from data tabulated by Figgis.[5]

Next we shall discuss temperature-independent paramagnetism. The magnetic field can cause a distortion of the electron distribution in a given orbital, and thereby can effect a mixing of high-lying excited states into the ground state, together with a lowering in energy of the ground state. One result of this mixing is the deviation of μ_{eff} from the spin-only value, as discussed. Another result, when the excited state lies above the ground state by an amount much greater than kT, is temperature-independent paramagnetism.

We write

$$\chi_M = \frac{C}{T} + \chi_{TIP} + \chi_{dia} \qquad (25.16)$$

The magnitude of χ_{TIP} for complexes having various ground terms may be estimated from the value of Δ (in cm^{-1}) using the following formulas.[5] For $^1A_{1g}$ (d^6) and all A_2 and A_{2g} complexes,

$$\chi_{TIP} = \frac{2.09}{\Delta} \text{ cgs units mole}^{-1} \qquad (25.17)$$

For E and Eg complexes,

$$\chi_{TIP} = \frac{1.04}{\Delta} \text{ cgs units mole}^{-1} \qquad (25.18)$$

There is no temperature-independent paramagnetism for $^1A_{1g}$ (d^0 and d^{10}) and $^6A_1$ and $^6A_{1g}$ (d^5) complexes.

Some Examples

Let us apply the systematics discussed in the preceding paragraphs to the magnetic data for Cs_3CoCl_5. For this salt, which contains the anionic complex $CoCl_4^{2-}$, Holm and Cotton[6] obtained the data given in the first two columns of Table 25.4. Using the data of Table 25.1, we estimate the diamagnetic contribution to the susceptibility to be -235×10^{-6} cgs mole^{-1}. From Table 22.4, we estimate Δ_t for $CoCl_4^{2-}$ to be 3310 cm^{-1}; hence, using Eq. (25.17), we estimate the T.I.P. contribution to be 630×10^{-6} cgs mole^{-1}.

Table 25.4   Magnetic Data for Cs_3CoCl_5*

T	*Molar Susceptibilities* $\times 10^6$		
(°K)	χ_M	χ_M^{corr}	μ_{eff}
294.8	8855	8460	4.50
195.1	13,315	12,920	4.54
74.1	32,775	32,380	4.50

* See ref. 6.

When the sum of the estimated diamagnetic and T.I.P. contributions (395×10^{-6} cgs mole^{-1}) is subtracted from the raw χ_M values, we get the χ_M^{corr} values given in the third column of Table 25.4. A plot of $1/\chi_M^{corr}$ vs. T yields a straight line that crosses the $1/\chi_M^{corr} = 0$ axis at $-4°$. Hence, we use the relation $\mu_{eff} = 2.83(\chi_M^{corr})^{1/2}(T + 4)^{1/2}$ to obtain the values of μ_{eff} given in the last column of the table. From the list on p. 377, we find $\lambda = -172$ cm^{-1}.

[6] R. H. Holm and F. A. Cotton, *J. Chem. Phys.*, **31**, 788 (1959).

Thus, we calculate

$$\mu_{\text{eff}} = \left(1 + \frac{4 \cdot 172}{3310}\right)(3.87)$$

$$= (1.21)(3.87) = 4.68$$

This estimated value of μ_{eff} is in fair agreement with the experimental values in Table 25.4 and is a much better approximation than the spin-only value of 3.87.

The magnetic moments of cobalt(II) complexes can be used as criteria of the type of coordination (tetrahedral or octahedral) around the Co^{2+} ion.[7] As we have seen in the preceding paragraph, *tetrahedral* Co(II) complexes have magnetic moments that deviate significantly from the spin-only value. As required by theory, the moments are essentially independent of temperature (if the susceptibility has been corrected for T.I.P.). In a wide variety of such tetrahedral complexes, the magnetic moment has been found to range from 4.40 to 4.88. On the other hand, *high-spin octahedral* cobalt(II) complexes have a $^4T_{1g}$ ground state, and, because of spin-orbit coupling, the magnetic moment not only deviates markedly from the spin-only value, but also varies with temperature. Such complexes usually have room-temperature magnetic moments around 5.0 to 5.1, in agreement with the theoretically estimated value of 5.1 of Table 25.3. Of course, it is an easy matter to identify *low-spin octahedral* cobalt(II) complexes, which have one unpaired electron. The compound $K_2PbCo(NO_2)_6$, with $\mu_{\text{eff}} = 1.8$, is in this category.

The uncorrected χ_M values for $(NH_4)_2Ni(SO_4)_2 \cdot 6 H_2O$ are given in Table 25.5. In this case, the calculated diamagnetism, -194×10^{-6} cgs mole^{-1}, is almost cancelled by the calculated T.I.P., 235×10^{-6} cgs mole^{-1}. The small correction of -41×10^{-6} cgs mole^{-1} is just barely significant in the

Table 25.5 Magnetic Susceptibility Data for $(NH_4)_2Ni(SO_4)_2 \cdot 6 H_2O$*

T (°K)	$\chi_M \times 10^6$ (cgs mole^{-1})
290	4180
78	15,500
20.4	59,200
14.6	82,200

* Landolt-Börnstein, Group II: *Atomic and Molecular Physics*, vol. 2: *Magnetic Properties of Coordination and Organo-metallic Transition Metal Compounds*, K.-H. Hellwege, ed., Springer-Verlag, New York, 1966, pp. 2–228.

[7] C. S. G. Phillips and R. J. P. Williams, *Inorganic Chemistry*, vol. 2, Oxford University Press, Inc., 1966, p. 413.

case of the 290°K susceptibility, and it can be ignored for the data at lower temperatures. The Curie Law is obeyed very precisely; therefore, we calculate $\mu_{eff} = 2.83 (4140 \times 290 \times 10^{-6})^{1/2} = 3.10$. This experimental value may be compared with the value estimated from Eq. (25.14):

$$\mu_{eff} = \left(1 + \frac{4 \cdot 315}{8900}\right)(2.83) = 3.23$$

Free Radicals

A number of relatively stable free radicals are known. These are generally molecules possessing an odd number of electrons, with one unpaired electron. In such compounds there is usually little contribution from the orbital moment, so that the susceptibilities are represented by

$$\chi_M = \frac{N\beta^2}{kT} \tag{25.19}$$

corresponding to the spin-only value of μ_{eff} for one unpaired electron $\{\mu_{eff} = 2[S(S+1)]^{1/2} = 1.73\}$. Such compounds include a wide variety of nitroso compounds, such as $(C_6H_5)_2NO$, alkali-metal-hydrocarbon complexes, such as $Na^+ \cdot [naphthalene]^-$, and hydrazyls, such as diphenylpicrylhydrazyl.

EXPERIMENTAL METHODS[8]

The Gouy Balance

Most measurements of magnetic susceptibility are made with a Gouy balance. If a uniform rod of material is suspended (as shown in Fig. 25.4) with the lower end in a uniform magnetic field of magnitude H and the upper end in a lower field of magnitude H_0, the material will be pulled down with a force given by the expression

$$\text{force} = \Delta wg = \tfrac{1}{2}A(\kappa - \kappa_0)(H^2 - H_0^2) \tag{25.20}$$

where Δw is the apparent increase in weight of the sample upon turning on the field, g is the gravitational constant, A is the cross-sectional area of the material, and κ and κ_0 are the volume susceptibilities, respectively, of the material and the displaced medium (usually air). Usually, κ_0 is negligible

[8] B. N. Figgis and J. Lewis, "Magnetochemistry," *Technique of Inorganic Chemistry*, vol. 4, H. B. Jonassen and A. Weissberger, eds., Interscience Publishers, New York, 1965, p. 137. For simplified discussions, see D. P. Shoemaker and C. W. Garland, *Experiments in Physical Chemistry*, McGraw-Hill Book Company, New York, 1962, pp. 293–97; S. Kirschner, M. J. Albinak, and J. G. Bergman, *J. Chem. Educ.*, **39**, 576 (1962); L. J. Brubacher and F. E. Stafford, *J. Chem. Educ.*, **39**, 574 (1962); T. G. Dunne, *J. Chem. Educ.*, **44**, 142 (1967).

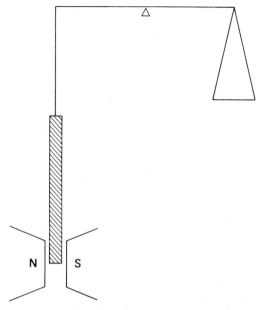

FIG. 25.4. Schematic representation of Gouy balance.

compared to κ, and, if the rod of material extends above the pole pieces a distance greater than eight times the pole gap width, H_0 is negligible compared to H. Then we may write

$$\kappa = \frac{2 \Delta w g}{H^2 A} \tag{25.21}$$

Generally, one does not determine the values of H and A, but rather calibrates the apparatus using a sample of known volume susceptibility. If the same cylindrical sample container and the same magnetic field strength are used for both the unknown sample and the standard compound, then we have the relation

$$\kappa_{unk} = \frac{\Delta w_{unk}\, \kappa_{std}}{\Delta w_{std}} \tag{25.22}$$

If the sample tube is filled to the same height with the unknown and standard substances, then we have the following useful relation between the gram susceptibilities:

$$\chi_{unk} = \frac{\Delta w_{unk}}{w_{unk}} \cdot \frac{w_{std}}{\Delta w_{std}} \cdot \chi_{std} \tag{25.23}$$

In accurate work, it is necessary to account for the diamagnetism of the sample tube:

$$\Delta w \text{ (corrected)} = \Delta w \text{ (sample + tube)} - \Delta w \text{ (empty tube)} \qquad (25.24)$$

The magnet should provide a field of 5000 G. or more, and the field should be reasonably homogeneous over a region considerably larger than the diameter of the sample tube. An electromagnet, the field of which can easily be turned on and off, is convenient, but a permanent magnet can be used if provision is made for moving it in and out of position under the balance.

The Gouy method may be used to determine the susceptibility of either a solution or a solid. In the case of a solid, the sample should be ground to a fine powder to avoid anomalies owing to anisotropy in the crystal and to facilitate uniform packing. When the gram susceptibility and molar susceptibility of a solid sample are calculated by Eqs. (25.2) and (25.3), the density used in the equations is the bulk density of the powdered material as it is packed in the tube.

In the study of aqueous solutions, the gram susceptibility is given by the relation

$$\chi \text{ (solution)} = \frac{\chi_M \text{ (solute)}}{M} p - 0.720 \times 10^{-6}(1 - p) \qquad (25.25)$$

where M is the molecular weight of the solute and p is the solute's weight fraction. Aqueous solutions of nickel(II) chloride are often used to calibrate Gouy balances. The gram susceptibility of such a solution is given by the relation

$$\chi = \left[\frac{10,030p}{T} - 0.720 \times 10^{-6}(1 - p) \right] \times 10^{-6} \qquad (25.26)$$

where p is the weight fraction of $NiCl_2$.

The crystalline substances $HgCo(NCS)_4$[9] ($\chi_{20°C} = 16.44 \times 10^{-6}$ cgs) and $[Ni(en)_3]S_2O_3$[10] ($\chi_{20°C} = 11.03 \times 10^{-6}$ cgs) are very good calibrating standards. They are easily prepared and are not hygroscopic.

The nmr Method

Another useful method for measuring magnetic susceptibilities is the nmr method.[11] The shifts of the proton resonance lines of inert reference molecules in solution caused by the presence of dissolved paramagnetic substances are given by the expression

$$\frac{\Delta H}{H} = \frac{2\pi}{3} \Delta \kappa \qquad (25.27)$$

[9] B. N. Figgis and R. S. Nyholm, *J. Chem. Soc.*, 4190 (1958).
[10] N. F. Curtis, *J. Chem. Soc.*, 3147 (1961).
[11] D. F. Evans, *J. Chem. Soc.*, 2003 (1959).

where $\Delta\kappa$ is the change in volume susceptibility. For aqueous solutions of paramagnetic substances, about 2 per cent of t-butanol is added as a reference substance, and an aqueous t-butanol solution of the same concentration is used as an external reference in a capillary tube in the nmr tube. Two resonance lines are generally observed for the methyl protons of the t-butanol because of the difference in volume susceptibilities of the two solutions, with the line for the paramagnetic solution at the higher frequency. The gram susceptibility of the solute is given by the expression

$$\chi = \frac{3}{2\pi C} \cdot \frac{\Delta H}{H} + \chi_0 + \frac{(d_0 - d_s)\chi_0}{C} \tag{25.28}$$

where C is the solute concentration in grams per milliliter of solution, χ_0 is the gram susceptibility of the solvent (-0.72×10^{-6} for dilute aqueous t-butanol solutions), d_0 is the density of the solvent, and d_s the density of the solution. For highly paramagnetic substances, the last term can often be neglected.

The Faraday Method

For the determination of the susceptibility of solid samples that are too small for the Gouy balance method, the Faraday method can be applied. In this method, a very small sample (even one small crystal can suffice) is suspended from a sensitive balance in a very nonhomogeneous magnetic field. The weight is determined with and without the magnetic field, and Δw is corrected for the diamagnetism of the sample container by use of Eq. (25.24). If κ_0 (the volume susceptibility of the displaced medium) is negligible, we may write

$$\kappa = \frac{g\Delta w}{V\ H(dH/dx)} \tag{25.29}$$

where V is the volume of the sample and (dH/dx) is the magnetic field gradient. Then for the gram susceptibility we have

$$\chi = \frac{g\Delta w}{w\ H(dH/dx)} \tag{25.30}$$

where w is the sample weight.

Just as in the case of the Gouy method, each of the constants in Eq. (25.30) is generally not determined, but rather the apparatus is calibrated with a standard substance, and Eq. (25.23) is employed.

Problems

1. A Gouy sample tube was filled to the same height with the calibrant [Ni(en)$_3$S$_2$O$_3$] and then with a sample of Cu(dimethylglyoxime)Cl$_2$. Pertinent data follow: temperature, 20°; Δw(empty tube) = -0.32 mg; Δw(Cu complex) = 3.52 mg;

$\Delta w[\text{Ni(en)}_3\text{S}_2\text{O}_3] = 9.30\,\text{mg}$; $w(\text{Cu}$ complex$) = 306.7\,\text{mg}$; $w[\text{Ni(en)}_3\text{S}_2\text{O}_3] = 415.0\,\text{mg}$. Calculate χ_M, correct for diamagnetism, and then calculate μ_{eff}, assuming that Curie's Law is obeyed.

2. Following are the diamagnetism-corrected χ_M values (in $10^{-6}\,\text{cgs mole}^{-1}$) for several compounds at $300°\text{K}$. Rationalize these values in terms of the structure and bonding in the compounds: $\text{K}_2\text{Ba Co(NO}_2)_6$, 1370×10^{-6}; VCl_4, 1200×10^{-6}; $\text{K}_3\text{Fe(CN)}_6$, 2100×10^{-6}; $[(\text{C}_2\text{H}_5)_4\text{N}]_2\text{NiCl}_4$, 6300×10^{-6}; $\text{Co(NCS)}_2\cdot 2\,\text{KSCN}\cdot 4\,\text{H}_2\text{O}$, 8400×10^{-6}; K_2NiF_6, 0.0 (diamagnetic compound); $\text{Co(NH}_3)_6\text{Cl}_3$, 100×10^{-6}; $\text{Cu(CH}_3\text{C}_6\text{H}_4\text{SO}_2\text{NC}_6\text{H}_4\text{SCH}_2\text{CH}_2\text{O)}$,[12] 858×10^{-6}.

3. Which would you expect to have the greater magnetic moment, $\text{CoCl}_4{}^{2-}$ or $\text{CoI}_4{}^{2-}$?

4. Suggest an electronic structure for the aqueous complex $(\text{H}_2\text{O})_5\text{FeNO}^{2+}$ (the "brown ring" test), for which $\mu_{\text{eff}} = 3.9$.

Reference for Further Reading

Earnshaw, A.; *Introduction to Magnetochemistry*, Academic Press, London, 1968.

---

[12] See S. Emori, *et al.*, *Inorg. Chem.*, **7**, 2419 (1968).

Electron

Spin

Resonance

26

The electron has a spin number of $\frac{1}{2}$. In an applied magnetic field, H, the electron may assume either of two orientations, corresponding to the energies $E = \pm\frac{1}{2}g\beta H$, where g is the electron gyromagnetic ratio or "g value," β is the Bohr magneton, and H is the field strength. If the system is subjected to radiation of the appropriate frequency, electrons will be "flipped" from their low energy orientation to their high energy orientation, and energy will be absorbed. This resonance frequency, v, may be calculated from the relation

$$\Delta E = hv = g\beta H$$

The electron spin resonance (esr) spectrum of a paramagnetic substance is generally determined by operating at a fixed frequency and varying the field strength. For technical reasons, the spectrum is generally not recorded as an absorption spectrum (as is usually done in nmr spectrometry), but as a derivative spectrum. That is, the spectrum is the first derivative (slope) of the corresponding absorption curve. In Fig. 26.1, two absorption curves and the corresponding derivative curves are plotted.

385

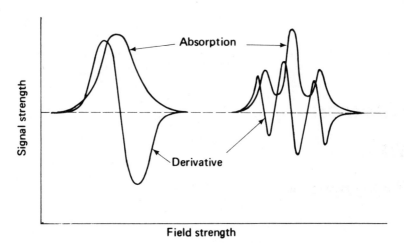

FIG. 26.1. ESR absorption curves and derivative curves.

One important characteristic of an esr spectrum is the field strength at which resonance occurs, or, what is more fundamental, the g value. (In the case of split signals, the g value corresponds to the "center of gravity" of the signal.)

Every paramagnetic substance is characterized by a unique g value, which is formally analogous to the chemical shift value of nmr spectrometry. Some representative g values are given in Table 26.1. For many crystalline sub-

Table 26.1   Some Representative g values*

Free electron	2.0023
Polycrystalline 1,1-diphenyl-2-picryl hydrazyl	2.0036
Organic radicals (in general)	2.0022–2.010
$B_{12}Cl_{11}$	2.011
Fe^{3+} in ZnO	2.0060
Fe^{2+} in MgO	3.4277
Co^{2+} in $ZnSO_4 \cdot 7\,H_2O$	6.90
Electron in NH_3	2.0012

*From M. Bersohn and J. C. Baird, *An Introduction to Electron Paramagnetic Resonance*, W. A. Benjamin, Inc., New York, 1966.

stances, the g value varies with the orientation of the crystal with respect to the applied magnetic field. Thus, a dilute solution of CoF_2 in ZnF_2 yields three values, $g_x = 2.6$, $g_y = 6.05$, and $g_z = 4.1$, corresponding to the g values observed by orienting a crystal in three mutually perpendicular directions with respect to the field.[1]

[1] M. Tinkham, *Proc. Roy. Soc. London*, **236A**, 535, 549 (1956).

LINE WIDTHS

In pure paramagnetic solids, the spin-spin interaction between neighboring magnetic dipoles sometimes causes the esr signal to be broadened. This broadening can be minimized by increasing the distance between the interacting dipoles. For this reason, esr spectra are frequently determined by using a paramagnetic material dissolved in a large amount of diamagnetic solid. It is usually convenient to dissolve the paramagnetic substance in an isomorphous host crystal. For example, a sharp esr spectrum can be obtained for a 1 per cent solution of $KCr(SO_4)_2 \cdot 12\,H_2O$ in the isomorphous salt $KAl(SO_4)_2 \cdot 12\,H_2O$. Sometimes the paramagnetic sites are "trapped" in a diamagnetic matrix. For example, esr spectra have been determined for NO_2 in a matrix of solid argon (the mixture was codeposited on a cold surface from the gas phase) and for OH radicals in ice (formed by irradiation of ice).

Transitions between the energy levels corresponding to different spins for a paramagnetic substance occur even in the absence of electromagnetic radiation. Energy is transferred between the electronic system and the thermal energy of the system by a mechanism called spin-lattice relaxation (here, "lattice" refers not only to crystalline lattices but also to molecular frameworks). When the relaxation time for an energy state is less than the period of the electromagnetic radiation, the transitions due to the radiation are lost among those due to the thermal relaxation. Therefore, the esr spectrum is broadened so much as to be essentially destroyed. Spin-lattice relaxation is facilitated by the presence of low-lying energy levels separated by about kT from the ground level. Thus, transition metal complexes having spin-orbit coupling (those having T ground terms; see p. 375) generally will not give esr spectra at room temperature. However, relaxation times increase with decreasing temperature, and many compounds that yield no spectrum at room temperature give good spectra at the temperatures of liquid helium, hydrogen, or nitrogen.

On p. 360, we briefly discussed nmr contact shifts of paramagnetic complexes and indicated that it is possible to relate electron spin-nuclear spin coupling constants to unpaired spin densities on atoms. This technique is possible whenever the lifetime of a complex in a given spin state is very short, that is, whenever the complex has a T ground state. In most other complexes, the electron spin relaxation occurs more slowly, and severe broadening is usually observed. Consequently, nmr contact shift measurement and esr spectroscopy are complementary techniques. When the electron spin relaxation time is long, the room-temperature esr spectrum can be obtained, but the nmr spectrum is severely broadened. When the relaxation time is short, the room-temperature esr spectrum is severely broadened, but the nmr spectrum is sharp and shows large contact shifts. Thus, octahedral high-spin complexes of Cr^{3+} ($^4A_{2g}$), Cr^{2+} (^5E_g), Fe^{3+} ($^6A_{1g}$), Ni^{2+} ($^3A_{2g}$), and Cu^{2+}

(^2E_g) can give room-temperature esr spectra, and tetrahedral complexes of Ni^{2+} $(^3T_1)$ and low-spin octahedral complexes of Mn^{3+} $(^3T_{1g})$ can give sharp, contact-shifted nmr spectra.

ZERO-FIELD SPLITTING

A species containing one unpaired electron can have its spin degeneracy removed only by application of a magnetic field. The same is true for a species containing more than one unpaired electron, as long as there is no appreciable magnetic interaction between the unpaired electrons. For example, consider a triplet species $(S = 1$ and $M_s = +1, 0, -1)$ in which there is no magnetic interaction. In the absence of a magnetic field, the three spin states are degenerate and are split upon application of a field as shown in Fig. 26.2A. In this case, the two allowed transitions $(\Delta M_s = 1)$ occur at the same field strength, and only one esr resonance is observed. However, if there is appreciable magnetic interaction between the two electrons (as can occur in an appropriate crystal field), the different M_s states have different energies even in the absence of a magnetic field. A possible energy level diagram for this case is shown in Fig. 26.2B, and it is clear that two different esr transitions are allowed.

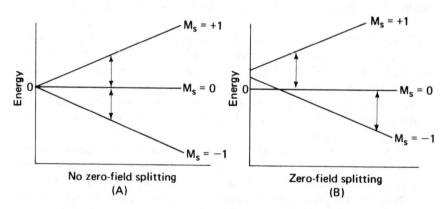

FIG. 26.2. Energy level diagrams for triplet species.

HYPERFINE SPLITTING

Electronic spin can couple with nuclear spin, and therefore esr signals will generally be split into "hyperfine" multiplets whenever the unpaired electrons are in the vicinity of magnetic nuclei. The splitting is quite analogous to the splitting of an nmr signal by interaction of the nuclear moment with a group of other nuclear moments. Each esr signal of an electronic system that interacts with a group of n equivalent nuclei of spin I is split into $2nI + 1$ lines. The relative intensities of the lines may be calculated by the method outlined in Chap. 24.

As a first example, consider the esr spectrum of the aqueous $ON(SO_3)_2{}^{2-}$ ion.[2] Here, the odd electron interacts with the nuclear magnetic moment of the ^{14}N nucleus ($I = 1$). We calculate $2nI + 1 = 2 \times 1 \times 1 + 1 = 3$ for the number of lines. The spectrum shown in Fig. 26.3 is in accord with this prediction.

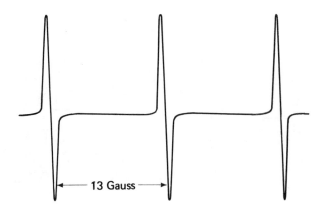

FIG. 26.3. The esr spectrum of an aqueous $ON(SO_3)_2{}^{2-}$ ion.

As a second example, consider the esr spectrum of hydrogen atoms trapped in calcium fluoride (see Fig. 26.4).[3] The spectrum consists of two main resonances attributable to the proton hyperfine splitting ($2 \times 1 \times \frac{1}{2} + 1 = 2$). Each of these main resonances consists of nine components. It seems

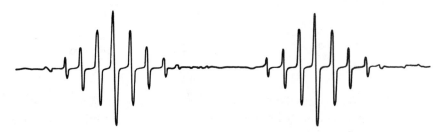

FIG. 26.4. The esr spectrum of H atoms in CaF_2.

[2] G. E. Pake, J. Townsend, and S. I. Weissman, *Phys. Rev.*, **85**, 682 (1952); *Phys. Rev.*, **89**, 606 (1953).

[3] J. L. Hall and R. T. Schumacher, *Phys. Rev.*, **127**, 1892 (1962).

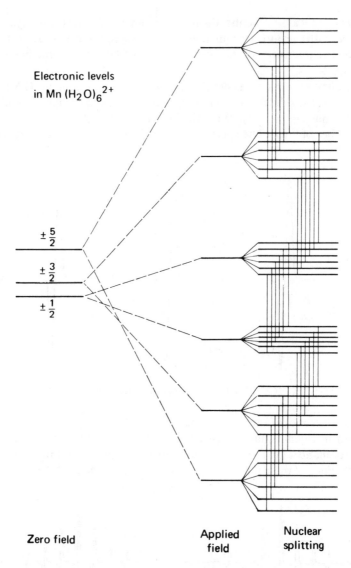

Electronic levels
in Mn $(H_2O)_6{}^{2+}$

$\pm \dfrac{5}{2}$

$\pm \dfrac{3}{2}$

$\pm \dfrac{1}{2}$

Zero field   Applied field   Nuclear splitting

FIG. 26.5. Electronic levels in $Mn(H_2O)_6{}^{2+}$.

most reasonable to ascribe this secondary splitting to the hyperfine interaction of eight equivalent fluorine nuclei ($I = \frac{1}{2}$). Such an interpretation is
reasonable because, in the CaF_2 crystal structure, there are holes bounded by
cubic arrangements of fluoride ions that can accommodate hydrogen atoms.
The spectrum of Fig. 26.4 was obtained with the magnetic field parallel to the
cube edge, so that the eight fluorine nuclei were magnetically equivalent.
Other crystal orientations yield more complicated spectra. The reader can,

as an exercise, show that the relative intensities expected for the multiplets of Fig. 26.4 are $1:8:28:56:70:56:28:8:1$.

As a third example, we shall consider the esr spectrum of the $Mn(H_2O)_6{}^{2+}$ ion.[4] As shown in Fig. 26.5, zero-field splitting gives three states, corresponding to $M_s = \pm\frac{1}{2}, \pm\frac{3}{2}, \pm\frac{5}{2}$. In a magnetic field, the degeneracy of each of these states is lifted, yielding six states. The Mn nucleus ($I = \frac{5}{2}$) causes hyperfine splitting of each of these states into six states. The esr spectrum consists of five peaks, each split into six components. One of the five sextets is shown in Fig. 26.6.

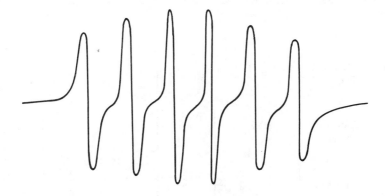

FIG. 26.6. The esr spectrum of aqueous Mn^{2+}

As a fourth example, we consider the esr spectrum of bis(salicylaldiimine)-copper(II).[5]

The spectrum of the complex made with ^{63}Cu ($I = \frac{3}{2}$) consists of four main resonances (hyperfine splitting by the copper nucleus), each resonance being split into what appears to be 11 peaks. It is believed that this secondary hyperfine splitting is caused by the two equivalent nitrogen atoms ($I - 1$) and the two equivalent hydrogen atoms ($I = \frac{1}{2}$) attached to carbon atoms (marked with asterisks), and thus we expect each multiplet to consist of a total of $(2 \times 2 \times 1 + 1)(2 \times 2 \times \frac{1}{2} + 1) = 15$ lines. The fact that only 11 lines are observed (see Fig. 26.7) is explained by assuming that four of the proton

[4] B. Bleaney and D. J. E. Ingram, *Proc. Roy. Soc. London*, **205A**, 336 (1951).
[5] A. H. Maki and B. R. McGarvey, *J. Chem. Phys.*, **29**, 35 (1958).

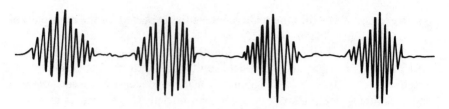

FIG. 26.7. The esr spectrum of bis(salicylaldiimine)copper(II).

hyperfine lines coincide with four of the nitrogen hyperfine lines, as shown below.

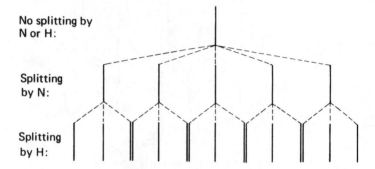

No splitting by
N or H:

Splitting
by N:

Splitting
by H:

Problems

1. Ordinary sulfur contains 0.74 per cent ^{33}S, which has a nuclear spin of $\frac{3}{2}$. Under high amplification, the esr spectrum of $ON(SO_3)_2{}^{2-}$ (Fig. 26.3) shows hyperfine splitting because of coupling to the ^{33}S nuclei. Sketch the expected spectrum.

2. At elevated temperatures, the esr spectrum of gamma-irradiated ammonium perchlorate consists of a triplet of quartets. Interpret.

3. The esr spectrum of the complex $[(NH_3)_5Co—O—O—Co(NH_3)_5]^{5+}$ consists of 15 lines ($I = \frac{7}{2}$ for Co). The cobalt hyperfine coupling constant is lower than that found for most other cobalt complexes. Discuss these results in terms of the electronic structure and bonding in the complex [see E. A. V. Ebsworth and J. A. Weil, *J. Phys. Chem.*, **63**, 1890 (1959)].

4. The esr spectrum of the product formed by depositing lithium, rubidium or cesium atoms on ice at $-196°$ (followed by warming to $-100°$) consists of seven lines with an approximate intensity ratio of $1:6:15:20:15:6:1$. Interpret the spectrum in terms of the structure of the species formed.

References for Further Reading

Atkins, P. W. and M. C. R. Symons, *The Structure of Inorganic Radicals*, Elsevier, New York, 1967.

Bersohn, M. and J. C. Baird, *An Introduction to Electron Paramagnetic Resonance*, W. A. Benjamin, Inc., New York, 1966.

Carrington, A. and A. D. McLachlan, *Introduction to Magnetic Resonance*, Harper & Row, Publishers, New York, 1967.

Drago, R. S., *Physical Methods in Inorganic Chemistry*, Reinhold Publishing Corp., New York, 1965, pp. 328–61.

Figgis, B. N., *Introduction to Ligand Fields*, Interscience Publishers, New York, 1966, pp. 293–310.

McGarvey, B. R., in *Transition Metal Chemistry*, vol. 3, R. L. Carlin, ed., Marcel Dekker, Inc., New York, 1966, p. 89.

Mass
Spectrometry

27

MASS SPECTROMETERS AND MASS SPECTRA

A mass spectrometer is a device for ionizing the molecules (and their fragments) in a gas and for determining the relative amounts of the positive ions having different mass/charge (m/e) ratios. The commonest method for ionizing a gas is the method of electron bombardment. The gaseous sample is allowed to flow through a capillary leak into a chamber where the stream of molecules is bombarded by a stream of electrons with a 90° angle of incidence. The positive ions formed are accelerated by a series of plates having high relative potentials, and a collimated beam of positive ions is formed. Several types of mass spectrometers are in use today, differing in the way in which the composition of the positive ion beam is determined. We shall not discuss the operating principles of mass spectrometers, inasmuch as an understanding of these principles is seldom required in the interpretation of mass spectra.[1]

---

[1] The interested reader is referred to R. W. Kiser, *Introduction to Mass Spectrometry and Its Applications*, Prentice-Hall, Inc., Englewood Cliffs, N. J., 1965.

Suffice it to say that a mass spectrum generally consists of a plot of ion intensity (as measured at a detector) against some function of m/e. Inasmuch as most of the ions are singly charged, the spectrum may be considered a means for identifying ions of different *mass*. A useful generalization (not always true) is that a mass spectrum shows the mass of a molecule and its fragments (formed by bond breaking without structural rearrangement).

A mass spectrum of hydrogen chloride is shown in Fig. 27.1 (the region of low m/e values is not shown). Another way of representing such a spectrum is as a histogram, illustrated in Fig. 27.2. Still another way, useful for showing very weak peaks, is the tabular representation given in Table 27.1.[2] Here the strongest peak is arbitrarily assigned an intensity of 100, and all the other peak intensities are given relative to 100.

Table 27.1   Mass Spectrum of HCl
(50 v Ionizing Voltage)*

$\dfrac{m}{e}$	Ion	Intensity
35	^{35}Cl	17
36	$H^{35}Cl$	100
37	^{37}Cl	5.4
38	$H^{37}Cl$	32.4

* See ref. 2.

Notice that the mass spectrum of HCl consists essentially of the sum of two spectra: the spectrum of $H^{35}Cl$ and the spectrum of $H^{37}Cl$. The peaks at masses 35 and 37, corresponding to Cl^+ ions, have an intensity ratio equal to the relative abundances of the ^{35}Cl and ^{37}Cl isotopes. The same is true of the peaks at masses 36 and 38, corresponding to HCl^+ ions. The situation is somewhat more complicated for molecules containing more than one polyisotopic atom. In the mass spectrum of BCl_3, the parent BCl_3^+ ion is represented by a band of eight lines, as shown in Fig. 27.3. Here one is concerned with the distribution of the following four isotopes, having the indicated abundances (in per cent): $^{10}B(19.6)$, $^{11}B(80.4)$, $^{35}Cl(75.5)$, and $^{37}Cl(24.5)$.

Table 27.2 lists the mass numbers and abundances of the naturally occurring isotopes.[3] These data can be very useful in interpreting mass spectra of compounds containing polyisotopic elements.

[2] American Petroleum Institute Research Project 44, Serial 93.
[3] See Appendix 1 in ref. 1.

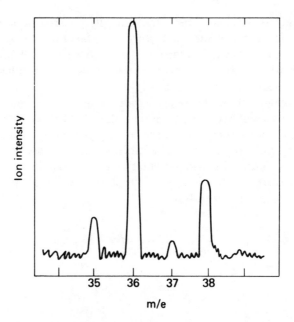

FIG. 27.1. Mass spectrum of HCl, as obtained with a strip chart recorder.

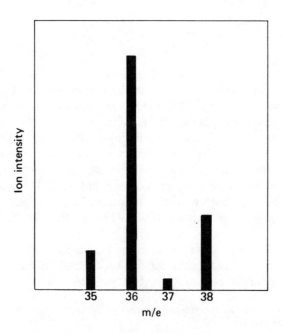

FIG. 27.2. Mass spectrum of HCl, plotted as a histogram.

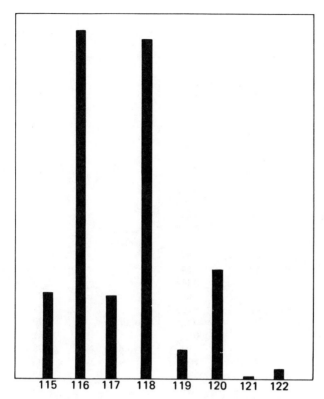

FIG. 27.3. Mass spectrum of BCl_3 in the m/e region 115–122.

Table 27.2   Mass Numbers and Abundances of the Naturally Occurring Isotopes*

Element	Mass Number	Relative Abundance	Element	Mass Number	Relative Abundance
Hydrogen	1	99.9855		17	0.0374
	2	0.0145		18	0.2039
Helium	3	trace	Fluorine	19	100.
	4	100.	Neon	20	90.92
Lithium	6	7.50		21	0.257
	7	92.50		22	8.82
Beryllium	9	100.	Sodium	23	100.
Boron	10	19.91	Magnesium	24	78.80
	11	80.09		25	10.15
Carbon	12	98.888		26	11.05
	13	1.112	Aluminum	27	100.
Nitrogen	14	99.633	Silicon	28	92.21
	15	0.367		29	4.70
Oxygen	16	99.759		30	3.09

Element	Mass Number	Relative Abundance	Element	Mass Number	Relative Abundance
Phosphorus	31	100.		66	27.81
Sulfur	32	95.018		67	4.11
	33	0.750		68	18.57
	34	4.215		70	0.62
	36	0.017	Gallium	69	60.16
Chlorine	35	75.7705		71	39.84
	37	24.2295	Germanium	70	20.51
Argon	36	0.337		72	27.40
	38	0.063		73	7.76
	40	99.600		74	36.56
Potassium	39	93.083		76	7.77
	40	0.012	Arsenic	75	100.
	41	6.905	Selenium	74	0.87
Calcium	40	96.97		76	9.02
	42	0.64		77	7.58
	43	0.145		78	23.52
	44	2.06		80	49.82
	46	0.003		82	9.19
	48	0.185	Bromine	79	50.537
Scandium	45	100.		81	49.463
Titanium	46	7.99	Krypton	78	0.354
	47	7.32		80	2.27
	48	73.98		82	11.56
	49	5.46		83	11.55
	50	5.25		84	56.90
Vanadium	50	0.24		86	17.37
	51	99.76	Rubidium	85	72.15
Chromium	50	4.352		87	27.85
	52	83.764	Strontium	84	0.560
	53	9.509		86	9.870
	54	2.375		87	7.035
Manganese	55	100.		88	82.535
Iron	54	5.82	Yttrium	89	100.
	56	91.66	Zirconium	90	51.46
	57	2.19		91	11.23
	58	0.33		92	17.11
Cobalt	59	100.		94	17.40
Nickel	58	67.77		96	2.80
	60	26.16	Niobium	93	100.
	61	1.25	Molybdenum	92	15.86
	62	3.66		94	9.12
	64	1.16		95	15.70
Copper	63	69.1		96	16.50
	65	30.9		97	9.45
Zinc	64	48.89		98	23.75

Element	Mass Number	Relative Abundance	Element	Mass Number	Relative Abundance
	100	9.62		128	31.79
Ruthenium	96	5.51		130	34.48
	98	1.87	Iodine	127	100.
	99	12.72	Xenon	124	0.096
	100	12.62		126	0.090
	101	17.07		128	1.919
	102	31.63		129	26.44
	104	18.58		130	4.08
Rhodium	103	100.		131	21.18
Palladium	102	0.96 .		132	26.89
	104	10.97		134	10.44
	105	22.23		136	8.87
	106	27.33	Cesium	133	100.
	108	26.70	Barium	130	0.101
	110	11.81		132	0.097
Silver	107	51.817		134	2.42
	109	48.183		135	6.59
Cadmium	106	1.215		136	7.81
	108	0.875		137	11.32
	110	12.39		138	71.66
	111	12.75	Lanthanum	138	0.089
	112	24.07		139	99.911
	113	12.26	Cerium	136	0.193
	114	28.86		138	0.250
	116	7.58		140	88.48
Indium	113	4.28		142	11.07
	115	95.72	Praseodymium	141	100.
Tin	112	0.95	Neodymium	142	27.11
	114	0.65		143	12.17
	115	0.34		144	23.85
	116	14.24		145	8.30
	117	7.57		146	17.22
	118	24.01		148	5.73
	119	8.58		150	5.62
	120	32.97	Samarium	144	3.15
	122	4.71		147	15.09
	124	5.98		148	11.35
Antimony	121	57.25		149	13.96
	123	42.75		150	7.47
Tellurium	120	0.089		152	26.55
	122	2.46		154	22.43
	123	0.87	Europium	151	47.82
	124	4.61		153	52.18
	125	6.99	Gadolinium	152	0.205
	126	18.71		154	2.23

Element	Mass Number	Relative Abundance	Element	Mass Number	Relative Abundance
	155	15.1		183	14.31
	156	20.6		184	30.66
	157	15.7		186	28.60
	158	24.5	Rhenium	185	37.07
	160	21.6		187	62.93
Terbium	159	100.	Osmium	184	0.02
Dysprosium	156	0.052		186	1.59
	158	0.090		187	1.64
	160	2.294		188	13.3
	161	18.88		189	16.1
	162	25.53		190	26.4
	163	24.97		192	41.0
	164	28.18	Iridium	191	38.5
Holmium	165	100.		193	61.5
Erbium	162	0.136	Platinum	190	0.013
	164	1.56		192	0.78
	166	33.41		194	32.9
	167	22.94		195	33.8
	168	27.07		196	25.2
	170	14.88		198	7.19
Thulium	169	100.	Gold	197	100.
Ytterbium	168	0.135	Mercury	196	0.146
	170	3.03		198	10.02
	171	14.31		199	16.84
	172	21.82		200	23.13
	173	16.135		201	13.22
	174	31.84		202	29.80
	176	12.73		204	6.85
Lutetium	175	97.41	Thallium	203	29.50
	176	2.59		205	70.50
Hafnium	174	0.18	Lead	204	1.40
	176	5.20		206	25.2
	177	18.50		207	21.7
	178	27.13		208	51.7
	179	13.75	Bismuth	209	100.
	180	35.24	Thorium	232	100.
Tantalum	180	0.012	Uranium	234	0.0056
	181	99.988		235	0.7205
Tungsten	180	0.14		238	99.274
	182	26.29			

* See Appendix 1 in ref. 1.

QUALITATIVE ANALYSIS

The mass spectrum of a compound is unique and can be used for the purposes of characterization and identification. Many compounds have their spectra tabulated in one or more of the following catalogs.

1. Mass Spectral Data, American Petroleum Institute Research Project 44, National Bureau of Standards, Washington, D. C.

2. Mass Spectral Data, Manufacturing Chemists Association Research Project, Agricultural and Mechanical College of Texas, College Station, Texas.

3. File of Uncertified Mass Spectra, American Society for Testing Materials, Committee E–14 on Mass Spectrometry.

Thus, one can often identify (or confirm the identification of) an unknown compound by simply comparing its mass spectrum with the tabulated spectrum of the compound.

The mass spectrum of a mixture of compounds is usually just the superposition of the spectra of the individual compounds. We shall show how this rule can be applied to the analysis of a binary mixture. In Table 27.3, the mass spectrum of dimethyl ether is given in tabular form in column A. The mass spectrum of a mixture of dimethyl ether and an unknown compound,

Table 27.3   Mass Spectra of $(CH_3)_2O$ and of $(CH_3)_2O$ plus an Unknown Compound

$\dfrac{m}{e}$	A $(CH_3)_2O$	B $(CH_3)_2O$ + unknown	$\dfrac{B}{A}$	$B - 0.231A$
12	2.41	0.82	0.340	0.26
13	5.60	1.38	0.246	0.09
14	12.33	7.73	0.627	4.89
15	58.05	21.45	0.369	8.04
16	2.44	29.80	>1	29.24
17	1.03	26.34	>1	26.10
27	0.52	0.40	0.770	0.28
28	3.08	22.11	>1	21.40
29	79.17	58.30	0.737	40.01
30	3.52	32.10	>1	31.29
31	7.05	48.92	>1	47.29
32	0.12	100.00	>1	99.97
33		0.85	>1	0.85
41	0.22	0.06	0.273	0.01
42	0.64	0.15	0.234	0.00
43	1.78	0.41	0.230	0.00
44	0.84	0.19	0.226	0.00
45	100.00	23.08	0.231	−0.02
46	45.61	10.53	0.231	0.00
47	1.20	0.27	0.225	−0.01

obtained with the same mass spectrometer under similar experimental conditions, is given in column B. In the fourth column, the ratio of intensities (B/A) is tabulated. For several mass values, the ratio of intensities is in the range from 0.225 to 0.234, and for no other mass value is the ratio smaller than 0.225. We conclude that, at these mass values, only the dimethyl ether contributes to the spectrum of the mixtures, and that, at all the other listed mass values, the unknown compound contributes. To calculate the contributions of the unknown compound at these latter mass values, we must subtract the estimated contributions of the dimethyl ether. These estimated contributions are made by multiplying each of the intensities in column A by 0.231 [assumed to be the most reliable value of the B/A factor because it corresponds to the most intense peak in the $(CH_3)_2O$ spectrum, at mass 45]. The calculated spectrum of the unknown is given in the last column. For simplicity in the interpretation of the calculated spectrum, it is best to ignore all the weak peaks with intensity values less than unity. We notice that there are five relatively strong peaks at mass 28, 29, 30, 31, and 32. Adjacent peaks often correspond to species differing by just one hydrogen atom; therefore, we should suspect that we have a compound containing at least four hydrogen atoms attached to a residue of mass 28. The possibilities of methanol, silane, and hydrazine should come to mind. The mass spectrum of CH_3OH might be expected to have fairly strong peaks at mass 12 (C^+) and mass 13 (CH^+). Only very weak peaks (probably attributable to impurities or experimental error) are found at these masses; therefore, we tentatively discard the CH_3OH possibility. The spectrum of SiH_4 would be expected to have peaks corresponding to the isotopic species $^{29}SiH_4$ and $^{30}SiH_4$, at masses 33 and 34, respectively. These peaks would, in accordance with the isotopic abundances, have intensities 5.1 and 3.4 relative to 100 for the $^{28}SiH_4$ peak at mass 32. Inasmuch as only a very weak peak is observed at mass 33, and no peak at mass 34, we discard the SiH_4 possibility. Thus, we are left with the N_2H_4 possibility. The peaks at masses 14 through 17 can be assigned to the species N^+, NH^+, $NH_2{}^+$, and $NH_3{}^+$. Only the latter peak is difficult to understand; its appearance implies a rearrangement of protons during the cleavage of N_2H_4. To make the identification of the unknown certain, we would have to compare the calculated spectrum with that of a sample of hydrazine. Such a comparison shows the spectra to be very similar.[4]

Mass spectrometry sometimes can be used to identify new compounds among the products of synthetic reactions. We may cite the example of the first identification of germyl phosphines. It was known that a series of polygermanes (of empirical composition Ge_nH_{2n+2}) could be prepared by passing germane (GeH_4) through an ozonizer type electric discharge.[5] When

[4] The procedures to be followed in analyzing more complicated mixtures are outlined by Kiser, *Introduction to Mass Spectrometry and Its Applications.* 1965.

[5] J. E. Drake and W. L. Jolly, *J. Chem. Soc.*, 2807 (1962).

a mixture of germane and phosphine was passed through such a discharge, a difficult-to-separate mixture of volatile products containing germanium, phosphorus, and hydrogen was formed. However, it was not immediately clear whether or not ternary hydrides (of both germanium and phosphorus) had formed, and if so, what the compositions of the hydrides were. By simply obtaining the mass spectrum of the mixture of products, it was possible to identify immediately the species GeH_3PH_2, Ge_2PH_7, and Ge_3PH_9.[6] These species are recognizable by the bands indicated in Fig. 27.4. Bands, rather than isolated peaks, are observed because of fragmentation by the successive loss of hydrogen atoms and because germanium has several isotopes of moderate abundance. Bands attributable to Ge_2H_6, Ge_3H_8, and Ge_4H_{10} are also shown in Fig. 27.4.

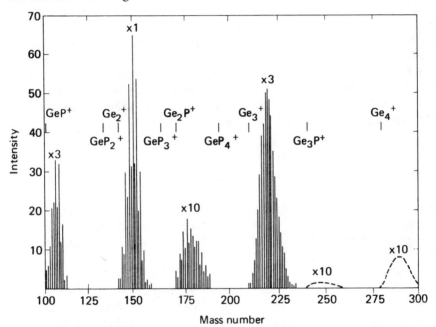

FIG. 27.4. Mass spectrum of products formed by passing GeH_4 and PH_3 through a silent electric discharge. (Individual peaks were not resolved beyond mass 240.)

Some high-resolution mass spectrometers permit the determination of m/e values to ± 0.0001 mass number. With such instruments, it is possible to distinguish between isobaric ions, i.e., ions that have the same weight to the nearest whole mass number but that differ slightly in weight because the atomic weights of the nuclides are not exact whole numbers on the basis of $^{12}C = 12.00000$. Accurate measurements of this type often are helpful in

[6] J. E. Drake and W. L. Jolly, *Chem. Ind. London*, 1470 (1962).

identifying the products of reactions. For example, a particular compound was suspected, on the basis of its method of synthesis, of being either $SiH_3-S-SiH_2-S-SiH_3$ (mass 155.9375) or $(SiH_3)_3PS$ (mass 155.9470).[6a] High-resolution mass spectrometry showed a parent peak with a mass of 155.9378, thus proving the first structure to be correct.

STRUCTURE DETERMINATION

Mass spectra may be used in structure determinations by applying the rule of thumb (really more of a hope than a rule) that ion fragments are formed without gross rearrangements of the atoms. Thus, the observation of a peak for the Mn_2^+ ion in the spectrum of $Mn_2(CO)_{10}$ is *consistent with* the structure having a manganese–manganese bond.[7] Similarly, spectra of the carbonyls $Co_4(CO)_{12}$ and $Ru_3(CO)_{12}$ show peaks corresponding to the metal clusters Co_4^+ and Ru_3^+, which are present in the parent molecules.[8] The ion MnH^+ has been observed in the mass spectrum of $HMn(CO)_5$, providing evidence for the manganese–hydrogen bond.[9] The mass spectra of $Fe_2Br_2(NO)_4$ and $CoBr_2(NO)_4$ are presented in Table 27.4.[10] These compounds have structures of the type

$$
\begin{array}{ccccc}
ON & & Br & & NO \\
\diagdown & \diagup & \diagdown & \diagup & \\
& M & & M & \\
\diagup & & \diagdown & \diagup & \diagdown \\
ON & & Br & & NO
\end{array}
$$

Notice that the cobalt bromide bridge is readily broken, whereas the iron bromide bridge is relatively stable. This increased stability of the bridge in the iron compound is probably attributable to a metal-metal bond, lacking in the cobalt compound.

Mass spectrometry was used to determine the molecular formula of the ruthenium carbonyl carbide $Ru_6(CO)_{17}C$, in which the "carbide" ion is at the center of an irregular octahedron of ruthenium atoms.[11] The differentiation between $Ru_6(CO)_{17}C$ and $Ru_6(CO)_{18}$ would have been difficult by the usual analytical techniques. However, the mass spectrum indicates the first formulation because the $Ru_6(CO)_{17}C^+$ ion is observed, followed by stepwise elimination of all 17 carbonyl groups giving finally the abundant ion Ru_6C^+.

The identification of isomers from their mass spectra is a less uncertain task than *a priori* structure determinations. Consider the mass spectra of

[6a] C. H. Van Dyke and E. W. Kifer, unpublished results.

[7] R. E. Winters and R. W. Kiser, *J. Phys. Chem.*, **69**, 1618 (1965).

[8] R. B. King, *J. Am. Chem. Soc.*, **88**, 2075 (1966).

[9] W. F. Edgell and W. M. Risen, Jr., *J. Am. Chem. Soc.*, **88**, 5451 (1966).

[10] B. F. G. Johnson, J. Lewis, I. G. Williams, and J. M. Wilson, *J. Chem. Soc.*, 338 (1967).

[11] B. F. G. Johnson, R. D. Johnston, and J. Lewis, *Chem. Commun.*, 1057 (1967).

Table 27.4   **Mass Spectra of $Fe_2Br_2(NO)_4$ and**
$Co_2Br_2(NO)_4$*

	$Fe_2Br_2(NO)_4$		$Co_2Br_2(NO)_4$	
Ion	$\dfrac{m}{e}$	*I*	$\dfrac{m}{e}$	*I*
$M_2Br_2(NO)_4{}^+$	394	45	398	1
$M_2Br_2(NO)_3{}^+$	364	95	368	2
$M_2Br_2(NO)_2{}^+$	334	100	338	2
$M_2Br_2NO^+$	304	54	308	1
$M_2Br_2{}^+$	274	90	278	4
M_2Br^+	193	18	198	0
$M_2{}^+$	112	0	118	0
$MBr(NO)_2{}^+$	197	0	198	90
$MBrNO^+$	167	0	168	60
MBr^+	137	23	138	38
$M(NO)_2{}^+$	116	0	119	0
MNO^+	86	0	89	0
M^+	56	63	59	100

     * See ref. 10.

three isomers of C_4F_6, presented in Table 27.5. In perfluoro-1,3-butadiene $(CF_2{=}CF{-}CF{=}CF_2)$, the middle carbon–carbon bond is weaker than the other two bonds; therefore, cleavage of the molecule into C_2F_3 fragments would be expected to be a fairly common process. Indeed, the $C_2F_3{}^+$ ion intensity is relatively much greater for perfluoro-1,3-butadiene than for the other two isomers. In perfluorocyclobutene, we may assume that fragmentation of the molecule will occur by successive cleavage of two carbon–carbon bonds. If the first cleavage occurs as follows:

$$\begin{array}{c} FC{=}CF \\ |\quad\ | \\ F_2C \nmid CF_2 \end{array}$$

then perfluoro-1,3-butadiene is formed, and we would expect subsequent fragmentation to occur as in perfluoro-1,2-butadiene. We note that although the $C_2F_3{}^+$ intensity is not as great as in the case of perfluoro-1,3-butadiene, it is much greater than in the case of perfluoro-2-butyne. If the first cleavage of perfluorocyclobutene occurs as follows:

$$\begin{array}{c} FC{=}CF \\ |\quad \\ F_2C{-}CF_2 \end{array}$$

the fragment $-CF_2-CF_2-CF{=}CF-$ is formed, and one would expect to obtain either CF_2 and C_3F_4 fragments or C_2F_4 and C_2F_2 fragments, depending on which C–C bond breaks in the second step. Actually, cleavage of the

Table 27.5   Mass Spectra of C_4F_6 Isomers*

Ion	Perfluoro-1,3-butadiene	Perfluorocyclobutene	Perfluoro-2-butyne
C^+	6.75	6.97	11.8
F^+	1.78	1.02	4.04
CF^+	49.6	64.0	53.4
CF_2^+	5.75	4.59	8.69
CF_3^+	5.52	7.68	42.1
C_2^+	1.85	1.77	5.14
C_2F^+	1.72	2.00	1.36
$C_2F_2^+$	1.10	6.48	2.45
$C_2F_3^+$	1.79	0.62	0.04
$C_2F_4^+$	0.23	3.73	0.00
C_3^+	2.13	1.55	6.38
C_3F^+	6.10	5.42	15.2
$C_3F_2^+$	11.5	8.56	21.5
$C_3F_3^+$	100	100	100
$C_3F_4^+$	10.5	3.77	10.8
$C_3F_5^+$	1.23	2.04	0.37
$C_4F_4^+$	4.31	1.35	14.2
$C_4F_5^+$	4.64	18.6	82.0
$C_4F_6^+$	29.9	26.2	39.5

* *J. Res. Natl. Bur. Std.*, **49**, 343 (1952).

middle carbon–carbon bond in the initial fragment would be expected to be energetically favored because it involves the formation of the relatively stable species, $CF_2{=}CF_2$ and $CF{=}CF$ (with a net increase in the number of bonds). Indeed, the $C_2F_4^+$ and $C_2F_2^+$ intensities are relatively much greater for perfluorocyclobutene than for the other two isomers. In perfluoro-2-butyne ($CF_3{-}C{\equiv}C{-}CF_3$), the terminal carbon–carbon bonds are weaker than the middle carbon–carbon bond, and cleavage into CF_3, C_2, and C_3F_3 fragments would be expected. We cannot compare the $C_3F_3^+$ intensities for the three isomers because they were arbitrarily taken to be 100; however, we observe that the CF_3^+ and C_2^+ intensities are the greatest for perfluoro-2-butyne.

Although isomer identification was fairly straightforward in the preceding case, it is not always so. Consider the mass spectra of *cis* and *trans* N_2F_2, shown in Table 27.6. It is not clear why the *trans* isomer should give a much higher yield of the $N_2F_2^+$ parent ion.[12]

[12] Possibly the explanation is to be found in the fact that less atomic rearrangement is required to convert *trans* N_2F_2 into a linear $N_2F_2^+$ than in the case of *cis* N_2F_2. The N-F bonds in $N_2F_2^+$ would be expected to have some double-bond character, requiring a linear configuration.

Table 27.6 Mass Spectra of *trans* and *cis* N_2F_2*

$\dfrac{m}{e}$	Ion	*trans* N_2F_2	*cis* N_2F_2
66	N_2F^+	25.3	0.5
47	N_2F^+	43.4	100.0
33	NF^+	5.0	6.0
28	N_2^+	100.0	84.5
23.5	N_2F^{2+}		1.4
19	F^+	1.8	5.3
14	N^+	11.6	10.5

* C. B. Colburn, *et al., J. Am. Chem. Soc.,* **81**, 6397 (1959).

The reader is referred to the text by McLafferty for a discussion of the interpretation of the mass spectra of organic compounds.[13] Many of the same principles apply to the interpretation of mass spectra of inorganic compounds.

Many pitfalls await the chemist who uses mass spectra to determine the structure and formulas of compounds. One might think that mass spectrometry would afford an ideal method for determining the molecular weight of a volatile compound. However, sometimes the peak of highest mass number in a spectrum corresponds to a fragment ion, rather than to a parent molecule ion. For example, the highest mass peak in the spectrum of B_8F_{12} corresponds to the fragment $B_7F_{10}^+$.[14] Similarly, the highest mass peak in the spectrum of C_2ClF_5 corresponds to $C_2ClF_4^+$.[13] There are various ways for recognizing situations of this type in the spectra of unidentified compounds. If the highest mass peak corresponds to an ion with an even number of electrons, it cannot be the parent ion peak except in the unusual case of a paramagnetic molecule. If, when the ionizing voltage is decreased, the intensity of the highest mass peak decreases relative to that of any of the other peaks, then it cannot be the parent ion peak.

Sometimes peaks occur at masses greater than that of the parent ion because of the combination of ions and molecules in the mass spectrometer. For example, the mass spectrum of ferrocene shows ions resulting from fragmentation of $(C_5H_5)_3Fe_2^+$.[15]

$$C_5H_5Fe^+ + (C_5H_5)_2Fe \longrightarrow C_5H_5{-}Fe{-}C_5H_5{-}Fe{-}C_5H_5^+$$

Such ion-molecule peaks can be recognized by their relative increase with increased sample pressure.

[13] F. W. McLafferty, *Interpretation of Mass Spectra*, W. A. Benjamin, Inc., New York, 1966.

[14] P. L. Timms, private communication.

[15] D. B. Chambers, F. Glockling, and J. R. C. Light, *Quart. Rev. London*, **22**, 317 (1968).

The rule that no rearrangements of bonds occur during the fragmentation of molecules in the mass spectrometer is often violated. For example, trifluoroboroxine has the structure

in which each boron atom is attached to only one fluorine atom. Yet the mass spectrum shows an intense band corresponding to $B_2OF_3^+$, a species that is clearly a consequence of rearrangement.[16] A particularly striking example of rearrangement is found in the mass spectrum of the molecule $[(CH_3)_2PBH_2]_3$, which has the structure[17]

The spectrum shows peaks for the ions $(CH_3)_2{}^{10}B^+$ and $(CH_3)_2{}^{11}B^+$ in the correct isotopic ratio but no evidence of any $(CH_3)_2P^+$ ions. It is believed that vibration brought the methyl groups close enough to boron, and hydrogen near enough to phosphorus, for an exchange to be part of the decomposition.[18]

Problems

1. The peak $m/e = 224$ appears in the mass spectrum of $Fe_2(CO)_9$. How could you show that this peak results from $Fe_2(CO)_4^+$ and not $Fe(CO)_6^+$?

2. In the mass spectra of $(C_5H_5)_2Fe$, $C_5H_5MnC_6H_6$, and $(C_6H_6)_2Cr$, the most intense peaks of metal-containing species are those due to $(C_5H_5)_2Fe^+$, $C_5H_5Mn^+$, and Cr^+. What do you conclude about the bonding?

3. The mass spectra (determined only for m/e values between 12 and 35) for silane and for a mixture of silane and an unknown material are given. Identify the unknown material and the species corresponding to all the listed m/e values. Note that the

[16] F. W. McLafferty and R. S. Gohlke, *Chem. Eng. News,* May 18, 1964, p. 96.

[17] A. B. Burg and R. I. Wagner, *J. Am. Chem. Soc.,* **75**, 3872 (1953).

[18] The spectrum was obtained by V. Dibeler and L. Wall; the interpretation was provided by A. B. Burg.

partial pressure of silane in the sample of mixture was greater than the pressure of silane in the sample of pure silane.

$\dfrac{m}{e}$	Pure SiH$_4$	Mixture of SiH$_4$ + ?
14	1.0	1.3
14.5	2.1	2.5
15	1.4	1.5
15.5	0.2	0.8
16	0.1	0.5
16.5		0.2
17		0.4
28	17.2	20.7
29	18.3	22.0
30	100.0	120.2
31	80.5	113.3
32	7.0	108.4
33	2.4	28.7
34	0.1	84.4

4. Identify the compound for which the following mass spectrum (MCA Serial No. 36) was obtained.

10	5.8	53	9.0	162	5.6
11	21.9	54	22.4	163	2.5
12	5.5	55	33.3	169	1.2
13	12.4	56	35.1	170	3.9
23	1.4	57	39.0	171	8.1
24	1.8	58	48.7	172	11.2
33	2.6	59	61.4	173	10.7
34	8.6	60	73.2	174	4.8
35	9.8	61	85.0	175	3.3
36	4.9	62	74.9	180	1.6
37	2.2	63	60.1	181	5.2
42	1.17	127	12.2	182	9.4
43	6.5	128	8.5	183	8.8
44	13.8	138	4.5	184	4.5
45	22.0	139	12.4	185	3.6
46	3.2	149	1.6	186	3.0
47	41.6	150	3.9	187	14.3
48	47.2	151	4.0	188	51.1
49	35.2	159	1.1	189	100.0
50	2.2	160	3.9	190	80.8
52	2.0	161	6.9		

References for Further Study

Bruce, M. I., "Mass Spectra of Organometallic Compounds," *Adv. Organometallic Chem.*, **6**, 273 (1968).

Chambers, D. B., F. Glockling, and J. R. C. Light, "Mass Spectra of Organometallic Compounds," *Quart. Rev. London*, **22**, 317 (1968).

Drago, R. S., *Physical Methods in Inorganic Chemistry*, Reinhold Publishing Corp., New York, 1965, pp. 374–88.

Kiser, R. W., *Introduction to Mass Spectrometry and Its Applications*, Prentice-Hall, Inc., Englewood Cliffs, N. J., 1965.

Lewis, J. and B. F. G. Johnson, "Mass Spectra of Some Organometallic Molecules," *Acc. Chem. Res.*, **1**, 245 (1968).

McLafferty, F. W., *Interpretation of Mass Spectra*, W. A. Benjamin, Inc., New York, 1966.

X-ray Diffraction

28

INTRODUCTION

The ultimate method for determining the structure of a molecule in a molecular crystal lattice is the determination of the crystal structure by X-ray diffraction.[1] No other technique, except possibly microwave spectroscopy,[2] can give detailed structural information with the reliability and accuracy of the X-ray diffraction method. Unfortunately, structure determination by X-ray diffraction is a fairly difficult, time-consuming process, and few synthetic chemists manage to determine the crystal structures of the compounds that they prepare. However, it is important that every synthetic chemist learn the capabilities and limitations of the X-ray method. For this purpose, various books, among those listed under References for Further Study, should be consulted.

[1] Neutron diffraction could also be included, but this technique is not as commonly employed as X-ray diffraction.

[2] Microwave spectroscopy is capable of giving very accurate structural parameters for relatively uncomplicated gaseous molecules.

411

As a general rule, to determine a structure by X-ray diffraction, it is necessary to have a suitable single crystal with a volume of at least 10^{-4} cc. A powder is ordinarily useless for a detailed structure determination. However, the X-ray diffraction pattern of a powdered substance is very characteristic of the substance and can be used for identification. In some cases, a "powder pattern" can give information about symmetry or the size of the molecular unit that eliminates certain possible structures from consideration.

In this chapter we shall briefly discuss the utility of the powder method of X-ray diffraction and shall describe procedures for indexing the diffraction lines for cubic crystals.

THE CRYSTALLINE STATE

In a crystal, the atoms are arranged in a regular three-dimensional pattern, the repeating units of which may be found at regular intervals in many different directions (the repeating unit is chosen so that it displays the full symmetry of the crystal lattice). We choose three directions (x, y, and z) corresponding to the three shortest repeating distances (a, b, and c) as the *crystal axes*. The repeating units may be atoms, molecules, or even groups of atoms or molecules. If we replace the repeating units by points, we form a *space lattice*.

There are seven different *crystal systems*, which differ from one another in terms of the parameters of the space lattices. These parameters are the three repeating distances, a, b, and c, and the three angles between the crystal axes. The angle between the x and y axes is γ, that between the y and z axes is α, and that between the x and z axes is β. The restrictions on the six parameters corresponding to the seven crystal systems are given in Table 28.1.

In Fig. 28.1 are shown the lattice points lying in one of the xy planes of a space lattice. The lines indicate the intersection of planes that are parallel to the z axis of the lattice. Notice that each of these planes is but one member of a family of parallel planes. Each family of planes is characterized by three integers, h, k, and l, called *Miller indices*. A Miller index of a family of planes is the number of such planes that are crossed on moving a distance of one lattice spacing along one of the axes of the lattice. The indices h, k, and l refer to the axes x, y, and z and the lattice spacings a, b, and c, respectively. For example, consider the planes labeled 130 in Fig. 28.1. One such plane is crossed on moving a distance a in the x direction; hence, $h = 1$. Three such planes are crossed on moving a distance b in the y direction; hence, $k = 3$. Because the planes are parallel to the z axis, no planes are crossed on moving in the z direction, and $l = 0$.

The 100, 010, and 001 planes form the boundaries of a set of parallelepipeds with side dimensions of a, b, and c. Each parallelepiped, called a unit cell, contains an integral number of structural units.

Table 28.1   The Seven Crystal Systems

System	Lattice Constants	Minimum Symmetry
Triclinic	$\alpha \neq \beta \neq \gamma$ $a \neq b \neq c$	only one-fold (identity or inversion) symmetry
Monoclinic	first setting: $\alpha = \beta = 90°; 90° < \gamma < 180°$ $a \neq b \neq c$ second setting (most common): $\alpha = \gamma = 90°; 90° < \beta < 180°$ $a \neq b \neq c$	one two-fold (rotation or inversion) axis is along z for first setting, and one is along y for second setting
Orthorhombic	$\alpha = \beta = \gamma = 90°$ $a \neq b \neq c$	three two-fold (rotation or inversion) axes, mutually perpendicular
Trigonal-hexagonal (two related systems)	**hexagonal** setting: $\alpha = \beta = 90°; \gamma = 120°$ $a = b \neq c$ **rhombohedral** setting: $90° \neq \alpha = \beta = \gamma < 120°$ $a = b = c$	three-fold or six-fold axis along z three-fold axis is body diagonal of the unit cell
Tetragonal	$\alpha = \beta = \gamma = 90°$ $a = b \neq c$	one four-fold axis
Cubic	$\alpha = \beta = \gamma = 90°$ $a = b = c$	four intersecting three-fold axes, 54° 44′ from the crystallographic axes

It is convenient (although not rigorously correct) to assume that a beam of X-rays impinging on a crystal is reflected from a family of crystal planes when the angle of incidence (and reflection) is such that the reflections from successive planes reinforce one another. It turns out that

$$\lambda = 2d_{hkl} \sin \theta \tag{28.1}$$

where λ is the X-ray wavelength, d_{hkl} is the interplanar spacing, and θ is the angle of reflection. The interplanar spacing is a function of the integers h, k, and l and the parameters required to describe the space lattice. If the space lattice belongs to the cubic system, the expression for d_{hkl} is very simple:

$$d_{hkl} = \frac{a}{\sqrt{h^2 + k^2 + l^2}} \tag{28.2}$$

where a is the lattice constant. More complicated expressions are required for space lattices of lower symmetry. It should be clear at this point that the angles of the X-ray reflections depend only on the dimensional parameters of

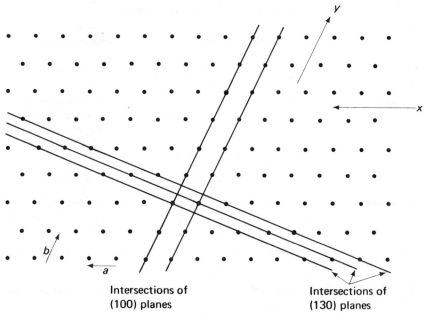

FIG. 28.1. An *xy* plane of a space lattice.

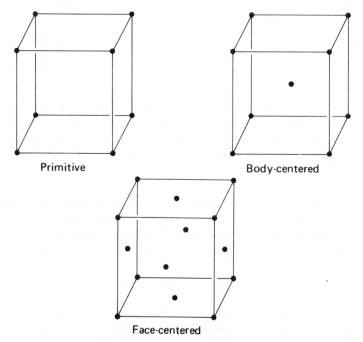

FIG. 28.2. The three cubic lattices.

the space lattice, and not on the arrangement of the atoms. Thus, one can find examples of quite chemically unrelated compounds having identical X-ray diffraction patterns with respect to the angles of reflections. Tin tetraiodide, SnI_4, and rubidium aluminum alum, $RbAl(SO_4)_2 \cdot 12 H_2O$, are examples of such compounds; both have primitive cubic space lattices with a lattice spacing of 12.2 Å. The relative *intensities* of the reflections depend on the arrangement of atoms in each repeating unit. However, the interpretation of the intensities is a fairly complicated subject, and we shall not be concerned with it in this chapter.

By combining Eqs. (28.1) and (28.2), we obtain, for a cubic crystal,

$$\sin^2 \theta = \frac{\lambda^2}{4a^2}(h^2 + k^2 + l^2) = \frac{\lambda^2}{4a^2} N \tag{28.3}$$

where

$$N = h^2 + k^2 + l^2$$

There are three types of cubic lattice: primitive cubic, face-centered cubic, and body-centered cubic (see Fig. 28.2). In a primitive cubic lattice, X-ray reflections can be obtained from all planes. In a face-centered cubic lattice,

Table 28.2 Miller Indices of Cubic Crystals for Which Reflections Are Possible

N	Primitive Cubic	Face-Centered Cubic	Body-Centered Cubic
1	100		
2	110		110
3	111	111	
4	200	200	200
5	210		
6	211		211
7			
8	220	220	220
9	300, 221		
10	310		310
11	311	311	
12	222	222	222
13	320		
14	321		321
15			
16	400	400	400
17	410, 322		
18	411, 330		411, 330
19	331	331	
20	420	420	420
21	421		
22	332		332
23			
24	422	422	422
25	500, 430		

reflections are obtained only from planes for which h, k, and l are all even or all odd. In a body-centered cubic lattice, reflections are obtained only from planes for which the sum $h + k + l$ is even. The planes from which reflections are possible, and the corresponding values of N, for the three types of cubic crystals are tabulated in Table 28.2. There are certain values of N (7, 15, 23, etc.) for which reflections are not obtained in any of the cubic lattices. These are the integers that cannot be expressed as the sum of three integers squared.

OBTAINING THE POWDER PATTERN
AND THE VALUES OF θ

The crystalline material is thoroughly ground to a fine, free-flowing powder with a mortar and pestle. This powder is then loaded into a thin-walled glass capillary tube about 0.5 mm in diameter in much the same way that an ordinary melting-point capillary is filled, except that particular care must be exercised not to break the X-ray capillary tube. It is only necessary to obtain a column of powder about 0.5 cm long in the capillary. The section of capillary containing the sample is then mounted at the center of a cylindrical X-ray camera, as shown in Fig. 28.3. A narrow sheet of X-ray film is placed on the inner wall of the camera (coaxial with the sample); the sample capillary is rotated about its axis, and a beam of X-rays is directed at the sample. The beam of X-rays is reflected by the sample into a large number of conical beams, with the sample at the common apex of all the cones. Each cone of reflected X-rays hits the film at two places; thus, each diffraction line has a mate an equal distance opposite the unreflected beam position. After a suitable exposure time, the film is removed from the camera, developed,

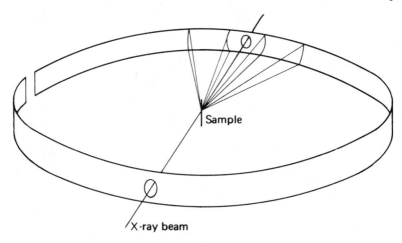

Sample

X-ray beam

FIG. 28.3.

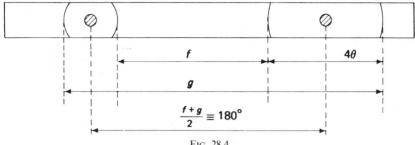

FIG. 28.4.

fixed, and dried. When flattened out, the film looks somewhat like the drawing in Fig. 28.4.

If the X-ray beam passed through the exact centers of the film holes, the distance between hole centers would correspond to an angle of 180°. By measuring this distance, one could thereby obtain a calibration factor in degrees per millimeter for the film. Because of variations in film dimensions, it is best to determine this calibration factor by taking the average of the two extreme distances of two pairs of lines, as shown in Fig. 28.4.

The distance between a corresponding pair of lines is a measure of 4θ for that particular reflection. Thus, by using the film calibration factor it is an easy matter to convert the distances between pairs of lines to values of θ.

The characteristic features of a diffraction pattern are the observed reflection angles, θ, and the intensities of the corresponding lines. To present these data in the customary manner (so that they may be compared with data for other powders) and to facilitate the indexing of the pattern, a table is constructed with the following headings: relative intensity of line; film distance (in millimeters); θ; $\sin^2 \theta$; and d. The relative intensity of a line is usually estimated visually, using a scale from 0 to 100. The film distance and θ are determined as previously discussed, and $\sin^2 \theta$ is obtained using tables found in various books. The d value is calculated from the relation $d = \lambda/2 \sin \theta$. Entries are put in the table for every pair of lines in the powder pattern.

Relative intensities and d values for the powder patterns of thousands of compounds are tabulated on file cards by the American Society for Testing Materials. These tabulations are very useful for identifying the powder patterns of materials for which the data have been tabulated.

INDEXING POWDER PATTERNS FOR CUBIC CRYSTALS

The *indexing* of a powder pattern is the assignment of Miller indices to the various lines of the pattern—a simple procedure for cubic crystals. For tetragonal, hexagonal, and rhombohedral crystals, the method is slightly troublesome and is outlined in various texts. However, triclinic, monoclinic, and orthorhombic crystals are difficult to index, and such crystals are usually indexed from single-crystal diffraction patterns.

A crystalline solid having a cubic space lattice may be readily identified if it is transparent and not too highly colored. Crystals of the material are placed on a microscope slide, wet with a nonvolatile liquid in which the crystals are insoluble (mineral oil can often be used), covered with a cover glass, and placed on the stage of a microscope having a polarizer and analyzer.

Cubic crystals are optically isotropic, and, no matter what their orientation, between crossed nicol prisms they appear dark, like the field of the microscope. Crystals of all the other crystal systems are optically anisotropic and, when rotated between crossed nicols, appear alternately light and dark, showing extinction at 90° intervals.

Tabular Method

The values of $\sin^2 \theta$ for the lines are tabulated in numerical order, and the differences between adjacent values are calculated. The smallest difference should be the largest common factor of the $\sin^2 \theta$ values. When the $\sin^2 \theta$ values are divided by this common factor, a series of integers, or numbers very close to integers, should be obtained. These integers are compared with the allowed N values for the three types of cubic crystals (Table 28.2) in order to determine the crystal type. In the case of the primitive and face-centered cubic lattices the integers correspond to N values, whereas in the case of body-centered cubic lattices, the integers must be multiplied by two to obtain the N values. The body-centered cubic lattice may be recognized by the occurrence of the integer 7, corresponding to $N = 14$.

Slide Rule Method

The d values for the lines are tabulated in decreasing numerical order. The slider of an ordinary slide rule is removed and replaced in an inverted position so that the A scale (which spans two decades) is in contact with the C scale (which spans one decade). The slider is moved until the various d values (as read on the C scale) are positioned opposite integers (or numbers very close to integers) on the A scale. As before, these integers are compared with the N values in Table 28.2 in order to identify the lattice type.

UNIT CELL VOLUME

After indexing a cubic powder pattern, the lattice parameter, a, may be calculated from the relation $a = dN^{1/2}$ for each d value. All the a values thus calculated should be in close agreement, and a weighted average may be taken as the best value of a for the crystal. If the chemical unit corresponding to the empirical formula of the compound has a molecular weight M, then the number of chemical units per unit cell, n, may be calculated from the relation

$$n = \frac{\rho a^3 N}{M} \tag{28.4}$$

where ρ is the density of the compound and N is Avogadro's number. The number n must be an integer, and thus even a rather approximate value of the density can suffice to determine n if n is small. For noncubic crystals, the quantity a^3 in Eq. (28.4) is replaced by a more general expression for the volume of the unit cell.

As an example of the application of a knowledge of n to a structural problem, we cite the case of the compound T_2SnCl_2 (T is the bidentate ligand tropolone, $C_7H_5O_2{}^-$).[3] From the density and lattice constants, it has been determined that $n = 8$; that is, there are eight T_2SnCl_2 units per unit cell. If the compound had an ionic structure, composed of T_3Sn^+ and $SnCl_6{}^{2-}$ ions, it would be necessary that the unit cell contain an integral number of $(T_3Sn^+)_2SnCl_6{}^{2-}$ units—that is, that the number of T_2SnCl_2 units per unit cell be either 3, 6, 9, or some higher multiple of 3. Such an ionic structure is therefore inconsistent with the observed value of 8 for the number of T_2SnCl_2 units per unit cell.

ISOMORPHISM

Two compounds having analogous formulas (e.g., the pairs of compounds NaF/CaO and CaF_2/ThO_2) and related crystal structures are said to be isomorphous. Isomorphous compounds give similar X-ray powder patterns; therefore, if a compound can be shown to have a powder pattern like that of another compound of known structure, the structure of the first compound is probably the same. For example, Donoghue and Drago[4] found that the powder pattern of $Ni[OP(NMe_2)_3]_4(ClO_4)_2$ is the same as that of the corresponding zinc compound. It was argued that because the $OP(NMe_2)_3$ groups are undoubtedly tetrahedrally grouped around the zinc atom in the zinc compound, the same sort of structure exists in the nickel compound. This technique must be used with caution, however. Thus, the zinc and copper(II) dithiocarbamates have the same unit cell and space group, but there are important differences in structure between these complexes.[5]

[3] E. L. Muetterties and C. M. Wright, *J. Am. Chem. Soc.*, **86**, 5133 (1964).

[4] J. T. Donoghue and R. S. Drago, *Inorg. Chem.*, **1**, 866 (1962).

[5] M. Bonamico *et al.*, *Acta Cryst.*, **19**, 887 (1965).

Problems

1. From the following powder pattern data for four different cubic crystals, calculate the lattice types and the lattice parameters (CuK_α radiation; $\lambda = 1.540$ Å).
   (a) d values (Å): 3.15, 1.93, 1.65, 1.37, 1.253, 1.115, 1.051, 0.966, 0.923, 0.910, 0.864, 0.833.
   (b) d values (Å): 2.23, 1.58, 1.29, 1.12, 1.002, 0.915, 0.848, 0.746, 0.708, 0.675, 0.623.
   (c) $\sin^2 \theta$ values: 0.0398, 0.0796, 0.119, 0.157, 0.198, 0.237, 0.316, 0.358, 0.397, 0.434, 0.477, 0.516, 0.550, 0.635, 0.670, 0.717.
   (d) θ values (degrees): 10.73, 15.32, 16.57, 17.75, 18.90, 19.93, 20.92, 21.90, 22.83, 24.58, 25.47, 26.30, 27.13, 28.03, 28.75.

2. A compound of empirical formula $(CH_3)_2PBH_2$ was shown to form crystals of the orthorhombic system, with $a = 11.1$, $b = 13.1$, and $c = 10.4$. The measured density was 0.98 g cc^{-1}. Now, crystallographically equivalent general positions in orthorhombic crystals occur in sets of 4, 8, 16, or 32. What may be said about the molecular structure of the compound, assuming that all of the molecules are crystallographically equivalent?

References for Further Study

Azároff, L. V., *Elements of X-ray Crystallography*, McGraw-Hill Book Company, New York, 1968.

Azároff, L. V. and M. J. Buerger, *The Powder Method in X-ray Crystallography*, McGraw-Hill Book Company, New York, 1958.

Buerger, M. J., *Crystal Structure Analysis*, John Wiley & Sons, Inc., New York, 1960.

D'Eye, R. W. M. and E. Wait, *X-ray Powder Photography*, Academic Press, Inc., New York, 1960.

Wheatley, P. J., *The Determination of Molecular Structure*, Oxford, 1959.

Other
Structural
Techniques

29

There are several physical tools that can be used to obtain structural information which are not as commonly used as the techniques discussed in the preceding chapters. It is important that synthetic chemists at least be aware of the fundamental principles of these tools, of the kinds of structural information they give, and of their limitations. Therefore, in the following paragraphs these methods are briefly discussed, and references to more detailed discussions are given.

RAMAN SPECTROSCOPY

In Raman spectroscopy, a sample is illuminated by monochromatic light, and the scattered light that emerges at right angles to the incident light is analyzed in a spectrometer. Most of the scattered light corresponds to photons that have undergone elastic collisions with the molecules of the sample; that is, the frequency of the scattered light is the same as that of the incident light. However, some of the photons undergo inelastic collisions in which they lose or gain energy corresponding to vibrational or rotational transitions. Consider a collision in which a vibration is excited from its ground state to its first excited state, with an energy difference corresponding to the frequency v. If the incident Raman light is of frequency v_0, the scattered light has a frequency $v_0 - v$, and the spectral line with this frequency is called a *Stokes line*. Now consider the collision of a photon with a molecule in the first excited vibrational state, in which the energy of frequency v is transferred to the photon. The scattered light has a frequency $v_0 + v$, and the spectral line is called an *anti-Stokes line*. Because there are fewer molecules in the excited vibrational state than in the ground vibrational state, the anti-Stokes lines are much weaker than the corresponding Stokes lines. A typical Raman spectrum is shown in Fig. 29.1.[1]

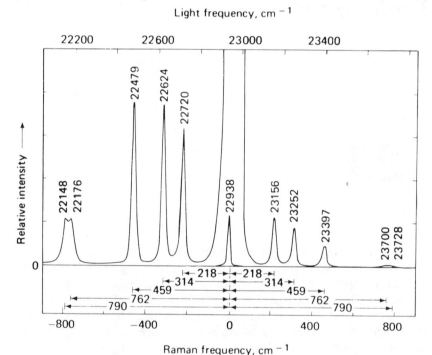

FIG. 29.1. Raman spectrum of CCl_4 excited with the blue line of the mercury arc, 4358Å ($22938 \ cm^{-1}$).[1] (Reproduced with permission of the Journal of Chemical Education.)

For a vibration to be Raman active, it is necessary that there be a change in the polarizability of the molecule during the vibration. This selection rule can be expressed in group-theoretical terms as was done for the infrared selection rule on p. 300. A vibration is Raman active if it belongs to the same

[1] R. S. Tobias, *J. Chem. Educ.*, **44**, 2, 70 (1967).

irreducible representation as one of the components of the polarizability tensor of the molecule. These components transform as x^2, y^2, z^2, xy, xz, yz, or a linear combination of these. Thus, the Raman-active vibrations can be readily identified from the character table corresponding to the molecular point group. Inasmuch as the infrared and Raman selection rules are different, these two forms of spectroscopy are more or less complementary. Indeed, a thorough study of the vibrational structure of a molecule should include both infrared and Raman work. An important corollary of the infrared and Raman selection rules is the *Rule of Mutual Exclusion*: For any molecule with a center of symmetry, the infrared-active vibrations are Raman inactive, and vice versa. This rule is often very valuable in structure determination.

Infrared spectra cannot be obtained over a wide frequency range for aqueous solutions because of the strong absorption by the water molecules. However, good Raman spectra of aqueous solutions can be obtained, and have been used to study the nature of many inorganic aqueous species. Vibrations with frequencies as low as 100 cm^{-1} are readily observed in Raman spectra, whereas relatively sophisticated apparatus is required to observe such low-frequency vibrations in infrared spectra.

A good discussion of inorganic applications of Raman spectroscopy is given by Tobias.[1]

MICROWAVE SPECTROSCOPY

Pure rotational transitions in molecules can be observed by the absorption of energy in the microwave region of the electromagnetic spectrum (about 5000 to 200,000 Mc, or $0.17 \times 10^{-6} - 6.7 \times 10^{-6}$ cm^{-1}). The quantized rotational energy of a molecule can be expressed in terms of the molecule's moments of inertia, and the microwave spectrum corresponds to transitions among these energy levels. Thus, the moments of inertia can be calculated from the spectrum, and, by the application of solid geometry, these yield bond angles and bond distances.

Although extremely accurate structural parameters can be obtained from microwave spectroscopy, there are several limitations of the method. First, the molecule must have a permanent dipole moment for a spectrum to be observable. Second, a vapor pressure of at least 1 μ (10^{-3} mm) must be available. Third, complex, polyatomic molecules usually yield spectra too difficult to interpret. Discussions of inorganic applications of microwave spectroscopy are given by Wilson,[2] Drago,[3] and Kirchhoff.[4]

[2] E. B. Wilson, Jr., *Science*, **162**, 59 (1968).

[3] R. S. Drago, *Physical Methods in Inorganic Chemistry*, Reinhold Publishing Corp., New York, 1965.

[4] W. H. Kirchhoff, *Chem. and Eng. News*, **47**, 88 (Mar. 24, 1969).

NEUTRON DIFFRACTION

A nuclear reactor can serve as the source of an intense beam of thermal neutrons. It can be shown that a beam of such neutrons has a de Broglie wavelength of the order of magnitude of interatomic distances. Therefore, after monochromatization by reflection from a single crystal, neutrons can be used in crystal diffraction studies in much the same way that X-rays are. However, whereas X-rays are scattered primarily by electrons, neutrons are scattered by nuclei and by unpaired electrons. Thus, neutron diffraction permits the location of light atoms, such as hydrogen, in crystals much more precisely than X-ray diffraction, and it is very helpful in elucidating the structure of ferromagnetic and antiferromagnetic compounds.[5,6]

ELECTRON DIFFRACTION

Electron diffraction is based on the wavelike character of a beam of electrons. The wavelength of an electron may be calculated from the de Broglie relation

$$\lambda = \frac{h}{mv}$$

where h is Planck's constant, m is the mass of the electron, and v is its velocity. If a beam of electrons, accelerated to about 40,000 v (corresponding to $\lambda \approx 0.06$ Å), is passed through a stream of gaseous molecules, the electrons are diffracted into a pattern that can be observed by allowing the electrons to fall on a photographic plate to be developed later. By tracing the developed plate with a microphotometer, it is possible to make a plot of scattered electron intensity against a function of the scattering angle. This plot is generally a falling curve with several poorly-resolved maxima and minima, from the positions of which it is possible to calculate interatomic distances. These distances can yield accurate molecular parameters if the shape of the molecule is known and if the molecule is relatively simple. However, there is considerable ambiguity in the interpretation of the data when the molecular shape is unknown, particularly when the symmetry is low and more than about six parameters are required to describe the molecule.[5]

NUCLEAR QUADRUPOLE RESONANCE

A nucleus with an electric quadrupole moment will interact with an inhomogeneous electric field caused by asymmetry in the surrounding electron

[5] P. J. Wheatley, *The Determination of Molecular Structure*, Oxford University Press, Inc., New York, 1959.

[6] W. C. Hamilton and J. A. Ibers, *Hydrogen Bonding in Solids*, W. A. Benjamin, Inc., New York, 1968.

distribution. In nuclear quadrupole resonance, transitions among the allowed quadrupole orientations are observed. These studies must be carried out on solids.

The frequency of a transition is related to the field gradient at the nucleus, and can be correlated with the degree of ionic character in the bond. For example, the nqr frequency for the free chlorine atom (which has a valence *p* electron "hole") is 54.9 Mc. In the covalent molecule C_6H_5Cl, the frequency is 34.6 Mc, and in the relatively ionic $CrCl_3$, 12.8 Mc. Applications of nqr to inorganic chemistry are discussed by Drago[3] and Sillescu.[7]

MÖSSBAUER SPECTROSCOPY

Mössbauer spectroscopy is based on the recoil-less emission and resonance absorption of low-energy γ radiation. A typical Mössbauer spectrometer consists of a recoilfree source that emits γ radiation, a γ-ray detector, and, between these, a sample containing the same radioactive nucleus in a different chemical state. The source or sample is mounted on a velocity drive to permit variation of the γ-ray energy by the Doppler effect. Because the γ-ray transitions have very narrow line widths, interactions between the nucleus and the orbital electrons cause measurable isomer shifts, quadrupole splitting, and magnetic hyperfine interactions. A great deal of chemical information can be derived from the first two effects. The absence of quadrupole splitting indicates that the local point symmetry of the nucleus is cubic or nearly cubic; its absence indicates distortion from cubic symmetry. Consider the ^{57}Fe Mössbauer spectra of Fig. 29.2. In $Na_4[Fe(CN)_6]$, the iron atom is situated in octahedral symmetry and a single-line spectrum is obtained. In $Na_2[Fe(CN)_5NO]\cdot 2 H_2O$, the distorted octahedral configuration causes the line to split into a well-resolved doublet.

Brief discussions of chemical applications of Mössbauer spectroscopy are given by Danon,[7] Drago,[3] Duncan and Golding,[8] and Fluck.[9]

X-RAY ABSORPTION SPECTROSCOPY AND PHOTOELECTRON SPECTROSCOPY

Various spectroscopic techniques are associated with electronic transitions of the inner shells of atoms. Two of these techniques are of particular interest to chemists because they yield structural information about compounds: X-ray absorption spectroscopy and photoelectron spectroscopy.

[7] H. A. O. Hill and P. Day, editors, *Physical Methods in Advanced Inorganic Chemistry*, Interscience Publishers, New York, 1968.

[8] J. F. Duncan and R. M. Golding, *Quart. Rev. London*, **19**, 36 (1965).

[9] E. Fluck, *Adv. Inorg. Chem. Radiochem.*, **6**, 433 (1964).

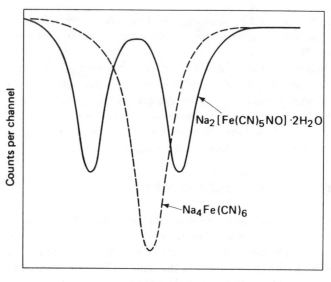

FIG. 29.2. Mössbauer spectra of an octahedrally symmetric complex $Na_4[Fe(CN)_6]$ and of a distorted octahedral complex $Na_2[Fe(CN)_5NO] \cdot 2H_2O$.

In X-ray absorption spectroscopy,[10] a beam consisting of a continuous spectrum of X-rays is passed through a thin sample, and the intensity of the transmitted X-rays is studied with an X-ray spectrometer. In this process, a bound electron can be either removed completely from the atom or promoted to one of the optical excitation levels. A series of overlapping absorption lines are formed, and most studies have involved the determination of the energy of the low-energy side of the first absorption line. These X-ray absorption edge energies have been correlated with, among other things, the oxidation state of the atom.

Photoelectron spectroscopy is a more promising technique, in which a sample is irradiated with a monochromatic beam of vacuum ultraviolet light or X-rays, and the energy spectrum of the ejected electrons is obtained. The sample can be either a gas[11] or a solid.[12] Examples of nitrogen 1s-electron spectra for two nitrogen compounds are given in Fig. 29.3.[13] The electrons from nitrogen atoms that are in different chemical environments have different

[10] C. Bonnelle, in ref. 7.

[11] D. W. Turner, *Chem. Britain*, 435 (1968) (see also ref. 7).

[12] K. Siegbahn, "ESCA. Atomic, Molecular, and Solid-State Structure Studied by Means of Electron Spectroscopy," *Nova Acta Reg. Soc. Scient. Uppsal., Ser. IV*, **20** (1967).

[13] J. M. Hollander, D. Hendrickson, and W. L. Jolly, unpublished work (1968); see *J. Chem. Phys.*, **49**, 3315 (1968).

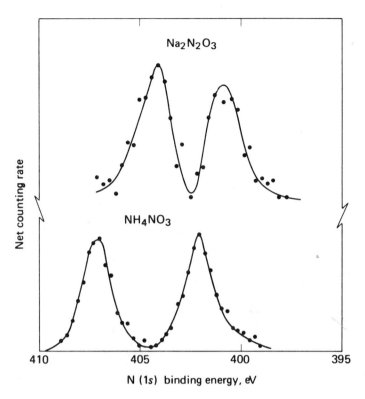

FIG. 29.3. Nitrogen 1s-electron spectra from two nitrogen compounds. Two lines in the lower spectrum arise because the two nitrogens in NH_4NO_3 are in different oxidation states, $+5$ and -3. Two lines in the $Na_2N_2O_3$ spectrum establish the inequivalence of these two nitrogens. (Reproduced with permission from *Science*, **161**, 745 (1968).)

binding energies. Consequently, two peaks are observed in the spectrum for ammonium nitrate. The two peaks in the $Na_2N_2O_3$ spectrum establish the nonequivalence of the nitrogen atoms in $N_2O_3^{2-}$.

SYNTHESES

Part
IV

The Literature of Synthetic Inorganic Chemistry

30

Before carrying out any synthetic work, you should survey the literature. The thoroughness of the literature survey depends upon the scope of the work. If you plan to embark on a systematic study of a class of compounds, to explore a new field of chemistry, or to develop a new method for preparing a compound, the literature survey should be extremely thorough. On the other hand, if you plan to prepare a compound that is fairly well characterized, the literature survey may be cursory—amounting to an hour or two in the library. We shall discuss the general procedure for making a cursory survey of the latter type.

If you are not familiar with one of the elements in the compound to be synthesized, your first step should be to read about the chemistry of that element in some reference book, such as one of those in Table 30.1. This reading should provide you with enough background so that you can understand other literature regarding the element and make modifications or improvements of synthetic methods that you will find in the literature.

Your next step should be to seek several synthetic methods, or "recipes," for the compound in question. There are several places in which you may search.

Table 30.1 Sources of Descriptive Material on Inorganic Chemistry*

F. Basolo and R. G. Pearson, *Mechanisms of Inorganic Reactions*, 2nd ed., John Wiley & Sons, Inc., New York, 1967.

F. Basolo and R. C. Johnson, *Coordination Chemistry*, W. A. Benjamin, Inc., New York, 1964.

G. E. Coates, M. L. H. Green, and K. Wade, *Organometallic Compounds*, vols. 1–2, Methuen & Co., Ltd., London, 1967.

F. A. Cotton and G. Wilkinson, *Advanced Inorganic Chemistry*, 2nd ed., Interscience Publishers, New York, 1966.

B. E. Douglas and D. H. McDaniel, *Concepts and Models of Inorganic Chemistry*, Blaisdell Publishing Co., New York, 1965.

H. J. Emeléus and J. S. Anderson, *Modern Aspects of Inorganic Chemistry*, 3rd ed., D. Van Nostrand Co., Inc., Princeton, N. J., 1960.

R. B. Heslop and P. L. Robinson, *Inorganic Chemistry. A Guide to Advanced Study*, 2nd ed., Elsevier, New York, 1963.

W. L. Jolly, *The Chemistry of the Non-Metals*, Prentice-Hall, Inc., Englewood Cliffs, N. J., 1966.

M. M. Jones, *Elementary Coordination Chemistry*, Prentice-Hall, Inc., Englewood Cliffs, N. J., 1964.

W. M. Latimer, *Oxidation Potentials*, 2nd ed., Prentice-Hall, Inc., Englewood Cliffs, N. J., 1952.

R. K. Murmann, *Inorganic Complex Compounds*, Reinhold Publishing Corp., New York, 1964.

L. E. Orgel, *Transition-Metal Chemistry*, 2nd ed., John Wiley & Sons, Inc., New York, 1966.

R. A. Plane and R. E. Hester, *Elements of Inorganic Chemistry*, W. A. Benjamin, Inc., New York, 1965.

C. S. G. Phillips and R. J. P. Williams, *Inorganic Chemistry*, vols. 1–2, Oxford University Press, Inc., New York, 1965.

E. G. Rochow, *Organometallic Chemistry*, Reinhold Publishing Corp., New York, 1964.

E. G. Rochow, *The Metalloids*, D. C. Heath & Company, Boston, 1966.

R. T. Sanderson, *Inorganic Chemistry*, Reinhold Publishing Corp., New York, 1967.

A. F. Wells, *Structural Inorganic Chemistry*, 3rd ed., Oxford University Press, Inc., New York, 1962.

* Books on very specialized areas of inorganic chemistry have been omitted.

1. *Inorganic Syntheses* (McGraw-Hill Book Company). The subject and formula indexes of volume 10 cover volumes 1 through 10, and the indexes of subsequent volumes are cumulative back to volume 11. This series contains reliable procedures for the preparation of inorganic and organometallic

compounds that were of current research interest at the time of publication of the volumes. Each synthesis was checked in an independent laboratory.

2. *Handbook of Preparative Inorganic Chemistry* (G. Brauer, ed., 2nd ed., vols 1–2, Academic Press Inc., New York, 1963). The syntheses in these volumes are not as reliable as those in *Inorganic Syntheses*.

3. *Preparative Inorganic Reactions* (Interscience Publishers, division of John Wiley & Sons, Inc.). In this series of volumes, the rationale and theory behind the preparation of various classes of compounds are discussed. Literature synthetic procedures are critically evaluated, and specific recommendations for syntheses are made.

4. **Gmelin's** *Handbuch der Anorganischen Chemie*, **Pascal's** *Nouveau Traité de Chimie Minérale*, or **Mellor's** *Comprehensive Treatise on Inorganic and Theoretical Chemistry*. You will seldom find any detailed recipes in these treatises, but you will find literature references to synthetic methods. These references are not listed in Table 30.1 because they are too encyclopedic for a preliminary, general study of an element. However, they are very comprehensive and are useful in the initial stages of a literature survey.

5. *Chemical Abstracts*. There exist both subject and formula indexes, which appear both annually and decennially. There is no need to search in the annual index for any year covered by a decennial index. The abstracts in *Chemical Abstracts* seldom include detailed synthetic procedures, but they can help to decide whether it would be profitable to look up the original articles.

6. If all the above methods fail, it is likely that the compound was discovered only in the last two or three years. If you know the journal in which it is likely that articles on the compound have appeared, it is sometimes fruitful to look through the **recent annual indexes** of that journal. Most articles on inorganic chemistry appear in the following journals:

Inorganic Chemistry

Journal of the American Chemical Society

Journal of the Chemical Society (London), Part A

Chemical Communications

Journal of Inorganic and Nuclear Chemistry

Inorganic and Nuclear Chemistry Letters

Zeitschrift für Anorganische und Allgemeine Chemie

Russian Journal of Inorganic Chemistry (English translation of *Zhurnal Neorganicheskoi Khimii*)

Journal of Organometallic Chemistry

7. You may be fortunate enough to know **somebody who is working with compounds like the one in which you are interested.** Ask him about the compound; he may be able to save you a great deal of time.

A number of textbooks, listed in Table 30.2, contain directions for the preparation of miscellaneous inorganic compounds. Students who are casting about for syntheses to use as laboratory exercises may find these textbooks useful.

Table 30.2 Inorganic Chemistry Texts Containing Synthetic Procedures

D. M. Adams and J. B. Raynor, *Advanced Practical Inorganic Chemistry*, John Wiley & Sons, Inc., New York, 1965.

R. J. Angelici, *Synthesis and Technique in Inorganic Chemistry*, W. B. Saunders Co., Philadelphia, 1969.

R. E. Dodd and P. L. Robinson, *Experimental Inorganic Chemistry*, Elsevier, New York, 1954.

A. King and A. J. E. Welch, *Inorganic Preparations, A Systematic Course of Experiments*, 2nd ed., George Allen & Unwin, London, 1950.

W. G. Palmer, *Experimental Inorganic Chemistry*, Cambridge University Press, London, 1954.

G. Pass and H. Sutcliffe, *Practical Inorganic Chemistry*, Chapman & Hall, Ltd., London, 1968.

G. G. Schlessinger, *Inorganic Laboratory Preparations*, Chemical Publishing Co., Inc., New York, 1962.

H. F. Walton, *Inorganic Preparations*, Prentice-Hall, Inc., Englewood Cliffs, N. J., 1948.

Techniques

Brauer's *Handbook of Preparative Inorganic Chemistry* and Dodd and Robinson's *Experimental Inorganic Chemistry* contain discussions of experimental techniques. More detailed material can be found in the two series of volumes *Technique of Inorganic Chemistry* and *Technique of Organic Chemistry* (edited by H. Jonassen and A. Weissberger, and A. Weissberger, respectively, Interscience Publishers, New York); these series are recommended as supplements to the material in this book.

A Course
Outline

31

The First Laboratory Period

After you have received your locker assignment, your first job is to check carefully the contents of the locker against the list of locker equipment. Missing items should be replaced, if possible, from the storeroom. During the remainder of the first laboratory period, you are to prepare a sample of chromium(II) acetate, following the directions of Synthesis 1 (p. 442). This synthesis will serve as a good test of your technique. If the product is pure and is stored in a tightly stoppered container, it will remain red indefinitely; otherwise, it will probably turn greenish-gray anywhere from a few minutes to a few weeks after its preparation. If time is available during the first laboratory period, you may practice the glass-blowing assignment to be described. During the remaining laboratory periods, you will be expected to complete several assignments, which are described.

Five Required Syntheses

Between the first laboratory period and the last three weeks of the term, you will be expected to prepare five compounds of your own choice. These preparations must be chosen with care because a number of requirements must be fulfilled. These requirements are as follows.

1. Three of the preparations (in addition to the chromium(II) acetate synthesis) must be taken from Chap. 32. Within a week or two after completing each of these preparations, you should give the instructor a report on your work. You are encouraged to suggest improvements in the procedure and to devise new methods for characterization.

2. Two of the preparations must be taken from the literature. Possible literature sources include the various journals and the *Inorganic Syntheses* series described in Chap. 30.

3. Each of the synthesized compounds must be of a different type. In other words, you should try to become familiar with a variety of inorganic compounds, and should not prepare, for example, two volatile halides or two organo-metallic compounds (except, of course, when one of these is an intermediate in the preparation of the other, or when the synthesis yields more than one product).

4. At least seven of the techniques or pieces of apparatus listed in Table 31.1 must be used in carrying out the preparations. Each preparation must be approved by the instructor, who will forbid preparations that are too dangerous, too difficult, or trivial.

5. Each product either must have its purity determined by a quantitative analysis (preferably volumetric) or must be characterized by one or two physical chemical methods. The text by A. I. Vogel (*A Textbook of Quantitative Inorganic Analysis Including Elementary Instrumental Analysis*, 3rd ed., John Wiley & Sons, Inc., New York, 1961) is a useful source of quantitative analytical methods. At least five of the techniques or pieces of apparatus listed in Table 31.2 must be used in the characterization of the five compounds. It is quite acceptable for two or more characterization techniques to be applied to one product.

In carrying out the preparations, avoid slavish adherence to published procedures. There are very few synthetic procedures that cannot be improved with respect to one or more of the following points: simplicity of procedure; purity of product; percentage yield of product; and cost of product.

Gas-Handling Equipment

One of the following experiments must be performed.

(a) Gas chromatographic analysis of a mixture of N_2 and O_2, according to the procedure outlined on pp. 188 to 190.

Table 31.1   Apparatus and Techniques Used in the Syntheses of Chapter 32*

Maintaining controlled atmosphere in reaction vessel near atmospheric pressure (1, 2, 7, 8, 12, 16, 18, 21, 22, 27–29, 31, 33, 36)

Nonaqueous solvent, b.p. > 25° (9, 11, 15–22, 24–31)

Nonaqueous solvent, b.p. < 25° (2, 33)

Vacuum-line manipulation (25–27, 37, 38)

Electrolysis (3–5)

Muffle furnace (23, 34, 35)

Tube furnace (6–8, 11, 29, 35)

Electric discharge (36–38)

Ion exchange (23, 34)

Autoclave (17)

Dean-Stark water trap (19, 21)

Alkali metal dispersion (18)

Photosynthesis (16, 25)

Inert atmosphere bag or box (2, 29, 33)

Crystal growing (35, 39)

Abderhalden drier (8, 15)

Distillation at atmospheric pressure (7, 22, 24, 30, 31)

Vacuum distillation (21, 30)

Sublimation (6, 8, 11, 17, 22, 33)

Thermal transport reaction (35)

Column chromatography (28)

Liquid-liquid extraction (9, 15, 24)

Continuous liquid-solid extraction (28, 29)

* Numbers refer to syntheses.

Table 31.2   Apparatus and Techniques Applicable to the Characterization of the Compounds of Chapter 32*

pH-meter titration (21, 23, 31)

Toepler pump (4, 12, 13)

Gas chromatography (18, 24)

Nuclear magnetic resonance (10, 13, 14, 18–24)

Infrared spectrometry

  KBr pellet or mull (3, 10–22, 28, 29, 31, 33)

  solution or liquid (7, 16–18, 20, 22, 24, 28–30, 38)

  gas (25–27, 36)

X-ray diffraction (1–4, 6, 8–23, 28–35, 38)

Mass spectrometry (7, 12, 13, 17, 21, 22, 24–27, 36, 38)

Polarimetry (14, 39)

Visible or ultraviolet spectrometry (3, 7–15, 17–22, 28–30, 32, 34, 36)

Magnetic susceptibility (1, 11, 12, 32–34)

Electron spin resonance (11, 32, 37, 39)

Vapor pressure and b.p. determination (7, 21, 24–27, 30)

Melting point determination (6, 7, 17–22, 24–30, 38)

Vapor density determination (7, 21, 24–27)

Cryoscopic molecular weight determination (6, 7, 17, 18, 20–22, 24, 28–30, 38)

* Numbers refer to syntheses.

(b) Determination of the equivalent weight of an acid in liquid ammonia, according to the procedure outlined in Appendix 13.

(c) Identification of the constituents of a mixture of volatile materials, according to the general procedure discussed in Chap. 8. This mixture will contain three components that can be cleanly separated by Toepler pumping through a $-112°$ trap followed by a $-196°$ trap.

The gas chromatographic analysis (a) is the simplest, and the analysis of the volatile mixture (c) is the most difficult, of the gas-handling experiments. This should be taken into account when choosing syntheses. Thus, a student would be expected to analyze the volatile mixture (c) if none of his five syntheses was particularly difficult.

Glass-blowing Assignment

As soon as possible, you should attain sufficient glass-blowing skill so that you can satisfactorily perform the following operations.

1. Make a closed round end on a tube.
2. Join, in a straight line, two tubes of unequal bore.
3. Make a "T" connection.

Directions for these operations are given in Appendix 1, and your instructor will demonstrate them. However, no amount of reading or instruction can replace actual practice with a torch if any dexterity is sought. Several torches will be available for use by the class, and you are advised to spend your spare laboratory time practicing glass blowing.

Special Research Problem

During the last three weeks of the term, you will work on a minor research problem that will involve the preparation of some entirely new compound or a new synthetic method for a known compound. Around the middle of the term, a list of research problems will be posted from which you may choose a problem. The results of this research should be described in a term paper having the style of a typical article from the journal *Inorganic Chemistry*. You should refer to the "Notice to Authors of Papers" that frequently appears in this journal immediately following the Author Index. It is also helpful to read the *Handbook for Authors of Papers in the Journals of the American Chemical Society*, which is available from the society.

The paper should not exceed one thousand words. An extensive review of the literature is not appropriate; the introduction should give only enough background material to show why the work was done. On the other hand, it is important that the experimental work be described in sufficient detail that reproduction of the results is possible. The conclusions and discussion of the work are presented in a separate section.

The manuscript should be typewritten, double spaced. If you wish to keep a copy of the paper, retain a carbon.

LABORATORY NOTEBOOK

A bound notebook, approximately eight by ten inches, should be used. Before proceeding with any preparation, write in it notes from the literature, the expected main and side reactions, the physical properties of the reactants and products, and yield calculations. At this time, show the notebook to the instructor; he may ask questions concerning the experiment and may make certain suggestions regarding the procedure.

During the course of a preparation, all significant observations should be recorded immediately. Drawings should be made of apparatus. At the conclusion of a preparation, the notebook and the prepared compound should be shown to the instructor.

SAFETY IN THE LABORATORY

No laboratory work will be permitted in the absence of the instructor or teaching assistant except with written approval of the instructor.

Safety goggles must be worn whenever you are in the laboratory.

Familiarize yourself with the location and the use of the first aid cabinet and the fire extinguishers. Appendices 9 and 10 (Safety Precautions and First Aid) should be studied.

Synthetic
Procedures

32

1. THE PREPARATION OF CHROMIUM(II) ACETATE HYDRATE

$$2\,Cr^{3+} + Zn \rightarrow 2\,Cr^{2+} + Zn^{2+}$$
$$2\,Cr^{2+} + 4\,OAc^- + 2\,H_2O \rightarrow [Cr(OAc)_2]_2 \cdot 2\,H_2O$$

Total time required: 3 hours

Actual working time: 2 hours

Preliminary study assignment:
   F. A. Cotton and G. Wilkinson, *Advanced Inorganic Chemistry*, 2nd ed., Interscience Publishers, New York, 1966, pp. 820–21.

Reagents required:
   15 g of CrCl$_3$·6 H$_2$O
   12 g of mossy zinc
   90 g of NaOAc·3 H$_2$O
   25 ml of concentrated HCl
   200 ml of ice-cold boiled water
   75 ml of alcohol (95 per cent)
   75 ml of ether

Procedure

Assemble the apparatus shown in Fig. 32.1. A small filter flask is fitted with a two-hole rubber stopper through which passes a dropping funnel and a glass tube that reaches to within 1 mm of the bottom of the flask. The side tube of the flask is connected with a short piece of rubber tubing to a tube that dips about 1 cm beneath the surface of water in a beaker.

   Place the mossy zinc and a solution of the chromium(III) chloride in 50 ml of water in the flask. Slowly add concentrated hydrochloric acid through the dropping funnel until the solution has a clear robin's egg blue color. The acid must not be added too rapidly or the solution may froth over through the side tube. Meanwhile, make a slurry of the sodium acetate in 80 ml of water. While hydrogen is still being rapidly evolved, pinch the rubber tubing so as to force the chromium(II) chloride solution over into the sodium acetate solution that is contained in a 250-ml Erlenmeyer flask. Dip the delivery tube under the sodium acetate solution during the transfer. A red precipitate of chromous acetate forms. Immediately stopper the flask and cool by swirling in running water. *NOTE:* Do not add the chromium solution to the acetate unless the solution is a pure blue. A muddy blue or greenish-blue color indicates the presence of chromium(III), which will cause the formation of difficult-to-filter chromic hydroxide.

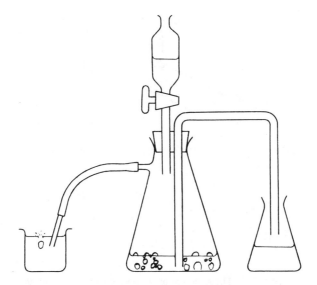

Fig. 32.1. Apparatus for the preparation of chromium(II) acetate hydrate.

Filter the chromous acetate on a sintered-glass funnel and wash it with four 50-ml portions of the cold air-free water. Do not stir up the precipitate with the wash water very much or the filtration may proceed slowly. Wash with the alcohol and then the ether, and finally suck the crystals as dry as possible on the funnel. Be careful not to suck air through the precipitate when it is wet with water or alcohol. Spread the product thinly on watch glasses, and allow it to dry thoroughly at room temperature.[1] Transfer to a test tube with a rubber stopper (tared), and seal tightly. Weigh, and calculate the percentage yield.

Characterization

A sample of chromium(II) acetate properly sealed in a test tube will remain brick red indefinitely. However, if air is admitted to the sample, it will gradually turn to the gray-green color characteristic of the oxidized material. Pure $[Cr(OAc)_2]_2 \cdot 2\,H_2O$ is diamagnetic because of electron-electron interaction between the chromium atoms in the dimeric molecule. Any paramagnetism in a sample is indicative of impurity. The structure of the compound is shown in Fig. 32.2. The chromium atoms essentially maintain a coordination number of six.

[1] Sometimes when the chromium(II) acetate is being dried on the funnel or on a watch glass, a portion of the material turns from the usual brick red to a gray-green color. This change in color will spread throughout the sample if the gray-green material is not isolated from the remaining red material. The color change [undoubtedly an oxidation of the Cr(II) to Cr(III)] is accompanied by a great deal of heat evolution, sometimes sufficient to crack a watch glass.

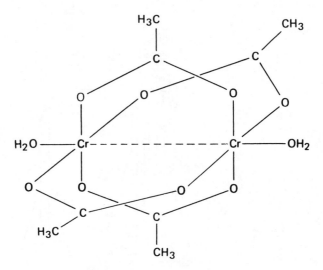

FIG. 32.2. The structure of chromium(II) acetate hydrate.

2. THE PREPARATION OF SODIUM AMIDE

$$Na + NH_3 \longrightarrow NaNH_2 + \tfrac{1}{2}H_2$$

Total time required: 2 to 3 days

Actual working time: 4 hours

Preliminary study assignment:
R. Levine and W. C. Fernelius, *Chem. Revs.*, **54**, 449 (1954), and, in this book, Chap. 7, Manipulation in an Inert Atmosphere, Chap. 5, Solvents, and Appendix 4, Compressed-Gas Cylinders.

Reagents required:
150 ml of liquid ammonia
10-g chunk of sodium
Crystal of $Fe(NO_3)_3 \cdot 9\,H_2O$

Special apparatus required:
Wide-mouthed, unsilvered Dewar flask (300 to 400 ml), equipped with special stopper and addition tube, as shown in Fig. 32.3
U-shaped drying tube with mercury, as shown in Fig. 32.3
Inert atmosphere box or inert atmosphere bag (argon or nitrogen filled)
Knife

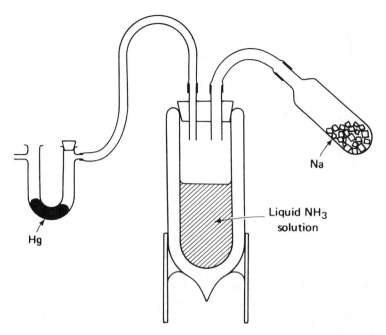

FIG. 32.3. Apparatus for the preparation of $NaNH_2$.

Procedure

In the inert-atmosphere box or inert-atmosphere bag, scrape the chunk of sodium clean of oxide and oil with a knife. Then cut about 5 g into pea-sized pieces and put them into the addition tube. After attaching a short piece of large-diameter tubing to the tube and clamping it closed with a pinch clamp, it may be brought out into the air.

Pass the outlet tube of an ammonia cylinder through the larger opening in the top of the Dewar and flush ammonia gas through the Dewar for a minute or so by barely opening the cylinder valves. Inasmuch as the Dewar is not covered with tape, particular care should be taken to avoid injury from flying glass in case the Dewar should implode. Then open the valves further and collect about 150 ml of liquid ammonia in the Dewar. Close the valves, remove the Dewar from the ammonia cylinder, and place the Dewar in a hood. Add a small crystal (about the size of a small rice kernel) of $Fe(NO_3)_3 \cdot 9$ H_2O to the ammonia. Attach the addition tube by means of the rubber tubing, and connect the smaller opening at the top of the Dewar to the mercury bubbler, which acts as an outlet for vaporized ammonia and prevents air from entering the Dewar when all the ammonia has vaporized. Make sure that all the connections are air tight. Now add the pieces of sodium at a rate such as to maintain a moderate effervescence of the ammonia solution.

About half an hour after adding all the sodium, the blue color of the solution should disappear. The addition tube with its rubber tube may then be removed and a rubber stopper placed in the tube at the top of the Dewar. Let the apparatus stand for a day or two while the ammonia evaporates. Then transfer the apparatus to the inert-atmosphere box or the inert-atmosphere bag, and, with the aid of a long spatula, transfer the sodium amide from the Dewar to a tightly capped bottle.

It is extremely important to prevent the access of air to the sodium amide. Oxygen reacts to form a yellow surface coating of various oxidation products. The partially oxidized material is explosive and can be detonated by friction or heat.

Characterization

Sodium amide reacts rapidly with moisture to give sodium hydroxide and ammonia. Consequently, sodium hydroxide is a common impurity in sodium amide, and a good way to check the purity of a sodium amide sample is to hydrolyze it in a known excess of acid and then to titrate the remaining acid with standard base. Each millimole of sodium amide will consume 2 mmoles of acid, whereas 1 mmole of sodium hydroxide present will consume only 1 mmole of acid:

$$NaNH_2 + 2\,H^+ \longrightarrow Na^+ + NH_4^+$$
$$NaOH + H^+ \longrightarrow Na^+ + H_2O$$

When the product is being transferred from the Dewar to the sample bottle, it may be convenient to transfer a small analytical sample of the product (about 0.05 to 0.08 g) to a previously weighed weighing bottle. The weighing bottle may then be removed from the box and reweighed to determine the weight of sample taken. (If the bottle was originally weighed while containing air, a correction must be made for the different weight of the inert gas.) Quickly add the entire analytical sample (weighing bottle and all) to excess standard 0.1 M acid, and back-titrate the solution with standard 0.1 M base, using methyl red as the indicator.

Questions

1. What is the purpose of the iron(III) nitrate?

2. How many milliliters of 0.1 M HCl are equivalent to 0.08 g of sodium amide?

3. Why cannot phenolphthalein be used as the indicator in the titration?

4. A 0.0766-g sample of sodium amide was added to 50.00 ml of 0.1000 M HCl that had been diluted to about 100 ml. The resulting solution was titrated with 0.1000 M NaOH; exactly 12.08 ml was required to reach the methyl red endpoint. Assuming that NaOH was the only impurity, calculate the percentage of $NaNH_2$ in the sample.

3. THE PREPARATION OF POTASSIUM PEROXYDISULFATE

$$2\,HSO_4^- \longrightarrow S_2O_8^{2-} + 2\,H^+ + 2\,e^-$$

Total time required: 2 days

Actual working time: 3 hours

Preliminary study assignment:
M. Ardon, *Oxygen*, W. A Benjamin, Inc., New York, 1965, pp. 72–75, and, in this book, Chap. 10, Electrolytic Synthesis.

Reagents required:
40 g of $KHSO_4$
About 1500 g of crushed ice
25 ml of 95 per cent alcohol
10 ml of ether

Special apparatus required:
d.c. power supply
Insulated wire for connections
Platinum-wire electrode (1 cm long; 0.64 mm in diameter)
Platinum-foil electrode
Large-diameter test tube (35 mm in diameter; 7 in. tall)

Procedure

Dissolve 40 g of potassium hydrogen sulfate in 90 ml of water, and then cool the solution to 5°. Decant 80 ml of the solution into the large test tube, and assemble the platinum-wire anode and platinum-foil cathode as shown in Fig. 32.4. Surround the test tube with an ice bath in an 800-ml beaker, and pass a current of about 1.5 amp for 2 hours. The ice must be replenished every half-hour or so. Collect the precipitate on a small sintered glass funnel or a sintered glass crucible without using any wash water. Dry the crystals with suction; wash, successively, with alcohol and ether, and dry in the desiccator for a day or two. A yield of about 1.6 g is generally obtained.

Characterization

The material may be analyzed by the following procedure. Dissolve a 0.25-g sample in 30 ml of water, add 4 g of potassium iodide, stopper the flask, and swirl to dissolve the iodide. After allowing the solution to stand for at least 15 min, add 1 ml of glacial acetic acid and titrate the liberated iodine with standard thiosulfate. At least two samples should be analyzed. Calculate

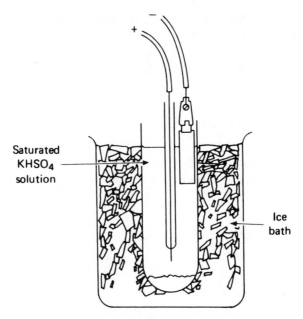

FIG. 32.4. Apparatus for electrolytic preparation of $K_2S_2O_8$.

the current efficiency, taking account of the fact that the product is not 100 per cent pure.

The infrared spectrum[2] of the product can be obtained.

Questions

1. By following the preceding procedure, a student prepared 1.6 g of potassium peroxydisulfate. A 0.2703-g sample of the material required 18.85 ml of 0.1020 N thiosulfate. What was the current efficiency?

2. Why must the solution of peroxydisulfate and iodide be allowed to stand for 15 min before adding acid and titrating?

3. How might one prepare peroxymonosulfuric acid? What are the structures of the peroxymonosulfate and peroxydisulfate ions?

4. THE PREPARATION OF SILVER MONOXIDE[3]

$$Ag^+ + H_2O \longrightarrow AgO + 2H^+ + e^-$$

Total time required: $1\frac{1}{2}$ days

[2] A. Simon and H. Richter, *Naturwiss.,* **44**, 178 (1957).

[3] See A. Noyes, D. de Vault, C. Coryell, and T. Deahl, *J. Am. Chem. Soc.*, **59**, 1326 (1937).

Actual working time: 4 hours

Preliminary study assignment:
J. A. McMillan, *Chem. Revs.*, **62**, 65 (1962); and, in this book, Chap. 10, Electrolytic Synthesis, and Chap. 8, The Vacuum Line.

Reagents required:
35 g of $AgNO_3$
50 ml of concentrated HNO_3
Ice

Special apparatus required:
d.c. power supply
Insulated wire for connections
Platinum-foil electrodes (2×10 cm and 1×8 cm)
Stirring motor and glass stirrer (or magnetic stirrer and stirring bar)
Tube with sintered-glass disk (cathode cup)
Crystallizing dish (7.5 in. in diameter)
Vacuum manifold (for analysis)

Procedure

Approximately 150 ml of 1.5 M HNO_3 containing 35 g of $AgNO_3$ is placed in a 250-ml beaker surrounded by an ice bath. A cathode cup containing about 30 ml of 10 M HNO_3 is partially immersed, and the platinum-foil electrodes and stirrer are assembled as shown in Fig. 32.5. The cathode cup should be tipped slightly to prevent the formation of a gas bubble beneath the sintered glass disk. A current of 3 amp is passed for 2 hours, after which the precipitate of $Ag_7O_8NO_3$ is filtered off on a fritted glass funnel and washed with water. This material is then converted to AgO by boiling in water for 2 hours; the AgO is filtered off and dried in a desiccator at room temperature.

Characterization

The AgO may be analyzed by measuring the oxygen evolved when a sample is heated in vacuo. Weigh out 0.1 g of the material into a glass tube with a ball joint; attach to the vacuum line, and evacuate. Connect the Toepler pump, and heat the AgO with a torch until no more oxygen is pumped by the Toepler pump. (It is best to have a liquid nitrogen trap between the sample and the Toepler pump.) From the amount of oxygen collected and the weight of the sample, the purity may be calculated.

Silver monoxide reacts with iodate in basic solution to form a periodate complex of Ag(III):

$$12 AgO + 10 OH^- + 4 IO_3^- \longrightarrow 5 Ag_2O + 2 Ag(IO_6)_2^{7-} + 5 H_2O$$

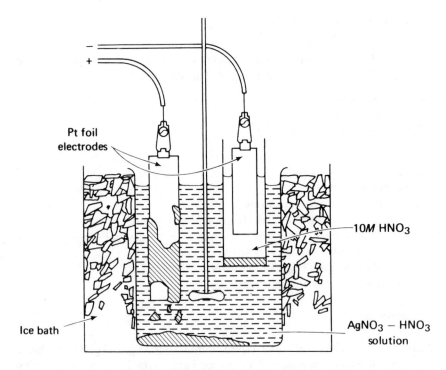

FIG. 32.5. Apparatus for the electrolytic preparation of AgO.

The salt $Na_5H_2Ag(IO_6)_2 \cdot 16 H_2O$ may be precipitated by subsequent treatment with sodium hydroxide.[4]

Questions

1. What is the reaction that takes place when silver monoxide is heated in vacuo?

2. A student obtained 55.0 ml of oxygen (at 158 mm pressure and 25°) from the decomposition of a 0.1239-g sample of silver monoxide. What was the empirical formula of the sample, on the assumption that the only impurity was silver(I) oxide?

3. Describe the structure of AgO.

5. *ELECTRODEPOSITION OF NICKEL-TIN ALLOY*

$$NiSnF_x^{4-x} + 4 e^- \longrightarrow NiSn + xF^-$$

Total time required: 4 to 5 hours

[4] G. L. Cohen and G. Atkinson, *Inorg. Chem.*, **3**, 1741 (1964).

Actual working time: 3 to 4 hours

Preliminary reading assignment:
N. Parkinson, *J. Electrochem. Soc.*, **100**, 107 (1953), and, in this book, Chap. 10, Electrolytic Synthesis.

Reagents required:
3 or 4 strips of copper, approximately 2.5 cm × 10 cm × 1 mm
75 g of $NiCl_2 \cdot 6 H_2O$
12.5 g of $SnCl_2 \cdot 2H_2O$
7 g of NaF
9 g of NH_4HF_2

Special equipment required:
d.c. power supply
Fine steel wool and emery paper
Magnetic stirrer-hot plate and plastic-enclosed stirring bar
Polyethylene beaker, 300 ml
Carbon rod
Large flat-bottomed crystallizing dish

Procedure

The copper strips are cleaned with the steel wool, polished with the emery paper, and then rubbed free of emery with a clean cloth.

The nickel(II) chloride, tin(II) chloride, and sodium fluoride are dissolved in 200 ml of water, and the solution is diluted to 250 ml. The solution is transferred to the polyethylene beaker, and then the ammonium bifluoride is dissolved while the mixture is stirred magnetically. The polyethylene beaker is placed in the crystallizing dish on the magnetic stirrer-hot plate; water is added to the crystallizing dish to a level equal to that of the electrolytic solution, and the hot plate is adjusted to maintain the temperature of the stirred electrolytic bath at 65 ± 3°. The carbon rod anode and one of the copper strip cathodes (previously weighed to ±0.5 mg) are immersed in the bath to a depth of approximately 6 cm such that the electrodes are about 6 cm apart. A d.c. potential is applied across the electrodes; the voltage is adjusted to obtain a cathode current density of 0.016 amp cm^{-2} (counting both sides of the cathode strip). The electrolysis is continued for 30 min; the copper strip should then have a bright, adherent coating of NiSn alloy. The strip is rinsed, dried, and weighed; the current efficiency is calculated. Using the other copper strips, the effects on the nature of the deposit of changing either the current density, the thickness of the deposit, or the temperature can be determined.

Characterization

The shininess, smoothness, and adherence of the plated alloy can be evaluated by visual inspection. The alloy can be analyzed for nickel and tin by X-ray emission spectrography.[5]

6. THE PREPARATION OF ALUMINUM(III) IODIDE

$$2 Al + 3 I_2 \longrightarrow Al_2I_6$$

Total time required: 7 hours

Actual working time: 7 hours

Preliminary study assignment:
F. A. Cotton and G. Wilkinson, *Advanced Inorganic Chemistry*, 2nd ed., Interscience Publishers, New York, 1966, pp. 440–41, and, in this book, Chap. 11, High-Temperature Processes.

Reagents required:
5 g of aluminum turnings
10 g of iodine
20 ml of acetone

Special apparatus required:
75 cm of 22-mm tubing and 15 cm of 8-mm tubing
Tube furnace and Variac
Thermocouple and potentiometer or pyrometer
Vacuum pump

Procedure

Clean the aluminum turnings with acetone, and then allow them to dry thoroughly in the air. Fashion the glass tubes into the shape shown in Fig. 32.6. Place the aluminum and iodine near the middle of the main tube and seal off the large end. Attach the 8-mm tube to a vacuum pump (protected with an intervening cold trap or tube containing NaOH pellets) and evacuate the apparatus to 0.5 mm or lower. Seal off the apparatus at a point some 10 cm from the collection bulb and place it in the tube furnace as shown, with the aluminum in the center of the furnace. Raise the furnace temperature to 500° and repeatedly sublime the iodine from one end of the tube to the other by means of a Fisher burner. When the iodine has completely reacted, remove the tube from the furnace and sublime all the Al_2I_6 into the collection

[5] C. Bonnelle, *Physical Methods in Advanced Inorganic Chemistry*, H. A. O. Hill and P. Day, eds., Interscience Publishers, New York, 1968, Chap. 2.

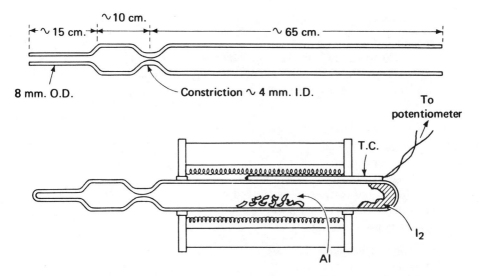

FIG. 32.6. Apparatus for the preparation of Al_2I_6.

tube (take care that the constriction does not become plugged). Seal off the collection tube and discard the main reaction tube.

Characterization

A portion of the Al_2I_6 may be sublimed into the narrow tube projecting from the collection tube, and a 5-cm length of this narrow tubing may be sealed off. This small sample of material may now be used for a melting-point determination and, if desired, for analysis. The reported melting point of aluminum iodide is 191°; the boiling point is 386°.

Questions

1. What is the structure of the Al_2I_6 molecule?

2. The cryoscopic molecular weights of aluminum(III) halides in benzene correspond to the dimers (Al_2X_6). However, the molecular weights in ether correspond to the monomers (AlX_3). Explain.

7. THE PREPARATION OF TITANIUM TETRACHLORIDE[6]

$$TiO_2 + 2C + 4Cl_2 \longrightarrow TiCl_4 + 2COCl_2$$

Total time required: 2 or 3 days

[6] See H. F. Walton, *Inorganic Preparations*, Prentice-Hall, Inc., Englewood Cliffs, N. J., 1948, pp. 107–109.

Actual working time: 6 hours

Preliminary study assignment:
    F. A. Cotton and G. Wilkinson, *Advanced Inorganic Chemistry*, 2nd ed.,
    Interscience Publishers, New York, 1966, p. 801, and, in this book, Chap.
    11, High-Temperature Processes.

Reagents required:
    24 g of TiO_2 powder
    8 g of carbon black
    Chlorine cylinder
    Nitrogen cylinder
    About 10 g of cottonseed or linseed oil
    Fibrous asbestos or glass wool
    A few copper turnings

Special apparatus required:
    Glass reaction tube
    Tube furnace and Variac
    Thermocouple and potentiometer or pyrometer
    8 cm × 8 cm iron sheet
    100-ml distilling flask

Procedure

Intimately mix the titanium dioxide and the carbon black in an iron crucible,
and add just enough cottonseed or linseed oil to make a stiff paste. After
thoroughly mixing these materials, cover the crucible with the flat piece of
sheet iron and slowly heat, in the hood, with the Fisher burner. Gradually
increase the temperature until no more fumes are evolved and the crucible is
red hot. Allow the crucible to cool while covered and then break up the porous
mass into pea-sized pieces. Discard any fine powder. Pack these pieces into
the Pyrex tube, using plugs of fibrous asbestos or glass wool to hold the
material in place. Attach a one-hole stopper with a glass tube to the large
end of the reaction tube and pass nitrogen through the tube, while heating it
to about 500°. When all the moisture has been driven from the mixture of
TiO_2 and carbon, stop the heating and allow the tube to cool while still
passing the nitrogen. If 2 hours of working time are not then available,
tightly stopper both ends of the tube to protect the reactants from moisture.
    Set up the apparatus in a hood as shown in Fig. 32.7. Pass chlorine at a
rate such that about three to four bubbles per second form in the sulfuric acid
bubbler. Hold the temperature of the furnace at 550° (no higher, or the glass
may melt). When it appears that no more titanium tetrachloride is condensing in the receiving flask, stop the flow of chlorine and turn off the furnace.

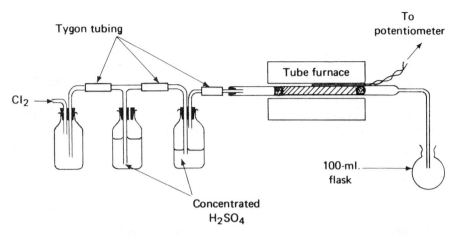

FIG. 32.7. Apparatus for the preparation of $TiCl_4$.

Add a few copper turnings to the liquid, stopper the flask,[7] and let it stand for a day or two. Then attach a distilling head and condenser to the flask; distill the liquid (b.p. = 136.4°), and collect and store the distillate in a glass-stoppered bottle.[7] A yield of about 30 g of product is usually obtained.

Characterization

Titanium tetrachloride is the starting material for a wide variety of titanium compounds. Dichlorobis(acetylacetonato)titanium(IV) is formed by the reaction of $TiCl_4$ with acetylacetone.[8]

$$TiCl_4 + 2 C_5H_8O_2 \longrightarrow Ti(C_5H_6O_2)_2Cl_2 + 2 HCl$$

Titanium tetraethoxide is formed by the ethanolysis of $TiCl_4$.[9]

$$TiCl_4 + 4 C_2H_5OH + 4 NH_3 \longrightarrow Ti(OC_2H_5)_4 + NH_4Cl$$

Titanium(III) chloride can be prepared by the reduction of $TiCl_4$ with hydrogen at high temperatures.[10]

$$TiCl_4 + \tfrac{1}{2} H_2 \longrightarrow TiCl_3 + HCl$$

Titanium tetrabromide is prepared by the reaction between $TiCl_4$ and hydrogen bromide.[11]

$$TiCl_4 + 4 HBr \longrightarrow TiBr_4 + 4 HCl$$

[7] Seal the stopper with a little Kel–F #90 grease.
[8] V. Doron, *Inorg. Syn.*, **7**, 50 (1963); *Inorg. Syn.*, **8**, 37 (1966).
[9] D. C. Bradley, R. Gaze, and W. Wardlaw, *J. Chem. Soc.*, 721 (1955).
[10] W. L. Groeneveld, G. P. M. Leger, J. Wolters, and R. Waterman, *Inorg. Syn.*, **7**, 45 (1963).
[11] R. B. Johannesen and C. L. Gordon, *Inorg. Syn.*, **9**, 46 (1967).

Dichlorobis(cyclopentadienyl)titanium(IV) is formed by the reaction of an alkali metal cyclopentadienyl with $TiCl_4$.[12]

$$2 NaC_5H_5 + TiCl_4 \longrightarrow (C_5H_5)_2TiCl_2$$

Ammonium and potassium hexachlorotitanates can be precipitated from solutions of $TiCl_4$ in aqueous hydrochloric acid.[13]

Solutions of $TiCl_4$ in aromatic solvents, such as benzene, toluene, xylene, diphenyl ether, anisole, and phenetole, are colored, probably because of charge-transfer interaction.[14]

Questions

1. Why is the crude titanium tetrachloride stored over copper metal for a day or two?

2. Name three other volatile chlorides that may be made by a method analogous to that used here for titanium tetrachloride.

3. What are the purposes of each of the bottles between the chlorine cylinder and the reaction tube?

4. Would it be possible to omit the carbon from the reaction mixture and to carry out the following reaction?

$$TiO_2 + 2 Cl_2 \longrightarrow TiCl_4 + O_2$$

8. THE PREPARATION OF MOLYBDENUM(II) CHLORIDE [OCTACHLOROHEXAMOLYBDENUM(II) CHLORIDE]

$$Mo + \tfrac{5}{2} Cl_2 \longrightarrow MoCl_5$$
$$3 MoCl_5 + 2 Mo \longrightarrow 5 MoCl_3$$
$$9 MoCl_3 \longrightarrow 3 MoCl_5 + (Mo_6Cl_8)Cl_4$$

Total time required: 8 hours

Actual working time: 5 hours

Preliminary study assignment:
   J. C. Sheldon, *J. Chem. Soc.*, 1007 (1960); F. A. Cotton and G. Wilkinson, *Advanced Inorganic Chemistry*, 2nd ed., Interscience Publishers, New York, 1966, pp. 949–50; and, in this book, Chap. 11, High-Temperature Processes.

[12] G. Wilkinson and J. M. Birmingham, *J. Am. Chem. Soc.*, **76**, 4281 (1954).
[13] G. W. A. Fowles and D. Nicholls, *J. Inorg. Nucl. Chem.*, **18**, 130 (1961).
[14] J. B. Ott, J. R. Goates, R. J. Jensen, and N. F. Mangelson, *J. Inorg. Nucl. Chem.*, **27**, 2005 (1965).

Reagents required:
  8 g of powdered molybdenum metal
  Chlorine cylinder
  Nitrogen cylinder

Special apparatus required:
  Tube furnace
  Vycor or silica tube, 20 to 25 mm O.D., 100 cm long
  Thermocouple and pyrometer
  Porcelain boat
  Vacuum pump with cold trap
  Abderhalden drier

Procedure

The apparatus shown in Fig. 32.8 is assembled. The molybdenum powder is placed in the boat, and the boat is placed in the middle of the Vycor tube, centered in the tube furnace. The rubber stoppers at each end of the Vycor tube are coated on the inside with paraffin wax or Kel–F grease.

A stream of chlorine gas is passed through the tube at the rate of about four bubbles per second. After several minutes, the tube furnace is turned on, and the temperature is adjusted to about 500° with the aid of a Powermite or Variac. Soon dark green and red-brown crystals will deposit in section *B* of the tube. After about 20 min or when the tube is opaque with crystals to within about 5 cm of the stopper on side *B*, the flow of chlorine is stopped and the Tygon tubing from the bubbler is disconnected from the glass tubing attached to side *A*. Immediately, the delivery tube from the nitrogen bubbler is attached to side *B*, and a stream of nitrogen (about four bubbles per second) is passed through the tube in a direction opposite to the original chlorine flow. When all the chlorine has been flushed from the tube, the furnace temperature is increased to 600–700° and side *B* is heated with the hot flame of a Fisher burner to drive the $MoCl_5$ through the furnace to side *A*. A residue of $MoCl_3$ (brown red) and $(Mo_6Cl_8)Cl_4$ (yellow) in side *B* will probably remain. The nitrogen delivery tube is then switched to side *A*, and the flaming procedure is repeated on side *A* to drive the $MoCl_5$ back to side *B*. The $MoCl_5$ is repeatedly passed over the hot molybdenum in this way until most of it has been reduced to the less volatile $MoCl_3$. Care must be taken not to allow the rubber stoppers to become hot. Finally, with nitrogen entering side *A*, side *A* is heated very strongly with the burner until practically all the material turns yellow. Then, with the nitrogen flowing in the same direction, side *B* is similarly heated, allowing the unreacted $MoCl_5$ to escape out the exit tube of side *B*.

The tube is cooled to room temperature; the stoppers are removed from the ends of the tube, and the boat containing the residual molybdenum is

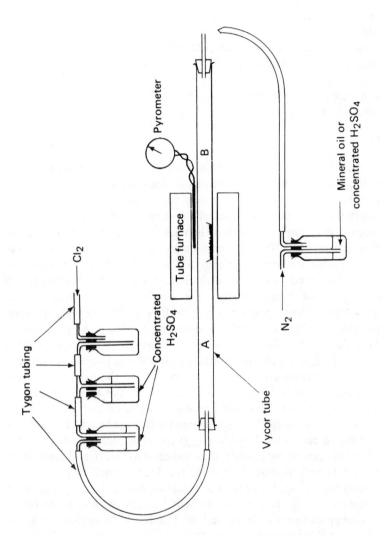

FIG. 32.8. Apparatus for the preparation of $(Mo_6Cl_8)Cl_4$.

pulled out of the tube with a long wire with a hook at the end. Then a solid rubber stopper is attached to one end of the tube, and about 100 ml of warm (50°) 6 M HCl is poured into the tube. The other end of the tube is stoppered, and the aqueous acid is sloshed around the tube to dissolve the yellow $(Mo_6Cl_8)Cl_4$. If all the $(Mo_6Cl_8)Cl_4$ does not dissolve, more aqueous HCl is added. The solution is filtered, evaporated on a hot plate to one-fourth the original volume of the solution, and then cooled in an ice bath. The precipitate of bright yellow $(H_3O)_2[(No_6Cl_8)Cl_6]\cdot 6\ H_2O$ is filtered off, transferred to a boat, and placed in an Abderhalden drying pistol. The pistol is evacuated to less than 1 mm through a liquid-nitrogen-cooled trap and, during pumping, is heated with refluxing nitrobenzene to 211° for 2 hours. Generally, about 8 g of purified $(Mo_6Cl_8)Cl_4$ is obtained.

Characterization

Molybdenum(II) chloride contains $Mo_6Cl_8^{4+}$ groups, having the structure shown in Fig. 32.9. This cluster persists in solutions and in a variety of derivatives, in which as many as six ligands can be coordinated to the six molybdenum atoms in the cluster. Thus, compounds such as $[(Mo_6Cl_8)(OH)_4(H_2O)_2]\cdot 12\ H_2O$, $[(CH_3)_4N]_2[(Mo_6Cl_8)Cl_6]$, and $\{(Mo_6Cl_8)[(CH_3)_2SO]_6\}(ClO_4)_4$ have been prepared.

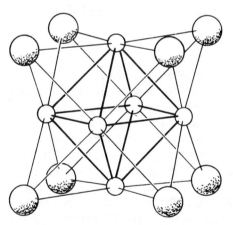

FIG. 32.9. The structure of the complex ion $[Mo_6Cl_8]^{4+}$. (Adapted from L. Pauling, "The Nature of the Chemical Bond," 3rd ed., Cornell Univ. Press, Ithaca, N.Y., 1960.)

Solutions of $Mo_6Cl_8^{4+}$ salts in aqueous acids have an ultraviolet absorption spectrum characterized by a maximum at approximately 300 mμ ($\varepsilon_{max} \approx 3000$) and a shoulder at about 330 mμ.[15]

[15] J. C. Sheldon, *J. Chem. Soc.*, 1007 (1960).

9. THE PREPARATION OF 12-TUNGSTOSILICIC ACID[16]

$$12\,WO_4^{2-} + SiO_3^{2-} + 26\,H^+ \longrightarrow H_4SiW_{12}O_{40}{\cdot}xH_2O + (11 - x)H_2O$$

Total time required: 2 or 3 days

Actual working time: 5 hours

Preliminary study assignment:
   H. J. Eméleus and J. S. Anderson, *Modern Aspects of Inorganic Chemistry*, 3rd ed., D. Van Nostrand Co., Inc., Princeton, N. J., 1960, pp. 325–36, and F. A. Cotton and G. Wilkinson, *Advanced Inorganic Chemistry*, 2nd ed., Interscience Publishers, New York, 1966, pp. 941–46.

Reagents required:
   50 g of $Na_2WO_4{\cdot}2\,H_2O$
   2.7 ml of sodium silicate solution (water glass of density 1.38 or 40° Bé)
   65 ml of concentrated HCl
   50 ml of diethyl ether

Special apparatus required:
   Magnetic stirrer-hot plate and stirring bar

Procedure

Dissolve the sodium tungstate(VI) 2-hydrate in 100 ml of water, and then add the sodium silicate solution. Vigorously stir the solution on the magnetic stirrer-hot plate and hold at incipient boiling while adding 30 ml of concentrated hydrochloric acid drop by drop from a dropping funnel. This operation should take at least 10 min. After filtering and cooling the solution, add 20 ml more of concentrated hydrochloric acid. Then shake the solution in a separatory funnel with 35 ml of ether. (If three liquid phases do not form, it will be necessary to use a little more ether.) Withdraw the bottom layer of oily ether complex into a beaker and discard the other two phases. Rinse out the separatory funnel and return the ether complex to the funnel along with a solution of 12 ml of concentrated hydrochloric acid in 38 ml of water and about 10 ml of ether. After shaking the mixture, run the lower phase into an evaporating dish and allow it to stand in a drafty hood for a day or two. Dry the remaining crystals in an oven at 70° for two hours. Avoid touching the moist crystals with anything metallic or they may turn blue. A yield of 32 g is generally obtained.

[16] This procedure is adapted from one given by E. North, *Inorg. Syn.*, **1**, 129 (1939).

Characterization

Weigh out two 2-g samples of the material and titrate their aqueous solutions with 0.1 M NaOH, using methyl orange as an indicator. The equivalent weight should agree closely with the value 751, corresponding to the tetrabasic acid $H_4SiW_{12}O_{40} \cdot 7 H_2O$.

Questions

1. How many structurally different kinds of oxygen atoms are there in the basic 12-tungstosilicate ion (exclusive of water of hydration)? How many oxygen atoms are there of each kind?

10. THE PREPARATION OF CHLOROPENTA-AMMINE COBALT(III) CHLORIDE, NITROPENTAAMMINECOBALT(III) CHLORIDE, AND NITRITOPENTAAMMINE-COBALT(III) CHLORIDE

Total time required: 2 days

Actual working time: 6 hours

Preliminary study assignment:
    R. B. Penland, T. J. Lane, and J. V. Quagliano, *J. Am. Chem. Soc.*, **78**, 887 (1956), and I. R. Beattie and D. P. N. Satchell, *Trans. Faraday Soc.*, **52**, 1590 (1956).

Reagents required:
    Ammonium carbonate[17]: 50 g (method A only)
    Concentrated aqueous ammonia: 160 ml (method A) or 75 ml (method B)
    $CoCl_2 \cdot 6 H_2O$: 20 g
    NH_4Cl: 5 g (method A) or 10 g (method B)
    Concentrated HCl: 340 ml (method A) or 150 ml (method B)
    95 per cent ethanol: 95 ml (method A) or 75 ml (method B)
    $NaNO_2$: 10 g
    30 per cent H_2O_2: 16 ml (method B only)

Special apparatus required:
    Magnetic stirrer-hot plate

---

[17] Commercial "ammonium carbonate" is really a mixture of ammonium bicarbonate and ammonium carbamate.

Preparation of $[Co(NH_3)_5Cl]Cl_2$ (method A)[18]:

$$Co^{2+} + 4NH_3 + HCO_3^- + \tfrac{1}{4}O_2 \longrightarrow [Co(NH_3)_4CO_3]^+ + \tfrac{1}{2}H_2O$$

$$[Co(NH_3)_4CO_3]^+ + 2H^+ + H_2O \longrightarrow [Co(NH_3)_4(H_2O)_2]^{3+} + CO_2$$

$$[Co(NH_3)_4(H_2O)_2]^{3+} + NH_3 \longrightarrow [Co(NH_3)_5H_2O]^{3+} + H_2O$$

$$[Co(NH_3)_5H_2O]^{3+} + 3Cl^- \longrightarrow [Co(NH_3)_5Cl]Cl_2 + H_2O$$

Prepare a solution of 50 g of ammonium carbonate in 250 ml of water by powdering the carbonate and stirring it with cold water until it has dissolved. Ignore any small amounts of undissolved material. Add 125 ml of concentrated ammonia, followed by a solution of 20 g of cobalt chloride in 50 ml of water. Put the solution in a filter flask fitted with a one-hole rubber stopper through which passes an 8 to 10 mm diameter glass tube reaching to the bottom of the flask. Connect the side arm of the flask to an aspirator (with trap) or house vacuum, and draw air through the solution for about 3 hours. During this time, the red color of the solution will deepen. Add 5 g of ammonium chloride and pour the solution into an evaporating dish. Gently heat the solution until it has evaporated to the first appearance of crystals. Then add just enough 6 *M* HCl to expel all the carbon dioxide. Neutralize with concentrated ammonia and add an additional 15 ml of ammonia. Heat on the steam bath for about 45 min; then add 120 ml of concentrated hydrochloric acid and continue the heating for another half-hour. Finally, cool the solution in ice, filter the $[Co(NH_3)_5Cl]$ Cl_2 on a sintered glass funnel, and wash with a few milliliters of ice water, followed by 20 ml of ethanol. Spread the product on a watch glass and air dry. The yield is about 12 g of purple-red $[Co(NH_3)_5Cl]Cl_2$.

Preparation of $[Co(NH_3)_5Cl]Cl_2$ (method B)[19]:

$$Co^{2+} + NH_4^+ + 4NH_3 + \tfrac{1}{2}H_2O_2 \longrightarrow [Co(NH_3)_5H_2O]^{3+}$$

$$[Co(NH_3)_5H_2O]^{3+} + 3Cl^- \longrightarrow [Co(NH_3)_5Cl]Cl_2 + H_2O$$

Dissolve 10.0 g of ammonium chloride in 60 ml of concentrated aqueous ammonia in a 500-ml Erlenmeyer flask. While continuously agitating the solution with a magnetic stirrer, add 20 g of finely powdered cobalt(II) chloride 6-hydrate in small portions. With continued stirring of the resulting brown slurry, slowly add 16 ml of 30 per cent hydrogen peroxide from a dropping funnel. When the effervescence has ceased, slowly add 60 ml of concentrated HCl. Continue the stirring on a hot plate, holding the temperature at about 85° for 20 min; then cool the mixture to room temperature and

[18] Adapted from W. E. Henderson and W. C. Fernelius, *A Course in Inorganic Preparations*, McGraw-Hill Book Company, New York, 1935, pp. 128–29.

[19] Adapted from G. Schlessinger, *Inorg. Syn.*, **9**, 160 (1967).

filter off the precipitated $[Co(NH_3)_5Cl]Cl_2$. Wash with 40 ml of ice water in several portions, followed by 40 ml of cold 6 M HCl. Dry the product in an oven at 100° for several hours. The yield is about 18 g of product.

Preparation of $[Co(NH_3)_5ONO]Cl_2$ and $[Co(NH_3)_5NO_2]Cl_2$[20]:

$$[Co(NH_3)_5Cl]^{2+} + H_2O \longrightarrow [Co(NH_3)_5H_2O]^{3+} + Cl^-$$

$$[Co(NH_3)_5H_2O]^{3+} + NO_2^- \longrightarrow [Co(NH_3)_5ONO]^{2+} + H_2O$$

$$[Co(NH_3)_5ONO]^{2+} \longrightarrow [Co(NH_3)_5NO_2]^{2+}$$

Dissolve 10.0 g of $[Co(NH_3)_5Cl]Cl_2$ in a solution of 15 ml of concentrated aqueous ammonia in 160 ml of water while stirring and heating. Filter off any slight precipitate of cobalt oxide that may form, and cool the filtrate to about 10°. Titrate the solution, with continued cooling, with 2 M HCl until it is just neutral to litmus. Then dissolve 10.0 g of sodium nitrite in the solution, followed by 10 ml of 6 M HCl. Allow the solution to stand in an ice bath for an hour or two, and then filter off the precipitated crystals of $[Co(NH_3)_5ONO]Cl_2$. Wash with 50 ml of ice water, followed by 50 ml of alcohol, and air dry at room temperature for 1 hour. The yield is about 9 g. Upon standing, isomerization to the nitro isomer occurs.

Dissolve 4.0 g of the $[Co(NH_3)_5ONO]Cl_2$ in 40 ml of hot water containing a few drops of aqueous ammonia, and then add, while cooling, 40 ml of concentrated HCl. Cool the solution thoroughly and filter off the $[Co(NH_3)_5NO_2]Cl_2$. Wash the product with 25 ml of alcohol and air dry at room temperature for several hours. The yield is about 3.5 g.

Characterization of $[Co(NH_3)_5NO_2]Cl_2$ and $[Co(NH_3)_5ONO]Cl_2$

As soon as the salts are dry, make a KBr pellet of each isomer and obtain its infrared spectrum.[21] After the pellets have stood at room temperature for a week or so, obtain the infrared spectra again. Compare all the spectra and explain the changes that have occurred. Note the color changes in the salts.

The effect of sunlight[22] or ultraviolet light[23] and of heat (a 100° oven) on the infrared spectra of KBr pellets of the salts can be studied. From the results, the student should be able to devise simple methods for interconverting the nitro and nitrito isomers.

[20] Adapted from S. M. Jorgensen, *Z. anorg. Chem.*, **5**, 147 (1894); *Z. anorg. Chem.*, **17**, 455 (1898).

[21] R. B. Penland, T. J. Lane, and J. V. Quagliano, *J. Am. Chem. Soc.*, **78**, 887 (1956).

[22] B. Adell, *Z. anorg. allgem. Chem.*, **279**, 219 (1955).

[23] W. W. Wendlandt and J. H. Woodlock, *J. Inorg. Nucl. Chem.*, **27**, 259 (1965); V. Balzani, R. Ballardini, N. Sabbatini, and L. Moggi, *Inorg. Chem.*, **7**, 1398 (1968).

11. THE PREPARATION OF MANGANESE (II) PHTHALOCYANINE

$$4\,C_6H_4(CN)_2 + Mn(OAc)_2 \cdot 4\,H_2O \longrightarrow Mn[C_6H_4(CN)_2]_4 + \tfrac{1}{2}O_2 + 3\,H_2O + 2\,HOAc$$

Total time required: 1 day

Actual working time: 2 hours

Preliminary study assignment:
   F. H. Moser and A. L. Thomas, *J. Chem. Educ.*, **41**, 245 (1964), and L. H. Vogt, Jr., H. M. Faigenbaum, and S. E. Wiberley, *Chem. Revs.*, **63**, 269 (1963).

Reagents required:
   8 g of $Mn(OAc)_2 \cdot 4\,H_2O$
   17 g of phthalonitrile
   40 ml of 1,2-propanediol (propylene glycol)
   150 ml of 95 per cent ethanol
   20 ml of pyridine
   Boiling chip

Special apparatus required (for sublimation):
   40-cm length of 25-mm Pyrex tubing, closed at one end
   Tube furnace and pyrometer
   Vacuum pump with McLeod gauge

Procedure[24]

Eight grams of powdered manganous acetate 4-hydrate, 17 g of phthalonitrile, 40 ml of 1,2-propanediol (propylene glycol), and a boiling chip are placed in a 100-ml round-bottom flask. The mixture is heated under reflux for about 45 min, cooled,[25] and then mixed with 100 ml of water. The black product is collected on a sintered-glass funnel,[26] washed with six 25-ml portions of hot alcohol, and air dried on a watch glass. The yield is about 9 g of manganese(II) phthalocyanine.

Characterization

One interesting characteristic of the compound is its remarkable thermal stability and its ability to be sublimed at high temperatures. The following

---

[24] Adapted from the procedure of H. A. Rutter, Jr., and J. D. McQueen, *J. Inorg. Nucl. Chem.*, **12**, 362 (1960).
   [25] The ground-glass joint should be loosened while it is still warm or it may "freeze."
   [26] The filtration proceeds very slowly.

procedure may be used for purifying the material by sublimation. By means of a small funnel attached to a long glass tube, place about 1 g of the manganese(II) phthalocyanine at the closed end of the 25-mm diameter Pyrex tube. Connect the open end to a vacuum pump. Hold the tube horizontally, with the end containing the sample inside a tube furnace. At least 20 cm of tubing should lie outside the furnace. While maintaining a vacuum of 50 μ or lower, gradually raise the temperature of the furnace. At first, volatile organic impurities will sublime from the sample and will appear near the opening of the tube furnace. These may be forced to the far end of the exposed tube, near the connection to the source of vacuum, by application of a flame. Later, when the tube reaches a temperature of approximately 420°, the manganese(II) phthalocyanine will sublime. Be careful not to heat the tube much over 500° or it may collapse. When the sublimation is judged complete, allow the tube to cool; crack it open, and remove the product.

Another interesting characteristic of the compound is its ability to form a complex with pyridine, which can reversibly react with molecular oxygen.[27] The structure of manganese(II) phthalocyanine is shown.

When the compound is dissolved in pyridine in the absence of air, the complex *A* (shown on p. 466, where the rhombs represent the square planar phthalocyanine ligand, $C_{32}H_{16}N_8$) forms. Upon introduction of air, complex *B* forms and then, more slowly, complex *C* forms. Complex *B* is green and has a characteristic absorption band at 716 mμ; complex *C* is blue and has a characteristic absorption band at 620 mμ. When a pyridine solution of complex *C* is boiled, it is rapidly converted to a solution of complex *B*, which, when cooled, slowly is reconverted to *C*. The cycle can be repeated many times. These transformations may be observed visually or spectrophotometrically, using approximately 10^{-4} *M* solutions.

[27] G. Engelsma, A. Yamamoto, E. Markham, and M. Calvin, *J. Phys. Chem.*, **66**, 2517 (1962); A. Yamomoto, L. K. Phillips, and M. Calvin, *Inorg. Chem.*, **7**, 847 (1968).

The magnetic susceptibility of the compound can be measured; previous measurements correspond to the presence of about three unpaired electrons per manganese atom. The esr spectrum can also be obtained.

12. THE PREPARATION OF COBALT(II) DI(SALICYLAL)ETHYLENEDIIMINE[28]

$$Co^{2+} + NH_2CH_2CH_2NH_2 + 2 C_6H_4(CHO)OH + 3 C_5H_5N \longrightarrow$$
$$CoC_{16}H_{14}N_2O_2 \cdot C_5H_5N + 2 C_5H_5NH^+ + 2 H_2O$$
$$CoC_{16}H_{14}N_2O_2 \cdot C_5H_5N \longrightarrow CoC_{16}H_{14}N_2O_2 + C_5H_5N$$

Total time required: 8 hours

Actual working time: 2 hours

Preliminary study assignment:
L. H. Vogt, Jr., H. M. Faigenbaum, and S. E. Wiberley, *Chem. Revs.*, **63**, 269 (1963); F. A. Cotton and G. Wilkinson, *Advanced Inorganic Chemistry*, 2nd ed., Interscience Publishers, New York, 1966, pp. 367–68; and, in this book, Chap. 8, The Vacuum Line.

Reagents required:
12.5 g of $Co(OAc)_2 \cdot 4 H_2O$
16 ml of pyridine
3.4 ml of ethylenediamine
10.5 ml of salicylaldehyde

Special apparatus required:
Vacuum desiccator
Vacuum line and reaction tube for removing pyridine and measuring oxygen uptake

[28] R. H. Bailes and M. Calvin, *J. Am. Chem. Soc.*, **69**, 1886 (1947).

Liquid nitrogen
Magnetic stirrer-hot plate
Oil bath

Procedure

A solution of the cobalt(II) acetate in 150 ml of water in a 300- or 500-ml Erlenmeyer flask is heated to 90 to 95° on a magnetic stirrer-hot plate. While the solution is stirred, the pyridine is added, followed by the ethylenediamine. The heater of the hot plate is then turned off, and the solution is vigorously stirred for about 2 min. The salicylaldehyde is added, and a stream of nitrogen is passed over the mixture in the flask. The mixture may become so thick that the stirring bar is ineffective; manual stirring is then required. Stirring is maintained for about 10 min and then stopped. The mixture is allowed to cool slowly to room temperature under a nitrogen atmosphere. The brown precipitate is collected on a sintered glass filter, washed with 20 ml of water, and dried overnight in a vacuum desiccator. The yield is about 17 g of cobalt di(salicylal)ethylenediimine 1-pyridinate.

About 0.4 g of the pyridinate is weighed out and transferred to the bottom of a reaction tube, which is then attached to the vacuum line and evacuated. While pumping through a −196° trap (liquid nitrogen) to collect the pyridine, warm the tube to 170° in an oil bath and maintain it at that temperature for 1.5 to 2 hours. The brown powder that remains [cobalt di(salicylal)ethylenediimine] is cooled to room temperature.

Characterization

The most significant characteristic of this compound is its ability to carry oxygen. To demonstrate this behavior, 1 atm pressure of oxygen is maintained over the product for at least 2 hours. During this time, $\frac{1}{2}$ mole of oxygen is absorbed per mole of cobalt complex. If the system is temporarily closed during the oxygen absorption period, the uptake of oxygen can be followed by the decrease in pressure, as indicated by a mercury manometer. The oxygenated material, $CoC_{16}H_{14}N_2O_2 \cdot \frac{1}{2} O_2$, is jet black. The oxygen content can be determined by heating the sample to 100° in an oil bath and by Toepler pumping the evolved oxygen through a −196° trap into a gas buret. The purity of the evolved oxygen can be determined, if desired, by mass spectrometry. The oxygenation-deoxygenation cycle can be repeated indefinitely.

CAUTION: The compound is reported to be quite toxic.[29] Inhalation of the finely divided dust should be avoided.

[29] H. Diehl and C. C. Hach, *Inorg. Syn.*, **3**, 196 (1950).

13. THE PREPARATION OF
NITROGENPENTAAMMINERUTHENIUM(II)
IODIDE[30]

$$RuCl_3(aq) \xrightarrow{N_2H_4} Ru(NH_3)_5N_2^{2+}$$

$$Ru(NH_3)_5N_2^{2+} + 2\,I^- \longrightarrow [Ru(NH_3)_5N_2]I_2$$

Total time required: 10 hours

Actual working time: 1 hour

Preliminary study assignment:
   A. D. Allen and F. Bottomley, *Acc. Chem. Res.*, **1**, 360 (1968).

Reagents required:
   1.0 g of "ruthenium trichloride"[31]
   10 ml of 85 per cent $N_2H_4 \cdot H_2O$
   10 g of NaI
   Ice

Special apparatus required:
   Magnetic stirrer and stirring bar

Procedure

The ruthenium chloride is dissolved in 12 ml of water in a 100-ml beaker. The solution is magnetically stirred, and 10 ml of 85 per cent hydrazine hydrate is added cautiously over a period of 2 min. The mixture undergoes vigorous effervescence during the addition of the hydrazine. Stirring is continued for 5 to 20 hours, and the resulting solution is filtered by gravity through a cone of filter paper. A solution of 10 g of sodium iodide in 6 ml of water is added to the filtrate and the mixture is stirred for about 5 min. The brown precipitate is collected on a small sintered-glass funnel and washed with 2 ml of ice-cold water. The product is dried for several hours in a vacuum desiccator. The yield is about 0.6 g of $[Ru(NH_3)_5N_2]I_2$.

Characterization

The product is contaminated with about 20 per cent $[Ru(NH_3)_6]I_2$; a purer product may be obtained by the reaction of $Ru(NH_3)_5H_2O^{2+}$ with azide.[30] The coordinated nitrogen may be liberated quantitatively by heating the solid in vacuo to about 400°. The liberated nitrogen should be Toepler

---

[30] A. D. Allen, F. Bottomley, R. O. Harris, V. P. Reinsalu, and C. V. Senoff, *J. Am. Chem. Soc.*, **89**, 5595 (1967).

[31] The commercial material contains both Ru(III) and Ru(IV) species.

pumped through a $-196°$ trap into a gas buret where it can be measured; the purity can be ascertained mass spectrometrically. The infrared spectrum of the solid shows the characteristic sharp band at $2129\ cm^{-1}$ due to $N{\equiv}N$ stretching. The ultraviolet absorption spectrum of the aqueous $Ru(NH_3)_5N_2^{2+}$ ion shows a peak at 221 mμ.

Question

1. How do you account for the fact that the $N{\equiv}N$ frequency in $Ru(NH_3)_5N_2^{2+}$ is reduced by $200\ cm^{-1}$ from that in free nitrogen ($2330\ cm^{-1}$)?

14. THE PREPARATION OF (−)- and (+)-TRIS(o-PHENANTHROLINE)IRON(II) PERCHLORATE TRIHYDRATE[32]

$$Fe^{2+} + 3(o\text{-phen}) \longrightarrow Fe(o\text{-phen})_3^{2+}$$

$$(-)\text{-}Fe(o\text{-phen})_3^{2+} + 2(+)\text{-}SbO{\cdot}C_4H_4O_6^- + 4\,H_2O \longrightarrow$$
$$[Fe(o\text{-phen})_3](SbO{\cdot}C_4H_4O_6)_2{\cdot}4\,H_2O$$

$$Fe(o\text{-phen})_3^{2+} + 2\,ClO_4^- + 3\,H_2O \longrightarrow [Fe(o\text{-phen})_3](ClO_4)_2{\cdot}3\,H_2O$$

Total time required: 6 hours

Actual working time: 2 hours

Preliminary study assignment:
   S. Kirschner, *Preparative Inorganic Reactions,* vol. 1, W. L. Jolly, ed., Interscience Publishers, New York, 1964, Chap. 2, p. 29, and F. Daniels, *et al., Experimental Physical Chemistry,* 6th ed., McGraw-Hill Book Company, New York, 1962, pp. 236–42.

Reagents required:
   1.0 g of *o*-phenanthroline monohydrate
   0.55 g of $FeSO_4{\cdot}7\,H_2O$
   1.15 g of potassium (+)-antimonyl tartrate
   25 ml of 0.05 *M* NaOH
   about 5 g of sodium perchlorate

Special apparatus required:
   Polarimeter (for characterization)

Procedure

One gram of *o*-phenanthroline monohydrate is stirred with a solution of 0.55 g of ferrous sulfate-7 hydrate in 45 ml of water until the *o*-phenanthroline

[32] Adapted from the procedure of F. P. Dwyer and E. C. Gyarfas, *J. Proc. Roy. Soc. N. S. Wales,* **83**, 263 (1950).

is completely dissolved. A solution of 1.15 g of potassium antimonyl tartrate in 15 ml of water is then added gradually over a period of 2 min. The mixture is rapidly cooled in an ice bath, and the dark red precipitate of (−)-*tris(o-phenanthroline)iron(II)* (+)-antimonyl tartrate is filtered off immediately. The product is sucked as dry as possible on the filter and is then spread on a watch glass to air dry. The filtrate is immediately cooled in an ice bath and about 10 ml of a moderately concentrated solution of sodium perchlorate is added. A red precipitate of (+)-[Fe(o-phen)$_3$](ClO$_4$)$_2$·3 H$_2$O forms, which is filtered off, washed with 10 ml of ice water, and air dried.

The (−)-*tris(o-phenanthroline)iron(II)*(+)-antimonyl tartrate is dissolved in about 25 ml of 0.05 *M* NaOH at about 5°, and the solution is filtered. About 10 ml of a moderately concentrated solution of sodium perchlorate is added, and the (−)-[Fe(o-phen)$_3$](ClO$_4$)$_2$·3 H$_2$O is isolated as previously described for the (+)-salt.

Characterization

Solutions of the salts in water may be examined polarimetrically; using the sodium D line, the absorbance of the solutions is so great that rather dilute solutions (with correspondingly weak rotations, in the neighborhood of 0.2°) must be used. The rotation should be followed for 1 or 2 hours; the half-life for the racemization at room temperature can be determined.

The proton nmr spectrum of the complex ion may be obtained.[33]

Question

1. How would the yields change if the initial precipitate of (−)-[Fe(o-phen)$_3$] (SbO·C$_4$H$_4$O$_6$)$_2$·4 H$_2$O were allowed to stand for several hours before filtration? Explain.

15. THE PREPARATION OF
MERCURY(II) DITHIZONATE

$$Hg^{2+} + 2 H_2C_{13}H_{10}N_4S \longrightarrow Hg(HC_{13}H_{10}N_4S)_2 + 2 H^+$$

Total time required: 5 hours

Actual working time: 1 hour

Preliminary study assignment:
  L. S. Meriwether, E. C. Breitner, *et al., J. Am. Chem. Soc.,* **87**, 4441, 4448 (1965).

[33] J. D. Miller and R. H. Prince, *J. Chem. Soc.,* 3185 (1965).

Reagents required:
  0.27 g of $HgCl_2$
  0.45 g of dithizone, $H_2C_{13}N_{10}N_4S$
  0.5 ml of concentrated aqueous HCl
  100 ml of $CHCl_3$
  100 ml of benzene

Special apparatus required:
  Abderhalden drier or vacuum drying oven

Procedure

A solution of the mercury(II) chloride and hydrochloric acid in 50 ml of water is shaken with a solution of the dithizone in the chloroform in a separatory funnel until the color of the aqueous phase does not change with continued shaking. The chloroform phase is separated off and allowed to dry in an evaporating dish in the hood. The solid residue is transferred to a tared weighing bottle or test tube, which is then placed in an Abderhalden drier and held in vacuo (< 10 mm) at 80° (boiling benzene) for 3 hours. A vacuum drying oven may be substituted for the Abderhalden drier.

Characterization

Mercury(II) dithizonate has the structure in which each ligand forms an N,S chelate with the metal atom.[34]

Solutions of the complex are photochromic.[34,35] In the dark, or under weak illumination, the solutions are orange (with an absorption maximum at 490 mμ); when exposed to sunlight or bright artificial light, they are blue (with $\lambda_{max} = 605$ mμ). When a blue solution is held in weak illumination, the color returns to orange in about 1 hour, although this time can be markedly

[34] L. S. Meriwether, E. C. Breitner, and N. B. Colthup, *J. Am. Chem. Soc.*, **87**, 4448 (1965).
[35] L. S. Meriwether, E. C. Breitner, and C. L. Sloan, *J. Am. Chem. Soc.*, **87**, 4441 (1965).

reduced by adding protonic species, such as water, or by raising the temperature. The return rate of a *ca* 10^{-5} *M* solution in benzene can be conveniently followed spectrophotometrically.

The photochromism is an inherent property of the ligand because it occurs in the presence of a variety of metals.[35] The central metal atom determines the photochemical stability, the return rate, and the color of the complex. A suggested mechanism for the return reaction involves a *cis-trans* isomerization of an azomethine group and a rate-determining proton shift.[34]

16. THE PREPARATION OF DIIRON ENNEACARBONYL[36]

$$2\,Fe(CO)_5 \xrightarrow{\;h\nu\;} Fe_2(CO)_9 + CO$$

Total time required: 26 hours

Actual working time: 1 hour

Preliminary study assignment:
   F. A. Cotton and G. Wilkinson, *Advanced Inorganic Chemistry*, 2nd ed., Interscience Publishers, New York, 1966, pp. 720–37.

Reagents required:
   50 ml of $Fe(CO)_5$
   150 ml of glacial acetic acid
   100 ml of 95 per cent ethanol
   50 ml of ether
   Cylinder of nitrogen

Special apparatus required:
   Mercury vapor ultraviolet lamp with water-cooled quartz sheath
   Reaction vessel for use with ultraviolet lamp

CAUTION: The iron carbonyls are very toxic materials. Great care must be taken to avoid breathing the fumes of $Fe(CO)_5$ and to avoid contact of the liquid with the skin. The carbon monoxide evolved during the photosynthesis is very poisonous. *All operations should be carried out in a hood.*

Procedure

The reaction vessel, with the water-cooled quartz sheath in position (see Fig. 32.10), is three-quarters filled with a solution prepared by dissolving 50 ml of

---

[36] Adapted from the procedures of E. Speyer and H. Wolf, *Ber.*, **60**, 1424 (1927), and E. H. Braye and W. Hübel, *Inorg. Syn.*, **8**, 178 (1966).

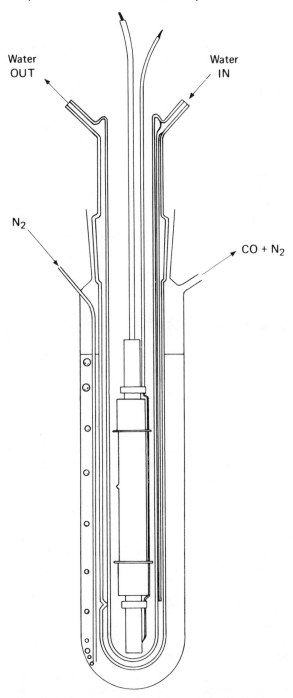

FIG. 32.10. Apparatus for the photosynthesis of $Fe_2(CO)_9$.

iron pentacarbonyl in 150 ml of glacial acetic acid. A stream of nitrogen gas is passed through the solution to eliminate oxygen and to stir the solution. The flow of cooling water is started, the ultraviolet lamp is inserted in the sheath, shielding is placed around the apparatus to protect the eyes, and the lamp is turned on. After about 24 hours, the lamp is turned off, and the precipitated $Fe_2(CO)_9$ is collected on a sintered-glass filter and washed with ethanol and then ether. The product is dried for 1 hour in a vacuum desiccator. A 75 per cent yield of $Fe_2(CO)_9$ is usually obtained.

Characterization

Diiron enneacarbonyl forms bronze mica-like platelets having a density of 2.08 gcc^{-1}. The compound is nonvolatile, almost insoluble in organic solvents, and decomposes at 100 to 120°. The infrared spectrum[37] may be readily interpreted in terms of the structure in which there are six terminal CO groups and three bridging CO groups.

17. THE PREPARATION OF DICOBALT OCTACARBONYL[38]

$$2\,CoCO_3 + 2\,H_2 + 8\,CO \longrightarrow Co_2(CO)_8 + 2\,H_2O + 2\,CO_2$$

Total time required: $1\frac{1}{2}$ days

Actual working time: 5 hours

Preliminary study assignment:
F. A. Cotton and G. Wilkinson, *Advanced Inorganic Chemistry*, 2nd ed., Interscience Publishers, New York, 1966, pp. 720–37, and, in this book, Chap. 14, High Pressure Apparatus.

Reagents required:
7.5 g of $CoCO_3$
75 ml of low-boiling petroleum ether
Cylinder of carbon monoxide with at least 400 P.S.I. pressure
Cylinder of hydrogen with at least 1700 P.S.I. pressure
Cylinder of nitrogen

Special equipment required:
300 ml stainless-steel autoclave and accessory equipment (described in Chap. 14 and pictured in Figs. 14.1 and 14.2)
Refrigerator with freezing compartment

[37] R. K. Sheline and K. S. Pitzer, *J. Am. Chem. Soc.*, **72**, 1107 (1950).

[38] Essentially, the same directions are given by I. Wender, H. W. Sternberg, S. Metlin, and M. Orchin, *Inorg. Syn.*, **5**, 190 (1957). Used by permission of the McGraw-Hill Book Company, New York.

Procedure

Place the petroleum ether and cobalt(II) carbonate in the autoclave. Following the general procedure outlined in Chap. 14, flush the autoclave three times with carbon monoxide and then fill it with carbon monoxide to a pressure of 1700 P.S.I. Turn the booster pump off and close all the valves. Replace the carbon monoxide cylinder with the hydrogen cylinder. With the valve attached to the gauge closed, flush the system with hydrogen. Fill the pressure chamber with hydrogen at cylinder pressure, and close the cylinder valve (valve 1). Open the gauge valve and boost the pressure to 3500 P.S.I. The autoclave now contains approximately equimolar amounts of hydrogen and carbon monoxide. Turn off the booster pump and close all the valves. Start the rocking mechanism and heat the autoclave to 150 to 160°. After holding the temperature in this range for 3 hours, stop the agitation and heating. When the autoclave has reached room temperature (after about 4 hours), carefully vent and open it in a hood. Pour the dark solution into a large beaker and filter it through filter paper into a 125-ml Erlenmeyer flask. Store the flask overnight in the freezing compartment of a refrigerator, whereupon large, well-formed crystals of the product will form. Decant the solvent and dry the crystals by passing a stream of dry nitrogen through the flask for several minutes. A yield of approximately 6 g of dark-red crystals is usually obtained.

Characterization

Dicobalt octacarbonyl melts at 50 to 51°, decomposing to black $[Co(CO)_3]_4$. If a sample is stored in an evacuated tube, it will slowly sublime, forming bright-orange crystals on the walls. The material is soluble in *n*-hexane and carbon tetrachloride, and the infrared and ultraviolet spectra of such solutions can be obtained.[39]

Questions

1. Describe two other synthetic methods that have been used for preparing metal carbonyls.

2. How may cobalt carbonyl hydride be prepared?

18. THE PREPARATION OF TIN TETRAPHENYL

$$2\,Na + C_6H_5Cl \longrightarrow C_6H_5Na + NaCl$$
$$4\,C_6H_5Na + SnCl_4 \longrightarrow (C_6H_5)_4Sn + 4\,NaCl$$

[39] J. W. Cable, R. S. Nyholm, and R. K. Sheline, *J. Am. Chem. Soc.*, **76**, 3373 (1954).

Total time required: 2 days

Actual working time: 7 hours

Preliminary study assignment:
   G. E. Coates, M. L. H. Green, and K. Wade, *Organometallic Compounds,*
   vol. 1, Methuen & Co. Ltd., London, 1967, pp. 414–21; T. P. Whaley,
   Inorg. Syn., **5,** 6 (1957); I. Fatt and M. Tashima, *Alkali Metal Dispersions,*
   D. Van Nostrand Co., Inc., Princeton, N. J., 1961, pp. 39–49; and, in this
   book, Chap. 5, Solvents, and Appendix 4, Compressed-Gas Cylinders.

Reagents required:
   275 ml of dry toluene (350 ml if wet)
   35 ml of chlorobenzene
   15 g of sodium
   10 ml of $SnCl_4$
   Dry ice

Special apparatus required:
   500-ml, three-neck, round-bottom flask
   Variac
   High-speed stirring motor
   Stirring rod with sharp metal blades[40]
   Cylinder of argon
   About 1000 ml of kerosene in pan
   Extra sintered-glass funnel and filter flask

Procedure

Fifteen grams of clean sodium chunks and 250 ml of dry toluene[41] are placed
in the flask. A thermometer and an argon inlet tube are inserted through one
of the side arms of the flask. The other side arm is stoppered. Insert the
stirrer through the main mouth of the flask, taking care that the stirring
blades cannot hit the thermometer and that they are above the chunks of
sodium. While stirring gently, and with a slow stream of argon flowing, heat
the contents slowly to 105°. Then lower the stirrer so that the blades are about
1 cm from the bottom of the flask and turn the stirrer on full power. It will be
found necessary to increase the power input to the heating mantle in order to
keep the temperature at 105°. After about 10 min of vigorous stirring at 105°,
remove the heating mantle from the flask. When the temperature has fallen
to 99°, stop the stirrer and allow the flask to cool to room temperature. The
sodium should now be in the form of a fine sand. Stir the solution gently to

---

[40] The Labline Stir-o-vac (Cat. No. 1280) and High Speed Motor (Cat. No. 1285) can be used.

[41] Toluene may be dried by simple distillation. Put about 350 ml in the distilling flask and
discard the initial cloudy distillate and the last 25 ml. Xylene may be substituted for toluene.

see if any of the particles have agglomerated. If so, the process must be repeated. If the sodium dispersion is not to be used immediately, thoroughly flush the flask with argon and tightly stopper it.

Using the heating mantle, heat the dispersion, with moderately vigorous stirring, to 45°. Attach a dropping funnel containing 35 ml of chlorobenzene to the unused side arm and add 2 to 3 ml of the chlorobenzene to the flask.

NOTICE: The flask should never contain more than 3 ml of unreacted chlorobenzene! If more than this amount is present, an uncontrollably vigorous reaction may take place, resulting in a fire.

Remove the heating mantle from the flask. The reaction should start, as evidenced by a rise in the temperature. If the reaction does not start at 45°, cautiously raise the temperature to 50° (no higher!). If the reaction starts at this temperature, the temperature may suddenly rise to as high as 55°, so be ready to cool the flask quickly with the kerosene bath. (If the reaction does not start at 50°, cool the flask to room temperature, cautiously hydrolyze the mixture with alcohol, and discard.)

Temperatures in excess of 50° will cause no great harm at the beginning of the synthesis, but thereafter the temperature must be kept below 45°; keep the flask partially immersed in the kerosene bath and cool the kerosene bath by occasionally adding pieces of Dry Ice to it. The temperature of the reaction mixture may be held between 40 and 45° by adjusting the rate of addition of chlorobenzene.

After all the chlorobenzene has been added (about 1 to 2 hours), place a solution of 10 ml of stannic chloride in 25 ml of toluene in the dropping funnel, and, over a period of 30 min, add this solution to the reaction flask. During this addition, it is necessary to cool the flask so as to keep the temperature below 45°. The flask may now be stored indefinitely (without protection from the air) until the tin tetraphenyl is extracted from the mixture.

Wipe the kerosene from the bottom of the flask, and, with moderate stirring, heat the mixture to incipient boiling and quickly filter through a sintered-glass funnel. It is best to keep most of the solid residue in the reaction flask. Cool the filtrate to room temperature and filter off the product on another sintered-glass funnel. Return the filtrate to the original flask and repeat the extraction two or three times until no more product precipitates on cooling the solution to room temperature. It is helpful to add another 100 ml of toluene to the mixture to reduce the necessary number of extractions. The final solution should be cooled in an ice bath before filtering. Suck the crystals of tin tetraphenyl as dry as possible on the filter and then let them air dry for 4 to 20 hours on a watch glass.

A yield of about 25 g of material melting at 226 to 228° should be obtained. A purer product (melting at 229°) may be obtained by recrystallization from benzene or toluene.

Characterization

Tin tetraphenyl may be characterized by its infrared spectrum[42] and its ultraviolet absorption spectrum.[43] The compound is an important intermediate for the preparation not only of phenyltin compounds but also of phenyl derivatives of other elements. Both Synthesis 18 (triphenyltin compounds) and Synthesis 20 (phenylboron compounds) require tin tetraphenyl as a starting material.

Questions

1. By what other methods may tin tetraphenyl be prepared, and in what ways is the present method superior?

2. What by-products will be formed if the reaction mixture is allowed to get too warm?

19. THE PREPARATION OF TRIPHENYLTIN CHLORIDE, TRIPHENYLTIN HYDROXIDE, AND BIS(TRIPHENYLTIN) OXIDE

$$3\,(C_6H_5)_4Sn + SnCl_4 \longrightarrow 4\,(C_6H_5)_3SnCl$$

$$(C_6H_5)_3SnCl + NH_3 + H_2O \longrightarrow (C_6H_5)_3SnOH + NH_4{}^+ + Cl^-$$

$$2\,(C_6H_5)_3SnOH \longrightarrow (C_6H_5)_3Sn-O-Sn(C_6H_5)_3 + H_2O$$

Total time required: 2 days

Actual working time: 6 hours

Preliminary study assignment:
  G. E. Coates, M. L. H. Green, and K. Wade, *Organometallic Compounds,* vol. 1, Methuen & Co. Ltd., London, 1967, pp. 437–44, and B. Kushlefsky, I. Simmons, and A. Ross, *Inorg. Chem.,* **2,** 187 (1963).

Reagents required:
  21 g of $(C_6H_5)_4Sn$
  2 ml of $SnCl_4$
  600 ml of 95 per cent ethanol
  60 ml of concentrated aqueous ammonia
  30 ml of toluene

Special apparatus required:
  Oil bath
  Mechanical stirrer
  Dean-Stark water trap
  Vacuum desiccator

[42] R. C. Poller, *J. Inorg. Nucl. Chem.,* **24,** 593 (1962).
[43] S. R. La Paglia, *J. Mol. Spectry.,* **7,** 427 (1961).

Preparation of $(C_6H_5)_3SnCl$[44]:

Twenty-one grams of tin tetraphenyl and 2.0 ml of tin tetrachloride are placed in a 100-ml round-bottom flask fitted with a reflux condenser. It is unnecessary to water cool the condenser. The flask is placed in an oil bath the temperature of which is gradually raised, over $1\frac{1}{2}$ hours, to 230°. The bath is held at 220 to 240° for 3 hours. Then the reaction vessel is allowed to cool to 100°, and the reaction mixture is poured into a beaker where it is cooled to room temperature. The solidified mass is broken up and very thoroughly shaken or stirred with 250 ml of warm (about 50°) 95 per cent ethanol. The ethanol dispersion is filtered, and about 500 ml of water is slowly added to the stirred filtrate. The precipitate is collected on a filter and is allowed to air dry on a large watch glass. About 20 to 21 g of crystals melting at 103 to 105° are usually obtained.

Preparation of $(C_6H_5)_3SnOH$[45] and $[(C_6H_5)_3Sn]_2O$[46]:

Sixteen grams of triphenyltin chloride is dissolved in about 300 ml of 95 per cent ethanol at 50°. Any undissolved material is filtered off. Sixty milliliters of concentrated aqueous ammonia is then added while the warm solution is stirred. After 1 hour of stirring, about 500 ml of water is slowly added to the mixture. The precipitate is collected on a sintered-glass filter and is washed thoroughly with boiling water. After the product is suction dried for several minutes, it is recrystallized from hot 95 per cent ethanol.[47] About 10 g of $(C_6H_5)_3SnOH$ is usually obtained. The material usually melts in the range 118 to 122°.

Prepare $[(C_6H_5)_3Sn]_2O$ by heating 8 g of the hydroxide with 35 ml of dry toluene and a few boiling chips in a reflux apparatus equipped with a Dean-Stark water trap. When water stops collecting in the trap, cool the solution to room temperature and filter immediately. After standing overnight or longer, about 4 g of the oxide is deposited. Filter this off and dry in a vacuum desiccator. The melting point is 121–123°.

Characterization

The hydroxide and oxide are easily distinguished and identified by their unique infrared spectra.[48] Triphenyltin hydroxide shows a strong doublet at

[44] Similar directions are given by K. A. Kozeschkow, M. M. Nadj, and A. P. Alexandrow, *Chem. Ber.*, **67**, 1348 (1934).

[45] O. Schmitz-Dumont, *Z. Anorg. Chem.*, **248**, 289 (1941).

[46] B. Kushlefsky, I. Simmons, and A. Ross, *Inorg. Chem.*, **2**, 187 (1963).

[47] Not all of the crude product will be found to be soluble in ethanol. The insoluble residue may be discarded.

[48] B. Kushlefsky, I. Simmons, and A. Ross, *Inorg. Chem.*, **2**, 187 (1963).

910 and 897 cm^{-1} that is completely absent in triphenyltin oxide, and the oxide shows a band at 774 cm^{-1} that is absent in the hydroxide. Poller[42] has discussed the systematics of the infrared spectra of phenyltin compounds.

Triphenyltin fluoride is essentially insoluble in water, and the chloride has been used as a reagent for the gravimetric determination of fluoride.[49]

20. THE PREPARATION OF DI- AND TRIBENZYLTIN CHLORIDES[50]

$$3 C_6H_5CH_2Cl + 2 Sn \longrightarrow (C_6H_5CH_2)_3SnCl + Sn^{2+} + 2 Cl^-$$
$$2 C_6H_5CH_2Cl + Sn \longrightarrow (C_6H_5CH_2)_2SnCl_2$$

Total time required: 3 days

Actual working time: 4 hours

Preliminary reading assignment:
G. E. Coates, M. L. H. Green, and K. Wade, *Organometallic Compounds*, vol. 1, Methuen & Co. Ltd., London, 1967, p. 413ff.; and K. Sisido, Y. Takeda, and Z. Kinugawa, *J. Am. Chem. Soc.*, **83**, 538 (1961).

Reagents required:
19 g of tin powder (150 to 200 mesh)[51]
4 g of NaOH
15 ml of methanol
18 ml of benzyl chloride
100 ml of ethyl acetate
75 ml of acetic acid

Special equipment required:
Magnetic stirrer and stirring bar
Melting-point apparatus

Procedure

The tin powder is added to 40 ml of a 10 per cent NaOH solution in a stoppered flask, and the mixture is shaken vigorously for 10 min. The powder is washed by decantation until the wash water shows no alkalinity to litmus; it is then rinsed with methanol. After air drying, the lumps are broken up with a mortar and pestle.

[49] A. I. Vogel, *A Textbook of Quantitative Analysis*, 3rd ed., John Wiley & Sons, Inc., New York. 1961, pp. 570–71.

[50] Adapted from the procedure of K. Sisido, Y. Takeda, and Z. Kinugawa, *J. Am. Chem. Soc.*, **83**, 538 (1961).

[51] Coarser powder can be used, but the yields suffer.

A mixture of 9 g of the purified tin powder and 75 ml of water is vigorously stirred with a magnetic stirrer in a 100-ml round-bottom flask equipped with a condenser and a heating mantle. The water is brought to a boil, and 9 ml of benzyl chloride is added from the top of the reflux condenser over a period of 2 min. Refluxing and stirring are continued for 1 or 2 hours, and, after cooling, the solid material is filtered off and air dried. The solid is extracted with a small volume of hot ethyl acetate; the hot solution is filtered, and the filtrate is evaporated at room temperature. The residue of crude product is recrystallized from hot glacial acetic acid and air dried overnight. The yield is about 8 g of tribenzyltin chloride (m.p. = 139 to 144°).

Two drops of water are added to 9 g of purified tin powder and the mixture is kneaded together. The moist tin is stirred vigorously with 75 ml of toluene in a 100-ml round-bottom flask, and the toluene is brought to a boil. Nine milliliters of benzyl chloride is added during a period of 3 min, and the stirring and refluxing are continued for 3 hours. After the slurry is cooled in an ice bath, the solid material is filtered off and extracted with a small volume of hot ethyl acetate. The hot solution is filtered and the filtrate is evaporated at room temperature. The residue of crude product is recrystallized from hot glacial acetic acid and air dried overnight. The yield is about 9 g of dibenzyltin chloride (m.p. = 160 to 163°).

Characterization

The treatment of tri- and dibenzyltin chloride with sodium hydroxide yields tribenzyltin hydroxide and dibenzyltin oxide, respectively.[52] The latter compounds undergo slow oxidation in air to give, among other things, benzaldehyde. In the case of tribenzyltin hydroxide, the reaction is

$$(C_6H_5CH_2)_3SnOH + O_2 \longrightarrow C_6H_5CHO + (C_6H_5CH_2)_2Sn(OH)_2$$

21. THE PREPARATION OF PHENYLDICHLORO-
BORANE, PHENYLDIHYDROXYBORANE
(PHENYLBORIC ACID), AND
TRIPHENYLBOROXINE[53]

$$Sn(C_6H_5)_4 + 4\,BCl_3 \longrightarrow 4\,C_6H_5BCl_2 + SnCl_4$$
$$C_6H_5BCl_2 + 2\,H_2O \longrightarrow C_6H_5B(OH)_2 + 2\,HCl$$
$$3\,C_6H_5B(OH)_2 \longrightarrow (C_6H_5BO)_3 + 3\,H_2O$$

[52] P. Pfeiffer, *Z. anorg. allgem. Chem.*, **68**, 102 (1910); T. A. Smith and F. S. Kipping, *J. Chem. Soc.*, **103**, 2034 (1913).

[53] Adapted from the procedure of J. E. Burch, W. Gerrard, M. Howarth, and E. F. Mooney, *J. Chem. Soc.*, 4916 (1960).

Total time required: 2 days

Actual working time: 5 hours

Preliminary study assignment:

    W. Gerrard, *The Organic Chemistry of Boron*, Academic Press Inc., New York, 1961, pp. 58–60, 67–70, and K. Torssell, *Progress in Boron Chemistry*, vol. 1, H. Steinberg and A. L. McCloskey, eds., The Macmillan Company, New York, 1964, Chap. 9, p. 369.

Reagents required:

    20 g of tetraphenyl tin
    50 ml of carbon tetrachloride
    Cylinder of boron trichloride
    Boiling chips
    50 ml of hexane
    50 ml of toluene (optional)
    About 200 ml of mineral oil

Special apparatus required:

    100-ml round-bottom flask with side arm
    Magnetic stirrer and stirring bar
    50-ml distillation flask with side arm
    Water aspirator with traps and manometer
    Dean-Stark trap (optional)
    Tygon tubing

Procedure

Twenty grams of tetraphenyl tin and 50 ml of carbon tetrachloride are placed in a 100-ml round-bottom flask equipped with a side arm gas inlet tube and a reflux condenser topped by a rubber stopper with a piece of glass tubing through it. A magnetic stirring bar is placed in the flask, and the flask is placed in an electric heating mantle on a magnetic stirrer (see Fig. 32.11). A stream of boron trichloride from a cylinder is passed into the stirred suspension, causing the reaction mixture to become warm. When there is no more exothermic reaction, and copious white fumes form at the exit tube at the top of the condenser, the flow of boron trichloride is stopped. The gas delivery tube is disconnected from the side arm and the latter is stoppered. The reaction mixture is very gradually heated to the boiling point, and reflux is maintained for 5 to 6 hours.

    After the reaction flask is cooled to room temperature, the reflux condenser is replaced by a simple distillation head and a few boiling chips are added to the flask. Using a one-hole rubber stopper, a 50-ml distillation flask with side arm is attached to the side arm of the distillation head. The

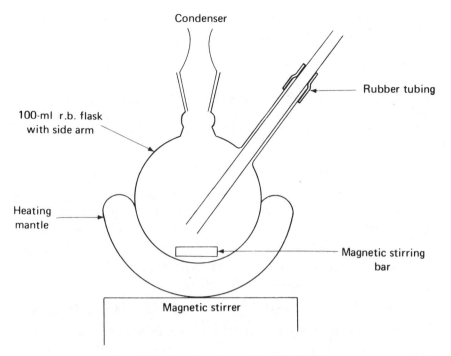

Fig. 32.11. Setup for the synthesis of $C_6H_5BCl_2$.

side arm of the distillation head should extend to 4 cm from the bottom of the 50-ml distillation flask. The side arm of the distillation flask is connected to a high-capacity trap cooled with Dry Ice; the trap is connected to a manometer and a water aspirator provided with a surge trap. The system is then evacuated to 10 to 30 mm pressure until most of the material volatile at room temperature has evaporated from the reaction flask. The flask temperature is then gradually raised by using an oil bath. A small amount of $SnCl_4$ that was not vaporized at room temperature will come off at bath temperatures below 60°; this material should be discarded. The phenyldichloroboron distills at approximately the following temperatures (vapor) and corresponding pressures: 63°/10 mm; 71°/15 mm; 76°/20 mm; 80°/25 mm; 84°/30 mm. The yield is about 19 g of $C_6H_5BCl_2$. The part of the product that is not to be used in subsequent steps should be stored in a sealed ampule.

About 10 g of phenyldichloroboron is dissolved in 50 ml of hexane in a beaker and magnetically stirred. Water is added drop by drop until the effervescence (evolution of HCl) ceases. The white precipitate of phenylboric acid, $C_6H_5B(OH)_2$, is filtered off, washed with about 25 ml of hexane, and air dried. The yield is about 8 g of $C_6H_5B(OH)_2$, melting at 216°. The acid may be converted to triphenylboroxine, $(C_6H_5BO)_3$, either by digestion with toluene using a Dean-Stark water trap or by heating overnight in an oven at 110°. The boroxine melts at 218°.

Characterization

Phenylboric acid (also referred to as benzeneboronic acid) is a weak acid (pK = 8.86), yet sufficiently strong to titrate in aqueous solution.[54] The carbon–boron bond is hydrolytically cleaved by aqueous solutions of the diamminesilver ion[55]:

$$C_6H_5B(OH)_2 + H_2O \xrightarrow{\text{Ag(NH}_3)_2{}^+} C_6H_6 + B(OH)_3$$

There is considerable question as to whether a meaningful melting point can be determined for phenylboric acid. The dehydration is quite rapid, and the reported melting point may be that of the anhydride, triphenylboroxine.

22. THE PREPARATION OF FERROCENE[56]
[Bis(cyclopentadienyl)iron(II)]

$$8\,KOH + 2\,C_5H_6 + FeCl_2 \cdot 4\,H_2O \longrightarrow Fe(C_5H_5)_2 + 2\,KCl + 6\,KOH \cdot H_2O$$

Total time required: 1 day (or 5 hours if product dried in vacuo)

Actual working time: 4 hours

Preliminary study assignment:
   J. Birmingham, "Synthesis of Cyclopentadiene Metal Compounds," *Adv. Organometallic Chem.*, **2**, 365 (1965); G. E. Coates, M. L. H. Green, and K. Wade, *Organometallic Compounds*, vol. 2, Methuen & Co. Ltd., London, 1968, pp. 90–115; W. L. Jolly, *Inorg. Chem.*, **6**, 1435 (1967); and, in this book, Chap. 5, Solvents, and Appendix 4, Compressed-Gas Cylinders.

Reagents required:
   120 ml of 1,2-dimethoxyethane
   About 60 g of KOH pellets (protected from moisture)
   11 ml of cyclopentadiene—obtained from the thermal cracking of dicyclopentadiene (3a,4,7,7a-tetrahydro-4,7-methanoindene)
   13 g of $FeCl_2 \cdot 4\,H_2O$
   50 ml of dimethyl sulfoxide (DMSO)
   90 ml of concentrated aqueous HCl
   About 200 g of ice
   Cylinder of nitrogen or argon

[54] G. E. K. Branch, D. L. Yabroff, and B. Bettman, *J. Am. Chem. Soc.*, **56**, 937 (1934).
[55] J. R. Johnson, M. G. Van Campen, and O. Grummitt, *J. Am. Chem. Soc.*, **60**, 111 (1938).
[56] Adapted from the procedure of W. L. Jolly, *Inorg. Syn.*, **11**, 120 (1968).

Special apparatus required:
   Fractional distillation apparatus
   300 ml three-necked (standard taper joints) round-bottomed flask
   Magnetic stirrer and large stirring bar
   100-ml dropping funnel, standard taper
   T-tube mercury bubbler with standard-taper connection to flask
   Pyrex Petri dish with cover, 150 × 20 mm

Procedure

The potassium hydroxide pellets are quickly ground with a mortar and pestle until the largest particles are less than 0.5 mm in diameter. Because it is very difficult to pulverize a large quantity of potassium hydroxide at one time, the pulverization should be carried out in batches of 15 g or less. It is important to minimize the exposure of the KOH powder to the atmosphere; therefore, it is stored in a tightly capped tared bottle.

The cyclopentadiene is prepared by the thermal cracking of dicyclopentadiene. Dicyclopentadiene is slowly distilled through a fractionating column, collecting only that material which refluxes below 44° (cyclopentadiene boils at 42.5°, and dicyclopentadiene at 170°). If several students plan to prepare cyclopentadiene, it is advised that they cooperate in this step and distill at least 50 ml of dicyclopentadiene from a flask having a capacity twice the volume of the dicyclopentadiene. The freshly distilled cyclopentadiene must be used within 2 or 3 hours, or stored at − 78° until use, because slow dimerization occurs at room temperature.

The magnetic stirring bar, 120 ml of 1,2-dimethoxyethane, and 50 g of powdered potassium hydroxide are placed in the three-necked flask. One side neck is stoppered and the other is connected to the T-tube mercury bubbler and the nitrogen cylinder (see Fig. 32.12). While the mixture is slowly stirred and the flask is being flushed with a stream of nitrogen, 11.0 ml of cyclopentadiene is added. The main neck is then fitted with the dropping funnel with its stopcock open. After about 99 per cent of the air has been flushed from the flask, the stopcock is closed, and a solution of 13.0 g of $FeCl_2 \cdot 4\,H_2O$ in 50 ml of dimethyl sulfoxide[57] is placed in the dropping funnel. The mixture is stirred vigorously.[58] After about 10 min, the T-tube is lifted above the mercury surface (to reduce the pressure in the flask to atmospheric), and drop-by-drop addition of the iron(II) chloride solution is begun. The rate of addition is adjusted so that the entire solution is added in 45 min. Then the dropping funnel stopcock is closed and vigorous stirring is continued for a further 30 min. Finally, the nitrogen flow is stopped, and the mixture is

[57] The rate of dissolution of the $FeCl_2 \cdot 4\,H_2O$ in DMSO is slow unless the solid is pulverized.
[58] Some magnetic stirrers are not adequate for vigorous stirring, and therefore should not be used.

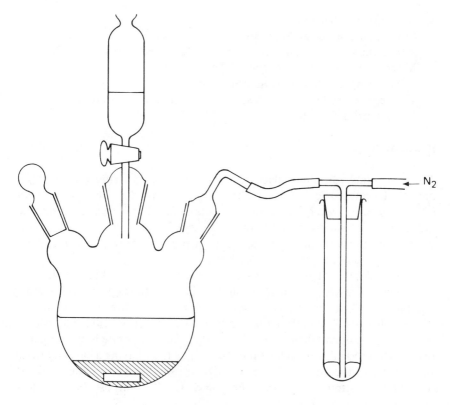

N₂

FIG. 32.12. Apparatus for the preparation of ferrocene.

added to a mixture of 180 ml of 6 *M* HCl and about 200 g of crushed ice. Some of the resulting slurry may be used to rinse the reaction flask. The slurry is stirred for about 15 min, and the precipitate is collected on a sintered-glass funnel and washed with four 25-ml portions of water. The moist solid is spread out on a large watch glass and dried in the air overnight. The yield is about 11.5 g of ferrocene. This product should be quite satisfactory as an intermediate for subsequent syntheses. An extremely pure product can be obtained by sublimation. The material to be sublimed is placed in the inverted cover of a Petri dish so that none of the material is within 2 mm of the side wall of the cover. The Petri dish itself (the smaller of the pair) is inverted and placed in the cover, and the apparatus is then placed on a hot plate, as shown in Fig. 32.13. The hot plate is gradually warmed up until the top surface of the apparatus is almost too hot to touch. After 4 to 10 hours, the ferrocene should be completely sublimed onto the upper glass surface and should be completely separated from the small amount of residue on the bottom by a gap of several millimeters. If any of the ferrocene crystals

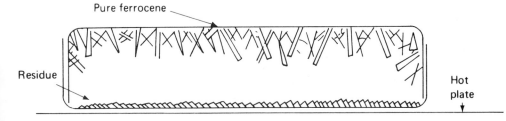

FIG. 32.13. Sublimation of ferrocene.

are touching the residue, the temperature of the hot plate should be increased, and more time allowed for complete sublimation. The yield is about 11.4 g of ferrocene, melting at 173–174°.

Characterization

The infrared spectrum of ferrocene may be determined by using either solutions in carbon tetrachloride and carbon disulfide or a KBr pellet. Absorption bands are observed at the following frequencies[59,60] (in cm^{-1}): 170(m), 478(s), 492(s), 782(w), 811(s), 834(w), 1002(s), 1051(w), 1108(s), 1188(w), 1411(s), 1620(m), 1650(m), 1684(m), 1720(m), 1758(m), 3085(s). A complete interpretation of the spectrum has been given by Lippincott and Nelson.[60]

The ultraviolet spectrum[61] in ethanol or hexane shows maxima at 325 mμ ($\varepsilon = 50$) and 440 mμ ($\varepsilon = 87$), and rising short-wavelength absorption ($\varepsilon = 5250$ at 225 mμ).

Ferrocene is readily oxidized to the blue ferricinium ion, $Fe(C_5H_5)_2{}^+$. Ferricinium tungstosilicate may be prepared by the following procedure. Dissolve $\frac{1}{2}$ g of ferrocene in 10 ml of concentrated sulfuric acid; allow the solution to stand for 15 min to 1 hour and then pour it into 150 ml of water. Stir the resulting blue aqueous solution for a few minutes and filter to remove the sulfur precipitate. Add a solution of 2.5 g of 12-tungstosilicic acid (Synthesis 9) in 20 ml of water, and filter off, wash with water, and air dry the pale blue tungstosilicate. Ferricinium perchlorate may be prepared by an electrolytic process described by Stranks.[62] Ferrocene may be readily converted to the complex $[(C_5H_5)Fe(C_6H_6)]PF_6$, in which a benzene ring has replaced one of the cyclopentadienide rings.[63]

[59] G. Wilkinson, P. L. Pauson, and F. A. Cotton, *J. Am. Chem. Soc.*, **76**, 1970 (1954).

[60] E. R. Lippincott and R. D. Nelson, *Spectrochim. Acta*, **10**, 307 (1958).

[61] G. Wilkinson, M. Rosenblum, M. C. Whiting, and R. B. Woodward, *J. Am. Chem. Soc.*, **74**, 2125 (1952); L. Kaplan, W. L. Kester, and J. J. Katz, *J. Am. Chem. Soc.*, **74**, 5531 (1952); A. T. Armstrong, F. Smith, E. Elder, and S. P. McGlynn, *J. Chem. Phys.*, **46**, 4321 (1967).

[62] D. R. Stranks, *Inorg. Syn.*, **7**, 203 (1963).

[63] R. B. King, *Organometallic Syn.*, **1**, 138 (1965).

A molecular orbital treatment of the bonding in ferrocene is described by Cotton.[64]

Questions

1. Describe five other methods for preparing ferrocene.

2. What is the structure of dicyclopentadiene?

3. X-ray diffraction data indicate that the ferrocene molecule is centrosymmetric. X-ray data indicate that this is not true for ruthenocene. What can be said about the structures of ferrocene and ruthenocene as a consequence of these observations?

4. What other types of organometallic compounds can be prepared by the use of potassium hydroxide? Why is much of the deprotonating power of KOH lost when it is used in solution?

23. THE PREPARATION OF SODIUM TRIPHOSPHATE

$$NaH_2PO_4 \cdot H_2O + 2\,Na_2HPO_4 \cdot 12\,H_2O \longrightarrow Na_5P_3O_{10} + 15\,H_2O$$
$$5\,Na^+ + P_3O_{10}^{5-} + 5\,H^+_{(resin)} \longrightarrow H_5P_3O_{10} + 5\,Na^+_{(resin)}$$

Total time required: 5 hours

Actual working time: 3 hours

Preliminary study assignment:
J. R. Van Wazer, *Phosphorus and Its Compounds*, vol. 1, Interscience Publishers, New York, 1958, pp. 638–56, and, in this book, Chap. 19, Structure from Chemical Data; Chap. 11, High-Temperature Processes; and Chap. 13, Ion Exchange.

Reagents required:
1.5 g of $NaH_2PO_4 \cdot H_2O$
7.8 g of $Na_2HPO_4 \cdot 12\,H_2O$ or 5.8 g of $Na_2HPO_4 \cdot 7\,H_2O$
500 ml of 1 *M* HCl
150 ml of standardized 0.1 *M* NaOH
About 50 ml of Dowex-50 Ion Exchange Resin, 20–50 mesh
Standard buffer solution (pH 4 or 7)

Special apparatus required:
Platinum crucible, 10 ml
Muffle furnace
pH meter
Magnetic stirrer and stirring bar

---

[64] F. A. Cotton, *Chemical Applications of Group Theory*, Interscience Publishers, New York, 1963.

Procedure

Place an intimate mixture of the two phosphates in a platinum crucible. (If hydrates different from those indicated are used, adjust the amounts so that the mole ratio of dihydrogen phosphate to monohydrogen phosphate is kept at 1:2.) Heat the mixture at 540 to 580° for 2 hours and then allow it to cool in the air. Knock the product out of the crucible by lightly tapping the bottom of the upturned crucible with a metal object. Grind the product with a mortar and pestle to a fine powder.

Characterization

The sodium triphosphate should be characterized by converting it to triphosphoric acid and titrating the "strong" and "weak" acidic hydrogens. Quantitative conversion to the acid is accomplished by passing a solution of the salt through the ion exchange resin in the hydrogen form. Wash the cation exchange resin into a 50-ml buret having a glass-wool plug at the bottom (see Fig. 13.2). Take care that the liquid level never falls below the top of the resin bed. Convert the resin to the hydrogen form by allowing about 500 ml of 1 M HCl to percolate through it. Remove the hydrochloric acid by washing the resin with distilled water until the eluate gives no test for chloride with silver ion. Weigh out about 0.25 g of the sodium triphosphate and dissolve it in 20 ml of water (dissolution is slow). Allow this solution, followed by 150 ml of water, to percolate slowly through the column into a 400-ml beaker. Standardize the pH meter, dip the rinsed electrodes into the triphosphoric acid solution, and titrate the solution with standardized 0.1 M NaOH while stirring magnetically. Record the pH of the solution after each 1-ml addition of base (more frequently at the endpoints). Plot the pH against the volume of base added and, from the endpoints, estimate the average chain length for the acid.

Phosphorus-31 nuclear magnetic resonance is another method for determining the relative number of phosphate "end groups" $\left(\begin{array}{c} O \\ \| \\ -O-P-O \\ | \\ O \end{array}\right)$

and phosphate "middle groups" $\left(\begin{array}{c} O \\ \| \\ -O-P-O- \\ | \\ O \end{array}\right)$ in a phosphate sample.

End groups and middle groups give separate resonances around $+5$ and $+19$ p.p.m. from 85 per cent H_3PO_4, respectively.[65] Classical chemical

[65] C. F. Callis, J. R. Van Wazer, J. N. Shoolery, and W. A. Anderson, *J. Am. Chem. Soc.*, **79**, 2719 (1957).

methods can also be used to quantitatively analyze for phosphate, diphosphate, and triphosphate.[66]

The higher polyphosphates (tetraphosphate, pentaphosphate, etc.) can be prepared as mixtures by employing a higher $H_2PO_4^-/HPO_4^{2-}$ ratio in the high-temperature synthesis. These polyphosphates can be separated by ion exchange chromatography on an analytical scale[67] and even for the isolation of macroscopic samples.[68] The book by Van Wazer may be consulted for paper chromatographic methods for analyzing polyphosphate mixtures.

Questions

1. Give two methods for separating a mixture of sodium polyphosphates of different chain length.

2. How would you distinguish cyclotriphosphate from a very high molecular weight straight-chain polyphosphate?

24. THE PREPARATION OF TRIMERIC AND TETRAMERIC DIMETHYLSILOXANE[69]

$$x(CH_3)_2SiCl_2 + x\,H_2O \longrightarrow [(CH_3)_2SiO]_x + 2x\,H^+ + 2x\,Cl^-$$

Total time required: 4 hours

Actual working time: 3 hours

Preliminary study assignment:

E. G. Rochow, *An Introduction to the Chemistry of the Silicones*, 2nd ed., John Wiley & Sons, Inc., New York, 1951, pp. 78–88, and J. Cason and H. Rapoport, *Laboratory Text in Organic Chemistry*, 2nd ed., Prentice-Hall, Inc., Englewood Cliffs, N. J., 1962, pp. 289–311 (discussion of fractional distillation).

Reagents required:

50 ml of $(CH_3)_2SiCl_2$[70]
50 ml of diethyl ether
5 g of Drierite
Ice

---

[66] R. N. Bell, *Anal. Chem.*, **19**, 97 (1947); A. G. Buyers, *Anal. Chim. Acta*, **19**, 118 (1958).

[67] J. A. Grande and J. Beukenkamp, *Anal. Chem.*, **28**, 1497 (1956).

[68] E. J. Griffith and R. L. Buxton, *J. Am. Chem. Soc.*, **89**, 2884 (1967).

[69] See W. Patnode and D. F. Wilcock, *J. Am. Chem. Soc.*, **68**, 358 (1946).

[70] Dimethyldichlorosilane is commercially available, or it may be prepared from silicon and methyl chloride [*Inorg. Syn.*, **3**, 56 (1950)].

Special equipment required:
   Magnetic stirrer and stirring bar
   Small fractional-distillation apparatus (with a 15-cm Vigreux column or
      a column with a heated jacket)
   Infrared lamp or heat gun

Procedure

Place a solution of the dimethyldichlorosilane and ether in a dropping funnel the tip of which reaches almost to the bottom of a 250-ml beaker containing 100 ml of water. Slowly add the solution through the funnel while vigorously stirring the water. Throughout the addition, hold the reaction mixture at 15 to 20° by surrounding the beaker with an ice bath. Separate the two phases with a separatory funnel; discard the lower phase, and wash the ether phase, first with 50 ml of water, then with 50 ml of a solution of sodium bicarbonate, and finally with another 50 ml of water. Allow the ether phase to stand in a flask containing about 5 g of Drierite (calcium sulfate) for 1 hour; then transfer the dry solution[71] to the pot of the fractional-distillation apparatus. The first fraction to come over is ether (b.p. = 34.6°).

CAUTION: Keep flames away while distilling the ether!

The second fraction is the trimer of dimethylsiloxane, hexamethyl-cyclotrisiloxane (b.p. = 134°). Inasmuch as this fraction is a solid at room temperature (m.p. = 64°), it is advisable to remove the water-cooled condenser from the distillation apparatus when practically all the ether has come over. The delivery tube of the distillation head will suffice as an air condenser. In fact, it may be necessary to warm this tube with an infrared lamp or a heat gun to keep it from clogging with solid trimer. The third fraction is the tetramer, octamethylcyclotetrasiloxane (b.p. = 175°). Usually, it is the last pure fraction that can be obtained from a distillation at atmospheric pressure. Attempts to collect further fractions usually result in such a high pot temperature that the high molecular weight diols of the type

$$HO[Si(CH_3)_2O]_xSi(CH_3)_2OH$$

are pyrolyzed to cyclic dimethylsiloxanes and water. Thus, a cloudy distillate (a mixture of the trimer, tetramer, and water) is obtained.

Excluding any material formed by pyrolysis, the usual yields of trimer and tetramer are 3 and 16 g, respectively. These yields may be augmented by a second fractionation of the crude pyrolysis product.

---

[71] If the products are not fractionally distilled soon after the synthesis, the procedure must be carried through to this point (at which the solution has been decanted from the Drierite) within a period of 3 to 4 hours, or the yield of trimer will be negligible.

Characterization

The determination of the boiling points of the trimer and tetramer and the melting point of the trimer serve to identify these compounds. The infrared spectra[72] (in CCl_4 and CS_2 solutions) also can be obtained.

Questions

1. If undiluted dimethyldichlorosilane is added to water, smaller yields of the low polymers of dimethylsiloxane are obtained than if the preceding directions are followed. Offer an explanation.

2. In what way does the preparation of "silicone oils" differ from the above synthesis? Explain the reason for the difference.

25. THE PREPARATION OF DIBORANE[73]

$$KBH_4 + H_3PO_4 \longrightarrow \tfrac{1}{2}B_2H_6 + H_2 + K^+ + H_2PO_4^-$$

Total time required: 4 hours

Actual working time: 4 hours

Preliminary study assignment:
   M. F. Hawthorne, *The Chemistry of Boron and Its Compounds*, E. L. Muetterties, ed., John Wiley & Sons, Inc., New York, 1967, Chap. 5, pp. 223–323, and, in this book, Chap. 8, The Vacuum Line, and Chap. 6, The Maintenance and Measurement of Low Temperatures.

Reagents required:
   0.5 g of KBH_4
   20 ml of 85 per cent H_3PO_4
   Liquid nitrogen
   Dry Ice, acetone, and CS_2 for cold baths

Special apparatus required:
   200-ml two- or three-necked flask with adapter, glass wool, and "tipping tube"
   Vacuum line
   Magnetic stirrer and stirring bar

   CAUTION: Diborane is an inflammable, highly toxic gas that may cause death after short exposure to small quantities. Care must be used in carrying out the following operations to prevent the release of diborane into the atmosphere.

---

[72] N. Wright and M. J. Hunter, *J. Am. Chem. Soc.*, **69**, 803 (1947).

[73] Adapted from procedures given by B. J. Duke, J. R. Gilbert, and I. A. Read, *J. Chem. Soc.*, 540 (1964), and A. D. Norman and W. L. Jolly, *Inorg. Syn.*, **11**, 15 (1968).

Procedure

The 200-ml flask, equipped as shown in Fig. 32.14, is connected to the vacuum line so that the volatile products can be pumped through a series of three U traps. The glass wool in the adapter to the vacuum line prevents solid material from being carried into the vacuum line. Twenty milliliters of 85 per cent H_3PO_4 is placed in the reaction vessel, and the "tipping tube" containing 0.5 g of KBH_4 is attached to it. The acid is stirred, and the reaction vessel is evacuated through the series of traps for a period of 20 min. Then the trap nearest the reaction vessel is cooled to $-78°$ (Dry Ice–acetone mixture), and the other two traps are cooled to $-196°$ (liquid nitrogen). The potassium hydroborate is slowly and steadily added to the acid (over a period of 20 to 30 min) so that the pressure in the reaction system (as measured by the mercury manometer) never exceeds 5 mm and so that the layer of foam never becomes thicker than 2.5 cm.

When all the hydroborate has been added, the system is thoroughly evacuated, and the reaction vessel is disconnected and placed in the hood. The $-78°$ trap (which contains principally water) is closed off from the rest

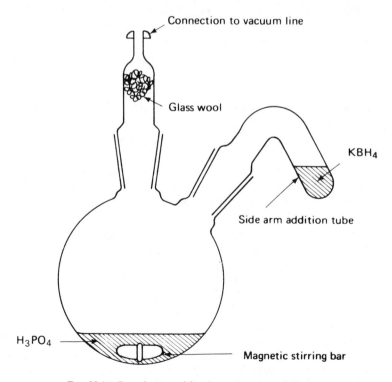

FIG. 32.14. Reaction vessel for the preparation of diborane.

of the system, disconnected, quickly placed in the hood, and replaced by a clean trap. The latter trap is then evacuated without allowing air to come in contact with the diborane in the $-196°$ traps. All the crude diborane is then combined in either the first or third trap of the series. Purification is achieved by passing the diborane in vacuo through a $-111.6°$ trap (CS_2 slush) into a $-196°$ trap. The traces of material that condense at $-111.6°$ are discarded.

The yield of purified diborane (about 2 mmoles) is determined by measuring the pressure and temperature of the gas in a system of known volume.

Characterization

The vapor pressure of diborane at the temperature of a CS_2 slush is 225 mm.[74] The infrared spectrum[75] shows absorption peaks at 3670(w), 2625(s), 2558(s), 2353(w), 1853(m), 1602(vs), 1197(s), 1178(s), 1154(s), and 973(m, sh) cm^{-1}.

If diborane is condensed in a reaction tube with an excess of anhydrous trimethylamine, and the mixture is slowly warmed to room temperature, the adduct trimethylamine-borane,[76] $(CH_3)_3NBH_3$, will form in essentially quantitative yield. The excess amine may be removed by holding the reaction mixture at $0°$ and evacuating. Trimethylamine-borane is a white volatile solid which can be sublimed in vacuo (vapor pressure = 0.8 mm at 23°) and which melts without decomposition at 94°. Complete hydrolysis is achieved only by treatment with strong mineral acids.

If a glass vessel containing diborane (at a pressure less than 250 mm) is exposed to sunlight or an ultraviolet lamp, decomposition occurs, with formation of hydrogen and several higher boron hydrides. Measurable amounts of B_4H_{10}, B_5H_9, and B_5H_{11} can be prepared by using the simple apparatus shown in Fig. 32.15. A quartz tube of about 150 ml volume is charged with about 200 mm pressure of diborane and exposed for 6 to 8 hours to a 450-watt mercury lamp. A little droplet of mercury is included in the tube to permit the formation of $Hg(^3P_1)$ atoms, which react with diborane.[77] The lamp emits large amounts of infrared radiation; therefore, the tube must be cooled with a strong blast of air and the stopcock must be shielded from the light (see Chap. 15 for a discussion of the safety precautions that must be taken when working with ultraviolet lamps). The unreacted diborane, the tetraborane, and the mixture of pentaboranes may be separated by passing the resulting reaction mixture through traps at $-112°$ (B_5H_9 and B_5H_{11}),

[74] A. B. Burg, *J. Am. Chem. Soc.*, **74**, 1340 (1952).

[75] R. C. Lord and E. Nielsen, *J. Chem. Phys.*, **19**, 1 (1951); also see *A.P.I. Res. Proj. 44*, Serial No. 742.

[76] A. B. Burg and H. I. Schlesinger, *J. Am. Chem. Soc.*, **59**, 780 (1939); Technical Bulletin C-200 of the Callery Chemical Co., Pittsburgh, Pa. This compound may be more readily prepared by the reaction of trimethylammonium chloride with sodium hydroborate. See J. Bonham and R. S. Drago, *Inorg. Syn.*, **9**, 8 (1967).

[77] T. Hirata and H. E. Gunning, *J. Chem. Phys.*, **27**, 477 (1957).

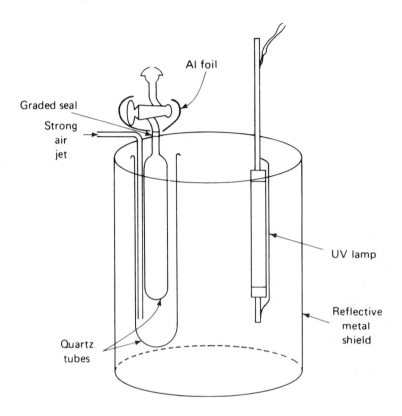

Fig. 31.15. Simple apparatus for gaseous photolysis.

$-130°$ (B_4H_{10}), and $-196°$ (B_2H_6). The hydrides may be identified from their infrared spectra[78] and their mass spectra.

Significant yields of tetraborane, the pentaboranes, and decaborane ($B_{10}H_{14}$) can be prepared by heating diborane at a pressure of about 300 mm in a 200-ml all-glass reaction vessel equipped with a breakseal for about 3 hours at 110°. The lower molecular weight hydrides can be separated and identified as previously described, but to prevent the decaborane from collecting with the pentaboranes, it is necessary to place the reaction vessel in an ice bath when the more volatile products are distilled out.[79] The decaborane remaining in the reaction vessel can be sublimed out at room temperature and can be identified by its melting point (100°), determined in a sealed capillary.

[78] L. V. McCarty, G. C. Smith, and R. S. McDonald, *Anal. Chem.*, **26**, 1027 (1954).

[79] This should be done a few minutes after the distillation has started, but before the reaction vessel has warmed up to 0°.

Diborane may be safely destroyed by slowly distilling it with pumping (to remove hydrogen) through a trap filled with NaOH (not KOH) pellets followed by a $-196°$ trap. Any diborane that survives this treatment should be passed through the NaOH trap again.

26. THE PREPARATION OF GERMANE[80]

$$HGeO_3^- + BH_4^- + 2H^+ \longrightarrow GeH_4 + H_3BO_3$$

Total time required: 3 hours

Actual working time: 3 hours

Preliminary study assignment:
W. L. Jolly and A. D. Norman, "Hydrides of Groups IV and V," *Prep. Inorg. Reactions,* **4**, 1 (1968), and, in this book, Chap. 8, The Vacuum Line, and Chap. 6, The Maintenance and Measurement of Low Temperatures.

Reagents required:
120 ml of glacial acetic acid
2 g of KOH pellets
1 g of GeO_2
1.5 g of KBH_4
Liquid nitrogen
$CHCl_3$ and CS_2 for cold baths
Ascarite

Special apparatus required:
500-ml three-necked round-bottomed flask with adapter for connection
    to vacuum line and stopper
100-ml dropping funnel
Magnetic stirrer and stirring bar
Vacuum line

Procedure

The reaction flask, fitted with the stopper, the dropping funnel, the adapter, and the magnetic stirring bar, is attached to the vacuum line so that the evolved gases can be pumped through a series of three U traps. The flask is

---

[80] Adapted from W. L. Jolly, *J. Am. Chem. Soc.,* **83**, 335 (1961), and W. L. Jolly and J. E. Drake, *Inorg. Syn.,* **7**, 34 (1963).

charged with 120 ml of glacial acetic acid, and the dropping funnel is charged with a solution prepared by dissolving, in order, the potassium hydroxide, the germanium dioxide, and the potassium hydroborate in 25 ml of water. While the acid is stirred, the system is evacuated through the series of traps for several minutes until only the vapor of acetic acid fills the system. Then the stopcock immediately before the first trap is closed and the traps are cooled to $-196°$ (liquid nitrogen). The solution in the dropping funnel is slowly added to the stirred acid over a period of 10 to 15 minutes. During the addition, the pressure in the system preceding the series of traps (as measured by a mercury manometer) is maintained at approximately 100 mm by adjustment of the stopcock immediately before the series of traps. In this way, the pressure of hydrogen gas in the traps will probably not exceed 10 mm, and the germane will be efficiently condensed. Then the stopcock immediately before the series of traps is completely opened for several minutes in order to pump most of the germane and hydrogen from the flask. Finally, the stopcock immediately above the reaction vessel is closed, and the reaction vessel is disconnected and quickly placed in a hood. The contents of the liquid-nitrogen traps (germane, digermane, acetic acid, water, and carbon dioxide) are combined in a trap at either end of the series, and the mixture is distilled through a $-63.5°$ trap ($CHCl_3$ slush) into a $-196°$ trap, to remove acetic acid and water. Then the carbon dioxide is removed by passing the material through a trap containing Ascarite, and finally the material is passed through a $-111.6°$ trap (CS_2 slush) to remove digermane and residual water.

The yield of germane (about 7 mmoles) can be determined by measurement of the pressure and temperature of the gas in a system of known volume.

Characterization

Germane has a vapor pressure at $-111.6°$ (CS_2 slush) of 182 mm.[81] The infrared spectrum[82] shows peaks at 2105, 943, and 815 cm^{-1}.

The reaction of germane with potassium hydroxide to form potassium germyl, followed by reaction with methyl iodide to form methylgermane, is an example of the use of germane in an organogermanium compound synthesis[83]:

$$2KOH + GeH_4 \longrightarrow K^+ + GeH_3^- + KOH \cdot H_2O$$

$$CH_3I + GeH_3^- \longrightarrow CH_3GeH_3 + I^-$$

[81] A. E. Finholt, A. C. Bond, Jr., K. E. Wilzbach, and H. I. Schlesinger, *J. Am. Chem. Soc.*, **69**, 2692 (1947).

[82] J. W. Straley, C. H. Tindal, and H. H. Nielsen, *Phys. Rev.*, **62**, 161 (1942).

[83] D. S. Rustad, T. Birchall, and W. L. Jolly, *Inorg. Syn.*, **11**, 128 (1968).

27. THE PREPARATION OF
DEUTERIUM CHLORIDE[84]

$$C_6H_5COCl + D_2O \longrightarrow C_6H_5COOD + DCl$$
$$C_6H_5COCl + C_6H_5COOD \longrightarrow (C_6H_5CO)_2O + DCl$$

Total time required: 4 hours

Actual working time: 4 hours

Preliminary study assignment:
Chapter 21, Infrared Spectrometry, in this book.

Reagents required:
1.0 ml of D_2O
40 ml of benzoyl chloride
Cylinder of helium or hydrogen
About 20 ml of concentrated H_2SO_4
Liquid nitrogen
2 boiling chips

Special apparatus required:
100-ml round-bottom flask with two side arms
U trap with stopcocks and ball joint
2 gas-washing bottles
Rubber septum; hypodermic needle and syringe

Procedure

The apparatus shown in Fig. 32.16 is assembled. The ground joints are greased with silicone grease; the flask is charged with 40 ml of benzoyl chloride and a few boiling chips; the apparatus is thoroughly flushed with a stream of helium or hydrogen (if hydrogen, the exit gas must be passed into a hood). The flow of helium (or hydrogen) is adjusted to a slow, steady rate (about one bubble per second in the gas-washing bottle), and the benzoyl chloride is gently refluxed for about 15 min. Then the benzoyl chloride is cooled to about 110° and liquid nitrogen is poured into the Dewar flask. At this point, the apparatus should be free of benzoic acid and water, and it is ready for the synthesis of DCl. One milliliter of deuterium oxide is added drop by drop to the reaction vessel, with a syringe and a hypodermic needle inserted through the rubber septum, over a period of 10 min. The temperature is then raised again, and the benzoyl chloride is refluxed for half an hour. The flow of helium (or hydrogen) is stopped; the stopcocks of the U trap are

---

[84] Adapted from the procedure of H. C. Brown and C. Groot, *J. Am. Chem. Soc.*, **64**, 2223 (1942). See G. Brauer, *Handbook of Preparative Inorganic Chemistry*, vol. 1, 2nd ed., Academic Press, Inc., New York, 1963, pp. 129–31.

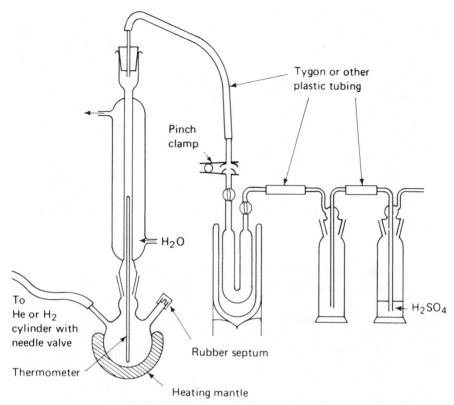

Tygon or other
plastic tubing

Pinch
clamp

H_2O

To
He or H_2
cylinder with
needle valve

Thermometer

Rubber septum

Heating mantle

H_2SO_4

FIG. 32.16. Apparatus for the preparation of DCl.

closed; the U trap is attached to the vacuum line (while the trap is cooled with liquid nitrogen), and the helium (or hydrogen) is pumped out of the trap. The DCl is purified by distilling it in vacuo into another liquid nitrogen-cooled trap through a trap cooled to $-111.6°$. The yield of DCl is determined by measuring its pressure in a system of known volume (3 to 5 l).

The same apparatus may be used to prepare HCl so that the properties of DCl and HCl can be compared. As soon as the DCl has been removed from the U trap in which it was originally collected, the U trap is reconnected to the synthetic apparatus. The apparatus is then flushed out with helium (hydrogen), and, by using a few drops of ordinary water, some HCl is prepared. The first fraction of HCl will be contaminated with DCl and should be discarded. The HCl should be purified in the same way as described for the DCl.

Characterization

The vapor pressures at $-111.6°$ (CS_2 slush) of DCl and HCl are approximately 116 and 122 mm, respectively. A simple, approximate method for determining the isotopic purity of the DCl (i.e., for determining the DCl/HCl ratio) is to examine the infrared spectrum of the gas at a pressure around 700 to 750 mm in a cell at least 7 cm long. Deuterium chloride's fundamental vibration frequency ($v = 0 \rightarrow v = 1$) is 2090 cm^{-1}; an absorption band with only P and R branches is observed. The corresponding band of hydrogen chloride occurs at 2885 cm^{-1}. It is a fairly easy matter to detect as little as 5 per cent HCl in DCl using these infrared absorption peaks. (Of course, a much more accurate and sensitive analysis can be obtained from a mass spectrum.) By using a high-resolution infrared spectrophotometer, the rotational fine structure of the P and R branches can be observed, and the data can be used to evaluate the bond distance.[85] If the first overtone bands are used (around 2.45 μ for DCl and 1.78 μ for HCl), a visible near infrared spectrophotometer, such as the Cary Model 14 recording spectrophotometer, can be used.

28. THE PREPARATION OF SULFUR NITRIDE, S₄N₃Cl, AND HEPTASULFUR IMIDE

When ammonia is passed into a solution of disulfur dichloride in an inert solvent, the mixture goes through a series of color changes, and a variety of products (including S_4N_4, S_7NH, $S_6(NH)_2$, S_4N_3Cl, and S_8) are formed. The relative amounts of these products can be varied markedly by changing the reaction conditions. By using an excess of ammonia at temperatures above 20°, the production of S_4N_4 is favored.[86] By using an excess of ammonia at temperatures below 0°, the production of S_7NH is favored.[87] By stopping the flow of ammonia at an intermediate point in the sequence of color changes, S_4N_3Cl can be obtained in good yield.[88]

A. Preparation of Tetrasulfur Tetranitride, S₄N₄

$$6 S_2Cl_2 + 16 NH_3 \longrightarrow S_4N_4 + 8 S + 12 NH_4Cl$$

or $\qquad 6 SCl_2 + 16 NH_3 \longrightarrow S_4N_4 + 2 S + 12 NH_4Cl$

Total time required: 2.5 days

Actual working time: 6 hours

[85] D. P. Shoemaker and C. W. Garland, *Experiments in Physical Chemistry*, McGraw-Hill Book Company, New York, 1962, pp. 309–15; F. E. Stafford, C. W. Holt, and G. L. Paulson, *J. Chem. Educ.*, **40**, 245 (1963); L. W. Richards, *J. Chem. Educ.*, **43**, 552 (1966).

[86] M. H. M. Arnold, U.S. Patent 2,372,046, March 20, 1945; M. Becke-Goehring, *Inorg. Syn.*, **6**, 123 (1960); M. Villena-Blanco and W. L. Jolly, *Inorg. Syn.*, **9**, 98 (1967).

[87] M. Becke-Goehring, H. Jenne, and E. Fluck, *Ber.*, **91**, 1947 (1958).

[88] M. Becke-Goehring and H. P. Latscha, *Z. anorg. allgem. Chem.*, **333**, 181 (1964).

Preliminary study assignment:

M. Becke-Goehring and E. Fluck, *Developments in Inorganic Nitrogen Chemistry*, vol. 1, C. B. Colburn, ed., Elsevier, New York, 1966, Chap. 3, see pp. 186 ff.

Reagents required:

1400 ml of CCl_4
50 ml of S_2Cl_2
Chlorine cylinder
Ammonia cylinder
450 ml of dioxane
500 ml of benzene
30 g of sodium-lead alloy ("dri-Na", J. T. Baker Chemical Co.)

Special apparatus required:

Paddle stirrer
Heavy-duty stirring motor
2-l three-necked round-bottomed flask
Soxhlet extractor and thimble

Procedure

Add 1400 ml of dry carbon tetrachloride[89] and 50 ml of disulfur dichloride to a 2-l three-necked flask. Insert a paddle stirrer through the main neck, a gas inlet tube through one of the side necks, and a thermometer through the other side neck. While stirring the mixture briskly, pass a stream of chlorine into the solution until a distinctly green layer of chlorine gas is observed over the solution[90] (the color of the solution changes from yellow to orange red). Stop the flow of chlorine and connect the gas delivery tube to a cylinder of ammonia. Immerse the flask to the level of the carbon tetrachloride solution in a water bath of running tap water, and pass ammonia through the stirred solution.[91] Pass the ammonia as rapidly as possible without causing material to splash from the flask or allowing the temperature of the reaction mixture to rise above 50°.

After approximately 2 hours, when the reaction has stopped and the entire mixture has turned a "golden poppy" color, stop the ammonia flow.

---

[89] Carbon tetrachloride may be dried by allowing it to stand over anhydrous calcium sulfate (Drierite) for several days.

[90] The preliminary chlorination of the S_2Cl_2 to SCl_2 may be omitted, if desired, but the total yield of S_4N_4 is thereby reduced by a factor of 2.

[91] The heat of reaction will prevent the temperature of the reaction mixture from falling below 20°, even though the water bath temperature is less than 20°. The water bath may be omitted, but then the flow of ammonia gas must be reduced to maintain the temperature below 50°.

It does no harm to pass an excess of ammonia; it is important that the "golden poppy" stage be reached.

Filter the reaction mixture on a large sintered-glass or Büchner funnel, and vigorously slurry the damp solid material with 1 l of water for 5 to 10 min. Filter off the remaining solid, and allow it to air dry thoroughly for 1 or 2 days. If the material is not yellow or orange at this time, it should be discarded and the synthesis should be repeated. Place the dry material in an extraction thimble and extract with 400 ml of dry dioxane[92] in a Soxhlet extractor until the eluate is only weakly colored orange yellow.[93] It is important that sufficient dioxane be used to avoid the possibility of concentrating the solution in the pot so much that the extracted S_4N_4 decomposes. Tetrasulfur tetranitride will explode if heated in a dry flask. When the eluate is cooled to room temperature, some of the S_4N_4 will crystallize out. Filter off this material and dry it in air, and evaporate the filtrate to dryness at a temperature lower than 60° (evaporation at room temperature at atmospheric pressure is slow, but satisfactory). The residue from the evaporation is recrystallized from hot benzene to remove sulfur. A combined yield of 16 g of S_4N_4 is generally obtained.

Characterization

Tetrasulfur tetranitride, as prepared by the preceding procedure, usually has a melting point of 178–179°. However, by repeated recrystallization from benzene, or by purification on an alumina chromatographic column, S_4N_4 with a melting point as high as 187–187.5° has been obtained.[94] The infrared spectrum of S_4N_4 is given by Lippincott and Tobin.[95]

CAUTION: Sulfur nitride should be handled with care, inasmuch as it will often explode when struck or when heated above 100°. Even the small amount of material contained in a melting-point capillary can explode violently.

B. Preparation of S_4N_3Cl

$$5\,S_2Cl_2 + 12\,NH_3 \longrightarrow S_4N_3Cl + 6\,S + 9\,NH_4Cl$$

Total time required: 2 days

Actual working time: 4 hours

---

[92] Dioxane may be dried by refluxing over sodium–lead alloy, followed by distillation.

[93] If the extraction is stopped too soon, S_4N_4 will be left in the thimble. If the extraction is prolonged, the product will be contaminated with sulfur.

[94] M. Villena-Blanco, "A Study of the Reactions Between Gaseous Ammonia and Sulfur Chlorides," M.S. Thesis, University of California, Berkeley, Calif., October, 1963.

[95] E. R. Lippincott and M. C. Tobin, *J. Chem. Phys.*, **21**, 1559 (1953).

Preliminary study assignment:

M. Becke-Goehring and E. Fluck, *Developments in Inorganic Nitrogen Chemistry*, vol. 1, C. B. Colburn, ed., Elsevier, New York, 1966, Chap. 3, see pp. 210–14, and D. A. Johnson, G. D. Blyholder, and A. W. Cordes, *Inorg. Chem.*, **4**, 1790 (1965).

Reagents required:
1400 ml of CCl_4
50 to 60 ml of S_2Cl_2
Ammonia cylinder
145 ml of concentrated H_2SO_4
850 ml of glacial acetic acid
180 ml of concentrated HCl
1125 ml of acetone
1225 ml of diethyl ether

Special apparatus required:
Paddle stirrer
Heavy-duty stirring motor
2-l three-necked round-bottomed flask

Procedure

Fifty ml of disulfur dichloride and 1400 ml of dry carbon tetrachloride[89] are added to a 2-l three-necked flask. A paddle stirrer is inserted through the main neck, a gas inlet tube is inserted through one of the side necks, and a thermometer is inserted through the other side neck. While the mixture is stirred briskly, a stream of ammonia gas is passed into the solution at a rate so as to maintain the temperature in the range 35 to 55°. If desired, the flask may be placed in a water bath during the passage of ammonia to carry off the heat of reaction, but the temperature should be maintained at 35 to 55°. As soon as the slurry begins to turn from an orange-tan to an olive-tan color (usually after $\frac{1}{2}$ to 1 hour),[96] the ammonia flow is stopped. Two of the necks of the flask are stoppered; one is fitted with a reflux condenser, and the slurry is refluxed for half an hour using a heating mantle. If the mixture does not turn bright yellow in this time, 10 ml of disulfur dichloride is added, and the refluxing is continued for another half-hour. The warm slurry is filtered through a sintered-glass funnel. The yellow solid is spread out on watch glasses and allowed to air dry at room temperature for 1 day. Sometimes a thin dark-brown coating forms on the mass of crystals, which may be disregarded. The yield is about 90 g of a mixture of S_4N_3Cl and NH_4Cl. The mixture is finely pulverized with a mortar and pestle.

[96] The entire sequence of color changes is usually the following: dark brown, orange-tan, olive-tan, olive-drab, dark brown, bright orange. No great harm is done if the olive-tan stage is passed. It is only necessary to add more S_2Cl_2 before refluxing.

CAUTION: Do not grind the material unless the bulk of it is bright yellow. Explosions have resulted from the grinding of mixtures containing S_4N_4, which is orange.

Fifty grams of the S_4N_3Cl–NH_4Cl mixture[97] is added, over a 10-min period, to 85 ml of concentrated sulfuric acid magnetically stirred in an 800-ml beaker. This latter operation should be carried out in a hood. After about 1 hour, effervescence will have ceased and all the material will have dissolved (except a small residue of sulfur). Glacial acetic acid (450 ml) is added; the mixture is stirred, and the slurry is filtered through a sintered-glass filter. The crystals of ammonium bisulfate on the filter are washed with 50 ml of acetic acid. The combined filtrate is transferred to a clean 800-ml beaker and cooled in an ice bath. During stirring, 100 ml of concentrated hydrochloric acid is added. The beaker is then removed from the ice bath, and a mixture of 625 ml of acetone and 625 ml of ether is added with stirring. Stirring is continued for 15 min. The crystals are collected on a sintered-glass filter, washed with 50 ml of ether, and air dried on a watch glass for half an hour.

The partially purified crystals from the preceding precipitation are dissolved in 60 ml of concentrated sulfuric acid in an 800-ml beaker, and 350 ml of glacial acetic acid is added. The solution is then treated as described for the filtrate in the preceding paragraph, except that only 80 ml of concentrated hydrochloric acid and 500 ml each of acetone and ether are used. After washing the crystals with ether, they are air dried for about 2 hours. The yield of purified S_4N_3Cl from 50 g of NH_4Cl–S_4N_3Cl mixture is about 5.4 g.

Characterization

The salt S_4N_3Cl is a bright canary yellow solid that is stable in dry air but is slowly attacked by moist air. It becomes dark green upon long exposure to light, and thus should be stored in darkness. In most organic solvents, S_4N_3Cl is insoluble, but it is appreciably soluble in anhydrous formic acid. The material has infrared absorption peaks at 8.6, 9.9, and 14.7 μ. It has no clean-cut melting point, but it melts with decomposition in the range 180–200°. By simple metathetical reactions, S_4N_3Cl may be converted to the corresponding iodide,[98] nitrate,[98] bromide,[98–100] thiocyanate,[98–100] bisulfate,[99] tetraphenylborate,[100] hexachloroantimonate(V),[100] tetrachloroantimonate(III),[100] pentachloroantimonate(III),[100] and fluoride.[101] X-ray diffrac-

---

[97] The amounts of reagents in this purification procedure may be scaled up or down to accommodate more or less of the S_4N_3Cl–NH_4Cl mixture.

[98] W. Muthmann and E. Seitter, *Ber.*, **30**, 627 (1897).

[99] A. Meuwsen and O. Jakob, *Z. Anorg. Chem.*, **263**, 200 (1950).

[100] M. Becke-Goehring and H. P. Latscha, *Z. Naturforsch.*, **17b**, 125 (1962).

[101] O. Glemser and E. Wyszomirski, *Chem. Ber.*, **94**, 1443 (1961).

tion studies[102,103] of the nitrate have shown that the $S_4N_3^+$ cation is a planar seven-membered ring:

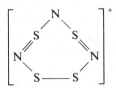

Nitrogen-15 nuclear magnetic resonance[104] has demonstrated the existence of two types of nitrogen atoms in the ion, in an abundance ratio of 1:2.

C. Preparation of Heptasulfur Imide

Total time required: 1.5 days

Actual working time: 4 hours

Preliminary study assignment:
  M. Becke-Goehring and E. Fluck, *Developments in Inorganic Nitrogen Chemistry*, vol. 1, C. B. Colburn, ed., Elsevier, New York, 1966, Chap. 3, see pp. 177–85.

Reagents required:
  1 l of dimethylformamide
  Ammonia cylinder
  Ice
  Salt
  100 ml of S_2Cl_2
  50 to 100 ml of concentrated HCl
  250 ml of tetrahydrofuran
  400 ml of methanol

Special apparatus required:
  Paddle stirrer
  Heavy-duty stirring motor
  2-l three-necked round-bottomed flask
  Magnetic stirrer and stirring bar

Procedure

Place 1 l of dry dimethylformamide in a 2-l three-necked flask. Insert a paddle stirrer through the main neck, a gas inlet tube through one of the side necks, and a thermometer through the other side neck. Immerse the flask in

[102] J. Weiss, *Z. Anorg. Allgem. Chem.*, **333**, 314 (1964).
[103] A. W. Cordes, R. F. Kruh, and E. K. Gordon, *Inorg. Chem.*, **4**, 681 (1965).
[104] N. Logan and W. L. Jolly, *Inorg. Chem.*, **4**, 1508 (1965).

an ice salt bath.[105] While stirring, pass a vigorous stream of ammonia gas through the solvent. When the solution is saturated with ammonia and the temperature has fallen to $-5°$, quickly add about 5 ml of disulfur dichloride to the flask (this addition is conveniently carried out using a pipet and a rubber bulb). A vigorous reaction will take place, and the temperature of the mixture will rise. When the temperature has again fallen to $-5°$, add another 5 ml of S_2Cl_2. Repeat this process until 100 ml of S_2Cl_2 has been added. Allow the ammonia flow to continue for another 15 min, and let the cold mixture stand at approximately $0°$ for about 1 hour. Slowly pour the slurry, with stirring, into a mixture of 2 l of ice-cold water, 50 ml of concentrated hydrochloric acid, and 500 g of granulated ice. Neutralize the resulting mixture with dilute hydrochloric acid, and allow the mixture to stand for about 2 hours (no longer than 10 hours, or the yield will suffer) in order to allow the precipitate to settle out. Decant off the supernatant, collect the precipitate on a sintered-glass funnel, and wash it with water. Spread the solid out on watch glasses and allow it to air dry for several hours. Add the crude product to a glass-stoppered flask containing 250 ml of peroxide-free tetrahydrofuran and a magnetic stirring bar, and vigorously stir the mixture for at least one half-hour. Filter the mixture (discard the insoluble material) and allow the filtrate to evaporate at a temperature below $60°$. By repeated recrystallization of the residue from hot methanol,[106] about 16 g of moderately pure S_7NH can be obtained. If very pure S_7NH is desired, this latter product may be dissolved in a minimum volume of carbon tetrachloride (about 1 l) and separated in several batches by elution chromatography on a silica gel column.[107] For example, about 150 ml of the saturated carbon tetrachloride solution may be passed through a column 24 cm long and 75 mm wide in each batch separation. Elution with carbon tetrachloride yields a band of sulfur followed closely by a band of S_7NH. S_7NH in the eluate may be detected by mixing a small volume of the eluate with an equal volume of methanolic KOH. A red-violet color forms in the presence of S_7NH. When heated, the mixture turns yellow, as it also does in the presence of sulfur alone.

Characterization

S_7NH is a pale yellow, almost colorless solid with a melting point of $113.5°$. The compound may be distinguished from sulfur by its infrared spectrum.[108] Absorption bands occur at 805, 1290, and 3300 cm^{-1}.

[105] A bath of alcohol or acetone to which chunks of Dry Ice are occasionally added can be used.

[106] The principal impurity, sulfur, is relatively insoluble in methanol. Hence, it is not necessary to dissolve the entire residue during the recrystallization.

[107] M. Villena-Blanco and W. L. Jolly, *J. Chromatog.*, **16**, 214 (1964); H. G. Heal and J. Kane, *Inorg. Syn.*, **11**, 184 (1968).

[108] M. Goehring and G. Zirker, *Z. anorg. Chem.*, **285**, 70 (1956).

29. THE PREPARATION OF $S_3N_2O_2$

$$S_2Cl_2 + 8\,SOCl_2 + 6\,NH_4Cl \longrightarrow 3\,S_3N_2O_2 + 24\,HCl + SO_2$$

Total time required: 9 hours

Actual working time: 3 hours

Preliminary study assignment:
    W. L. Jolly and M. Becke-Goehring, *Inorg. Chem.*, **1**, 76 (1962).

Reagents required:
    150 g of NH_4Cl
    40 ml of S_2Cl_2
    30 ml of $SOCl_2$
    30 ml of benzene
    150 g of Drierite

Special equipment required:
    Tube furnace
    Glass reaction tube (28 mm O.D., 60 cm long)
    Air condenser (20 mm O.D., 60 cm long)
    2 gas-washing bottles
    Glass wool

Procedure

The granular ammonium chloride is packed into the reaction tube, as shown in Fig. 32.17, so that the length of the bed of ammonium chloride is about

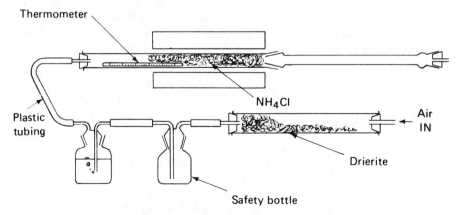

FIG. 32.17. Apparatus for the preparation of $S_3N_2O_2$.

6 cm shorter than the tube furnace. The joint connecting the reaction tube and the air condenser is positioned just inside the tube furnace so that it does not become clogged with sublimed ammonium chloride. The rest of the apparatus is set up as shown in Fig. 32.17, except that at first nothing is put in the second gas-washing bottle. Kel–F grease is used on the joint of the second bottle; ordinary grease can be used on the first bottle. Springs or rubber bands should be used to hold the joints on the bottles closed. A stream of dry air is passed through the apparatus,[109] and the reaction tube is heated to 170 to 180°. After dry air is passed for about 20 min at this temperature, the second gas-washing bottle is opened and a mixture of 40 ml of S_2Cl_2 and 30 ml of $SOCl_2$ is added. The gas-washing bottle is reassembled, and the flow of air and the temperature are maintained for 10 to 20 hours. During this time, the inside of the air condenser becomes coated with an orange solid. Then the air flow and heating are stopped, and the air condenser is removed from the reaction vessel and quickly attached to a 100-ml flask containing 30 ml of dry benzene. A water-cooled condenser is attached to the top of the air condenser, and the benzene is refluxed for 3 hours, or until only a pale yellow residue remains in the air condenser. The solution is allowed to cool to below 40°; then a stream of dry air is passed over the solution in the flask until a fairly thick slurry of $S_3N_2O_2$ forms. The orange crystals are filtered off and sucked dry on a sintered glass filter flushed with argon. In a dry atmosphere (preferably in a dry bag), the product is transferred to a tared sample bottle. The yield is 2 to 4 g.

Characterization

Pure $S_3N_2O_2$ melts at 100.7°. The crude orange-yellow product prepared by this method usually melts at 97 to 100°; however, recrystallization from benzene will yield a bright yellow material melting above 100°. The compound is extremely sensitive toward moisture, traces of which cause it to decompose, principally according to the following reaction:

$$2\,S_3N_2O_2 \longrightarrow S_4N_4 + 2\,SO_2$$

$S_3N_2O_2$ has the structure:

---

[109] The flow rate should be adjusted so that when an 8-mm exit tube is immersed in water, bubbles form at the rate of about 4 sec^{-1}.

30. THE PREPARATION OF TRIMERIC
AND TETRAMERIC PHOSPHONITRILIC
CHLORIDE[110]

$$PCl_5 + NH_4Cl \longrightarrow PNCl_2 + 4\,HCl$$

Total time required: 3 days

Actual working time: 9 hours

Preliminary study assignment:
  N. L. Paddock, *Quart. Rev. London*, **18**, 168 (1964); R. A. Shaw, R. Keat, and C. Hewlett, *Prep. Inorg. Reactions*, **2**, 1 (1965); and, in this book, Chap. 5, Solvents.

Reagents required:
  300 ml of dry 1,1,2,2-tetrachloroethane
  120 g of PCL_5
  40 g of NH_4Cl
  About 25 ml of diethyl ether
  About 25 ml of benzene
  Drierite desiccant
  Boiling chips

Special apparatus required:
  1-l round-bottom flask with ground glass joint
  1-l size heating mantle
  Magnetic stirrer and stirring bar
  Drying tube
  Special wide-bore distillation head and vacuum receiving adapter
  Aspirator, manometer, water trap, and drying tube
  Vigreux column, 30 cm long, 15 mm in diameter
  Electrically heated jacket for Vigreux column[111]
  2 Variacs
  Electrical heating tape
  Infrared lamp
  Glass-blowing torch
  Oil bath
  Hot plate

[110] The procedure is essentially that of R. Schenck and G. Römer, *Ber.*, **57B**, 1343 (1924), including some modifications introduced by M. L. Nielsen and T. J. Morrow, *Inorg. Syn.*, **6**, 97 (1960).

[111] Methods for the construction of a heated jacket for a fractionating column are described by J. Cason and H. Rapoport, *Laboratory Text in Organic Chemistry*, 2nd ed., Prentice-Hall, Inc., Englewood Cliffs, N. J., 1962, pp. 291–93, and K. B. Wiberg, *Laboratory Technique in Organic Chemistry*, McGraw-Hill Book Company, New York, 1960, pp. 50–52.

CAUTION: The phosphonitrilic chlorides and their derivatives are toxic and should be handled only in a hood. Avoid contact with the skin.

Procedure

Add the ammonium chloride, phosphorus pentachloride, tetrachloroethane, and magnetic stirring bar to the 1-l flask. Attach a reflux condenser topped by a drying tube containing a little Drierite. While magnetically stirring, heat the mixture with the heating mantle to the boiling point in a hood. Maintain a moderate reflux for 10 to 15 hours. After bringing the reaction products to room temperature, filter off the unreacted ammonium chloride and transfer the solution, a few boiling chips, and the magnetic stirring bar to the cleaned and dried 1-l flask.

Set up for a vacuum distillation (using a 500-ml flask as receiver), and, while stirring, remove the solvent at a pressure of 15 mm or less. A water bath should be used for the heating; its temperature should be raised to 75° (no higher). When no more solvent comes over at 75°, stop the distillation and transfer the hot liquid residue to a 250-ml round-bottom flask.

Set up for a vacuum distillation using an oil bath and the jacketed Vigreux column (see Fig. 32.18). Place a short thermometer or thermocouple in the space between the Vigreux column and the jacket. No condenser is necessary. The wide-bore distillation head is connected directly to the vacuum receiving adapter; a 100- or 250-ml flask is used as a receiver. The joints should be lubricated with silicone grease; the distillation head should be wrapped with electrical heating tape. The electrical current supplied to the column jacket and the heating tape should be controlled by Variacs. The receiving adapter should be heated with the infrared lamp or a gentle flame. A pressure of about 10 mm is maintained throughout the distillation. With the oil bath temperature around 100° and the column temperature around 90°, a few drops of tetrachloroethane may come over. These may be forced out of the receiving flask by application of a flame. About 20 g of crude trimeric phosphonitrilic chloride will come over at column temperatures between 120 and 135°. When no more material distills over at 135°, the vacuum is broken with dry air and the receiving flask is replaced by a clean dry flask. About 2 g of an intermediate fraction will come over at column temperatures between 135 and 185°. Again, the receiving flask is changed. Finally, about 10 g of crude tetrameric phosphonitrilic chloride is collected at column temperatures between 185 and 220°. Do not raise the oil bath temperature above 250°. This fractional distillation lasts about 5 hours.

By recrystallization of the first fraction from a small amount of warm ether, about 16 g of trimer (m.p. = 113.5–114.5°) can be obtained. By recrystallization of the final fraction from a small amount of hot benzene, about 7 g of tetramer (m.p. = 123–124°) can be obtained.

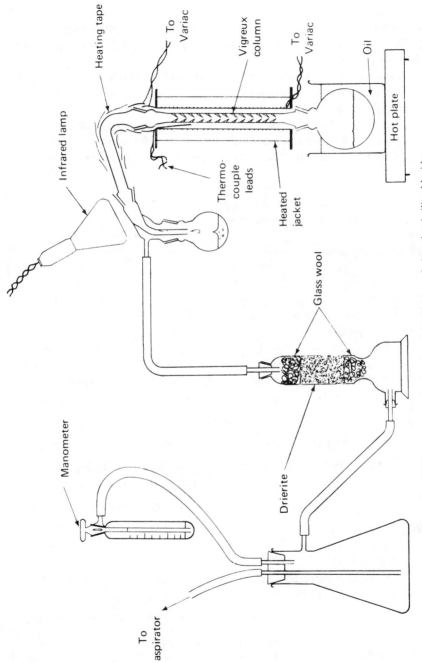

FIG. 32.18. Apparatus for fractional distillation of phosphonitrilic chlorides.

Characterization

The infrared spectra of the two isomers may be determined by using solutions in CCl_4 and CS_2 or by using KBr pellets. The spectra may be compared with those in the literature.[112]

Various derivatives of the phosphonitrilic chlorides can be prepared. The reader is referred to the chapter by Shaw, Keat, and Hewlett (cited in the preliminary study assignment) for examples.

The X-ray diffraction patterns for the two isomers can be determined and compared with those reported in the literature.[113] Lines at $d = 7.63\,\text{Å}$ and $d = 5.66\,\text{Å}$ are characteristic of the trimer and tetramer, respectively.

31. THE PREPARATION OF
DIAMIDOPHOSPHORIC ACID

Total time required: 2 days

Actual working time: 4 hours

Preliminary study assignment:
   M. L. Nielsen, *Developments in Inorganic Nitrogen Chemistry*, vol. 1, C. B. Colburn, ed., Elsevier, New York, 1966, Chap. 5, see pp. 369–78; and, in this book, Chap. 1, pp. 9–10, and Appendix 6, Compressed-Gas Cylinders. Method A: H. N. Stokes, *Am. Chem. J.*, **15**, 198 (1893). Method B: R. Klement and O. Koch, *Ber.*, **87**, 333 (1954), and *Angew. Chem.*, **65**, 266 (1953).

Method A

$$C_6H_5OH + POCl_3 \longrightarrow C_6H_5OPOCl_2$$
$$C_6H_5OPOCl_2 + 4\,NH_3 \longrightarrow C_6H_5OPO(NH_2)_2 + 2\,NH_4^+ + 2\,Cl^-$$
$$C_6H_5OPO(NH_2)_2 + 2\,OH^- \longrightarrow C_6H_5O^- + PO_2(NH_2)_2^- + H_2O$$
$$PO_2(NH_2)_2^- + HOAc \longrightarrow HOPO(NH_2)_2 + OAc^-$$

Reagents required:
   25 g of phenol
   27 ml of $POCl_3$[114]
   80 ml of concentrated aqueous NH_3
   670 ml of 95 per cent ethanol

---

[112] L. W. Daasch, *J. Am. Chem. Soc.*, **76**, 3403 (1954); S. Califano, *J. Inorg. Nucl. Chem.*, **24**, 483 (1962).

[113] A. H. Herzog and M. L. Nielsen, *Anal. Chem.*, **30**, 1490 (1958).

[114] The $POCl_3$ must be pure. If it cannot be obtained from a recently opened bottle, it should be distilled to free it of hydrolysis products.

12 g of KOH (85 per cent)
50 ml of glacial acetic acid
10 ml of ether or absolute ethanol

Special apparatus required:
15-cm Vigreux column with heated jacket

Procedure

Twenty-five grams of phenol and 27 ml of $POCl_3$[114] are placed in a 100-ml round-bottom flask fitted with a reflux condenser. The mixture is heated with a heating mantle at reflux in a good hood for about 12 hours. The mixture is cooled and the reflux condenser is replaced with a distillation head. The material is distilled, using a short (about 15-cm) Vigreux column with a heated jacket. Only the fraction that boils at 235–255° is collected. The yield is about 36 g of crude phenyl dichlorophosphate.

The phenyl dichlorophosphate is added slowly through a dropping funnel to 80 ml of vigorously stirred concentrated aqueous ammonia (28 per cent), cooled in an ice bath. The phenyl dichlorophosphate is added at a rate such that the temperature of the reaction mixture never exceeds 10°. The white precipitate of phenyl diamidophosphate is collected on a sintered-glass funnel and washed with 100 ml of cold water. The crude product is partially dried on the funnel and then recrystallized from hot 95 per cent alcohol. The product is air dried for several hours. The yield is about 20 g of phenyl diamidophosphate melting at 187–191°.

Twelve grams of potassium hydroxide is dissolved in 20 ml of water in a 250-ml beaker. Fifteen grams of phenyl diamidophosphate is added and the mixture is stirred while the temperature is kept at 40 to 55°. When the phenyl diamidophosphate has dissolved, the solution is cooled to about 5° in an ice bath, and, with stirring, 50 ml of glacial acetic acid is added at such a rate that the temperature never exceeds 10°. The mixture is held between 0 and 5° for 5 min, and the precipitated diamidophosphoric acid is collected on a sintered-glass funnel and washed with 60 ml of cold 95 per cent alcohol followed by 10 ml of absolute alcohol or ether. The material is spread out on a watch glass or filter paper and allowed to dry in a vacuum desiccator for at least 1 hour. The yield is about 7.9 g of crude product.

Purification is achieved by precipitation from aqueous solution by the addition of alcohol. The crude product is dissolved in the least amount of water required to dissolve it at 30° (no higher); the solution is filtered if necessary, and then cooled to about 20°. A volume of 95 per cent ethanol twice that of the aqueous solution is added with stirring; the mixture is cooled to between 10 and 15° and allowed to stand for about 30 min. The precipitated acid is collected on a sintered-glass funnel and dried in a vacuum desiccator. The product from method A will yield about 5.5 g of purified diamidophosphoric acid.

Method B

$$POCl_3 + 6 NH_3 \longrightarrow PO(NH_2)_3 + 3 NH_4Cl$$

$$PO(NH_2)_3 + OH^- \longrightarrow PO_2(NH_2)_2^- + NH_3$$

$$PO_2(NH_2)_2^- + HOAc \longrightarrow HOPO(NH_2)_2 + OAc^-$$

Reagents required:
  720 ml of freshly distilled, dry chloroform[115]
  50 ml of POCl$_3$[114]
  96 g of KOH pellets (85 per cent pure)
  400 ml of glacial acetic acid
  About 2.5 l of 95 per cent ethanol
  50 ml of absolute ethanol or diethyl ether
  Cylinder of ammonia
  Dry Ice

Special apparatus required:
  Magnetic stirrer and stirring bar
  Large evaporating dish

Procedure

A 1-l three-necked flask is placed in a large crystallizing dish on a magnetic stirrer, and the side necks are fitted with a low-temperature thermometer and a gas inlet tube that reaches almost to the bottom of the flask. A magnetic stirring bar and 700 ml of dry chloroform are placed in the flask, and a moderately fast stream of ammonia is passed through the stirred liquid. Alcohol is placed in the crystallizing dish, and particles of Dry Ice are added. The temperature of this cold bath is maintained at -25 to $-45°$ by the continuous addition of Dry Ice. When the chloroform has reached a temperature of $-15°$, the drop by drop addition of a solution of 50 ml of POCl$_3$[114] and 20 ml of chloroform from a dropping funnel is begun. The bottom tip of the dropping funnel should be about 5 cm above the top of the middle neck of the flask. Throughout the addition (which may last about 1 hour), the temperature of the chloroform should be maintained at from -15 to $-20°$. If a stalactite forms on the tip of the dropping funnel, it should be knocked off. Use a stirring rod to dislodge any crust of solid material that forms in the neck of the flask. When all the POCl$_3$ solution has been added, continue the ammonia flow for another 15 min; then stop the stirring, remove the flask

[115] Shake 850 ml of reagent grade chloroform with 400 ml of water; separate, dry with CaCl$_2$, reflux with P$_2$O$_5$, and finally distill.

from the cold bath, and loosely stopper the middle neck of the flask. If desired, at this point the reaction mixture may be stored at room temperature for a few days. When the flask has reached room temperature, the precipitate is collected on a sintered-glass funnel and air dried on watch glasses for several hours. The yield is about 90 g of a mixture of $PO(NH_2)_3$ and NH_4Cl.

Ninety-six grams of potassium hydroxide is dissolved in 160 ml of water in a 1-l beaker, and 80 g of the $PO(NH_2)_3$–NH_4Cl mixture is added to the stirred solution. Heat the slurry to 95°, and maintain it at this temperature until ammonia evolution ceases (about $\frac{1}{2}$ hour). If all the material does not dissolve, add a little water. Cool the solution to about 5° in an ice bath, adding just enough water to prevent precipitation, if necessary. Then add 400 ml of glacial acetic acid at such a rate that the temperature never exceeds 10°. Hold the mixture at 0 to 5° for 5 min, and collect the precipitate on a sintered-glass funnel and wash it with 100 ml of 95 per cent ethanol followed by 25 ml of absolute ethanol or ether. Air dry the material. The yield is about 35 g of crude diamidophosphoric acid. The purification scheme described under method A will yield about 12 g of the pure acid.

Characterization

Weigh out about 0.2 g of the purified acid on an analytical balance and dissolve the sample in 20 ml of water. Using a pH meter, titrate the solution with standardized 0.1 M NaOH. From the titration curve, estimate the purity of the acid and its ionization constant.

The infrared spectrum of the acid can be determined, using a KBr pellet. Absorption bands have been observed at the following frequencies (in cm^{-1}): 675(s), 905(sh), 915(m), 975(w), 1020(m), 1030(sh), 1090(m), 1130(s), 1255(s), 1415(s), 1565(m), 1625(w), 2065(sh), 2150(m), 2445(m), 2615(w), 2780(sh), 2920(s), 3280(m), and 3430(m). Assign the bands to particular groups in the molecule.

Aqueous solutions of diamidophosphoric acid are unstable; at room temperature decomposition to ammonium monohydrogen phosphate is almost complete in one week. Hydrolysis is greatly retarded, however, in alkaline solutions. The presence of phosphate in an aqueous solution of the acid can be ascertained by adding aqueous barium chloride to the solution. A freshly prepared solution of the pure acid will give no precipitate, but on heating or long standing, a precipitate of barium phosphate forms.

Questions

1. Why cannot diamidophosphoric acid be prepared by the partial ammonolysis of $POCl_3$ to $POCl(NH_2)_2$, followed by hydrolysis to $HOPO(NH_2)_2$?

2. How can one prepare $PO(NH_2)_3$ free from ammonium chloride?

32. THE PREPARATION OF
POTASSIUM NITROSODISULFONATE[116]

$$HNO_2 + 2\,HSO_3^- \longrightarrow HON(SO_3)_2^{2-} + H_2O$$
$$12\,K^+ + 6\,HON(SO_3)_2^{2-} + 2\,MnO_4^- \longrightarrow K_4[ON(SO_3)_2]_2 + 2\,MnO_2$$
$$+ 2\,OH^- + 2\,H_2O$$

Total time required: 3 hours

Actual working time: 2 hours

Preliminary study assignment:
   D. M. Yost and H. Russell, Jr., *Systematic Inorganic Chemistry*, Prentice-Hall, Inc., Englewood Cliffs, N. J., 1946, pp. 93–94.

Reagents required:
   14 g of $NaNO_2$
   21 g of $NaHSO_3$
   8 ml of glacial acetic acid
   8 ml of concentrated aqueous ammonia
   5 g of $KMnO_4$
   About 150 g of KCl
   15 ml of 95 per cent ethanol
   15 ml of diethyl ether
   Crushed ice

Special equipment required:
   None

Procedure

The synthesis is carried out in an 800-ml beaker in an ice bath. The reaction mixture is continuously stirred with a large magnetic stirring bar. The various solutions called for in the following procedure should all be prepared immediately before starting the synthesis.

   A solution of 14 g of sodium nitrite in 40 ml of water is placed in the beaker in the ice bath. After stirring for several minutes, 80 g of crushed ice is added, and a solution of 21 g of sodium bisulfite in 40 ml of water is added over a period of 1 min. Then 8 ml of glacial acetic acid is added; after 3 min, 8 ml of concentrated aqueous ammonia is added. Then, during 5 min, a solution of 5 g of potassium permanganate in 160 ml of water is added. After being stirred for about 2 min, the solution is suction filtered into a 1-l filter flask containing 400 ml of a saturated solution of potassium chloride

[116] Adapted from the method of H. J. Teuber, as reported by D. J. Cram and R. A. Reeves, *J. Am. Chem. Soc.*, **80**, 3094 (1958).

(saturated at 0°), cooled in an ice bath. If during the course of the filtration the unfiltered reaction mixture begins to decompose, as evidenced by effervescence and bleaching of the supernatant purple solution, the filtration should be stopped. It is usually possible to filter at least 200 ml of the solution. A beautiful precipitate of yellow crystals forms in the filter flask. This precipitate is collected on a funnel and is washed, successively, with 15 ml of ice water, 15 ml of alcohol, and 15 ml of ether. The compound is thoroughly dried by evacuation at less than 0.1 mm for several hours; it is best stored in an evacuated ampoule in a refrigerator.

Characterization

Potassium nitrosodisulfonate is interesting because, although it is bright yellow as a solid, its aqueous solutions are bright violet. In the solid state it is dimeric and diamagnetic; the aqueous $ON(SO_3)_2^{2-}$ ion is paramagnetic.[117] The esr spectrum of the solution consists of three lines of equal intensity; the splitting corresponds to coupling of the electron spin with the ^{14}N nuclear spin. The absorption spectrum of the solution shows maxima at 248 mμ and 545 mμ.[118] In acidic solutions, the nitrosodisulfonate ion rapidly decomposes:

$$4\,ON(SO_3)_2^{2-} + 3\,H_2O \longrightarrow 2\,HON(SO_3)_2^{2-} + N_2O + 4\,HSO_4^-$$

In basic solutions, the decomposition is very slow, and nitrite and hydroxylamine trisulfonate are found among the decomposition products.[118] The solutions are good oxidizing agents and may be analyzed iodimetrically:

$$2\,ON(SO_3)_2^- + 3\,I^- + 2\,H^+ \longrightarrow 2\,HON(SO_3)_2^{2-} + I_3^-$$

33. THE PREPARATION OF ANHYDROUS COPPER(II) NITRATE

$$Cu + 3\,N_2O_4 \longrightarrow Cu(NO_3)_2 \cdot N_2O_4 + 2\,NO$$
$$Cu(NO_3)_2 \cdot N_2O_4 \longrightarrow Cu(NO_3)_2 + N_2O_4$$

Total time required: 9 hours

Actual working time: 4 hours

[117] An orange-brown form of potassium nitrosodisulfonate has been shown to be paramagnetic and to contain essentially monomeric $ON(SO_3)_2^{2-}$ ions. [W. Moser, *et al.*, *J. Chem. Soc.* (A), 3039, 3043 (1968).] Violet paramagnetic salts are formed with a number of cations larger than potassium. [D. L. Fillmore and B. J. Wilson, *Inorg. Chem.*, **7**, 1592 (1968).]

[118] J. H. Murib and D. M. Ritter, *J. Am. Chem. Soc.*, **74**, 3394 (1952); B. J. Wilson and D. M. Ritter, *Inorg. Chem.*, **2**, 974 (1963).

Preliminary study assignment:

C. C. Addison and N. Logan, *Preparative Inorganic Reactions*, vol. 1, W. L. Jolly, ed., Interscience Publishers, New York, 1964, Chap. 6, p. 141.

Reagents required:

Strip of copper metal (about 1.2 cm × 15 cm × 1 mm)
20 ml of ethyl acetate
About 70 ml of N_2O_4 (obtained directly from cylinder)
P_2O_5 for drying tubes

Special apparatus required:

Special apparatus shown in Fig. 32.19
Vacuum pump and McLeod gauge, with liquid nitrogen trap to protect pump and gauge
Oil bath
Dry bag, with Drierite tube for drying flushing air

Procedure

Twenty milliliters of N_2O_4 is transferred from the cylinder to reaction vessel *A*, and then the copper strip and 20 ml of ethyl acetate are added. The reaction vessel is attached to the filtration apparatus, as shown in Fig. 32.19, and the reaction mixture is allowed to stand at room temperature for 4 hours. Then reaction vessel *A* is momentarily disconnected from the apparatus while any unreacted copper metal is removed and another 20 ml of N_2O_4 is added. The apparatus is inverted, and suction[119] is applied to the side arm below the fritted disk until all the liquid has been sucked from the precipitate of $Cu(NO_3)_2 \cdot N_2O_4$. Then reaction vessel *A* is momentarily removed while about 15 ml of N_2O_4 is poured onto the precipitate, and the liquid is drawn through the frit. This washing operation is repeated, and the crystals are sucked dry. Reaction vessel *A* is replaced with a similar, but clean, vessel, and the vessel containing the filtrate is removed and replaced with a cap. The apparatus is reinverted, allowing the crystals to fall to the bottom of the clean vessel.

In the dry bag, the vessel containing the crystals of $Cu(NO_3)_2 \cdot N_2O_4$ is removed from the filtration apparatus, and a small wad of glass wool is pressed down onto the crystals in the bottom of the vessel. The vessel is connected to a high-vacuum source (protected by a liquid nitrogen cold trap) and is immersed in an oil bath so that the oil surface is about 2 cm above the glass wool. While a vacuum of 10^{-2} mm or less is applied, the bath temperature is slowly raised to 120°. During this process, the compound decomposes to anhydrous copper(II) nitrate. The bath temperature is then raised to 200°, and the copper nitrate slowly sublimes to the cooler walls of the vessel above

[119] Through a Dry-Ice trap to protect the house vacuum line from the corrosive N_2O_4.

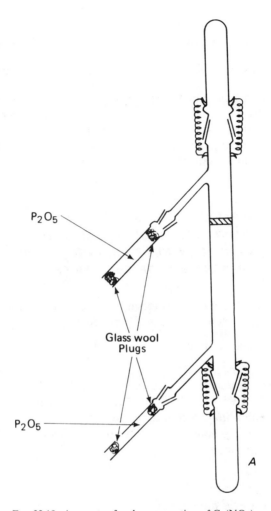

FIG. 32.19. Apparatus for the preparation of $Cu(NO_3)_2$.

the oil surface. In the dry bag, the anhydrous, sublimed product is transferred to a bottle which is then sealed with a greased glass stopper.

Characterization

Anhydrous copper(II) nitrate is one of a group of covalently bonded nitrates[120]
which have sufficient volatility that they can be sublimed at pressures of 0.1
to 1.0 mm Hg at temperatures in the range of 100 to 200°. A recent addition
to this group is anhydrous cobalt(III) nitrate, which has been prepared

[120] C. C. Addison and N. Logan, *Adv. Inorg. Chem. Radiochem.*, **6**, 71 (1964).

by the reaction of CoF_3 with N_2O_5.[121] Most of these volatile covalent nitrates are characterized by bidentate nitrate groups. In solid $CuNO_3$, the Cu atoms are linked together by

$$\begin{array}{c} O \\ | \\ -O-N-O- \end{array}$$

groups in such a way that each copper atom is surrounded by a distorted octahedron of oxygen atoms. In the vapor phase, two NO_3 groups are bonded to a copper atom so as to surround the copper atom by an elongated tetrahedron of oxygen atoms. Infrared spectra, in conjunction with Raman spectra, can be used to determine the mode of bonding of nitrate groups.[120]

34. THE SEPARATION OF SAMARIUM, NEODYMIUM, PRASEODYMIUM, AND LANTHANUM BY ION EXCHANGE[122] AND THE PREPARATION OF THE OXIDES

Total time required: 5 days (It is recommended that two students share the work of this experiment)

Actual working time: 12 hours

Preliminary study assignment:
F. H. Spedding and A. H. Daane, eds., *The Rare Earths*, John Wiley & Sons, Inc., New York, 1961; T. Moeller, *The Chemistry of the Lanthanides*, Reinhold Publishing Corporation, New York, 1963; and, in this book, Chap. 13, Ion Exchange, and Appendix 6, Theoretical Plates and the Binomial Expansion.

Reagents required:
0.5 g of didymium oxide[123]
265 g of citric acid monohydrate
0.5 g of HgI_2
About 1000 ml of concentrated aqueous ammonia
100 g of ammonium oxalate, $(NH_4)_2C_2O_4 \cdot H_2O$
125 g of Dowex–50 Resin (7.5 per cent cross linking; 200 to 400 mesh)
100 ml of concentrated HCl
Standard buffer solution, pH 4.00

---

[121] R. J. Fereday, N. Logan, and D. Sutton, *Chem. Commun.*, 271 (1968).

[122] This separation employs elution chromatography. For a description of a simple separation by displacement chromatography, see J. E. Powell, F. H. Spedding, and D. B. James, *J. Chem. Educ.*, **37**, 629 (1960).

[123] Didymium oxide (codes 420 or 422, from the Lindsay Chemical Division of the American Potash and Chemical Corporation, West Chicago, Ill.) has proved very satisfactory.

Special apparatus required:

Glass column (2.5 cm in diameter; about 35 cm long)
6-l bottle
Automatic sample collector (with interval timer)
About 36 50-ml beakers
Muffle furnace
pH meter
pHydrion paper

Preparation of Rare Earth Solution

Dissolve the didymium oxide in the smallest possible volume of 6 *M* HCl (50 volume per cent water; 50 volume per cent concentrated HCl), and then add concentrated aqueous ammonia drop by drop until the pH (as measured with pHydrion paper) is between 2 and 4. Keep the total volume of the solution below 25 ml.

Preparation of Eluant

Dissolve the citric acid in 5 l of water and titrate with concentrated ammonium hydroxide to a pH of *precisely* 3.05. (This operation must be done with the aid of a pH meter that has been standardized with a pH 4.00 buffer.) Transfer the solution to the 6-l bottle and add about 0.5 g of HgI_2 (mold inhibitor).

Preparation of Resin Column

Slurry the resin with two 200-ml portions of 6 *M* ammonia (2 volumes of concentrated ammonia plus 3 volumes of water) and then rinse with three 200-ml portions of water. Put a plug of glass wool in the bottom of the glass column, and add sufficient sand to make a flat supporting surface for the resin bed. Then slowly add the resin (as a thin slurry in water) to the column, always keeping a layer of water over the resin that has settled in the column. (Make the resin column 31 to 32 cm high; on standing, the resin will pack to a height of approximately 30 cm.)

The resin particles undergo partial size separation as they settle from the slurry, giving the column a banded appearance. This banding will not interfere with the proper functioning of the column. The final top surface of the column should be as flat as possible; it helps to add the last 10 to 20 cc of slurry from an inverted pipet. *Never allow the liquid level to fall below the surface of the resin.* Finally, add a 1-cm layer of sand to protect the top layer of resin.

Procedure

Set up the apparatus as shown in Fig. 32.20 and pass 300 to 400 ml of eluant through the column. Then remove the stopper from the column and allow the liquid level to fall just to the top of the resin bed. Add the didymium chloride solution with a pipet, taking care that the resin surface is not disturbed. Again allow the liquid level to fall to the surface of the resin. Carefully add a little eluant and restopper the column. Adjust the flow rate of eluant to 1.0 ml min^{-1} with the screw clamp.

The eluate is collected in successive 40-ml fractions. This is accomplished by setting the interval timer so that the turntable of the sample collector turns to a new beaker every 40 min. Thus, if the turntable is started with 24 empty beakers at 4:30 P.M., you must be back in the laboratory at 8:00 the next

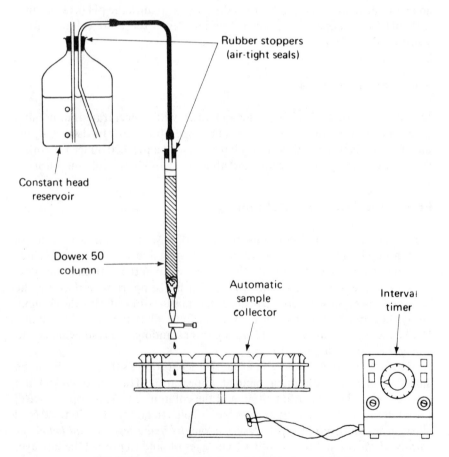

FIG. 32.20. Ion-exchange column for separating rare earths.

morning to replace the filled beakers with empty ones. To each 40-ml fraction add 2 ml of 1 M HCl and 5 ml of saturated ammonium oxalate solution. If a rare earth is present in a sample, a precipitate should form within 30 min. Make a rough visual estimate of the amount of precipitate in each beaker and plot it against sample number (see Fig. 32.21). It will be found that three distinct rare earth fractions are eluted in the first 2000 to 4000 ml. These are most likely samarium, neodymium, and praseodymium. At this point, the only rare earth remaining on the column is lanthanum, and, rather than wait for it to be eluted with the same eluant, it is better to hasten its elution by adding about 50 ml of concentrated ammonia to each liter of eluant remaining.

Combine the precipitates corresponding to a given rare earth fraction. Digest each group of precipitate in its mother liquor on the steam bath for several hours. Separately filter off each of the four groups of precipitate on ashless filter paper. Ignite the four samples in the muffle oven at 900° for 1 hour and then weigh each of the rare-earth oxide samples. The rare earths may be identified by the colors of the oxides (Sm_2O_3 is cream colored, Nd_2O_3 is pale blue, Pr_6O_{11} is brown black, and La_2O_3 is white).

Characterization

All the rare earths except lanthanum and lutecium can be distinguished from one another and analyzed in the presence of one another by means of the absorption spectra of the aqueous ions in the ultraviolet, visible, and near

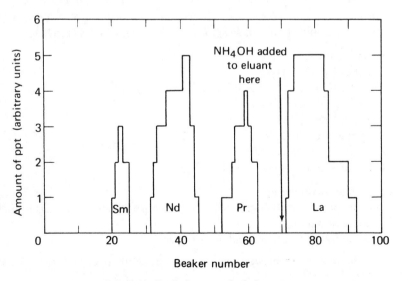

Fig. 32.21. Typical rare-earth elution curve.

infrared regions.[124] The ions have characteristic sharp absorption bands due to f orbital transitions. Solutions of the separated rare earths satisfactory for spectral analysis can be prepared by dissolving the oxides in aqueous hydrochloric acid.

Questions

1. Why is the pH of the eluant very critical?

2. What would be the effect of using a larger amount of didymium oxide (say, 5 g) in the above procedure?

35. THE PREPARATION OF TRIIRON TETRAOXIDE (MAGNETITE) AND ZINC FERRITE[125]

$$12\,Fe^{2+} + 23\,OH^- + NO_3^- \longrightarrow 4\,Fe_3O_4 + NH_3 + 10\,H_2O$$

$$Zn^{2+} + 2\,Fe^{2+} + 3\,C_2O_4^{2-} \longrightarrow ZnFe_2(C_2O_4)_3$$

$$ZnFe_2(C_2O_4)_3 \longrightarrow ZnFe_2O_4 + 2\,CO_2 + 4\,CO$$

Total time required: 2 days

Actual working time: 6 hours

Preliminary study assignment:
   C. Kittel, *Introduction to Solid State Physics*, 2nd ed., John Wiley & Sons, Inc., New York, 1956, pp. 443–48; F. A. Cotton and G. Wilkinson, *Advanced Inorganic Chemistry*, Interscience Publishers, New York, 1962, pp. 708–10; and R. Ward, *Prog. Inorg. Chem.*, **1**, 479–88 (1959).

Reagents required:
   27.8 g of $FeSO_4 \cdot 7\,H_2O$
   0.84 g of KNO_3 or 0.71 g of $NaNO_3$ } for Fe_3O_4
   15 g of 85 per cent KOH
   13.9 g of $FeSO_4 \cdot 7\,H_2O$
   7.2 g of $ZnSO_4 \cdot 7\,H_2O$
   1 ml of 10 per cent H_2SO_4 } for $ZnFe_2O_4$
   11.0 g of ammonium oxalate 1-hydrate

Special apparatus required:
   Centrifuge with centrifuge bottles or fine-porosity sintered-glass funnel
   Permanent magnet
   X-ray diffraction camera and accessories

[124] T. Moeller and J. C. Brantley, *Anal. Chem.*, **22**, 433 (1950); D. C. Stewart and D. Kato, *Anal. Chem.*, **30**, 164 (1958).
   [125] The suggestions of Mr. William Carlson are gratefully acknowledged.

Procedure for Preparation of Magnetite

The ferrous sulfate is dissolved in 200 ml of water, and the potassium or sodium nitrate and potassium hydroxide are dissolved in 100 ml of water. Each solution is heated to about 75° and the two solutions are mixed with vigorous stirring. A thick gelatinous green precipitate forms. After being stirred at 90 to 100° for 10 min, the precipitate turns to a finely divided dense black substance. The mixture is cooled to room temperature and is made acidic with a little 6 M HCl. The precipitated magnetite is centrifuged or filtered on a medium- or fine-porosity sintered-glass filter and washed with water until the wash water gives no test for sulfate with barium chloride. The product is dried at 110° for 1 or 2 hours. The yield is about 7.5 g of Fe_3O_4.

Procedure for Preparation of Zinc Ferrite

The zinc sulfate, ferrous sulfate, and sulfuric acid are dissolved in 100 ml of water, and the ammonium oxalate is dissolved in 150 ml of warm water. Heat the solutions to about 75° and, while stirring vigorously, add the oxalate solution to the metal sulfate solution. Stir the mixture while maintaining the temperature at 90 to 95° for about 5 min. The yellow mixed oxalate precipitate is filtered off on a sintered-glass funnel and washed with water until sulfate can no longer be detected in the wash water. The oxalate is dried for several hours at 110°.

The mixed oxalate is transferred to a 50-ml crucible *with a cover* and the covered crucible is placed in a muffle furnace and heated to 600 to 800°. After remaining 3 hours at this temperature, the material is allowed to cool to room temperature. The yield is about 5.7 g of $ZnFe_2O_4$.

Characterization

The magnetic characteristics of the oxides may be compared by sprinkling small portions (about 50 mg) onto a piece of paper and moving a magnet under it. An X-ray powder pattern may be obtained for each of the compounds, and the lines may be indexed. (The crystals are cubic; the student can easily determine the lattice type and the lattice parameters.)

Relatively large single crystals (octahedra several millimeters across) of Fe_3O_4 can be prepared from the powdered product of this synthesis by a high-temperature transport process employing the following reversible reaction.[126]

$$Fe_3O_4(s) + 8 HCl(g) \rightleftharpoons FeCl_2(g) + 2 FeCl_3(g) + 4 H_2O(g)$$

[126] R. Kershaw and A. Wold, *Inorg. Syn.*, **11**, 10 (1968).

A quartz tube (15 cm × 15 mm O.D.) is sealed at one end and joined at the other to a 10-mm piece of quartz tubing joined through a graded seal to a Pyrex 18/9 ball joint.

The tube is attached to the vacuum line, evacuated, and then filled with pure HCl (from a cylinder) to a pressure such that if all the HCl below the stopcock connecting the vacuum line to the tube were forced into the 15-mm O.D. portion of the tube, the pressure would be about 15 mm. The stopcock is closed, the HCl in the vacuum line is pumped away, and the HCl below the stopcock is transferred to a trap for temporary storage. About 0.5 g of Fe_3O_4 powder is placed in the bottom of the quartz tube; the tube is re-evacuated, and the Fe_3O_4 is heated with a flame until it is completely degassed. Then the measured-out HCl is condensed in the bottom of the tube with liquid nitrogen and the quartz tube is sealed off with a very hot flame immediately above the 15-mm O.D. section. The tube is placed in two adjacent tube furnaces, as shown in Fig. 32.22, the tube being supported on pieces of unglazed porcelain so that it is concentric with the furnaces. The ends of the

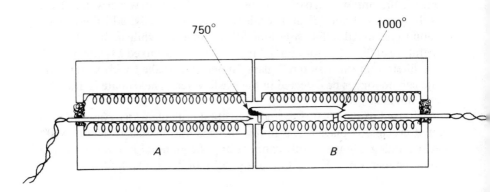

FIG. 32.22. Setup for Fe_3O_4 crystal growing. (Tube is positioned for initial cleanup of growth end.)

furnaces are loosely plugged with asbestos fiber. Furnace *B* is adjusted so that the temperature of the empty end of the tube is 1000 ± 20°, and then furnace *A* is adjusted so that the temperature of the Fe_3O_4-containing end is 750 ± 20°. The tube is kept in this temperature gradient for 15 to 24 hours in order to clean the growth end of stray nuclei by transporting them to the charge zone. The position of the tube is then reversed, so that the charge zone is at 1000° and the growth end is at 750°. These temperatures are maintained for about 10 days, after which the temperature of the charge zone is reduced to 750°, and the furnaces are then turned off. The crystals of Fe_3O_4 may be removed from the growth end by cracking open the quartz tube.

36. THE PREPARATION AND STUDY OF OZONE

$$\tfrac{3}{2} O_2 \longrightarrow O_3$$

Total time required: 3 to 7 hours

Actual working time: 3 to 7 hours

Preliminary study assignment:

W. L. Jolly, *The Chemistry of the Non-Metals*, Prentice-Hall, Inc., Englewood Cliffs, N. J., 1966, pp. 56–58, or M. Ardon, *Oxygen*, W. A. Benjamin, Inc., New York, 1965, pp. 48–67, and, in this book, Chap. 12, Electric Discharge Synthesis.

Reagents required:

Oxygen cylinder (a house oxygen line will suffice)
Other reagents, as required by the study undertaken

Special apparatus required:

High-voltage (a.c.) power supply
Ozonizer (see Fig. 12.3)

NOTICE: The high-voltage power supply is not to be connected to the power line except by the instructor, and it must be used under the direct supervision of the instructor. Great care must be taken to avoid electrocution. When in use, the ozonizer and the electrical connections should be barricaded so as to prevent other laboratory personnel from touching the apparatus. Suitable warning signs should be posted. Ozone is a poisonous gas, and either the apparatus should be assembled in a hood, or provision should be made to prevent the escape of ozone into the air. Under no circumstances should ozone be condensed by, for example, passing it through a liquid nitrogen-cooled trap. Many violent explosions have occurred during the handling of liquid ozone, and special apparatus and shielding are required for such work.

Experimental

Ozone may be prepared by passing oxygen through the ozonizer while a potential of about 15 kv is applied. It is important that no rubber tubing be used in the flow system beyond the ozonizer. In fact, even the use of plastic tubing should be minimized by making most of the connections with glass tubing. A variety of experiments can be carried out using ozone.

1. It is instructive to determine the yield and concentration of the ozone as a function of either the flow rate of oxygen or the applied potential. Ozone

may be analyzed by allowing it to react with a potassium iodide-ammonium chloride solution:

$$2 NH_4^+ + O_3 + 3 I^- \longrightarrow I_3^- + O_2 + H_2O + 2 NH_3$$

The ozone may be completely removed from an oxygen stream by allowing it to pass through a gas-washing bottle containing a solution that is 0.1 M in NH_4Cl and 0.1 M in KI. The resulting triiodide solution may be titrated with standard thiosulfate solution.

2. Ozone has been used to oxidize elements to their highest oxidation states in aqueous solution without chemical contamination. For example, the Pu^{3+} ion may be oxidized by ozone in acid solution to the plutonyl ion, PuO_2^{2+}. (This oxidation cannot be carried out by hydrogen peroxide, because the latter reagent is oxidized by the plutonyl ion.) The oxidizing power of ozone has been used to advantage in the quantitative analyses of manganese and iodine. Manganese can be quantitatively oxidized to permanganate by ozone if the oxidation is carried out in 1.5 M $HClO_4$, using Ag^+ as catalyst (about 50 mg $AgNO_3$/100 ml). Iodide is quantitatively oxidized to periodate in concentrated NaOH or KOH solutions.

3. Various oxides of elements in high oxidation state can be prepared by treatment of lower oxides with ozone. The following syntheses are examples of this technique:

(a) N_2O_5 (from NO_2)[127]

(b) NO_2Cl (from NOCl)[128]

(c) AgO (from Ag)[129]

(d) NiO_2 (?) [from Ni(II)][130]

4. The ultraviolet absorption spectrum of ozone can be obtained by using a 10-cm flow-through silica absorption cell, such as that pictured in Fig. 32.23. The ozone–oxygen mixture can be passed directly from the ozonizer through the cell. There are three principal absorption regions in the near ultraviolet and visible: a set of strong bands peaking at 2550 Å ($\varepsilon_{max} \approx 3000$), a set of bands in the range 3100–3500 Å ($\varepsilon < 20$), and a weak absorption system peaking at 6020 Å ($\varepsilon_{max} \approx 1.2$).[131] With the concentration of ozone that is usually obtained from an ozonizer, the bands in the 3000–3500 Å range can be easily studied. To measure the intense 2550-Å absorption band using a 10-cm cell, the ozone must be diluted by blowing air or oxygen into the cell. The 6020-Å absorption band is too weak to measure without concentrating the ozone or using a much longer path. The absorption spectrum

[127] A. D. Harris, J. C. Trebellas, and H. B. Jonassen, *Inorg. Syn.*, **9**, 83 (1967).

[128] H. J. Schumacher and G. Sprenger, *Z. Anorg. Chem.*, **182**, 139 (1929).

[129] J. Selbin and M. Usategui, *J. Inorg. Nucl. Chem.*, **20**, 91 (1961).

[130] J. S. Belew and C. Tek-Ling, *Chem. Commun.*, 1100 (1967).

[131] E. C. Y. Inn and Y. Tanaka, *J. Opt. Soc. Am.*, **43**, 870 (1953). For a theoretical discussion of the electronic spectrum, see A. D. Walsh, *J. Chem. Soc.*, 2266 (1953).

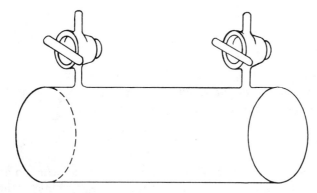

Fig. 32.23. Silica absorption cell for gases, for use in the ultraviolet, visible, and near infrared ranges.

of ozone in aqueous solutions is essentially the same as that of gaseous ozone, except that no fine structure is observed in the solution spectrum.[132] A solution of satisfactory absorbance in the 2550-Å region can be prepared by bubbling the ozone–oxygen mixture through the aqueous solution. Ozone is particularly stable toward decomposition in aqueous acetic acid[132] and in 7 M sodium hydroxide.[133]

Question

1. A stream of oxygen was passed at a rate of 20 cc min^{-1} at S.T.P. for 1 hour through a silent electric discharge followed by a gas-washing bottle containing aqueous ammonium iodide. The resulting triiodide required 85 ml 0.15 N thiosulfate. What was the mole percentage of ozone in the gas coming from the discharge tube?

37. THE PREPARATION AND STUDY OF ATOMIC NITROGEN

$$N_2 \longrightarrow 2N$$

Total time required: 4 hours

Actual working time: 4 hours

Preliminary study assignment:

W. L. Jolly, *Technique of Inorganic Chemistry*, vol. 1, H. B. Jonassen and A. Weissberger, eds., Interscience Publishers, New York, 1963, Chap. 5, p. 179; A. N. Wright and C. A. Winkler, *Active Nitrogen*, Academic Press Inc., New York, 1968; B. A. Thrush, *Science*, **156**, 470 (1967); and, in this book, Chap. 12, Electric Discharge Synthesis.

[132] H. Taube, *Trans. Faraday Soc.*, **53**, 656 (1957).
[133] L. J. Heidt and V. R. Landi, *J. Chem. Phys.*, **41**, 176 (1964).

Reagents required:
  Nitrogen cylinder
  Nitric oxide cylinder

Special apparatus required:
  Microwave generator and resonance cavity
  Vacuum line and apparatus illustrated in Fig. 32.24

Procedure

The apparatus shown in Fig. 32.24 is assembled and attached to a vacuum line in a room that can be completely darkened. The upper inlet of the reactor is connected to a system of 50 to 150 ml volume containing nitric oxide at a pressure of 400 to 700 mm. The reactor is evacuated, the traps are cooled with liquid nitrogen, and the stopcock leading to the source of nitrogen is adjusted so that nitrogen is pumped through the apparatus, with a pressure in the reactor of 2 to 4 mm. The microwave cavity is attached, and the dis-

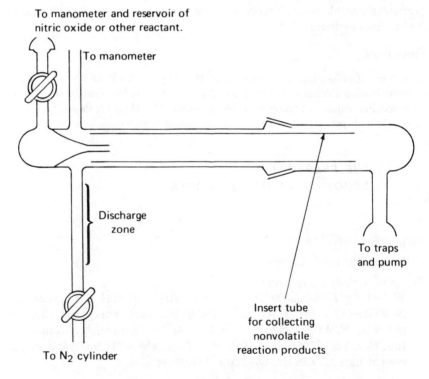

To manometer and reservoir of
nitric oxide or other reactant.

To manometer

Discharge
zone

To traps
and pump

Insert tube
for collecting
nonvolatile
reaction products

To N$_2$ cylinder

FIG. 32.24. Apparatus for studying the reaction of atomic nitrogen with nitric oxide or other volatile species.

charge is started. (It may be necessary to touch a Tesla coil to the tubing to start the discharge.) A yellow color (active nitrogen afterglow) will fill the apparatus between the discharge and the pump. The stopcock to the nitric oxide reservoir is then barely opened. The emitted color of the gas beyond the nozzle should then change from yellow to blue. The blue color is caused by the emission of excited NO molecules, formed by the reactions

$$N + NO \longrightarrow N_2 + O$$
$$N + O + M \longrightarrow NO^* + M$$
$$NO^* \longrightarrow NO + h\nu$$

If the stopcock is carefully opened to the point where the rate of introduction of nitric oxide is equal to the flow rate of atomic nitrogen at the nozzle, the reactor will become dark because of the complete consumption of atomic nitrogen by the first reaction.[134] However, if the nitric oxide is introduced too rapidly, the gas will glow white or greenish yellow because of the following reaction between the atomic oxygen and the excess nitric oxide:

$$NO + O \longrightarrow NO_2 + h\nu$$

To determine the flow rate of atomic nitrogen, the nitric oxide is introduced at a rate such as to maintain the reactor dark for a period of 5 or 10 min. The rate of nitric oxide introduction, and hence the atomic nitrogen flow rate, is calculated from the change in pressure of the reservoir during the period and the known volume of the reservoir.

Using the apparatus of Fig. 32.24, one can study the reactions of atomic nitrogen with many volatile compounds. For example, H_2S reacts[135,136] to give $(SN)_x$ and S_7NH (which collect on the insert tube), S_2Cl_2[136] gives NSCl (which collects in the cold trap), and ethylene[137] gives HCN (which collects in the cold trap).

Questions

1. In a "titration" of N atoms with NO, the pressure of NO in a 100-ml reservoir fell from 700 mm to 620 mm during 10 min. Room temperature was 20°. Calculate the flow rate of N atoms, in millimoles per minute.

2. In another run, N_2 was introduced into the system at a rate of 3.5 mmoles min[-1] and the flow rate of N atoms in the reaction tube was 0.07 mmole min[-1]. The pressure in the reaction tube was 2.0 mm, and the inside diameter of the reaction tube was 20 mm. What percentage of the N_2 was dissociated, and what was the flow velocity in the tube?

[134] P. Harteck, R. Reeves, and G. Mannella, *J. Chem. Phys.*, **29**, 608 (1958).
[135] R. A. Westbury and C. A. Winkler, *Can. J. Chem.*, **38**, 334 (1960).
[136] J. J. Smith and W. L. Jolly, *Inorg. Chem.*, **4**, 1006 (1965).
[137] G. G. Mannella, *Chem. Revs.*, **63**, 1 (1963).

38. THE PREPARATION OF
HEXACHLORODIGERMANE[138]

$$2\,GeCl_4 \longrightarrow Ge_2Cl_6 + Cl_2$$

Total time required: 4 hours

Actual working time: 4 hours

Preliminary study assignment:
W. L. Jolly, *Technique of Inorganic Chemistry*, vol. 1, H. B. Jonassen and
A. Weissberger, eds., Interscience Publishers, New York, 1963, Chap. 5,
p. 179, and, in this book, Chap. 12, Electric Discharge Synthesis.

Reagents required:
2 ml of $GeCl_4$
About 300 ml of concentrated HCl

Special apparatus required:
Microwave discharge apparatus (Fig. 12.5)

Procedure

Assemble the apparatus as pictured in Fig. 12.5 (use Kel–F grease on the
ground joints and stopcocks). Place the germanium tetrachloride in a glass
ampoule and connect it below one of the needle valves. Evacuate the main
part of the apparatus, cool the $GeCl_4$ with a dry ice bath, and, by momentarily opening the needle valve, pump the air from the ampoule. Remove the
dry ice bath from the $GeCl_4$ ampoule. Immerse the first trap in an ice-HCl
slush (about $-18°$), the second trap in a dry ice-acetone slush ($-78°$), and
the third trap in liquid nitrogen.

By turning the needle valve, adjust the pressure in the system to about
0.5 mm of Hg (about 6 mm of dioctyl phthalate). Turn on the air jet (to cool
the resonance cavity), turn on the fans (to cool the magnetron), and turn on
the power to the magnetron. If a discharge does not form spontaneously,
momentarily turn on the Tesla coil. When the discharge is established,
increase the pressure (by further opening the needle valve) to as high a value
as possible without causing the discharge to go out. When almost all of the
germanium tetrachloride has vaporized (after about 1 hour), close the needle
valve and turn off the power to the magnetron.

The glass tubing immediately beyond the discharge will contain polymeric
germanium chlorides. The $-18°$ trap will contain crystals of Ge_2Cl_6; the
$-78°$ trap will contain unreacted $GeCl_4$; and the liquid nitrogen trap will

---

[138] D. Shriver and W. L. Jolly, *J. Am. Chem. Soc.*, **80**, 6692 (1958).

contain chlorine. With a torch, seal off the entrance side (left side) of the $-18°$ trap and then seal off the exit side (right side). The trap containing the Ge_2Cl_6 will now be free. Keep it partially immersed in the cold bath until the ends of the trap cool to room temperature. Close all the cocks of the apparatus, remove all the cold baths, and, finally, remove the traps.

The Ge_2Cl_6 may be sublimed from one end of the sealed trap to another by simply cooling one end in an ice bath. Sublime half of the material into one end and half into the other end. Cool one end in an HCl-ice bath and the other end in liquid nitrogen, and let the tube stand for several minutes. Melt the bend in the tubing with a torch and separate the two ends (keep each end in its cold bath until the seals have cooled). The end cooled in the HCl-ice bath should contain pure Ge_2Cl_6, uncontaminated with $GeCl_4$. Determine the melting point of this sample. Pure Ge_2Cl_6 melts at 41–42°.

Questions

1. What are the physical properties of C_2Cl_6 and Si_2Cl_6? How are these compounds prepared?

2. Suggest a new method for preparing digermane.

39. THE PREPARATION OF
LARGE SINGLE CRYSTALS OF
KAl(SO$_4$)$_2$·12H$_2$O, NaClO$_3$ AND NaNO$_3$

Total time required: 3 to 14 days

Actual working time: 3 hours

Preliminary study assignment:
Chapter 16, Growing Crystals from Aqueous Solutions, in this text.

Reagents required:
85 g of $KAl(SO_4)_2$·12 H_2O and (optional) 30 g of $KCr(SO_4)_2$·12 H_2O
or 600 g of $NaClO_3$ and (optional) 35 g of $Na_2B_4O_7$·10 H_2O
or 550 g of $NaNO_3$

Special apparatus required:
2 Polaroid sheets (for examining $NaClO_3$ and $NaNO_3$)
Wooden cover for beaker used in growth by cooling (see Fig. 16.1)
Gallows-shaped support and cloth used in growth by evaporation (see Fig. 16.2)
About 20-cm length of fine monofilament thread

Procedure

By following the general procedure outlined in Chap. 16, prepare about 500 ml of a saturated solution of potassium aluminum sulfate dodecahydrate, sodium chlorate, or sodium nitrate. In Table 32.1, the solubilities of these

Table 32.1   Solubilities of Salts in Grams per 100 g of Water

$t(°C)$	$KAl(SO_4)_2 \cdot 12 H_2O$	$NaClO_3$	$NaNO_3$
20	11.40	101	88
25	14.14	106	92
30	16.58	113	96

salts at 20, 25, and 30° are given. From these data, you can estimate the amount of salt required to make a saturated solution at a particular growing temperature. When making the saturated solution, it is best to use about 20 per cent more salt than can dissolve at the growing temperature. Using one of the crystals formed in the preparation of the saturated solution as a seed, you may grow a large crystal either by evaporation of the solution or by the growth-by-cooling method. In the latter method, the amounts of salt to be added to the saturated solutions are as follows:

$KAl(SO_4)_2 \cdot 12 H_2O$:   4 g per 100 g of water

$NaClO_3$:   4 g per 100 g of water

$NaNO_3$:   3 g per 100 g of water

These amounts have been found to be satisfactory for a growing temperature of about 24°. If the average growing temperature differs appreciably from this figure, it may be necessary to use a different amount of added salt.

Characterization

Potassium alum forms crystals of the cubic system shaped like octahedra. By adding a saturated solution of the chromium alum, $KCr(SO_4)_2 \cdot 12 H_2O$, to a growing solution of the aluminum alum, you may grow purple, homogeneous, "mixed" crystals of these salts. If the solid solution of the chromium alum in the host $KAl(SO_4)_2 \cdot 12 H_2O$ is sufficiently dilute, a sharp esr spectrum can be obtained from the crystal.

Sodium chlorate forms crystals of the cubic system ordinarily shaped like cubes; however, if 6 g of borax for every 100 g of sodium chlorate is added to the growing solution, the crystals will be tetrahedral. Single crystals of sodium chlorate are optically active. If a crystal is placed between crossed

Polaroid sheets in front of a light source, it will be necessary to rotate the Polaroid sheets to extinct the crystal. Some crystals rotate light to the left, others rotate light to the right. With calibrated Polaroid disks, you can verify that the degree of rotation is proportional to the crystal thickness.

Sodium nitrate forms crystals of the hexagonal system that are shaped like rhombohedra. A crystal can be cleaved with a razor blade along planes parallel to the rhombohedral faces. By pressing a knife against the edge of a sodium nitrate crystal where the faces meet in an obtuse angle, gliding may be observed. A sodium nitrate crystal exhibits double refraction. Thus, if you look through a sodium nitrate crystal at a dot on a piece of paper, you will see two images of the dot. By using Polaroid sheets, you can show that sodium nitrate resolves light into two mutually perpendicular polarized beams.

Appendices

33

It is practically essential for a synthetic inorganic chemist to be able to blow glass. Even if a professional glass blower is available, a chemist will save much time by making simple apparatus himself.

Practically all laboratory glassware is made of borosilicate glass (Pyrex or Kimax).[1] This glass is remarkably resistant to thermal shock and does not soften until about 560°. A gas-oxygen flame is required when working this glass; a torch of the type pictured in Fig. 33.1 is convenient.[2] This torch may be clamped to the bench top so that the hands are free to manipulate glassware in the flame, or the torch may be held in the hand so that glassware which is rigidly mounted may be blown.

---

[1] Borosilicate glass can be distinguished from soft glass and silica by immersion in any one of the following liquids, in which it is invisible: (a) a mixture of 16 parts of methanol and 84 parts of benzene; (b) a mixture of 41 parts of benzene and 59 parts of carbon tetrachloride; (c) pure trichloroethylene; or (d) pure dimethyl sulfoxide.

[2] National Welding Equipment Company, Richmond, Calif., No. 3–A Blowpipe.

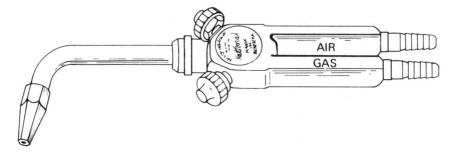

FIG. 33.1. A gas-oxygen hand torch for glass blowing.[2]

Cutting Glass Tubing

To cut tubing (up to about 15 mm in diameter), it is first scratched with the sharp edge of a file or diamond-tipped pencil, care being taken that the scratch, a few millimeters long, is perpendicular to the length of the tubing. Then the tubing is grasped so that the scratch is between the hands and opposite the thumbs. The break is made by a combined bending and pulling force, such that tension is created on the scratched side.

Tubing of diameter greater than 15 mm requires a different technique. After it is scratched with the file, the tip of a small piece of glass cane, made incandescent by heating in a hot flame, is touched to one end of the scratch. Upon cooling, a crack should form, permitting separation of the tubing into two pieces.

These cutting techniques should yield straight, clean breaks. If jagged ends should form, the sharp edges may be somewhat smoothed off by vigorously stroking with a wire gauze.

Closed Round Ends

The procedure for forming a closed rounded end on a piece of glass tubing is illustrated in Fig. 33.2. First, a piece of cane is fused to the end of the tube for a handle, as in Part (B). Then the tubing is rotated evenly in a small, hot flame, so that a short section near the end is heated. A thickened, constricted section will form, as shown in Part (C). While the glass is still hot, the two ends are pulled apart, as shown in Part (D). The tapered end is then rotated in the flame to complete the seal, and the narrow connecting tubing is pulled off. If this operation is properly performed, careful heating and blowing will yield a hemispherical end, as shown in Part (F). If too much glass is left on the end, the end wall will be too thick, as shown in Part (G). Excess glass may be removed by briefly touching a cold piece of glass to the hot end, as shown in Parts (H) and (I).

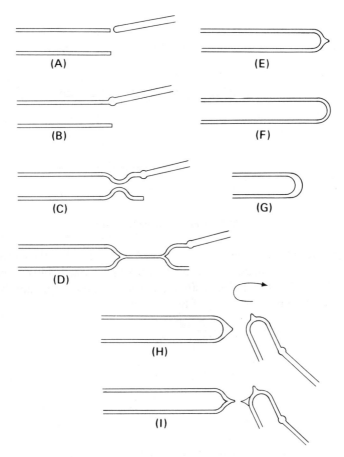

FIG. 33.2. Forming a closed rounded end on a piece of tubing.

End-to-End Seals

The simplest end-to-end seal is that between tubes of equal bore. The outer end of one tube is closed with a cork, or otherwise sealed. The ends to be sealed are heated in the flame until they are soft, and are then gently pressed together, as in Fig. 33.3A. It is important that good contact be made around the entire joint. The joint is then heated with occasional blowing to make a leak-proof seal, as in Part (B). While the glass is still hot, the seal is blown and elongated slightly. If properly done, the result is a single straight tube of uniform diameter and bore, as shown in Part (C).

The procedure for joining two tubes of unequal bore is the same as that described above, except that the diameter at the end of the larger tube must first be made equal to that of the smaller tube. This is accomplished by first

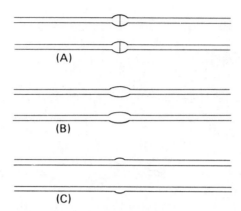

FIG. 33.3. Making a simple end-to-end seal.

making a closed round end, as in Fig. 33.4A. The center of the closed end is then heated with a small, hot flame, and a small bulge is blown, as in Fig. 33.4B. The end of this bulge is similarly heated, and a large, thin bubble is blown out and broken off, as in Fig. 33.4C and D.

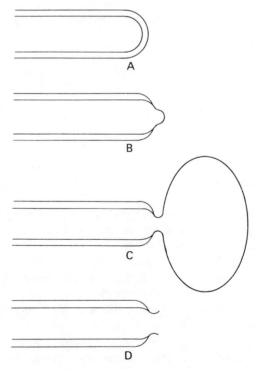

FIG. 33.4. Reducing the diameter at the end of a glass tube.

T Connections

A small area on one side of the tube that is to form the cross piece is heated with a small, hot flame, and a bulge is blown, as shown in Fig. 33.5A. The very end of the bulge is similarly heated and blown out to a thin bubble which is broken off, as in Fig. 33.5B and C. The resulting hole and the end of the other piece of tubing are heated and joined, as in Part (D). By alternately heating and blowing various points of the seal, a strong, smooth union is obtained, as in Part (E).

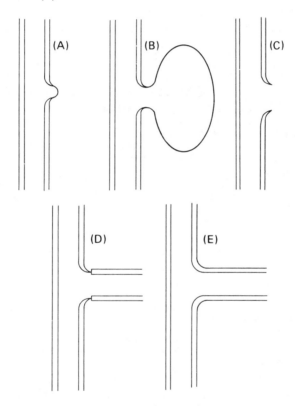

Fig. 33.5. Making a T connection.

General Instructions

When blowing glass, remember that the hottest portions of the glass respond to pressure or movement most readily. Thus, if glass is blown while in the flame, the thin parts become still thinner. Soon after glass is removed from the flame, the thin portions, which cool more rapidly, become inflexible, and blowing works out the thick parts. It is convenient to blow glassware through a piece of rubber tubing, one end of which is connected to the only open end of the glassware.

After blowing a piece of glassware, it should be cooled down to room temperature slowly, particularly so through the annealing region (500 to 600°). This gradual cooling may be crudely accomplished by heating the glassware in a bushy flame while gradually reducing the oxygen supply.

The two commonest troubles encountered by the beginning glassblower are that he does not heat the glass to a high enough temperature, and that he has difficulty, when making seals, in holding the two pieces as if they were one. The first trouble usually results from the uncomfortably bright light that glass imparts to the flame when it is at the temperature required for blowing. This difficulty may be eliminated by simply wearing goggles made of didymium glass. The second trouble may be eliminated by clamping the pieces to be joined on a ring stand. In other words, the torch, and not the pieces of glass, is moved (however, one of the glass pieces should be loosely clamped so that it may be slightly moved during the start of the seal).

References for Further Study

Dodd, R. E., and P. L. Robinson, *Experimental Inorganic Chemistry*, Elsevier, New York, 1954, pp. 90–97.

Heldman, J. D., *Techniques of Glass Manipulation in Scientific Research*, Prentice-Hall, Inc., Englewood Cliffs, N. J., 1946.

Sanderson, R. T., *Vacuum Manipulation of Volatile Compounds*, John Wiley & Sons, Inc., New York, 1948, Chap. 3, pp. 14–29.

Strong, J., *Procedures in Experimental Physics*, Prentice-Hall, Inc., Englewood Cliffs, N. J., 1944.

Wright, R. H., *Manual of Laboratory Glass-blowing*, Chemical Publishing Co., Inc., Brooklyn, N. Y., 1943.

2. HYGROSTATS

A saturated aqueous solution of a salt in contact with an excess of the solid phase at a given temperature will maintain constant humidity in a closed space. A number of salts suitable for this purpose are listed in Table 33.1. The relative humidity over a saturated solution is nearly independent of temperature, and therefore the relative humidities given in the table can be used with fair accuracy over the temperature interval of 12 to 28°. The vapor pressure of water over the solutions may be calculated from the data in the table and the following vapor pressure data for pure water (in mm Hg): 12°(10.5), 14°(12.0), 16°(13.6), 18°(15.5), 20°(17.5), 22°(19.8), 24°(22.4), 26°(25.2), 28°(28.3).

**Table 33.1   Relative Humidities over Saturated Aqueous
Solutions at 20°***

Solid Phase	Relative Humidity (per cent)
$Na_2HPO_4 \cdot 12\,H_2O$	95
K_2HPO_4	92
$KHSO_4$	86
KBr	84
$(NH_4)_2SO_4$	81
NH_4Cl	79.5
$NH_4Cl + KNO_3$	72.6
$Mg(C_2H_3O_2)_2 \cdot 4\,H_2O$	65
$NaBr \cdot 2\,H_2O$	58
$NaHSO_4 \cdot H_2O$	52
$KSCN$	47
$Zn(NO_3)_2 \cdot 6\,H_2O$	42
$CaCl_2 \cdot 6\,H_2O$	32.3
$KC_2H_3O_2$	20
$LiCl \cdot H_2O$	15

* *International Critical Tables*, vol. 1, McGraw-Hill
Book Company, New York, 1926, pp. 67–68.

3. DESICCANTS

Some of the common desiccants, with their approximate gas-drying efficiencies, are listed in Table 33.2. The use of the solid desiccants in the drying of
solvents is discussed on pp. 114–115.

Table 33.2   The Efficiencies of Desiccants

Desiccant	Residual Water (P_{mm} or $mg\,l^{-1}$)
Efficient trap at $-196°$	10^{-23}
CaH_2	$<10^{-5}$
Efficient trap at $-100°$	1×10^{-5}
P_2O_5	2×10^{-5}
$Mg(ClO_4)_2$	5×10^{-4}
Efficient trap at $-78°$	5×10^{-4}
BaO	7×10^{-4}
Linde molecular sieves	0.001
SiO_2 gel; active Al_2O_3; KOH	~ 0.002
$CaSO_4$	0.005
CaO; 96 per cent H_2SO_4	~ 0.01
$CaCl_2$	~ 0.2

4. COMPRESSED-GAS CYLINDERS

Many gases can be bought compressed in cylinders. Unfortunately, there is very little standardization as to type of valve outlet and cylinder color. Some of the common compressed gases and recommended fittings are listed in Table 33.3. Chemical supply house catalogs should be consulted for alternate fittings and for lists of other available gases.

Table 33.3   Fittings for Gas Cylinders

Substance	Cylinder Valve Outlet (in.)	Recommended Control	Full Cylinder Pressure (P.S.I.G., 70°F)
A, He, N_2 (water pumped)	0.965 RH INT (accepting bullet-shaped nipple)	automatic pressure regulator plus needle valve	1600–2200
A, He, N_2 (oil pumped)	0.965 LH INT (accepting bullet-shaped nipple)	automatic pressure regulator plus needle valve	1600–2200
H_2, CO	0.825 LH EXT (accepting round nipple)	automatic pressure regulator plus needle valve	1600–2000 800–1000
O_2	0.903 RH EXT (accepting round nipple)	automatic pressure regulator plus needle valve	1600–2200
Cl_2, SO_2	1.030 RH EXT (flat seat)	needle valve (monel, brass)	84, 35
CO_2	0.825 RH EXT (flat seat)	brass needle valve	830
HCl	0.825 LH EXT (flat seat)	monel needle valve	613
NH_3	$\frac{3}{8}$ RH INT IPS	steel needle valve	114

In most laboratory applications, gases are bled from cylinders into systems that are near atmospheric pressure and that are provided with some sort of safety outlet in case too high a pressure is reached. The main valve of a cylinder cannot be used for bleeding gas into such a system, inasmuch as this valve is simply an on-or-off valve. A needle valve (see Fig. 33.6) must be connected to the cylinder to control the flow of gas. With this arrangement, the main cylinder valve must never be opened unless the needle valve is closed. After opening the main valve, the needle valve is opened sufficiently to obtain the desired flow rate. This arrangement is unsatisfactory for gases having critical points below room temperature, because, as the cylinder pressure falls, the needle valve needs frequent adjustment to maintain a constant flow rate. For these gases, one should interpose an automatic pressure regulator (see Fig. 33.6) between the cylinder and the needle valve.

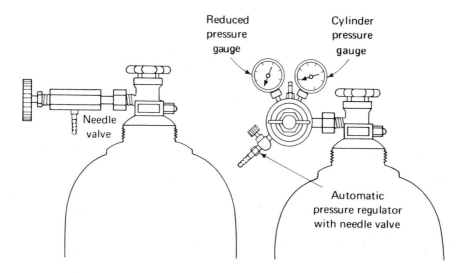

FIG. 33.6. Compressed-gas controllers.

The automatic regulator is used to reduce the cylinder pressure to the range of 5 to 50 P.S.I. This control pressure will remain essentially constant, despite a drop in cylinder pressure. Note that the pressure regulator valve is turned clockwise to increase the pressure and counterclockwise to decrease the pressure. One should never use an automatic regulator without a needle valve.

Many gases may be purchased in "lecture bottles" (2 × 15 in. cylinders) requiring special needle-valve fittings. These bottles are recommended whenever small amounts of gases are needed.

5. DRY SOURCES OF GASES

Few laboratories are stocked with more than a dozen kinds of compressed gases. When the occasional need arises for a gas that is not readily available, it is best to prepare it in the laboratory. Directions for the preparation and purification of various gases are described in the literature.[3] Certain gases may be very conveniently prepared in small amounts by simply heating certain solids. The latter technique is particularly useful in vacuum-line work. A list of gases and the solids heated to produce them is given in Table 33.4.

[3] See, for example, R. E. Dodd and P. L. Robinson, *Experimental Inorganic Chemistry*, Elsevier, New York, 1954, Chap. 3; or A. Farkas and H. W. Melville, *Experimental Methods in Gas Reactions*, The Macmillan Company, New York, 1939, Chap. 3.

Table 33.4   Dry Sources of Gases

Gas	Material to be Heated
O_2	$KMnO_4$, HgO, PbO_2, $KClO_3$, or $K_2Cr_2O_7$
N_2	$(NH_4)_2Cr_2O_7$ or NaN_3
N_2O	NH_4NO_3
NO	intimate mixture of 63.8 g KNO_2, 25.2 g KNO_3, 76 g Cr_2O_3, and 120 g Fe_2O_3
SO_3	$Fe_2(SO_4)_3$
BF_3	$CaF(BF_4)$ or $B_2O_3 + KBF_4$
H_2	UH_3
$H_2S(H_2Se)$	S(Se) + paraffin wax
CO_2	Dry Ice, $MgCO_3$, or $NaHCO_3(+H_2O)$
$(CN)_2$	$Hg(CN)_2$
SO_2	$Na_2S_2O_5$
Cl_2	$CuCl_2$
SiF_4	$BaSiF_6$
GeF_4	$BaGeF_6$
C_2F_4	Teflon

6. THEORETICAL PLATES AND THE BINOMIAL EXPANSION

Many processes involving the distribution of a chemical species between two phases may be discussed in terms of theoretical plates. Let us consider the situation in which one phase moves in a step by step manner through a series of batch equilibrations, or plates, in each of which equilibrium is established between the two phases.

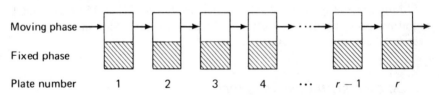

The concentrations of the chemical species in the two phases are related by the distribution coefficient

$$K = \frac{C_M}{C_F} \tag{33.1}$$

where C_M and C_F are the concentrations in the moving and fixed phases, respectively. Hence, the fraction f of a species in the moving phase during any equilibration is given by

$$f = \frac{1}{1 + (V_F/V_M K)} \tag{33.2}$$

where V_F and V_M are the volumes of the fixed and moving phases, respectively, in each plate. After one transfer (the moving phase has advanced to plate 2), the first plate will contain the fraction $1 - f$ of the species and the second plate will contain the fraction f of the species. After two transfers, the first plate will contain $(1 - f)^2$ of the species, the second plate will contain $2f(1 - f)$ of the species, and the third plate will contain f^2 of the species. The fraction in the rth plate is the rth term of the binomial expansion of

$$[(1 - f) + f]^n = 1 \tag{33.3}$$

where n is the number of transfers. The general formula for the fraction in the rth plate after n transfers is

$$F_{n,r} = \frac{n!}{(r - 1)!(n - r + 1)!} f^{r-1}(1 - f)^{n-r+1} \tag{33.4}$$

Note that $r \leqslant n + 1$; $F_{n,r}$ has a maximum value when $n = r/f$. The distribution of concentrations about the maximum value will be symmetrical (Gaussian) when r, n, and $n - r$ are large. The distribution is given by the equation[4]

$$F_{n,r} = \frac{1}{\sqrt{2\pi r}} e^{-f^2(n-r/f)^2/2r} \tag{33.5}$$

The maximum value of $F_{n,r}$ is $1/\sqrt{2\pi r}$.

It should be emphasized that the above treatment assumes the constancy of the distribution coefficient (Eq. 33.1). In actual practice, this coefficient is more or less a function of the concentration of the chemical species undergoing distribution, and even of the concentrations of other species. Hence, the binomial expansion treatment is only a rough approximation.

7. BUFFER SOLUTIONS

The following solutions have fairly high buffer capacity and may be used as primary standards of pH.

pH	Solution
1.68	$0.05\ M$ $KH_3(C_2O_4)_2 \cdot 2\ H_2O$
3.56	KH tartrate solution (saturated at 25°)
4.66	solution $0.1\ M$ in both HOAc and NaOAc
5.34	solution $0.05\ M$ in both NaHsuccinate and Na_2succinate
6.86	solution $0.025\ M$ in both KH_2PO_4 and Na_2HPO_4
9.18	$0.01\ M$ $Na_2B_4O_7$
9.93	solution $0.05\ M$ in both $NaHCO_3$ and Na_2CO_3
12.04	$0.05\ M$ Na_3PO_4

[4] The function $(1/\sqrt{2\pi})e^{-x^2/2}$ is tabulated in many mathematical handbooks.

Buffer solutions covering almost all pH values between 1 and 12 can be prepared from appropriate combinations of acids and their conjugate bases, as indicated:

pH Range	Acid/Conjugate Base Pair
1–3	HSO_4^-/SO_4^{2-}
3.7–5.7	$HOAc/OAc^-$
5.8–7.8	$H_2PO_4^-/HPO_4^{2-}$
8–10	$H_3BO_3/B(OH)_4^-$
9–11	HCO_3^-/CO_3^{2-}
11–12	HPO_4^{2-}/PO_4^{3-}

8. ACID-BASE INDICATORS

The color-change intervals for acid-base indicators in aqueous solutions are indicated graphically in Fig. 33.7.[5]

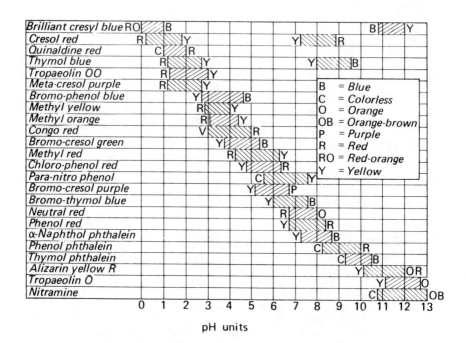

FIG. 33.7. Acid-base indicators for aqueous solutions.[5] (Reproduced with permission of John Wiley & Sons, Inc.)

[5] A. I. Vogel, *A Textbook of Quantitative Inorganic Analysis*, 3rd ed., John Wiley & Sons, Inc., New York, 1961, p. 54.

9. SAFETY PRECAUTIONS

1. Glasses (either ordinary eyeglasses or safety glasses) must be worn by all persons who enter the laboratory. When especially hazardous work is being done, a face shield should be worn.

2. Nobody may work in the laboratory in the absence of the instructor or teaching assistant, unless written approval has been obtained. Nobody should ever work alone.

3. Never work in the open with inflammable solvents within 5 ft of a flame. Be extremely careful of carbon disulfide and ether; these solvents should be heated only with warm-water baths. Inflammable gases, such as hydrogen and carbon monoxide (also poisonous), must be vented into fume hoods, and all connections must be carefully made to ensure the absence of leaks.

4. Operations involving the release of poisonous fumes (e.g., CCl_4, H_2S, O_3, PH_3, Hg, OsO_4, etc.) must be performed in a fume hood. Obnoxious liquids should not be poured into an open sink. If the liquid does not react violently with water, it may be slowly flushed down a drain in a fume hood with large quantities of water. The advice of the instructor should be sought regarding the disposal of all dangerous chemicals.

5. It is best never to eat, drink, or smoke in the laboratory.

6. All electrical equipment should be well insulated to prevent the possibility of shock. If this is not feasible, the equipment should be barricaded so as to prevent other people from accidentally approaching, and warning signs should be posted. Motors and the like should be grounded. Currents through the body as low as 0.1 amp are lethal; a current of this magnitude is easily produced by an ordinary 110-v line.

7. Any chemical which could—even as a remote possibility—explode, must be handled behind a suitable protective screen (e.g., armor plate, bullet-proof glass, or Plexiglas).

10. FIRST AID

Chemical Burns

Immediately flood the affected area with large quantities of water. Chemicals should be washed off the skin with soap and water. If corrosive material enters the eyes, as soon as the outsides of the eyes have been washed, the eyes should be forced open and thoroughly rinsed with water. A doctor should be consulted as soon as possible.

Cuts

If poisonous or corrosive material has contacted the area of the cut, the entire area should be flushed with water. If the rate of bleeding indicates a cut artery, a hand-applied tourniquet may be advisable; a doctor should be sought immediately. Any deep cut or a cut that is sensitive to pressure (indicating the possible presence of glass fragments) should be examined by a doctor.

Poison Gas

The exposed person should be removed to an area free of the poison gas and should be kept warm and as inactive as reasonably possible. A doctor should be sought. Watch for symptoms of shock. If breathing has stopped, apply artificial respiration.

Electrical Shock

Turn off the electrical apparatus or remove the victim from contact as soon as possible—without endangering your safety. If breathing has stopped, apply artificial respiration.

11. GROUP THEORY

An understanding of the principles of molecular symmetry is often of great help in interpreting the spectral properties of compounds. Even a superficial understanding of the language of symmetry and the working methods of group theory can aid a synthetic chemist who wishes to characterize his compounds by spectroscopic methods. Symmetry considerations aid the following procedures:

1. Classifying structures.
2. Classifying normal modes of vibration.
3. Classifying electronic states of molecules.
4. Identifying equivalent and nonequivalent atoms in molecules.
5. Predicting allowed transitions in electronic and vibrational spectra.
6. Predicting splitting of electronic levels in electric fields of various symmetry.
7. Constructing hybrid orbitals.
8. Identifying optical isomers.

In this section, we shall very briefly introduce the important concepts of symmetry and the application of group theory to chemical problems.

Symmetry

If a transformation of coordinates (a reflection, rotation, or a combination of these) produces no distinguishable change in the orientation of a molecule, the transformation is a *symmetry operation,* and the molecule possesses a *symmetry element.* A symmetry element is a point, line, or plane about which the symmetry operation is carried out. For single molecules, there are five symmetry elements and corresponding symmetry operations.

1. *Identity, E.* The identity operation consists in doing nothing to the molecule. All molecules possess this symmetry element, inasmuch as the aspect of a molecule is unchanged after doing nothing to it.

2. *Plane of symmetry (mirror plane), σ.* If reflection of all parts of a molecule through a plane yields an indistinguishable structure, the plane is a plane of symmetry. The water molecule (Fig. 33.8A) possesses two planes of symmetry, and the staggered Si_2H_6 molecule (Fig. 33.8B) possesses three planes of symmetry.

3. *Center of symmetry (inversion center), i.* If reflection through the center yields an indistinguishable configuration, the center is a center of symmetry. The staggered Si_2H_6 molecule (Fig. 33.8B) has a center of symmetry; that is, it is *centrosymmetric.*

4. *n-Fold axis of symmetry (rotational axis), C_n.* If rotation of a molecule by an angle $2\pi/n$ about an axis gives an indistinguishable structure, the axis is an axis of symmetry. The water molecule (Fig. 33.8A) possesses one two-fold axis of symmetry, and staggered Si_2H_6 (Fig. 33.8B) possesses one three-fold axis and three two-fold axes.

5. *n-Fold rotation-reflection axis (improper rotation), S_n.* If rotation by the angle $2\pi/n$ about an axis, followed by reflection through a plane perpendicular to the axis, yields an indistinguishable configuration, the axis is a rotation-reflection axis. Staggered Si_2H_6 (Fig. 33.8B) possesses a six-fold

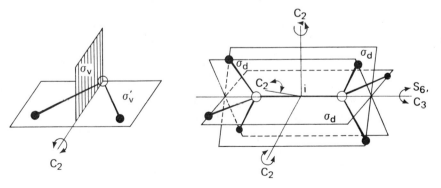

FIG. 33.8.

rotation-reflection axis. The highest-fold rotation axis of a molecule is taken as the vertical z axis. Mirror planes containing the z axis are *vertical planes*, σ_v, and a mirror plane perpendicular to the z axis is the *horizontal plane*, σ_h. *Diagonal planes*, σ_d, are vertical planes that bisect the angles between successive two-fold axes.

Point Groups

When a molecule possesses two or more symmetry elements, the elements always possess at least one point in common. The complete set of symmetry operations that can be carried out on a molecule constitutes a *point group*.

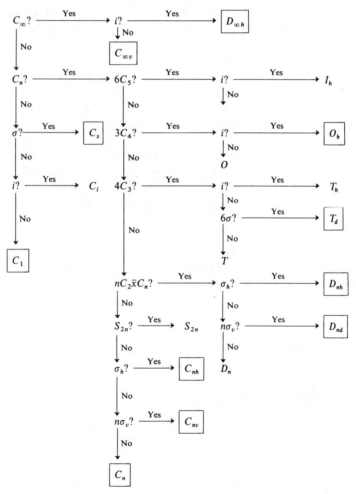

FIG. 33.9. Procedure for determining the point group of a molecule.

Although there is an infinite number of possible point groups, about 40 suffice to classify all known molecules. Point groups have labels, such as C_2, C_{2v}, D_{3h}, O_h, T_d, and so on. A dichotomous system for determining the point group of a given molecule is outlined in the diagram[6] of Fig. 33.9. We shall demonstrate the use of this diagram with several examples.

1. HCl. The HCl molecule possesses an ∞-fold axis of rotation along the H—Cl bond, but it possesses no inversion center; therefore, the molecule belongs to the $C_{\infty v}$ point group.

2. BFClBr. This planar triangular molecule possesses only one symmetry element—the plane of symmetry in which the atoms lie. Consequently, the point group is C_s.

3. $trans$-$N_2O_2{}^{2-}$, 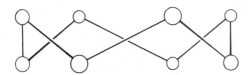 . This planar ion possesses only one symmetry axis, the two-fold axis of symmetry perpendicular to the plane of the ion. There is no S_4 axis but there is a horizontal plane. Consequently, the point group is C_{2h}.

4. S_8. The S_8 molecule consists of a zig-zag closed chain of sulfur atoms, as pictured in Fig. 33.10. Notice that the molecule has one four-fold axis and four two-fold axes perpendicular to the four-fold axis. It has no horizontal plane, but it has four vertical planes. Thus, the point group is D_{4d}.

FIG. 33.10. Side view of S_8 ring.

Further examples are given in Table 33.5. The student should become adept at recognizing symmetry elements and at determining the point groups of molecules. The following list of compounds can be used for practice:

1. $SiHDBr_2$
2. NH_3
3. CO_2
4. SiH_2Br_2
5. P_4O_6
6. BF_3
7. SiFClBrI
8. $POCl_3$
9. B,B,B,-trichloroborazine, $B_3N_3H_3Cl_3$
10. S_4N_4
11. $Mn(CO)_5I$
12. B_2H_6
13. the isomers of $Co(NH_3)_3Cl_3$
14. cis and $trans$ $Co(NH_3)_4Cl_2{}^+$
15. tennis ball (including seam)

[6] Similar diagrams are given by R. L. Carter, *J. Chem. Educ.*, **45**, 44 (1968), and J. Donohue, *J. Chem. Educ.*, **46**, 27 (1969).

The structures of unfamiliar compounds can be found in textbooks of inorganic chemistry. *Answers:* 1, C_s; 2, C_{3v}; 3, $D_{\infty h}$; 4, C_{2v}; 5, T_d; 6, D_{3h}; 7, C_1; 8, C_{3v}; 9, D_{3h}; 10, D_{2d}; 11, C_{4v}; 12, D_{2h}; 13, C_{3v} and C_{2v}; 14, C_{2v} and D_{4h}; and 15, D_{2d}.

Table 33.5   Examples of Some Common Point Groups

Point Group	Description	Examples
C_1	no symmetry (except E)	SiFClBrI, SOFCl
C_2	only C_2	H_2O_2
C_s	only σ	ONCl, HOCl
C_{2v}	C_2, $2\sigma_v$	H_2O, SO_2F_2, SCl_2, ClO_2^-
C_{3v}	C_3, $3\sigma_v$	SiH_3Cl, NH_3, PF_3
C_{4v}	C_4, C_2, $2\sigma_v$, $2\sigma_d$	BrF_5
$C_{\infty v}$	$C_\infty \equiv C_p$, $\infty\sigma_v$	N_2O, HCN, HCl
C_{2h}	C_2, σ_h, i	*trans*-$N_2O_2^{2-}$, *trans*-N_2F_2
D_{2h}	$3C_2$, 3σ, i	N_2O_4, $C_2O_4^{2-}$
D_{3h}	C_3, S_3, $3C_2 3\sigma_v$, σh	BCl_3, NO_3^-, SO_3
D_{4h}	C_4, S_4, C_2, $4C_2$, 4σ, σ_h, i	$PtCl_4^{2-}$
$D_{\infty h}$	S_∞, C_∞, C_p, S_p, ∞C_2, $\infty\sigma_v$, σh, i	H_2, CO_2, $HgCl_2$
D_{2d}	S_4, C_2, $2C_2$, $2\sigma_d$	B_2Cl_4, $H_2C{=}C{=}CH_2$
D_{3d}	S_6, C_3, $3C_2$, i, $3\sigma_d$	staggered Si_2H_6
T_d	tetrahedral	$GeCl_4$, ClO_4^-
O_h	octahedral	UF_6, SF_6, PF_6^-

Character Tables

For each point group there is a corresponding *character table*. Some of the more important character tables are given at the end of this section. At the top of the table are listed horizontally the various classes of symmetry operations of the point group. These classes are indicated by the symbols for the operations of the point group. For example, the classes of symmetry operations of the C_{2v} point group are E, C_2, $\sigma_v(xz)$, and $\sigma_v'(yz)$.[7] In this case, there are four classes of symmetry operations, and there is one operation for each class. In some cases, a point group may possess more than one symmetry operation of a given class. For example, the C_{3v} point group has three different σ_v planes, and one finds $3\sigma_v$ in the top row of the character table. For the same point group, there are two possible C_3 rotations (through 120 and 240°) other than rotation through 360° (the identity operation); thus, one finds $2C_3$ in the table.[8] The number preceding the symmetry operator symbol

---

[7] For planar C_{2v} molecules, the x axis is generally chosen perpendicular to the plane.

[8] The effect of carrying out a rotation m times may be written C_n^m. Thus, the operation C_3 or C_3^1 corresponds to rotation by 120°; C_3^2 corresponds to rotation by 240°; and C_3^3 corresponds to rotation by 360° ($C_3^3 = E$). Often, there are equivalent ways of representing an operation—for example, $C_6^3 = C_2$; $C_6^2 = C_3^1$. Similar remarks apply to the other symmetry operations, σ, i, and S.

is the number of distinct operations of that particular class and is represented by the general symbol g_R, where R refers to the fourth column in the table (absence of such a number implies that $g_R = 1$). The *order* of the point group, g, is the total number of operations in the group, or simply the sum of the g_R terms, $g = \sum_R g_R$.

Below the listing of the symmetry operations, there is a matrix of terms, such as $1, -1, 0, 2$, and so forth. Each term is a *character*, and each horizontal row contains the characters of an *irreducible representation* of the point group. Each irreducible representation is labeled by a symbol, such as A_1, B_2, and so forth; these symbols are presented in a column on the left side of the table. They are used for labeling normal vibrational modes, electronic transitions, molecular orbitals, and the like.

The symbols A and B refer to nondegenerate species that are symmetric and antisymmetric, respectively, about the principal axis. The symbols E and T refer to doubly and triply degenerate species, respectively.[9] It will be noted from the character tables that the number of irreducible representations of a group is always equal to the number of classes of symmetry operations of the group. On the right side of the table are listed various symbols, such as x, y, z, R_x, and so on. The symbols x, y, and z are the Cartesian coordinates (which can stand for vectors of translation or dipole moment) and the R terms represent rotations about the axes specified in the subscripts. Each of these functions can serve as a *basis* for the particular irreducible representation the characters of which are given on the same line.

We say that the function *transforms* under the symmetry operations of the point group according to a particular irreducible representation.[10] For example, consider the C_{2v} character table. A unit vector lying on the x axis, when subjected to the identity operation E, remains unchanged. The corresponding character is 1. When subjected to 180° rotation around the C_2 axis (z axis), the sign of the vector is reversed, corresponding to the character -1. Reflection through the xz plane produces no effect (because the vector lies in this plane), yielding the character 1. Reflection through the yz plane reverses the vector, and the corresponding character is -1. Thus, by successive application of the symmetry operators E, C_2, $\sigma_v(xz)$, $\sigma_v'(yz)$, we obtain the set of characters $1, -1, 1, -1$. It can be seen that these are the characters of the B_1 irreducible representation; therefore, the symbol x is found to the right of this row. A rotation about the x axis can be visualized as a curved arrow going around the x axis. If we carry out the C_{2v} symmetry operations on this arrow, we obtain the set of characters $1, -1, -1, 1$, which corresponds to the

---

[9] E is not to be confused with the identity operator.

[10] The procedures described in the following discussion are readily applied to any nondegenerate irreducible representation (e.g., A or B). The situation is a little more complicated for degenerate irreducible representations (e.g., E or T), and the reader may wish to consult one of the general references given at the end of this appendix for more details.

B_2 irreducible representation. (In the case of a rotation, a reversal of the direction of rotation corresponds to a character of -1.) Thus, the symbol R_x is found to the right of the B_2 characters. The character of the irreducible representation of the direct product of two vectors is obtained by multiplying together the characters of the irreducible representations for the individual vectors.[11] From the C_{2v} character table, we see that the vectors x and y transform separately as B_1 and B_2, respectively. Thus, we calculate that the product xy transforms according to $B_1 \times B_2$. To obtain the characters of the irreducible representation corresponding to this product, we simply multiply the corresponding pairs of characters under the same symmetry operation. We obtain $(1)(1)$, $(-1)(-1)$, $(1)(-1)$, $(-1)(1)$, or $1, 1, -1, -1$, which corresponds to the A_2 irreducible representation. Hence, $B_1 \times B_2 = A_2$, and xy is placed to the right of the A_2 representation. The reader should verify the placement of y, z, R_y, R_z, x^2, y^2, z^2, xz, and yz in the C_{2v} character table.

Reducible Representations

There is no problem in determining the representation of the direct product of two nondegenerate irreducible representations (e.g., $A_1 \times B_2$) or of a non-degenerate and a degenerate irreducible representation (e.g., $A_2 \times E$). But a difficulty arises in determining the direct product of two degenerate irreducible representations. In this case, we do not obtain, by multiplying the individual characters, an irreducible representation, but rather a *reducible representation*. For example, in the C_{3v} point group,

$$E \times E = \Gamma = 4, 1, 0$$

The product is obviously not any one of the irreducible representations. It may be considered the sum of several irreducible representations. In this case, Γ is the sum of A_1, A_2, and E,

$$\Gamma = A_1 + A_2 + E$$

This can be verified by adding the characters for these three symmetry species for each of the operations in the character table. Thus, for E (the identity operator), $1 + 1 + 2 = 4$; for $2C_3$, $1 + 1 - 1 = 1$; for $3\sigma_v$, $1 - 1 + 0 = 0$. These sums are the characters of the reducible representation. In many cases, a reducible representation may be reduced, or decomposed, to its component irreducible representations by inspection of the character table. However, there is a foolproof analytical method for doing this. The number of times a particular irreducible representation contributes to the reducible

---

[11] As shown in the following paragraph, this method is applicable only when one of the two irreducible representations is nondegenerate.

representation may be calculated from the expression

$$N = \frac{1}{g}\sum_R g_R\chi(R) \cdot \gamma(R)$$

where $\chi(R)$ is the character of the symmetry operation R in the irreducible representation (taken from the character table), and $\gamma(R)$ is the character of the symmetry operation R in the reducible representation. Thus, the reducible representation $\Gamma = 4, 1, 0$ of the C_{3v} point group may be reduced as follows:

$$N_{A1} = \tfrac{1}{6}(1 \cdot 1 \cdot 4 + 2 \cdot 1 \cdot 1 + 3 \cdot 1 \cdot 0) = 1$$
$$N_{A2} = \tfrac{1}{6}[1 \cdot 1 \cdot 4 + 2 \cdot 1 \cdot 1 + 3 \cdot (-1) \cdot 0] = 1$$
$$N_E = \tfrac{1}{6}[1 \cdot 2 \cdot 4 + 2 \cdot (-1) \cdot 1 + 3 \cdot 0 \cdot 0] = 1$$

A set of points or figures that includes all those points or figures which may be generated by successive application of all the symmetry operations can serve as the basis for a reducible representation. Two or more such sets together can also serve as the basis of a reducible representation.

In chemistry, it is commonly necessary to determine the reducible representation corresponding to a set of points (e.g., atoms), lines (e.g., bonds), or atomic orbitals (e.g., p orbitals). The following procedure may be followed. For each member of the set, carry out, in turn, one operation for each class of symmetry operation in the point group. A number is assigned to each of these operations. If the operation does not move or change the aspect of the point or figure, the number is $+1$. In the case of a figure having directional character (as a p orbital), if the operation causes a change in sign, the number is -1. For any other change the number is 0. After determining this series of numbers for each member of the set (always taking care to use the same operation for a given class of operation), add the numbers corresponding to the same class of symmetry operation. The resulting sums constitute the reducible representation sought.

As a first example, we shall determine the reducible representation for the P—F bonds in the trigonal bipyramidal PF_5 molecule. We make a table, using the headings of the D_{3h} point group. The sets of numbers corresponding to the two sets of equivalent bonds are indicated

D_{3h}:	E	$2C_3$	$3C_2$	σ_h	$2S_3$	$3\sigma_v$
axial $\Big\{$	1	1	0	0	0	1
bonds	1	1	0	0	0	1
equatorial $\Big\{$	1	0	1	1	0	1
bonds	1	0	0	1	0	0
	1	0	0	1	0	0
$\Gamma =$ 5	5	2	1	3	0	3

By adding the vertical columns, we obtain the reducible representation, which is readily reduced to $2A_1{}' + A_2{}'' + E'$.

As a second example, we shall determine the reducible representation for the p_π atomic orbitals of the two oxygen atoms in the nitrite ion. These orbitals are perpendicular to the plane of the ion. We obtain the following table.

C_{2v}:	E	C_2	$\sigma_v(xz)$	$\sigma_v'(yz)$
	1	0	0	-1
	1	0	0	-1
$\Gamma =$	2	0	0	-2

The reducible representation reduces to $A_1 + B_2$.

Character Tables for some Important Point Groups

C_s	E	σ_h	
A'	1	1	$x, y, R_z, x^2, y^2, z^2, xy$
A''	1	-1	z, R_x, R_y, yz, xz

C_{2h}	E	C_2	i	σ_h	
A_g	1	1	1	1	R_z, x^2, y^2, z^2, xy
B_g	1	-1	1	-1	R_x, R_y, xz, yz
A_u	1	1	-1	-1	z
B_u	1	-1	-1	1	x, y

C_{2v}	E	C_2	$\sigma_v(xz)$	$\sigma_v'(yz)$	
A_1	1	1	1	1	z, x^2, y^2, z^2
A_2	1	1	-1	-1	R_z, xy
B_1	1	-1	1	-1	x, R_y, xz
B_2	1	-1	-1	1	y, R_x, yz

C_{3v}	E	$2C_3$	$3\sigma_v$	
A_1	1	1	1	$z, x^2 + y^2, z^2$
A_2	1	1	-1	R_z
E	2	-1	0	$(x, y) (R_x, R_y), (x^2 - y^2, xy)(xz, yz)$

C_{4v}	E	$2C_4$	C_2	$2\sigma_v$	$2\sigma_d$	
A_1	1	1	1	1	1	$z, x^2 + y^2, z^2$
A_2	1	1	1	-1	-1	R_z
B_1	1	-1	1	1	-1	$x^2 - y^2$
B_2	1	-1	1	-1	1	xy
E	2	0	-2	0	0	$(x, y)(R_x, R_y), (xz, yz)$

Character Tables for some Important Point Groups—*(contd.)*

D_{2h}	E	$C_2(z)$	$C_2(y)$	$C_2(x)$	i	$\sigma(xy)$	$\sigma(xz)$	$\sigma(yz)$	
A_g	1	1	1	1	1	1	1	1	x^2, y^2, z^2
B_{1g}	1	1	-1	-1	1	1	-1	-1	R_z, xy
B_{2g}	1	-1	1	-1	1	-1	1	-1	R_y, xz
B_{3g}	1	-1	-1	1	1	-1	-1	1	R_x, yz
A_u	1	1	1	1	-1	-1	-1	-1	
B_{1u}	1	1	-1	-1	-1	-1	1	1	z
B_{2u}	1	-1	1	-1	-1	1	-1	1	y
B_{3u}	1	-1	-1	1	-1	1	1	-1	x

D_{3h}	E	$2C_3$	$3C_2$	σ_h	$2S_3$	$3\sigma_v$	
A_1'	1	1	1	1	1	1	$x^2 + y^2, z^2$
A_2'	1	1	-1	1	1	-1	R_z
E'	2	-1	0	2	-1	0	$(x, y), (x^2 - y^2, xy)$
A_1''	1	1	1	-1	-1	-1	
A_2''	1	1	-1	-1	-1	1	z
E''	2	-1	0	-2	1	0	$(R_x, R_y), (xz, yz)$

D_{4h}	E	$2C_4$	C_2	$2C_2'$	$2C_2''$	i	$2S_4$	σ_h	$2\sigma_v$	$2\sigma_d$	
A_{1g}	1	1	1	1	1	1	1	1	1	1	$x^2 + y^2, z^2$
A_{2g}	1	1	1	-1	-1	1	1	1	-1	-1	R_z
B_{1g}	1	-1	1	1	-1	1	-1	1	1	-1	$x^2 - y^2$
B_{2g}	1	-1	1	-1	1	1	-1	1	-1	1	xy
E_g	2	0	-2	0	0	2	0	-2	0	0	$(R_x, R_y),$ (xz, yz)
A_{1u}	1	1	1	1	1	-1	-1	-1	-1	-1	
A_{2u}	1	1	1	-1	-1	-1	-1	-1	1	1	z
B_{1u}	1	-1	1	1	-1	-1	1	-1	-1	1	
B_{2u}	1	-1	1	-1	1	-1	1	-1	1	-1	
E_u	2	0	-2	0	0	-2	0	2	0	0	(x, y)

D_{6h}	E	$2C_6$	$2C_3$	C_2	$3C_2'$	$3C_2''$	i	$2S_3$	$2S_6$	σ_h	$3\sigma_d$	$3\sigma_v$	
A_{1g}	1	1	1	1	1	1	1	1	1	1	1	1	$x^2 + y^2, z^2$
A_{2g}	1	1	1	1	-1	-1	1	1	1	1	-1	-1	R_z
B_{1g}	1	-1	1	-1	1	-1	1	-1	1	-1	1	-1	
B_{2g}	1	-1	1	-1	-1	1	1	-1	1	-1	-1	1	
E_{1g}	2	1	-1	-2	0	0	2	1	-1	-2	0	0	$(R_x, R_y), (xz, yz)$
E_{2g}	2	-1	-1	2	0	0	2	-1	-1	2	0	0	$(x^2 - y^2, xy)$
A_{1u}	1	1	1	1	1	1	-1	-1	-1	-1	-1	-1	
A_{2u}	1	1	1	1	-1	-1	-1	-1	-1	-1	1	1	z
B_{1u}	1	-1	1	-1	1	-1	-1	1	-1	1	-1	1	
B_{2u}	1	-1	1	-1	-1	1	-1	1	-1	1	1	-1	
E_{1u}	2	1	-1	-2	0	0	-2	-1	1	2	0	0	(x, y)
E_{2u}	2	-1	-1	2	0	0	-2	1	1	-2	0	0	

Character Tables for some Important Point Groups—*(contd.)*

D_{2d}	E	$2S_4$	C_2	$2C_2'$	$2\sigma_d$	
A_1	1	1	1	1	1	x^2+y^2, z^2
A_2	1	1	1	-1	-1	R_z
B_1	1	-1	1	1	-1	x^2-y^2
B_2	1	-1	1	-1	1	z, xy
E	2	0	-2	0	0	$(x, y), (R_x, R_y), (xz, yz)$

D_{3d}	E	$2C_3$	$3C_2$	i	$2S_6$	$3\sigma_d$	
A_{1g}	1	1	1	1	1	1	x^2+y^2, z^2
A_{2g}	1	1	-1	1	1	-1	R_z
E_g	2	-1	0	2	-1	0	$(R_x, R_y), (x^2-y^2, xy), (xz, yz)$
A_{1u}	1	1	1	-1	-1	-1	
A_{2u}	1	1	-1	-1	-1	1	z
E_u	2	-1	0	-2	1	0	(x, y)

D_{4d}	E	$2S_8$	$2C_4$	$2S_8^3$	C_2	$4C_2'$	$4\sigma_d$	
A_1	1	1	1	1	1	1	1	x^2+y^2, z^2
A_2	1	1	1	1	1	-1	-1	R_z
B_1	1	-1	1	-1	1	1	-1	
B_2	1	-1	1	-1	1	-1	1	z
E_1	2	$\sqrt{2}$	0	$-\sqrt{2}$	-2	0	0	(x, y)
E_2	2	0	-2	0	2	0	0	(x^2-y^2, xy)
E_3	2	$-\sqrt{2}$	0	$\sqrt{2}$	-2	0	0	$(R_x, R_y), (xz, yz)$

T_d	E	$8C_3$	$3C_2$	$6S_4$	$6\sigma_d$	
A_1	1	1	1	1	1	$x^2+y^2+z^2$
A_2	1	1	1	-1	-1	
E	2	-1	2	0	0	$(2z^2-x^2-y^2, x^2-y^2)$
T_1	3	0	-1	1	-1	(R_x, R_y, R_z)
T_2	3	0	-1	-1	1	$(x, y, z), (xy, xz, yz)$

O_h	E	$8C_3$	$6C_2$	$6C_4$	$3C_2(=C_4^2)$	i	$6S_4$	$8S_6$	$3\sigma_h$	$6\sigma_d$	
A_{1g}	1	1	1	1	1	1	1	1	1	1	$x^2+y^2+z^2$
A_{2g}	1	1	-1	-1	1	1	-1	1	1	-1	
E_g	2	-1	0	0	2	2	0	-1	2	0	$(2z^2-x^2-y^2,$ $x^2-y^2)$
T_{1g}	3	0	-1	1	-1	3	1	0	-1	-1	(R_x, R_y, R_z)
T_{2g}	3	0	1	-1	-1	3	-1	0	-1	1	(xz, yz, xy)
A_{1u}	1	1	1	1	1	-1	-1	-1	-1	-1	
A_{2u}	1	1	-1	-1	1	-1	1	-1	-1	1	
E_u	2	-1	0	0	2	-2	0	1	-2	0	
T_{1u}	3	0	-1	1	-1	-3	-1	0	1	1	(x, y, z)
T_{2u}	3	0	1	-1	-1	-3	1	0	1	-1	

Problems

1. Determine the following direct products in the indicated point groups:
   (a) $B_g \times A_u (C_{2h})$
   (b) $A_2 \times E (C_{4v})$
   (c) $A_2 \times B_1 (D_{2d})$
   (d) $A_2 \times T_1 (T_d)$

2. Express the following direct products as sums of irreducible representations:
   (a) $E_g \times E_g (D_{3d})$
   (b) $E_1 \times E_2 (D_{4d})$
   (c) $E_g \times T_{2g} (O_h)$

3. The three P—Cl bonds in PCl_3 are the basis of a reducible representation. Reduce it into its irreducible representations.

4. In $PtCl_4{}^{2-}$, the four chlorine p orbitals that are perpendicular to the plane of the molecule are the basis of a reducible representation. Reduce it into its irreducible representations.

References

Cotton, F. A., *Chemical Applications of Group Theory*, Interscience Publishers, New York, 1963.

Jaffé, H. H. and M. Orchin, *Symmetry in Chemistry*, John Wiley & Sons, Inc., New York, 1965.

Schonland, D. S., *Molecular Symmetry*, D. Van Nostrand Co., Inc., Princeton, N. J., 1965.

12. MOLECULAR ORBITALS

Molecular orbitals may be considered as being generated by the overlap of atomic orbitals; indeed, as a fair approximation they may be represented as linear combinations of atomic orbitals. Now, the identification of suitable combinations of atomic orbitals for the formation of molecular orbitals would, in general, be a very difficult task were it not for the fact that most molecules have symmetry. In symmetrical molecules, there exist sets of equivalent atoms having equivalent atomic orbitals. By combining these atomic orbitals into *group orbitals*, a considerable simplification can be achieved. In the following paragraphs, we shall outline a procedure for determining linear combinations of atomic orbitals (and the corresponding irreducible representations) suitable for the formation of group orbitals. We shall then briefly show how molecular orbitals may be formed by the combination of group orbitals and/or atomic orbitals.

Follow procedures 1 through 7 to identify and determine qualitatively the energy levels of the molecular orbitals of a molecule.

1. Determine the point group of the molecule.

2. Identify the set (or sets) of atoms that are sterically equivalent—that is, the atoms that can be transposed into each other's position by suitable symmetry operations.

3. Each of a set of equivalent atoms has one or more atomic orbitals, or hybrid atomic orbitals, that can be used in bonding. Identify those orbitals suitable for forming σ bonds and those suitable for forming π bonds.

4. For each group of atomic orbitals, calculate the reducible representation.

5. Decompose the reducible representation of each group of atomic orbitals into its irreducible representations. They are the irreducible representations of the various allowed combinations of the atomic orbitals.

6. The atomic orbitals of any unique atoms that lie on all the symmetry elements of the molecule transform as irreducible representations of the point group. The irreducible representations of such orbitals should be determined, an easy matter for s, p, and d orbitals. An s orbital on a unique atom always transforms totally symmetrically; all the characters of its irreducible representation are $+1$; p and d orbitals transform in the same way as their subscripts. Hence, one need only look up the appropriate subscripts [x, y, and z for p orbitals, and $2z^2 - x^2 - y^2$ (usually called z^2), $x^2 - y^2$, xy, xz, and yz for d orbitals] in the character table.

7. Those group orbitals and atomic orbitals that are of the same symmetry species and that overlap significantly form antibonding and bonding molecular orbitals. Those group orbitals and atomic orbitals that do not overlap significantly yield nonbonding molecular orbitals. By considering the relative energy levels of the atomic orbitals, their degree of overlap, and the relative electronegativities of the atoms, construct a qualitative energy level diagram for the molecular orbitals.

If it is desired to determine the linear combinations of atomic orbitals corresponding to a particular group orbital, procedures 8 through 11 should be followed. Procedures 10 and 11 should be followed for each of the irreducible representations of the group orbital (determined in step 5).

8. Construct a separate coordinate system about each atom in the group, so that plus and minus directions have meaning for each atom. The choice of coordinate systems is completely arbitrary, but it is wise to choose analogous axes in as symmetric a way as possible.

9. If none of the irreducible representations of the group orbital has a degeneracy greater than one (i.e., only A and/or B species are involved), perform each symmetry operation of the point group on one of the atomic orbitals of the group. (Every distinguishable operation must be carried out.

Thus, the number of operations equals g, the order of the group.) If the highest degeneracy of the irreducible representations is two (E species), perform the symmetry operations on two of the orbitals. If a T species is included among the irreducible representations, perform the operations on three of the orbitals. Make a table in which the horizontal headings are the symmetry operations and the vertical headings are the atomic orbitals, (ϕ_1, ϕ_2, etc.). Each term in the table corresponds to the *orbital* (ϕ_1, ϕ_2, etc.) into which the orbital being operated on transforms. A change in sign of an orbital is indicated by a minus sign.

10. Multiply each of the entries in the table by the corresponding character of one of the irreducible representations found in step 5. These multiplications should be carried out for only one row if the symmetry species is of type A or B, for two rows if of type E, and for three rows if of type T. Make a table of the resulting terms. Each horizontal series of entries is a linear combination of atomic orbitals (of the type $a_1\phi_1 + a_2\phi_2 + a_3\phi_3 + \cdots$) that can represent a group orbital. For nondegenerate symmetry species (A and B), there is only one linearly independent combination. For degenerate symmetry species (e.g., E and T), the number of linearly independent combinations is equal to the degeneracy. Write out the appropriate linear combinations of atomic orbitals, expressing each as simply as possible.

11. Molecular orbitals may be formed by the linear combination of group orbitals (or group orbitals and atomic orbitals) which are of the same symmetry species and which overlap adequately. In their simplest form, the molecular orbitals are the sum and difference of two orbitals, each multiplied by an appropriate factor.

$$\psi_{MO} = a\psi_{GO}{}^A + b\psi_{GO}{}^B$$
$$\psi_{MO}{}^* = a\psi_{GO}{}^A - b\psi_{GO}{}^B$$

The sum corresponds to the bonding orbital, and the difference to the antibonding orbital.

The Nitrite Ion

As a first example of the application of these rules, we shall consider the molecular orbitals of the nitrite ion, $NO_2{}^-$.

1. The point group of the nitrite ion is C_{2v}.
2. The two oxygen atoms are sterically equivalent.
3. Inasmuch as the O—N—O bond angle is 115°, it is a convenient (and also plausible) approximation to think of the in-plane nitrogen atomic orbitals as three sp^2 hybrid orbitals. Two of these orbitals may be oriented approximately along the bond axes and can form σ bonds with oxygen p orbitals aligned with the bonds, as shown in Fig. 33.11A. The third sp^2 hybrid

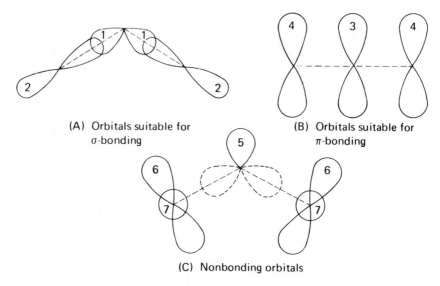

(A) Orbitals suitable for
σ-bonding

(B) Orbitals suitable for
π-bonding

(C) Nonbonding orbitals

FIG. 33.11. Orbitals and group orbitals in the valence shells of the nitrite ion. Equivalent orbitals are labeled with the same numbers.

orbital (shown in Fig. 33.11C) is then nonbonding. The three p orbitals perpendicular to the plane of the ion and suitable for forming π bonds are shown in Fig. 33.11B. We shall assume that the oxygen s orbitals and the in-plane oxygen p orbitals perpendicular to the bond axes are nonbinding; these orbitals are shown in Fig. 33.11C.

4, 5, 6. The irreducible representations of the orbitals shown in Fig. 33.11 are listed as follows:

group orbital 1 (σ bonding on N): $\Gamma = 2, 0, 0, 2 = A_1 + B_2$
group orbital 2 (σ bonding on O's): $\Gamma = 2, 0, 0, 2 = A_1 + B_2$
atomic orbital 3 (π bonding on N): from character table, B_1
group orbital 4 (π bonding on O's): $\Gamma = 2, 0, 0, -2 = A_2 + B_1$
hybrid orbital 5 (nonbonding on N): $\Gamma = 1, 1, 1, 1 = A_1$
group orbital 6 (nonbonding on O's): $\Gamma = 2, 0, 0, 2 = A_1 + B_2$
group orbital 7 (nonbonding on O's): $\Gamma = 2, 0, 0, 2 = A_1 + B_2$

7. An approximate energy-level scheme and correlation diagram for NO_2^- is given in Fig. 22.2. For detailed discussions of the relative energy levels, the reader is referred to the papers by Strickler and Kasha,[12] and McEwen.[13] The reader should notice that if the ion is considered to lie in the xz plane, the b_1 and b_2 notations are interchanged. Some authors use this convention.

8. We construct coordinate systems for each atom as shown in Fig. 33.12.

[12] S. J. Strickler and M. Kasha, *J. Am. Chem. Soc.*, **85**, 2899 (1963).
[13] K. L. McEwen, *J. Chem. Phys.*, **34**, 547 (1961).

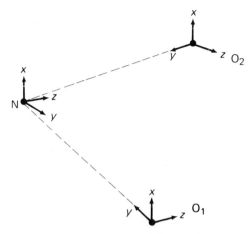

FIG. 33.12. Coordinate systems for the nitrite ion.

9. The following table gives the results of applying the C_{2v} symmetry operations to the three types of oxygen p orbitals shown in Fig. 33.11. The oxygen atoms are identified by the numbers 1 and 2.

		E	C_2	$\sigma_v(xz)$	$\sigma_v'(yz)$
σ orbitals:	$\sigma_1(\sigma)$	ϕ_1	ϕ_2	ϕ_2	ϕ_1
π orbitals:	$\phi_1(\pi)$	ϕ_1	$-\phi_2$	ϕ_2	$-\phi_1$
nonbonding orbitals: $\phi_1(n)$		ϕ_1	ϕ_2	ϕ_2	ϕ_1

The following table gives similar results for the two nitrogen sp^2 orbitals shown in Fig. 33.11A. The orbitals are identified by the subscripts y and z.

	E	C_2	$\sigma_v(xz)$	$\sigma_v'(yz)$
ϕ_y	ϕ_y	ϕ_z	ϕ_z	ϕ_y

10. The following table shows the functions obtained by multiplying the terms of the preceding tables by the characters of the appropriate irreducible representations (found in 5 above). Simplified functions (obtained by multiplying the terms in the table by $\frac{1}{2}$) are shown on the right side.

		E	C_2	$\sigma_v(xz)$	$\sigma_v'(yz)$	
O	σa_1	$+\phi_1$	$+\phi_2$	$+\phi_2$	$+\phi_1$	$\phi_1 + \phi_2$
	σb_2	$+\phi_1$	$-\phi_2$	$-\phi_2$	$+\phi_1$	$\phi_1 - \phi_2$
	πa_2	$+\phi_1$	$-\phi_2$	$-\phi_2$	$+\phi_1$	$\phi_1 - \phi_2$
	πb_1	$+\phi_1$	$+\phi_2$	$+\phi_2$	$+\phi_1$	$\phi_1 + \phi_2$
	$n_o a_1$	$+\phi_1$	$+\phi_2$	$+\phi_2$	$+\phi_1$	$\phi_1 + \phi_2$
	$n_o b_2$	$+\phi_1$	$-\phi_2$	$-\phi_2$	$+\phi_1$	$\phi_1 - \phi_2$
N	σa_1	$+\phi_y$	$+\phi_z$	$+\phi_z$	$+\phi_y$	$\phi_y + \phi_z$
	σb_2	$+\phi_y$	$-\phi_z$	$-\phi_z$	$+\phi_y$	$\phi_y - \phi_z$

11. The various allowed combinations of atomic orbitals are indicated both algebraically and schematically in Fig. 33.13.

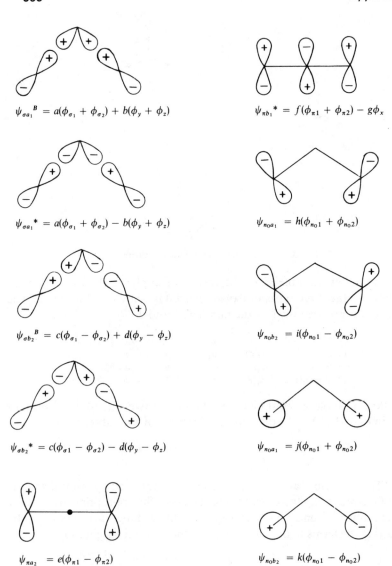

$$\psi_{\sigma a_1}{}^B = a(\phi_{\sigma_1} + \phi_{\sigma_2}) + b(\phi_y + \phi_z)$$

$$\psi_{\pi b_1}{}^* = f(\phi_{\pi 1} + \phi_{\pi 2}) - g\phi_x$$

$$\psi_{\sigma a_1}{}^* = a(\phi_{\sigma_1} + \phi_{\sigma_2}) - b(\phi_y + \phi_z)$$

$$\psi_{n o a_1} = h(\phi_{n o 1} + \phi_{n o 2})$$

$$\psi_{\sigma b_2}{}^B = c(\phi_{\sigma_1} - \phi_{\sigma_2}) + d(\phi_y - \phi_z)$$

$$\psi_{n o b_2} = i(\phi_{n o 1} - \phi_{n o 2})$$

$$\psi_{\sigma b_2}{}^* = c(\phi_{\sigma 1} - \phi_{\sigma 2}) - d(\phi_y - \phi_z)$$

$$\psi_{n o a_1} = j(\phi_{n o 1} + \phi_{n o 2})$$

$$\psi_{\pi a_2} = e(\phi_{\pi 1} - \phi_{\pi 2})$$

$$\psi_{n o b_2} = k(\phi_{n o 1} - \phi_{n o 2})$$

$$\psi_{\pi b_1}{}^B = f(\phi_{\pi 1} + \phi_{\pi 2}) + g\phi_x$$

$$\psi_{n N a_1} = \phi_{N s p^2}$$

Fig. 33.13. Algebraic and schematic representations of allowed combinations of atomic orbitals.

An Octahedral Transition-Metal Complex

A brief outline of the procedure for determining the molecular orbitals of an octahedral transition-metal complex, ML_6, will serve as a second example of the application of the rules.

    1. The complex possesses O_h symmetry.

    2. All six ligands are sterically equivalent.

    3. We shall only consider σ bonding. Hybrid atomic orbitals on the ligand atoms, directed toward the central metal atom, can be used for forming six σ bonds.

    4. For the group of six σ orbitals we calculate the reducible representation, as shown in the following table.

	E	$8C_3$	$6C_2$	$6C_4$	$3C_2'$	i	$6S_4$	$8S_6$	$3\sigma_h$	$6\sigma_d$
ϕ_1	1	0	0	0	0	0	0	0	1	0
ϕ_2	1	0	0	0	0	0	0	0	1	0
ϕ_3	1	0	0	0	0	0	0	0	1	0
ϕ_4	1	0	0	0	0	0	0	0	1	0
ϕ_5	1	0	0	1	1	0	0	0	0	1
ϕ_6	1	0	0	1	1	0	0	0	0	1
Γ =	6	0	0	2	2	0	0	0	4	2

    5. This reduces to $A_{1g} + E_g + T_{1u}$.

    6. As seen from the O_h character table, the d_{xy}, d_{xz}, and d_{yz} orbitals of the metal are a basis for the irreducible representation T_{2g}. The d_{z^2} and $d_{x^2-y^2}$

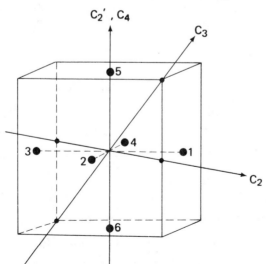

FIG. 33.14. Some of the O_h symmetry operations.

orbitals are a basis for E_g; the p_x, p_y, and p_z orbitals are a basis for T_{1u}, and the s orbital transforms as A_{1g}.

7. The energy level diagram is shown in Fig. 22.3.

8. The coordinate system for each ligand is chosen such that the direction toward the metal atom is positive.

By carrying out procedures 9 and 10, the combinations of the ligand σ orbitals that transform as A_{1g}, E_g, and T_{1u} can be determined. The procedures are fairly laborious, and we shall only show the calculations required to obtain the two E_g combinations. In the following table, we show the results of applying the 48 O_h symmetry operations to ϕ_1 and ϕ_2. Figure 33.14 shows some of the symmetry operations and the ligand numbering used. At the bottom of the table are the functions which result from multiplying the terms by the characters of the E_g irreducible representation.

	E	$8C_3$	$6C_2$	$6C_4$	$3C_2{}'$	i	$6S_4$	$8S_6$	$3\sigma_h$	$6\sigma_d$
ϕ_1	ϕ_1	ϕ_2	ϕ_2	ϕ_1	ϕ_1	ϕ_3	ϕ_2	ϕ_2	ϕ_1	ϕ_1
		ϕ_2	ϕ_3	ϕ_1	ϕ_3		ϕ_3	ϕ_2	ϕ_1	ϕ_1
		ϕ_4	ϕ_3	ϕ_2	ϕ_3		ϕ_3	ϕ_4	ϕ_3	ϕ_2
		ϕ_4	ϕ_4	ϕ_4			ϕ_4	ϕ_4		ϕ_4
		ϕ_5	ϕ_5	ϕ_5			ϕ_5	ϕ_5		ϕ_5
		ϕ_5	ϕ_6	ϕ_6			ϕ_6	ϕ_5		ϕ_6
		ϕ_6						ϕ_6		
		ϕ_6						ϕ_6		
ϕ_2	ϕ_2	ϕ_1	ϕ_1	ϕ_1	ϕ_2	ϕ_4	ϕ_1	ϕ_1	ϕ_2	ϕ_1
		ϕ_1	ϕ_3	ϕ_2	ϕ_4		ϕ_3	ϕ_1	ϕ_2	ϕ_2
		ϕ_3	ϕ_4	ϕ_2	ϕ_4		ϕ_4	ϕ_3	ϕ_4	ϕ_2
		ϕ_3	ϕ_4	ϕ_3			ϕ_4	ϕ_3		ϕ_3
		ϕ_5	ϕ_5	ϕ_5			ϕ_5	ϕ_5		ϕ_5
		ϕ_5	ϕ_6	ϕ_6			ϕ_6	ϕ_5		ϕ_6
		ϕ_6						ϕ_6		
		ϕ_6						ϕ_6		
after multiplying by E_g characters	$+2\phi_1$	$-2\phi_2$ $-2\phi_4$ $-2\phi_5$ $-2\phi_6$	0	0	$+2\phi_1$ $+4\phi_3$	$+2\phi_3$	0	$-2\phi_2$ $-2\phi_4$ $-2\phi_5$ $-2\phi_6$	$+4\phi_1$ $+2\phi_3$	0
	$+2\phi_2$	$-2\phi_1$ $-2\phi_3$ $-2\phi_5$ $-2\phi_6$	0	0	$+2\phi_2$ $+4\phi_4$	$+2\phi_4$	0	$-2\phi_1$ $-2\phi_3$ $-2\phi_5$ $-2\phi_6$	$+4\phi_2$ $+2\phi_4$	0

We obtain two allowable combinations of σ orbitals:

$$8\phi_1 - 4\phi_2 + 8\phi_3 - 4\phi_4 - 4\phi_5 - 4\phi_6$$

and

$$-4\phi_1 + 8\phi_2 - 4\phi_3 + 8\phi_4 - 4\phi_5 - 4\phi_6$$

Of course, any two independent linear combinations of these functions are also acceptable. If we multiply both functions by $-\frac{1}{4}$ and then add, and multiply both functions by $-1/12$ and then subtract, we obtain the results

$$2\phi_5 + 2\phi_6 - \phi_1 - \phi_2 - \phi_3 - \phi_4$$

and

$$\phi_1 + \phi_3 - \phi_2 - \phi_4$$

These functions are the usual ways of representing the E_g combinations. It should be obvious that these combinations combine with the metal d_{z^2} and $d_{x^2-y^2}$ orbitals, respectively, to form molecular orbitals. Similar procedures lead to the A_{1g} combination (which combines with the metal s orbital):

$$\phi_1 + \phi_2 + \phi_3 + \phi_4 + \phi_5 + \phi_6$$

and the T_{1u} combinations, $\phi_1 - \phi_3$, $\phi_2 - \phi_4$, and $\phi_5 - \phi_6$ (which combine with the metal p_x, p_y, and p_z orbitals, respectively). All these combinations can be obtained by a more "intuitive" approach described by Ballhausen and Gray.[14]

Problems

1. Construct qualitative molecular orbital energy level diagrams for the following species. Give the irreducible representation, the bonding character (bonding, non-bonding, or antibonding), and the electron occupancy for each level.
   (a) NO_3^-
   (b) B_2H_6
   (c) COF_2
   (d)

$$\left[\begin{array}{c} O \diagdown \\ \quad C - C \diagup O \\ | \quad || \\ \diagup C - C \diagdown \\ O \qquad\quad O \end{array} \right]^{2-} \qquad (\pi \text{ orbitals only})$$

13. DETERMINATION OF THE EQUIVALENT WEIGHT OF AN UNKNOWN ACID IN LIQUID AMMONIA BY REACTION WITH SODIUM

Your instructor will give you a small amount of a crystalline compound which is an acid in liquid ammonia. You will determine the equivalent weight of the acid by measuring the hydrogen evolved when the acid is allowed to react with excess sodium-ammonia solution. In the following directions, it is

[14] C. J. Ballhausen and H. B. Gray, *Molecular Orbital Theory*, W. A. Benjamin, Inc., New York, 1964, pp. 95–99.

assumed that the vacuum line (pictured in Fig. 8.14) has been pumped down with all the cocks closed except cocks 1, 2, 3, 12, and 13.

The unknown acid will be in the form of nonhygroscopic crystals and will require no drying. Using the metal "pill press" (see Fig. 33.15), prepare a pellet of the acid weighing 80 to 100 mg. Weigh the pellet to ± 0.2 mg and place it in the side arm of the reaction vessel (see Fig. 33.16). Cut a clean, pea-sized piece of sodium (about 0.1 g) and put it in the bottom of the vessel. Grease the ball joint; connect it to the socket below cock 5, and evacuate the vessel by opening cock 5. This last operation should be performed quickly to minimize the reaction of air with the sodium.

Connect an ammonia cylinder to the socket below cock 6 with Tygon tubing, and open cock 6. When the pressure has decreased to about 30 μ, close cock 1 and *carefully* open the ammonia-cylinder valves (the main valve first) so that ammonia bubbles at a moderate rate through the mercury bubbler. Cool the reaction vessel with liquid nitrogen until about 5 to 10 cc of solid ammonia has condensed in the vessel; then close the ammonia-cylinder valves and close cock 6. Open cock 1 until the pressure has decreased to about 10 μ; then close cock 1. Remove the liquid-nitrogen bath from the reaction vessel until ammonia begins bubbling through the mercury bubbler; then freeze the sodium-ammonia solution with liquid nitrogen and open cock 1 until the pressure has again reached about 10 μ.

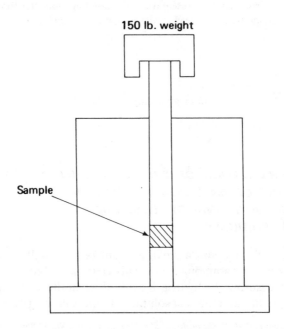

FIG. 33.15. Pill press.

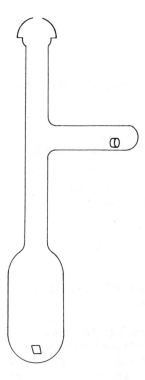

Fig. 33.16. Reaction vessel suitable for studying the reaction of a weak protonic acid with excess sodium in liquid ammonia.

Remove the liquid nitrogen bath from the reaction vessel and wait until the vapor pressure of the ammonia reaches about 10 cm; then immerse the vessel in a chloroform slush (see Chap. 6). Tip the reaction vessel slightly, and, by tapping, knock the acid sample into the sodium-ammonia solution. After about 10 min, the reaction is complete and the vessel is again immersed in liquid nitrogen. The hydrogen may now be Toepler pumped through a liquid-nitrogen trap (to remove the small amount of uncondensed ammonia) into bulb Y.

Problems

1. Why is the main ammonia cylinder valve opened first?

2. Why is the freshly distilled ammonia allowed to melt and then refrozen before the acid sample is added?

3. What are the equivalent weights, in liquid ammonia, of:
   (a) ammonium bromide?
   (b) phosphonium iodide?

(c) acetamide?

(d) phthalimide?

(e) sulfamic acid?

4. Why would the following materials make poor "unknown" acids?

(a) ammonium nitrate

(b) ammonium sulfate

5. What pressure will be exerted in a 50-ml bulb at 25° by the hydrogen evolved from the reaction of 0.100 g of diphenylamine with excess sodium-ammonia solution?

6. Exactly 0.105 g of an unknown acid was added to excess $Na-NH_3$ solution, and the evolved hydrogen was Toepler pumped into a system of volume 298 ml. The pressure and temperature were 115 mm and 25°, respectively. Calculate the equivalent weight of the acid.

14. PROBLEMS

The following problems are of broad scope and require knowledge of several of the chapters in this book.

1. Lithium hydroaluminate ($LiAlH_4$) and dimethylammonium chloride were allowed to react in ether to give hydrogen, lithium chloride, and compound A. Compound A was a solid melting at 90° with a cryoscopic molecular weight of 220. Compound A was dissolved in benzene and treated with a phenyllithium solution. A solid B precipitated, which was centrifuged off and found to be violently reactive with water. The supernatant liquid was concentrated by evaporation and cooled, whereupon compound C precipitated. Compound C melted at 204° and had an ebullioscopic molecular weight of 470. Mercuric chloride reacted vigorously with an ether solution of compound A to give mercury, hydrogen, and a solution of compound D, which was isolated by evaporation of solvent and then purified by vacuum sublimation. Compound D contained chlorine (49.8 per cent) and had an ebullioscopic molecular weight of 300. Identify the lettered compounds.

2. Aubke, Cady, and Kennard[15] allowed a large excess of peroxydisulfuryl difluoride, $FSO_2-O-O-SO_2F$, to react with 1.1056 g of I_2O_5. The I_2O_5 was completely consumed. The volatile products were pumped out of the reaction vessel, leaving 1.6995 g of solid product A. That portion of the volatile products not condensed by a liquid nitrogen trap occupied 37 cc (S.T.P.). Product A was found to contain 48.9 per cent iodine. When 0.1000 g of product A was dissolved in excess of dilute 0.1 M HI, 23.4 ml of 0.1000 N thiosulfate was required to titrate the liberated iodine. Identify compound A.

3. When 32 mmoles of $(CH_3)_3NBH_3$ and 32 mmoles of $(CH_3)_3NH^+I^-$ were heated together, 30 mmoles of H_2 was evolved. The principal product was a white compound that dissolved in water to form a conducting solution. The ^{11}B nmr spectrum of the solution was a $1:2:1$ triplet. Suggest a structure for the product.

[15] F. Aubke, G. H. Cady, and C. H. L. Kennard, *Inorg. Chem.*, 3, 1799 (1964).

4. Decaborane, $B_{10}H_{14}$, was dissolved in dimethyl sulfide and the solution was allowed to stand at room temperature for a week. During this time, 1 mole of H_2 was evolved per mole of $B_{10}H_{14}$, and colorless crystals (*A*) separated. The crystals contained 44.3 per cent boron, 19.7 per cent carbon, and 9.8 per cent hydrogen. They were dissolved in liquid ammonia, and the ammonia was then allowed to evaporate from the open flask. A solid (*B*), containing 22.1 per cent NH_3 (liberated

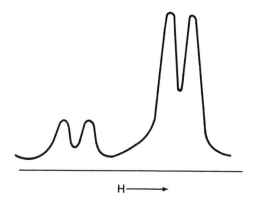

H⟶

on treatment with alkali), remained in the flask. The ^{11}B nmr spectrum of a solution of *B* is shown in the figure. A large excess of chlorine was passed into a solution of *B*; then addition of a concentrated solution of cesium chloride caused the precipitation of a white solid (*C*), which, after drying, contained 36.5 per cent cesium and no hydrogen. A solution of *C* was passed through a cation exchange column in the acid form. The effluent was concentrated by vacuum evaporation of the solvent with the formation of a white solid, *D*. This material contained no cesium and was acidic, with a neutral equivalent of 295. Give the formulas for *A*, *B*, *C*, and *D*, and suggest a structure for *D*.

5. A stream of fluorine, diluted with N_2, was passed through an aqueous solution of urea. Two moles of fluorine were consumed per mole of urea. By extraction of the solution with ether, evaporation of the ether layer, and sublimation of the residue, a white product *A* (melting at 43°) was obtained. At 80 to 90°, solutions of *A* in dilute mineral acids yielded equimolar amounts of CO_2 and a gas *B*. The mass spectrum of *B* showed peaks at m/e values of 53, 52, 34, 33, 20, 19, 15, and 14. When aqueous solutions of *A* were treated at 0° with aqueous KOH, a mixture of gases, in which compound *C* predominated, was obtained. Compound *C* contained only fluorine and nitrogen, and was observed to exist in two isomeric forms. Give structures for *A*, *B*, and the two isomers of *C*.

6. Dicobalt octacarbonyl, $Co_2(CO)_8$, and excess $H_3B:N(CH_3)_3$ in benzene were heated for 3 hours at 60 to 65°. Workup yielded crystals melting at 124 to 125°. The analysis was C, 29.6; H, 2.2; N, 2.7; Co, 33.0; and B, 1.7. The ^{11}B nmr spectrum showed a triplet. The mass spectrum showed peaks at the following m/e values (relatively strong peaks are italic): *529, 501, 473*, 472, 471, *445*, 444, *443*, 432, 417, *416, 415*, 389, *388, 387*, 373, 361, 360, *359*, 345, 332, *331*, 317, 316, 304, 289, 277, 276,

275, 261, 249, 247, 246, 245, 244, *243, 233,* 218, 217, 216, *205,* 202, *190, 189, 177.* Explain the nmr spectrum, identify the eight highest mass peaks of the mass spectrum, and suggest a structure for the compound that is consistent with the "effective atomic number" principle.[16]

7. A compound melting at $-44°$ and distilling at $35°$ (20 mm Hg) can be isolated from the reaction of $(CH_3)_2NPF_2$ with BrCN. The compound gives a mass spectrum in which the peak of highest m/e occurs at 158. This same compound gives a ^{19}F nmr spectrum consisting of two doublets, one twice as intense as the other. Each member of the more intense doublet is a doublet, and each member of the less intense doublet is a 1:2:1 triplet. Give the formula and structure of the compound.[17]

8. The reaction of NF_3 with O_2 in an electric discharge gives a volatile material of formula NF_3O, which could be formulated as

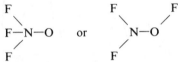

Explain how infrared, nmr, and mass spectroscopy could be used to distinguish between these possibilities.[18]

9. A chemist isolated a compound H_4PNO_3 from the hydrolysis products of $PNH(NH_2)_3$. He believed the structure to be

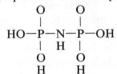

In fact, the salt Ag_3PNHO_3 was prepared.
(a) What do you think the structure was? Why?
(b) Give another method for preparing the compound.

10. Suggest a procedure for the preparation of imidodiphosphoric acid:

$$
\begin{array}{ccc}
 & O & \quad O \\
 & \| & \quad \| \\
HO\!-\!P\!-\!N\!-\!P\!-\!OH \\
 & | \quad H \quad | \\
 & O & \quad O \\
 & H & \quad H
\end{array}
$$

11. If calcium phosphide (Ca_3P_2) were treated with an alcohol (ROH), which of the following sets of products would you expect? Why?
(a) PR_3 and $Ca(OH)_2$
(b) PH_3 and $Ca(OR)_2$
(c) $P(OR)_3$ and CaH_2
(d) H_3PO_3 and CaR_2

[16] See *Inorg. Chem.,* **7,** 2265 (1968).
[17] See *Inorg. Chem.,* **7,** 2067 (1968).
[18] See *Inorg. Chem.,* **7,** 2064 (1968).

12. How can you account for the fact that nitryl chloride reacts with water to form hydrochloric acid and nitric acid, whereas nitryl chloride reacts with ammonia to form chloramine (NH_2Cl) and ammonium nitrite?

13. Can you account for the puzzling fact that phosphorus trichloride hydrolyzes to give phosphorous acid and hydrochloric acid, whereas nitrogen trichloride hydrolyzes to give ammonia and hypochlorous acid?

14. A solution of equimolar amounts of $Ni(CO)_4$ and $(CF_3)_2P(NCH_3)P(CF_3)_2$ in hexane liberated, after 24 hours, 2.45 moles of CO per mole of the initial phosphorus compound consumed. The molecular weight of the product (obtained after evaporation of the hexane) in ether was 922. Heated in a stream of CO, it liberated 82 per cent of the original phosphorus compound. The infrared spectrum of the product showed peaks at 2083, 2072, and 1917.5 cm^{-1}. Suggest a plausible formula and structure for the product.[19]

15. At 400°, $K_3Cr(CN)_6$ reacts with hydrogen to form ammonia, HCN, and an olive-green solid containing about 43 per cent potassium and 19.06 per cent chromium. The infrared mull spectrum shows two sharp peaks in the C–N stretching region (2178 and 2077 cm^{-1}), whereas $K_3Cr(CN)_6$ shows only one peak in this region (2130 cm^{-1}). The compound is insoluble in water and is unaffected by boiling aqua regia.[20] What may be said about the structure of the solid?

16. Exactly 11.2 cc (S.T.P.) of a colorless material is found to weigh 23 mg. The vapor pressure at 0° is measured as 11 ± 1 mm. When this material is distilled onto excess sodium-ammonia solution and allowed to react, 5.8 cc (S.T.P.) of hydrogen is evolved. What is the material?

17. What products would you expect if an equimolar mixture of $LiBH_4$ and $(CH_3)_3NHCl$ were heated?

18. Suggest an explanation for the fact that better yields of silanes are obtained from the ammonolysis of magnesium silicide in liquid ammonia than from the corresponding hydrolysis in water. On the other hand, why can boron hydrides be obtained from the aqueous hydrolysis of magnesium boride and not from the corresponding liquid ammonia ammonolysis?

19. Why cannot silanes be prepared by the reaction of silicon with molecular hydrogen, analogously to the preparation of lithium hydride?

15. LOCKER EQUIPMENT

1 Fisher burner, with tubing
1 Chromel triangle
1 wire gauze

[19] See A. B. Burg and R. A. Sinclair, *J. Am. Chem. Soc.*, **88**, 5354 (1966).
[20] See D. F. Banks and J. Kleinberg, *Inorg. Chem.*, **6**, 1849 (1967).

2 iron extension rings (3 and 4 in.)

1 iron support

2 extension clamps with asbestos-covered jaws

2 extension clamps with three-prong, asbestos-covered jaws

4 clamp holders

2 boxes of matches

1 test-tube rack

6 test tubes (20 × 150 mm)

7 beakers (1–100, 2–250, 1–400, 1–600, 1–800, and 1–1500 ml)

4 glass rods (2–7 × 280 mm, 2–6 × 180 mm)

4 Erlenmeyer flasks (1–125, 2–250, and 1–500 ml)

3 watch glasses (2–90, 1–150 mm)

1 polyethylene squeeze bottle, for water (500 ml)

1 weighing bottle (30 ml)

1 buret (50 ml)

3 graduated cylinders (1–10, 1–50, and 1–250 ml)

1 pipet (25 ml)

1 filtering funnel with short stem (90 mm)

1 cylindrical dropping funnel (125 ml)

1 pear-shaped separatory funnel (500 ml)

2 fritted-glass crucibles (15 ml)

1 fritted-glass funnel, medium porosity (6 cm)

1 neoprene filter adapter (for fritted-glass crucibles and funnel)

1 filter flask (500 ml)

1 filter trap, equipped

2 evaporating dishes (10 and 12 cm)

1 porcelain crucible with cover (50 ml)

1 iron crucible (50 ml)

1 pair of crucible tongs

2 round-bottomed flasks with 24/40 joints and heating mantles (1–100, 1–500 ml)

1 distilllng head, with 24/40 joint

1 condenser with 24/40 joint at receiver end (30 cm)

1 adapter with unground 24/40 joint

1 Powermite

2 thermometers (− 10–110° and − 10–360°)

1 mortar and pestle (10 cm)

1 file

1 pair of safety goggles

1 desiccator

1 "Scoopula"

1 Monel spatula   (continued next page)

2 towels
1 test-tube brush
1 bottle brush
rubber tubing (3 ft suction; 3 ft regular)
3 glass-stoppered bottles (1–125, 1–250, and 1–500 ml)

Index

579

SAFETY PRECAUTIONS AND FIRST AID: See page 549.

CONCENTRATIONS OF COMMERCIAL REAGENTS

Reagent	Density	Weight %	Molarity
Acetic acid	1.05	100%	17.5
Aqueous ammonia	0.90	28	15
Hydriodic acid	1.5	47	5.5
Hydrobromic acid	1.5	48	9
Hydrochloric acid	1.19	37	12
Hydrofluoric acid	1.15	48	28
Hydrogen peroxide	1.11	30	10
Nitric acid	1.42	69	16
Perchloric acid	1.69	72	12
Perchloric acid	1.54	60	9.2
Phosphoric acid	1.69	85	14.5
Sulfuric acid	1.84	96	18

SOME USEFUL CONSTANTS

$$R = 1.9872 \text{ cal deg}^{-1} \text{ mole}^{-1}$$
$$= 82.056 \text{ ml-atm deg}^{-1} \text{ mole}^{-1}$$
$$= 62,362 \text{ ml-mmHg deg}^{-1} \text{ mole}^{-1}$$
$$(PV)_{0°C} = 22,414 \text{ ml-atm mole}^{-1}$$
$$N = 6.0225 \times 10^{23} \text{ molecules mole}^{-1}$$
$$F = 96,487 \text{ coulomb equiv}^{-1}$$
$$= 23,061 \text{ cal volt}^{-1} \text{ equiv}^{-1}$$
$$k = 1.3805 \times 10^{-16} \text{ erg deg}^{-1} \text{ molecule}$$
$$0°C = 273.150 \text{ °K}$$
$$e = 4.8030 \times 10^{-10} \text{ e.s.u.}$$
$$h = 6.6256 \times 10^{-27} \text{ erg sec}$$
$$c = 2.9979 \times 10^{10} \text{ cm sec}^{-1}$$
$$g = 980.66 \text{ cm sec}^{-2}$$